Turbulent Multiphase Flows with Heat and Mass Transfer

Turbulent Multiphase Flows with Heat and Mass Transfer

Roland Borghi
Fabien Anselmet

iSTE

WILEY

First published 2014 in Great Britain and the United States by ISTE Ltd and John Wiley & Sons, Inc.

ISTE Ltd
27-37 St George's Road
London SW19 4EU
UK

www.iste.co.uk

John Wiley & Sons, Inc.
111 River Street
Hoboken, NJ 07030
USA

www.wiley.com

Library of Congress Control Number: 2013951424

British Library Cataloguing-in-Publication Data
A CIP record for this book is available from the British Library
ISBN: 978-1-84821-617-4

Table of Contents

Acknowledgments

The idea for this book emerged from several thesis projects at the University of Rouen where Roland Borghi has been working for more than fifteen years. Its writing owes much to our many interactions and discussions with doctoral candidates at the university as well as with our colleagues at the CORIA laboratory.

We must also make special mention of our many unfailingly open and stimulating discussions with Olivier Simonin, over a period of more than 30 years, and in the process of completing this book. Likewise, the arduous study of probability densities was greatly helped by years of exchange with Vladimir Sabelnikov, who even volunteered to proofread and critique the corresponding part of this book. Nevertheless, as usual, any error left in the text would only be our responsibility. We extend our gratitude to one and all, especially Olivier and Vladimir.

Fabien Anselmet offers warm personal thanks to a few colleagues who are his elders, including Roland Borghi of this book, for having introduced him to the fascinating world of turbulence, multiphase flows, drops and droplets. This book can consequently be considered as the result of thirty years of collaboration and discussion.

Introduction

I.1. The significance of multiphase flows and their modeling

Many industrial systems bring into play, in one way or the other, multiphase media involving the combination of liquids and gases, non-miscible liquids, fluids and solids.

Nuclear reactors (whether they use boiling or pressurized water) possess a cooling circuit where, in certain parts of this circuit, a mixture of water and vapor circulates, with water vapor forming on contact with hot walls needing to be cooled and drops of liquid water forming on contact with cold walls needing to be heated. Numerous other thermal engineering facilities possess this type of circuit for transferring heat, in order either to use this energy elsewhere or simply to prevent the machinery from being destroyed by the heat.

The extraction and transportation of oil products is done using conduits within which media are flowing with two or more phases: liquids of different densities and viscosities, gases and even solids. Problems of icing in aeronautics (on the leading edges of wings or ailerons or in Pitot tubes, etc.) also necessitate studying a humid air medium with drops of water flowing in the immediate vicinity of the wall. The short-distance transport of pulverulent materials such as wheat, sawdust and grain is done by blowing air loaded with these solid particles through ducts.

In liquid-fuel rocket engines used in space launchers, as well as in diesel engines, the combustion chamber contains a mixture of vaporizing droplets and combusting gases that give off a considerable release of heat in an astonishingly small volume. A combustible or oxidant liquid, or a mixture of both, is injected in tiny droplets into the combustion chamber, where these drops vaporize and the vapors can burn together, in a steady regime in a rocket engine and in a periodic regime in cycles of a diesel engine.

Fuel burners in glassworks furnaces, or vapor generators in thermal power stations, also inject jets of droplets of fuel into the zone of reacting gases. They produce not only heat and burned gases, but also smoke in which very small particles of carbon are dispersed, and the control of these particles is critical; they allow high heat transfers via radiation in furnaces, but can lead to significant air pollution from chimney exhaust.

Chemical engineering uses several types of gas–liquid reactors at controlled temperature, which are intended to produce specific chemical products rather than heat. Liquid and gaseous reactants are mixed as effectively as possible in order to be able to allow various chemical reactions at the interface between phases. Many chemical reactors also use a catalyst, which is most often in the form of a solid dispersed phase, and these reactors therefore bring multiphase flows into play.

Fluidized beds are currently the most effective devices for burning coal: air is blown forcefully through a highly dense dispersed solid phase composed of particles of dolomite and small particles of coal, enabling exchanges of energy among these three phases, which then causes and maintains chemical reactions. The energy released by combustion produces water vapor by the intermediary of a heat exchanger, the tubes of which can be closer to the combustion zone. The system not only enables adequate homogeneity of the temperature field but also maintains this temperature at around 1,300 K by avoiding the overproduction of NO, while stabilizing combustion at the same time. In addition, dolomite absorbs sulfur and reduces SO_2 emissions. There are also recirculating fluidized beds in which solid phases are entrained by the gas phase, recovered at the exit and reinjected at the entry to perfect combustion, even in the highest mass flow rate conditions. These are also easily transposable for the combustion of different types of combustibles, ranging from gas mixtures to various types of waste. Fluidized beds are also used in chemical engineering, or simply to dry solid particles, or to manufacture various types of powders.

Multiphase media often play in nature as well. Clouds contain tiny droplets of (non-pure) water along with particles of ice or snowflakes; agriculture utilizes jets of drops or droplets for the watering or treatment of plants. The dispersion of smoke or other natural or artificial aerosols into the atmosphere or confined gaseous environments, and the possible deposit of the solid or liquid particles they contain, is a source of ongoing problems that are difficult to solve or control. Landslides, avalanches, and flows of sand or various types of debris are also natural examples of flows of multiphase media, the behavior of which is difficult to predict. Soil, even stable soil, is always a multiphase medium containing at least one liquid phase, which is usually water, but air for backfill and ballast for railroads, and in civil engineering we must always have control over the strains of these media that is as complete as possible.

In all of these industrial or natural situations, the overall medium behaves very often like a fluid. This is obviously the case when there is no solid phase, but as soon as one fluid phase exists the medium can be in flow, at least in some parts. This makes it desirable to be able to study these media in the same way as classic fluids, for which we have nearly 100 years of accumulated knowledge and experience. Moreover, these flows almost always display a highly random nature, both in the positions of phases and because they show that velocity fluctuations develop, quite often, similarly to turbulent flows. Experience in modeling turbulent flows should, therefore, prove extremely useful. This experience is not confined solely to questions of motion; we can also represent phenomena in fluids such as heat exchange, diffusion and mixing of various constituents, and chemical reactions, all out of equilibrium. The short descriptions provided above for multiphase situations make it clear that they also include all these phenomena, and that these phenomena give them important specificities. Therefore, it is very useful to generalize the approach used for out of equilibrium continuous fluid media to these multiphase flows.

Since the early years of the use of multiphase devices in various fields of application, a notable body of results and knowledge of an empirical or even occasionally theoretical nature has been accumulated; this has been used to develop simplified and practical approaches for the study and prediction of their characteristics. The design of industrial machinery and the interpretation and control of natural multiphase phenomena have been aided by the various separate branches of this knowledge set, each of them directly linked to a particular application. However, for the last 20 years, the desire for increasingly detailed predictions has resulted in the increasingly frequent use of the methods of continuous fluid mechanics, thus showing a certain community of approach in the different cases. Even without necessarily seeking to build theoretical representations, when we wish to understand how the various basic physical phenomena that occur on a small scale combine, we find a strong similarity of description in numerous different multiphase situations. With regard to the specific aspect of mixing, this is well emphasized in [GUY 97].

It now seems possible and interesting to attempt a unified explanation of the basic theoretical concepts used in the modeling of all multiphase media of these various applications, even though the particularities of the various different situations must explicitly be involved at a certain point. This is the aim of this book. Seeking such a unified methodology has a purpose, even a threefold purpose: first, it may provide a more complete physical understanding of each situation by bringing together information and analyses of situations that are different but using similar phenomena. Second, it may serve as a motivation to use successfully certain modeling or study tools from one field of application in another. Finally, it may render a new field of application accessible that might seem too complex at first glance. However, it is not our objective here to develop the approach to the extent of completely addressing the issues posed by the various applications, even simply those referred to earlier. First, there are too many possible applications, but above all, it would be necessary at times to enter into more details of the modeling, and at a level that would lack interest for the non-specialist. In addition, many particular aspects of certain problems are not yet sufficiently known, and they still require critical discussions, a fact that further pushes away discussions too specific for our global purposes. We wish simply to show how a unified approach can establish a common basis of representation for all of these situations, how questions of modeling emerge, which aspects are general and which are more specific to different applications, etc. To answer these questions, the first result of this unified approach will be to make certain suggestions that are the outcome of comparisons with the issues of another type of multiphase medium.

We cited granular media and even soils above as examples of multiphase media containing solid phases. Landslides and avalanches of debris are obviously multiphase flows. Soil is generally considered as a solid medium, but it is particularly interesting to study the threshold beyond which this medium flows, completely or, most often, in part, which also poses the problem of determining the expansion of the zone which is flowing. We will show how the unified method presented here contributes a new and useful point of view for granular media. When this medium is in flow, it is easy to see how it might be seen as a particular turbulent multiphase flow. Just before the threshold of flow, the medium shifts and deforms; qualifying it as a turbulent medium is not absolutely appropriate, but the motions are sudden and random (as is the initial structure itself), even if they are of limited amplitude. However, it would be interesting to push the approach toward this limit. We will not be examining every situation here in which the medium is a deformable solid, or those in which one or more liquid or gas phases flow across a porous solid. The fields of soil mechanics and porous media mechanics are well established. It would be interesting to show how certain aspects of these studies are close to the topics we will discuss here, but space does not permit it.

I.2. Modeling and its related issues

The perspective that enables a common theoretical approach begins, of course, with the acknowledgment that multiphase media are similar to continuous media when they are viewed from far enough away that their detailed structure is no longer discernable. These detailed structures, moreover, are not the objective of studies; we are dealing simply with the phenomena that we witness and undergo at our level. However, the details of these structures have consequences for these phenomena. In these conditions, we must seek to represent the media and phenomena that occur by means of a certain number of characteristic macroscopic, or averaged, variables, and pinpoint the various types of laws they must satisfy. The first task is to adequately define the way to establish these averages. These averaged variables, like all expansive physical variables, satisfy balance equations that must then be written. The detailed structures of the medium and the small-scale phenomena that enter into play within them must be taken into account, or must be modeled, using certain adapted constitutive laws. These laws are usually found with local formulations, but this will not always be the case here; it will be necessary to create new partial differential equations, and in any case they will have to be validated in a fairly general manner through experimentation. In some cases, we will see the usefulness of suggesting several equations for the same medium but that will be valid in more or less general conditions.

Of course, the situation is more complicated for a multiphase medium than for a fluid medium composed of identical molecules. First, it is necessary to be able to monitor within the medium the way in which the different phases move, i.e. to understand the composition of the medium in different places. This is similar to a situation in which the fluid is composed of different molecules. However, these different phases are also liable to have different velocities, temperature and specific agitation energies, and these variables must be clearly defined in an appropriate manner. It will, therefore, be necessary to characterize the medium, even if it is only at a macroscopic level, by more characteristic variables, some of these being new types; for example, we will see the broad utility of the mean interface area per unit of volume. It should be remembered, however, that even if we use more variables, modeling will be only an approximation; the variables we are considering will never contain the whole complexity of the media, but only the aspects that we consider are the most important, and it will be necessary in any case to model complex behaviors that remain hidden.

The methodology of modeling is, of course, similar to that of classic fluid media, which covers the detailed movements of molecules. To avoid straying too far from this example, it is helpful to consider that the characteristic scale of the multiphase structure of the medium is much smaller than the macroscopic scale in which we are interested, that is of the device containing it, as is the case for classic flows. This is the hypothesis effectively advanced for the first attempts at modeling, and it is indeed generally the case when the medium is composed of solid particles widely diluted within a continuous fluid phase. But when we are examining the fragmentation or atomization of a liquid jet into drops in a gaseous flow, the liquid core and the first ligaments or parcels of liquid are of a size comparable to the conduits within which they are moving, while the drops being produced are expected to be much smaller. For flows in which bubbles of various sizes are transported within a liquid, or flows of an emulsion of two non-miscible liquids, there are conditions in which the bubbles, drops or inclusions are very small, but there are also others in which they extend over the entire diameter of the conduits. It is clear that these latter conditions, which cause the appearance of multiphase structural scales intermingled with the scales of the device being considered, and thus of the flow occurring, exceed the simple situation of very small heterogeneities. Even though these conditions complicate modeling, we must, nevertheless, in our theoretical description, position ourselves so as to address them. The methodology that we have chosen gives us a good perspective from which to approach this problem. The first models resulting from it, which are relatively classical in form, are not immediately well adapted to these situations but we will see how to improve them in this regard.

We have not only talked about multiphase structures but also about a certain degree of randomness; these are not well-ordered structures. The medium permanently displays disordered fluctuations of phase positions and velocity fluctuations, which very often resemble those existing in a turbulent flow. This situation originally occurs with the fact that we cannot completely control the preparation of the medium at the initial instant, or at entry of the flow field, either in terms of phases or in terms of the field of velocities. It is only when a relatively viscous liquid phase exists that this random character of preparation is not amplified during passage through the device. In the vast majority of cases, the initial disorder increases greatly during the flowing of the medium, especially when velocity is high; the spatial and temporal scales of the fluctuations grow to a certain level of saturation that ceases to be controlled solely by the initial and inlet conditions. Phenomena of amplification and saturation of fluctuations are in part the same as those that occur in single-phase turbulent flows. The multiphase character causes additional instabilities in the presence of trajectories of drops or particles (even in a Stokes flow, if more than two particles interact), and instabilities linked to the production or coalescence of drops or to collisions between solid particles enable the different inertias of the different phases to be fully at play. The resulting random and fluctuating medium can be referred to as an extended turbulent medium. It is similar to the classic single-phase turbulent medium, but shows fluctuations of new variables associated with additional phenomena and requires similar methods of modeling.

In addition to the averaged variables we have already mentioned, we will need to define and study the variances of fluctuations of these variables, and even their probability density functions, and find the balance equations they satisfy. In principle, these equations do not directly involve the physical nature of the phenomena that regulate these fluctuations; therefore, they can be used much more widely than for turbulent flows. The closure hypotheses of these equations will necessarily take into account the nature of the phenomena involved, and it is here that classic hypotheses for turbulence will have to be adapted. Moreover, it will also be necessary to take into account the fact that the small scales of fluctuations, which still retain the particularities of each type of medium being considered, may in this case retain an influence more often than in classic turbulent flows, where the so-called high Reynolds number situations are very common. Obviously, it will be necessary to complicate the representations of these fluctuating flows and the equations to which they lead, that is to build more complex models, but it will be interesting to be able to do this with the same spirit, and perhaps the same tools, as for classical turbulent flows.

We have also stated that various physical or chemical phenomena may occur, particularly, in the internal structure, giving it its specific utility. Momentum exchanges occur between various constituents via the intermediary of forces; exchanges of heat and mass occur as well (phase changes), most often coupled together. Chemical phenomena can also be used beneficially inside a given phase or at the interface between two phases. The approach we are seeking must be general enough to adapt to these diverse physical phenomena.

Part 1 of this book deals with the definition and description of the *basic approach*, which can be presented simply as the generalization to several phases concerning the usual approach used in the mechanics of continuous media to the thermomechanics of fluids in turbulent flow. Its objective is to provide a context of theoretical representation, its hypotheses, its possibilities and its limits, and to identify the different types of equations necessary to make this approach useful, i.e. to draw from it predictions concerning specific flows. The general description includes four aspects: the theoretical description of a piecewise continuous flow; the definition of an averaged or filtered description, usable in practice, and it is statistical averages that will first be used here; the writing of balance equations for averaged quantities that can be chosen to represent the medium in its evolution; and the definition of the necessary constitutive laws, which include equations of state and laws of irreversible processes, and the problem of choosing these. These basic developments are the subject of Chapters 1–4.

Part 2 of the book begins by discussing the principles of modeling the effect of irreversible processes that are active within the flows. There are, in Chapter 5, the irreversible processes linked to exchanges between phases, masses, momentum or energy. We will show the significance of defining and studying exchanges in the form of exchange fluxes per unit of interface area. It is not possible to calculate these exchange fluxes without understanding the detail of the phenomena occurring on a small scale in the immediate vicinity of the interfaces, and it is necessary to suggest approximate models. The specificities of the different multiphase flows must explicitly be involved in this; we will not attempt to examine each interesting case in detail, or even a single interesting case, but will study a certain number of specific *local exchange configurations* and discuss how they can serve to construct the desired models.

Equally important, there are also additional dispersion fluxes inside a phase, for momentum or energy, that are due to fluctuations, often turbulent in nature, the modeling of which is discussed in Chapter 6. We will see that they may be modeled by generalizing approaches that are usual in single-phase turbulent flows, but there remains at least one major problem to be studied: the influence of the presence of phases on the spectrum of the characteristic length scales. We will also discuss simplification in which the mass of one phase is considered simply as diffusing in the overall multiphase mixture itself, in a manner analogous to a mixture of truly miscible gaseous or liquid components, without examining its own convection velocity in detail.

In Chapter 7, we then focus on the mean interface area per unit of volume, which plays a critical role in all of these questions related to exchanges between phases. The generalization of the approach for filtered averages, such as the large eddy simulation, is discussed in Chapter 8. This approach is, particularly, indicated for more precise study of the presence of large scales of multiphase or dynamic structures, which are avoided in statistical approaches. There is still a lot of work to be done in these areas. To date, models and their approximations have been presented and discussed in relation to the available experimental data. Chapter 9 reconsiders most of these questions with the general viewpoint of thermodynamics of irreversible processes, which will require a generalization of the classic thermodynamic approach. Classically used approximations will be reinforced for the most part, but there will also be interesting openings. Chapters 10 and 11 focus on the experimental methods and results available to date, and on the new aspects to be studied further, after a brief description of the current and new experimental methods.

Part 3 analyzes in more detail the *granular flows and media*, a type of multiphase media that has generated recent interest. In Chapter 12, fluidized beds, which are widely used in industry, are considered as a typical example, and we give a summary of the various models proposed and show how the developments in the preceding chapters are useful to explain them, as well as to emphasize the improvements that are necessary. In Chapters 13–15, we discuss an improvement for dense granular media in which lasting contacts between grains must be taken into account. There is, of course, a regime of flow in such a media, where only short collisions are occurring, but this is a limited case, when the flows are relatively rapid. At the other limit of small displacements, the media can be studied like elastic solid media, and between these two limits one or more other regimes of motion may exist due to the fact that the contacts between two or more grains may last for more or less time, small deformations and displacements of grains take place, and sudden sliding is possible locally. The multiphase approach is still possible, but it is necessary to consider that each grain is a different solid phase for which we know the constitutive laws, and the search for constitutive laws for a solid phase that would be the ensemble of all of the grains is the center of the modeling issue. These situations are at the limit of the domain of soil mechanics, and we will not go deep into this area. We will confine ourselves to demonstrate how the approach can lead to a form of model in which the stress tensor of the medium is given by a constitutive law in the form of a differential balance equation, like a kind of viscoelastic medium with a threshold.

Part 4 examines the improvement of the representation of all these media and adds the possibility of better understanding the fluctuations around mean values. The fact that we have defined mean variables from the beginning does not actually mean that it is impossible to understand the fluctuations around them. This is possible, first, by simply studying the variances of the fluctuations of variables, which has already been discussed in the preceding chapters for velocity fluctuations. This will now be extended to enable knowledge of the *probability density functions, even joint probability density functions, of the various variables in certain phases*, which is an interesting improvement in several situations. When relatively rapid chemical reactions occur in a gas in the vicinity of liquid droplets that supply one of the reactants, it is absolutely necessary to take into account the strong and fluctuating heterogeneities of composition and temperature that are found in the gas, precisely due to the exchange of chemical species between phases and the chemical reactions locally occurring. In problems where the vaporization of an ensemble of drops takes place, the distribution of temperatures of different drops, if wide enough, can have a significant effect on the mean rate of vaporization. Finally, for the dispersion of particles or droplets, a better understanding of the distribution of fluctuations of velocities and diameters of droplets is very useful.

In Chapter 16, we examine the modeling of the probability density of compositions for the gas phase of a two-phase medium in combustion. Chapter 17 discusses the possibility of better understanding the temperature fluctuations of the liquid phase or of drops or solid particles, which can have significant consequences for phase changes; finally, in Chapter 18, we study the possibility of better understanding the distribution of velocity fluctuations and parcel sizes of a phase. To obtain a better understanding of the distribution of sizes of a phase, we must define different categories of parcels in this phase, with each one behaving like a phase itself, and the treatment of all these virtual phases can thus become quite unwieldy. For situations in which the phases are always highly dispersed, the classically used stochastic Lagrangian–Eulerian simulation method is more appropriate. We will show how this approach connects to the Eulerian method discussed in this book.

Since the field contains problems that are still confined largely to the research level (not only in the last part), we have been careful throughout to emphasize the main subjects in which modeling still has critical flaws and that must be the subject of the most promising effort.

Literature on the domain, less general and more focused on one of the applicative aspects, provides additional information and interesting developments. In particular, we will refer more than once to the following works: [SOO 89, DEL 81, NIG 91, OES 06, CRO 01, JAK 08, PRA 06, AND 11].

Finally, the basics of the thermo mechanical description of continuous media used here were taken from [BOR 08a] and [BOR 08b].

PART 1

Approach and General Equations

Part 1 of this book deals with the definition and description of the basic approach, which can be simply presented as the generalization to several phases of the usual approach of continuous media in the thermomechanics of fluids in turbulent flows. We thus often speak of "piecewise continuous media". Its objective is to give a framework for the theoretical representation, its hypotheses, possibilities and also its limits, to identify the different types of equations that are necessary to make this approach useful, i.e. by taking the "predictions" concerning specific flows. It should be noted that this approach is the only one that allows us to provide exact, instantaneous equations of the complete medium, from which we can then develop equations (which require closure assumptions, such as the more classical equations for turbulent flows with a single phase) of variances and probability density functions (PDF). The general description consists of four aspects, which are each the object of a chapter: the theoretical description of a "piecewise continuous flow or medium"; the definition of an averaged or filtered description, usable in practice, and in particular the statistical averages that will be mainly used here; the writing of balance equations for averaged quantities that can be chosen to represent the medium in its evolution; and finally, the definition of the necessary constitutive laws, which include "equations of state" and laws of irreversible processes, and the problem of choosing them.

Chapter 1

Towards a Unified Description of Multiphase Flows

1.1. Continuous approach and kinetic approach

In classic fluid mechanics theories, fluid is usually considered as a "continuous medium", described locally and at each instant using a certain number of characteristic variables, and its evolution is represented by "balance equations". These balance equations are partial differential equations in three-dimensional space and time, whose original writing uses the Eulerian point of view: a geometric point in space is designated, and at this point, the characteristics of the fluid at each instant are observed, in particular its velocity. This leads to the so-called continuity equation and then to Euler and Navier–Stokes equations and necessitates the definition of the Cauchy stress tensor, containing the pressure and tensor of viscous stresses. The description is, therefore, not complete until these new variables can be given by specific "constitutive laws", which represent the nature and small-scale properties of the fluid in question. These constitutive laws are not necessarily based solely on a theory; they can, with more generality, be of empirical origin, but in any case they must follow the principles of thermodynamics.

There is another way to find these same equations and laws by considering the fluid medium from the beginning as a set of atoms or molecules in steady motion and colliding frequently in the void space, under the action of the laws of classic mechanics for material points. This approach, often called the kinetic approach, was introduced by Boltzmann and Maxwell and has been used to recover classical Eulerian continuity equations and Navier–Stokes equations since the work of Chapman and Enskog in the early 20th Century. In this context, the characteristic variables that we might call "macroscopic" variables are defined in a small volume around each point in space, and in a small interval of time around a given instant. When the volume and the interval of time considered are very small in relation to the spatial and temporal scales of variation expected for macroscopic variables, in the macroscopic experiments and situations with which we are concerned, these macroscopic variables no longer depend on the real size of the volume nor the time interval, and have continuity properties with respect to space and time. It is in these conditions that we can consider the medium as continuous, on the macroscopic scale and with our limited view.

The base equation of the kinetic approach is the Boltzmann equation, which concerns the distribution function of the velocities of molecules at a given point M (x) and a given instant t. To be exact, and for a gas medium containing only one kind of molecule, this function, $f(u,x,t)$, is defined as the number of molecules that can be found, in the medium, in a small volume around point x approximately dx, at an instant t approximately dt, and which have velocity u approximately du (here, of course, the variables x and u are three-component vectors, as are dx and du). The number of molecules per unit of volume of the medium is, thus, defined as the integral in space of the values of velocities (from minus infinity to plus infinity) of this distribution function, and what we might call the macroscopic velocity of the medium is the integral, in this same space of velocities, of the product uf, divided by the integral of f itself. All the molecules of the gas being identical and of known mass, the volumetric mass of the medium is the product of the number of molecules per unit of volume and the mass of a molecule, and the velocity of the medium thus appears as the ratio between the momentum of the medium and its mass, both per unit of volume. The historical development of this approach is discussed in [CHA 60] and a very complete recitation of the developments in [HIR 54]. In the framework of the approximation of continuous media, this approach allows us to find the same forms of the balance equations of mass and momentum by using the first continuous approach discussed above. However, in addition, the fact that we can describe in more detail the microscopic structure of the material enables us to obtain the constitutive laws in a theoretical manner, that is expressions of all the variables that appear in the balance equations, namely the Cauchy stress tensor, pressure, viscous stress tensor, etc. These expressions are calculable only because of certain hypotheses on the microscopic characteristics of the gas being considered (for example the approximation of molecules as hard spheres of a certain mass). The energy balance equation of the continuous medium can also be found by defining the heat conduction flow in particular. When the approximation of the continuous medium is no longer valid, as is the case for rarefied gases, the Boltzmann equation remains valid and the balance equations of mass, momentum and energy still exist with the same overall form, but they contain additional terms, and the usual terms can no longer be calculated using the same approximations. In certain conditions, it is no longer useful to define a macroscopic velocity for the medium, and the calculation of the velocity distribution function itself, by using adequate approximations to represent collisions between molecules in the Boltzmann equation, is recommended.

When undertaking a theoretical study of multiphase fluid media, we might ask ourselves which of the two approaches is of more interest, especially when we wish to focus on a case in which one of the phases is dispersed into numerous parcels of different sizes.

In fact, it seems that the first studies involving clouds of drops in a gaseous medium used the kinetic approach to study the liquid phase composed of the set of drops by examining a distribution function where the variables were both the velocities and sizes of the drops, or $f(u,v,x,t)$, where v is the volume of the drops. At a given point and at a given instant, the number per unit of volume (of the medium) of drops of volume v, approximately dv, is still the integral of f on the space of velocities, and the total number of drops of all sizes is the integral of f both on the space of velocities and that of volumes. The function f follows a type of Boltzmann equation here again, in which the most difficult term to approximate is the term due to the collisions of particles. We may refer, for example to, [WIL 58]. However, it has never been suggested to represent the gaseous phase in which these drops are dispersed using a Boltzmann equation that, when the quantity of drops is relatively large, might include the influence of these drops on the motion of the fluid. The description of this gaseous phase has been kept in the form of a continuous medium, where the influence of the drops is represented by additional interaction terms.

For less well-defined multiphase media such as those containing bubbles and gas pockets that can be of various shapes, a “macroscopic continuous medium” approach has been applied from the beginning, by focusing on the characteristic variables for a small volume around each point in space. The basic hypothesis has been made that this volume is very large in relation to the size of the inclusions of the dispersed phases, and these variables therefore represent an “equivalent continuous medium” corresponding to our multiphase medium. The fictive volume within which the macroscopic variables are defined is often called the “representative elementary volume”. We may refer, for example to, [SOO 89] and [DEL 81]. In these conditions, the balance equations for these variables have the classic form but the necessary constitutive laws are more complicated; they cannot be calculated and must be obtained through experimentation. This approach is valid, but the definition of the characteristic variables of the medium, which must have a certain continuity, as well as those of the various constitutive laws, must be precisely made. Obtaining these laws is (too) highly dependent on experimentation, and the domain of validity of these laws is difficult to know with precision. Moreover, the condition that each inclusion must be much smaller than the representative elementary volume has very often been invalidated, or noted visually as soon as modern viewing methods were able to be used. Does this flaw seriously affect the approach, or is it responsible for only "small" errors? This remains an ongoing debate and, most likely, must depend on each of the various cases considered.

More recently, in the 1980s, it appeared possible to develop an approach that might be called intermediary, which defined the equivalent continuous medium more accurately and thus made it possible to obtain more information about the form of the constitutive laws, which in turn made it possible, importantly, to better examine situations with large inclusions, and could therefore be adapted to widely diverse practical situations. In fact, if the size of each inclusion in a phase is large enough to include a large number of molecules that interact on the interface with extremely numerous molecules of the other phase in contact, then it is possible to treat all of the phases, fluid or solid, like classic continuous media. The multiphase medium is, thus, a "piecewise continuous medium" in which each piece of continuous medium is separated from the other by an interface where the properties can show a discontinuity. Using this perspective, we enable the shape and size of the inclusions in each phase to vary in the medium in a totally arbitrary manner. However, the balance equations for phases in continuous media are unusable as such for representing the full multiphase medium, for the same reason that equations of material points mechanics cannot be used to represent a gas, which is a group of molecules. We do not precisely know the initial conditions of each molecule in the gas, nor the initial position or shape of each inclusion in each phase, even if these inclusions are not too small. To establish some macroscopic characteristics at the scale of a multiphase medium, and to deduce the macroscopic equations that they must satisfy, it is necessary to define, through an averaging operation that will be specified later, a "mean continuous" medium that will be the equivalent continuous medium. This approach has grown slowly since its introduction. It was presented for the first time in [DRE 83]. However, it is very similar to the approach used for flows in porous media by Marle in the beginning of 1967 [MAR 67]. At roughly the same time, Nigmatulin began using the same approach in the foundation chapters of his book [NIG 91], which also presents highly useful and detailed developments for several different applications. General equations for piecewise continuous media have been studied in detail by Kataoka [KAT 86], taking into account mass exchanges as well as exchanges of momentum or energy between these phases. This generality is one of the strong points of the approach.

This averaged piecewise continuous medium approach is the approach that we will follow in this book. In this way, we will avoid defining a distribution function for the sizes and velocities of parcels, but it will be important to know other macroscopic variables of the medium besides its volumetric mass and velocity. We will be interested at least by the "volume fraction" occupied in the medium by each of the phases, and also by the mean temperature or internal energy of each phase, which are to be defined precisely. It will also be necessary to have a thorough understanding of at least one macroscopic variable related to the size of the inclusions of a phase that is dispersed in another carrier phase. First, and in the most general of situations, it is through the "mean area of interface per unit of volume" of the medium that this knowledge will be introduced and studied. For the case of dispersed phases with "particles" of various sizes, if necessary, we can also understand the distribution of sizes and velocities by defining and studying the number of particles of volume v (approximately dv). This will come down to considering a multiphase medium with an infinite number of phases, with each category of particle in the given volume able to be considered as a different phase.

In Chapter 2, we will describe the basic equations that represent a piecewise continuous medium instantaneously and locally. The definition of averages and the obtaining of equations related to equivalent continuous media will be discussed in the following chapters.

1.2. Eulerian–Lagrangian and Eulerian formulations

If we wish to study only situations in which, in a fluid, small quasi-spherical particles of another fluid or a solid are dispersed, one practical method that comes immediately to mind is simply to study the motion of the center of gravity of these particles, taking into account the friction between them and the fluid that surrounds them, which depends on their volume and shape, using specific formulas that experiments can validate. If it is not only their motion that interests us, but also their temperature, for example, we can also study this quantity by supposing that it is homogeneous in the particle in question. This brings us back to consider the total internal energy of the particle, and its dependence on exchanges of heat and possibly of mass with the outside environment, without necessarily having to specify whether these changes occur more on one side or the other side of the particle, and using more or less empirical formulas.

This point of view, which separates the carrier fluid, still considered as a continuous medium, from a certain number of isolated objects, has been highly developed. It is not too complicated to implement when the particles have a well-defined geometric shape and retain this shape permanently. It has been used to construct an Eulerian representation of a mean particulate medium (for example [JAC 00]) and also, especially, for the so-called Eulerian-Lagrangian approach, in which the continuous fluid medium is still described by Eulerian equations but the particles of the other phase are observed in a Lagrangian manner (see [OES 06]). Suspensions containing liquid drops or solid particles can be studied using this perspective, as long as the droplets or particles do not break apart or cluster together too much. Exchanges between particles and the continuous medium have been studied in great detail, and these studies constitute a body of knowledge that is quite generally useful; we recommend, for example, [SIR 92] and [SIR 99]. The book by Crowe also provides a broad recapitulation of the knowledge and practices using this approach [CRO 01].

This particular point of view is not easy to generalize for types of multiphase media other than highly dispersed environments, and it is not inclined to allow a general theoretical approach, not least because it immediately introduces asymmetry into the description. Moreover, Lagrangian–Eulerian representation introduces significantly greater difficulties of application in situations where, as the number of particles is higher, these particles interact more with each other than with the fluid, and in situations where these particles do not have a constant and simple shape. There are, of course, methods for calculating the Lagrangian motion of solid parcels for granular media, taking into account the multiple contacts between these “grains”, but these contacts are firmly localized; the fluid between the grains is not taken into account, and increasing the number and complicating the shape of the grains greatly increases the calculation time. The Lagrangian point of view has certain advantages, not the same as the Eulerian point of view does; it is beneficial to make use of these advantages, but the theoretical separation of a Lagrangian part and an Eulerian part has no general justification.

The unified approach we are looking for will be completely Eulerian: the macroscopic continuous medium equivalent to the multiphase medium, which is non-continuous by nature, will be described by Eulerian balance equations. The necessary models will, therefore, be studied first of all in a Eulerian context. Nevertheless, it is of interest, as we will see in Chapter 18, to use a Lagrangian translation of certain equations and models, which is closer to the Lagrangian–Eulerian representation, often used in practice. We will then show how to link the balance equations of the two approaches. In addition, the possibility of coupled calculations, in the various geometric zones of the space, between a model expressed in Eulerian form and a model expressed in Lagrangian–Eulerian form is also a promising practical solution.

Chapter 2

Instant Equations for a Piecewise Continuous Medium

If we consider any multiphase medium at a given instant in any type of experimental situation, this medium is a "piecewise continuous" medium with different non-miscible phases because the local density of the medium undergoes sudden jumps on certain surfaces in the medium. Other local characteristics per unit of mass can also display these kinds of discontinuities, but it is not always the case. We will discuss this further later on. Here, we are considering a medium with two or more phases, at least one of which is fluid (gas or liquid); the others can be solid or fluid. A phase is usually defined by its physical nature, but it is also possible to define phases using other criteria, provided that several phases cannot be in the same place at the same instant; we will return to this point later as well.

We define an index, or indicative function, of phase i, as Φ_i, which is 1 in phase i and 0 in the others. Obviously, these indices are not independent, since their sum is 1 (at the interface, it is not necessary to define precisely the value of the index provided that it remains finite). In these conditions, if ρ_i is the density of phase i, defined only where phase i is found, $\rho_i\Phi_i$ will be defined everywhere, equal to ρ_i in phase i and zero elsewhere. Likewise, we can extend the definition of various interesting characteristics of phase i (velocity, internal energy, etc.) everywhere by simply multiplying them by Φ_i. We can thus write balance equations in integral and differential forms using characteristics of phase i that will be valid even in other phases (becoming $0 = 0$), and at interfaces (giving algebraic interface balance relationships). Since some variables are non-continuous, the derivatives used in differential equations must be defined in the sense of distributions, specifically by using Dirac distributions. This approach seems to have been presented for the first time by Kataoka [KAT 86]. Our discussion of the approach here will be relatively brief but will include the most important results.

2.1. Integral and differential forms of balance equations

The simplest form of a balance equation concerns a given control volume of space, defined by the experimentalist and set in relation to the chosen reference coordinates. For each extensive variable of the medium (total mass, mass of a certain type, total momentum, momentum of mass of a certain type, total energy, etc.), we simply write that the temporal variation rate is either due to processes of exchange with the exterior environment via the surface of the volume, or due to phenomena of production (positive or negative) internal to the control volume. If we write the extensive variables per unit of mass that can be defined locally as g, we have:

$$\frac{d}{dt}\iiint_{vol} \rho g dv + \iint_{surf} \rho g v_g^\alpha n_S^\alpha dS = \iiint_{vol} \rho \dot{w}^g dv$$

The second term to the left of the equals sign is on the left because the normal to the surface of the volume is defined, as is usual, as the outgoing normal. This is the flux of quantity g exiting from the volume (and thus exchanged with the exterior), and we can write it formally by defining a velocity v_g^α associated with this exchange flux per unit of surface in direction α. The quantity $\rho \dot{w}^g$ is the rate of production (or disappearance) of the quantity in question, per unit of volume, due to any physical or chemical process possible (which must then be identified). The dot above is simply a notation and does not indicate a temporal derivative; however, this notation refers to the fact that, without an exchange with the exterior, the previous balance equation shows that the rate is equal to a temporal derivative. The presence of the volume mass factor is simply a convention that is widely but not always used.

For the mass balance, $g = 1$ and the velocity associated with the exchange flux is simply the classic convection velocity $v_g^\alpha = v^\alpha$, but it is different for the other variables. For the mass of a substance in a fluid mixture, the variable g is the mass fraction (which we will write later, as is usual, as Y_j for substance j), and the velocity of this substance must add to the velocity of the total mass, the velocity of (molecular) diffusion $v_j^\alpha = v^\alpha + v_{Dj}^\alpha$. For the momentum, the flux per unit of surface (which in this case is a second-order tensor since the momentum is a vector) also includes the convection flux of the momentum vector plus the opposite of the

Cauchy stress tensor, and is therefore $\rho v^{\alpha}v^{\beta}-\sigma^{\alpha\beta}$. We know this is necessary in order to be able to represent the fact that the momentum of a system that does not exchange mass with the exterior is modified by the forces of contact with this exterior (as well as remote forces, but these, as it is not a surface-related phenomenon, must be taken into account in the rate of production). For the total energy, this flux per unit of surface also includes heat flux and the power of contact forces, in agreement with what is called the first law of thermodynamics.

For a continuous medium, using the so-called divergence theorem, which transforms the surface integral into a volume integral, we deduce from an integral balance, for a fixed control volume of any shape and position, a balance partial differential equation, as follows:

$$\frac{\partial}{\partial t}(\rho g)+\frac{\partial}{\partial x_{\alpha}}(\rho g v_{g}^{\alpha})=\rho\dot{w}^{g}$$

We are mainly interested here in the classic balances of total mass, momentum and total energy, which we rewrite showing the fluxes and rates of production:

$$\frac{\partial}{\partial t}(\rho)+\frac{\partial}{\partial x_{\alpha}}(\rho v^{\alpha})=0$$

$$\frac{\partial}{\partial t}(\rho v^{\beta})+\frac{\partial}{\partial x_{\alpha}}(\rho v^{\beta}v^{\alpha})=\frac{\partial}{\partial x_{\alpha}}(\sigma^{\alpha\beta})+\rho g^{\beta}$$

$$\frac{\partial}{\partial t}(\rho e_{t})+\frac{\partial}{\partial x_{\alpha}}(\rho e_{t}v^{\alpha})=\frac{\partial}{\partial x_{\alpha}}(-j_{Q}^{\alpha}+\sigma^{\alpha\beta}v^{\beta})+\rho g^{\beta}v^{\beta}$$

The mass balance of a substance j in the fluid medium is written, with the mass fraction of this substance, as:

$$\frac{\partial}{\partial t}(\rho Y_{j})+\frac{\partial}{\partial x_{\alpha}}(\rho Y_{j}v^{\alpha})=\frac{\partial}{\partial x_{\alpha}}(-j_{Dj}^{\alpha})+\rho\dot{w}_{j}$$

The diffusion flux of this substance in the medium is related to the diffusion velocity previously defined: $j_{Dj}^{\alpha}=\rho Y_{j}v_{Dj}^{\alpha}$, and the term of production, or disappearance, of the substance in question is zero unless it is caused by chemical reactions.

These differential balance equations are valid in each phase of a multiphase medium, whether these phases are gaseous, liquid or solid, with differences in the physical meaning and expression of the values defined in each of these phases. Here, we have applied classic principles, stipulating that the production rate of the total mass is always zero, that the production rate of momentum is the sum of the remote forces and that the total energy production rate is equal to the power of the remote forces. It is supposed here that the remote forces are composed solely of gravity due to Earth, and thus are linked to g^{β}, but this is transposable to other types of forces.

Integral balances can also be written for an environment in which two or more phases are present in the volume, *Vol*. By using a phase index as stated above (1 in phase *i* and 0 elsewhere) Φ_i, we have:

$$\frac{d}{dt}\iiint_{vol}\sum_{i}\rho_i\Phi_i g_i dv+\iint_{surf}\sum_{i}\rho_i\Phi_i g_i v_{gi}^{\alpha}n_S^{\alpha}dS=\iiint_{vol}\sum_{i}\rho_i\Phi_i\dot{w}_i^{g}dv \qquad [2.1a]$$

The production rate can include volumetric phenomena, as well as phenomena that occur only at the interface, and in this case, the part of the rate that represents this surface phenomenon has a Dirac peak structure placed at this interface (and its volumetric integral is therefore well defined).

A particularly interesting application of this balance concerns a very small volume that only contains a small portion of the interface between two phases *i* and *i'*, and then moves along with this interface. We can also write an integral balance equation for a mobile volume; and in this case, the second term has to use the velocity relative to the surface of the volume in motion. Here, the volume is moving with the interface, so we will have $v_{gi}^{\alpha}-v_s^{\alpha}$. When a very small volume is considered, the first term of the above equation is zero, the second term is simply the sum of the fluxes of *g* exiting from the side of phase *i* and the side of phase *i'* and the third term is reduced to the integral of the rate of production at the interface. In this case, by defining n_i^{α} as the normal at the interface exiting from *i* (and $n_{i'}^{\alpha}$ as the normal exiting from *i'*, $n_{i'}^{\alpha}=-n_i^{\alpha}$), we get the general algebraic equation valid at all points of the interface:

$$\rho_{is}g_{is}(v_{gis}^{\alpha}-v_s^{\alpha})n_i^{\alpha}+\rho_{i's}g_{i's}(v_{gi's}^{\alpha}-v_s^{\alpha})n_{i'}^{\alpha}=\Gamma_{ii'}^{sg} \qquad [2.1b]$$

The index *s* associated to *i* or *i'* indicates that the quantities within the phase are taken immediately at the interface. The term on the right-hand side represents the overall production rate per unit of surface of the quantity considered by surface processes at the interface between phases. It is zero unless such processes can exist and must be provided by studying the involved physics.

For masses, momentum and total energy, in the classic case that does not consider surface tension, these surface terms are zero and this interface balance relationship is written simply as:

$$\rho_{is}(v_{is}^{\alpha} - v_{s}^{\alpha})n_{i}^{\alpha} + \rho_{i's}(v_{i's}^{\alpha} - v_{s}^{\alpha})n_{i'}^{\alpha} = 0$$

$$\rho_{is}v_{is}^{\beta}(v_{is}^{\alpha} - v_{s}^{\alpha})n_{i}^{\alpha} - \sigma_{is}^{\alpha\beta}n_{i}^{\alpha} + \rho_{i's}v_{i's}^{\beta}(v_{i's}^{\alpha} - v_{s}^{\alpha})n_{i'}^{\alpha} - \sigma_{i's}^{\alpha\beta}n_{i'}^{\alpha} = 0$$

$$\rho_{is}e_{tis}(v_{is}^{\alpha} - v_{s}^{\alpha})n_{i}^{\alpha} + j_{Qi}^{\alpha}n_{i}^{\alpha} - \sigma_{is}^{\alpha\beta}v_{is}^{\beta}n_{i}^{\alpha} + \rho_{i's}e_{ti's}(v_{i's}^{\alpha} - v_{s}^{\alpha})n_{i'}^{\alpha} + j_{Qi'}^{\alpha}n_{i'}^{\alpha} - \sigma_{i's}^{\alpha\beta}v_{i's}^{\beta}n_{i'}^{\alpha} = 0$$

These relationships are used directly, leaving out the irreversible effects of viscosity and heat transfer, when we study shock waves or detonation or deflagration waves. We will see their general usefulness later on, as well as how to include the effects of surface tension in them.

As is often done in this book, we use, for the sake of simplicity, indexed notations with Einstein's notation, but these indexed notations are valid only in orthonormal Cartesian coordinates. Indices in Greek letters represent a direction in space, and if an index is repeated in a monomial, it implies a summation of the monomial with all possible values of this index. This is not the case with index i, which indicates a phase, or with other indices in Roman letters. The high or low position of the indices has no particular significance here.

In the general case, in which we consider any volume filled with a multiphase medium, it is not possible to move directly from an integral balance [2.1a] to a differential balance. However, since the medium remains piecewise continuous, we can obtain generalized differential balance equations, for each characteristic variable of each phase, by using derivatives in the sense of distributions. We will do this successively for masses, momentum and total energy, as we have stated, following the essential reasoning of Kataoka.

2.2. Phase mass balance equations in a piecewise continuous medium

We will show that the differential balance equation of the mass of phase i in the entire space can be written as:

$$\frac{\partial}{\partial t}(\rho_i\Phi_i) + \frac{\partial}{\partial x_\alpha}(\rho_i\Phi_i v_i^{\alpha}) = -\rho_{is}(v_{is}^{\alpha} - v_{s}^{\alpha})n_i^{\alpha}\sigma_i^{s} \qquad [2.1c]$$

This writing must be taken in the sense of distributions, since the phase index is non-continuous, and we will also see that the right-hand term has a Dirac peak structure.

The quantity σ_i^s is *the instantaneous area of interface per unit of volume*, for an infinitely small volume, which can be seen in Figure 2.1.

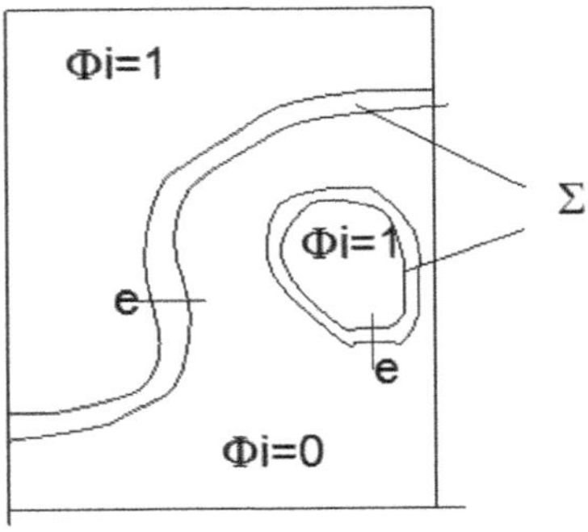

Figure 2.1. *Cross-section of a parallelepipedic volume with a two-phase medium*

Here, we see a cross-section of a parallelepipedic volume with a two-phase environment. The phase index is the limit of a function that is zero in one part of this volume, equal to one in another part and varies linearly between 0 and 1 in a small band of constant thickness *e* between the preceding parts, a band whose thickness tends progressively toward zero. In these conditions, it is clear that the algebraic value of the gradient of this function is nil in the first two parts of the volume, and constant and equal to $-1/e$ in the middle band (the gradient itself is a vector normal to the limits of this band, and directed toward the side where the phase index is 1). The integral of the opposite of this algebraic value in the whole volume is therefore $1/e$ multiplied by the portion of the volume in which the quantity is non-nil, which is the product of *e* by the total area of interface Σ in the volume, which gives Σ and is independent of *e* when *e* tends to be smaller. Thus:

$$\iiint_{Vol} -n_i^\alpha \frac{\partial \Phi_i}{\partial x_\alpha} dv = \Sigma = \iiint_{vol} \sigma_i^s dv .$$

The volume can be of any value; then we have: $-n_i^\alpha \dfrac{\partial \Phi_i}{\partial x_\alpha} = \sigma_i^s$, even when *e* is zero.

Thus, we can see that σ_i^s has a Dirac peak structure, meaning that it is proportional to such a Dirac peak. We can also call this quantity: "instantaneous interface density" (the notation is slightly similar to that of a Cauchy tensor, but the higher index *s* distinguishes it from the two Greek letters in the higher index for this tensor).

If the interface is represented by a relationship $F(x_1,x_2,x_3,t)=0$, with *F* being a dimensionless function, positive in phase *i*, $n_i^\alpha=-(\frac{\partial F}{\partial x_\alpha}/\left|\frac{\partial F}{\partial x_\alpha}\right|)_{F=0}$ and $\frac{\partial \Phi_i}{\partial x_\alpha}=\delta(F)\frac{\partial F}{\partial x_\alpha}$, which implies that $\sigma_i^s=\delta(F)\left|\frac{\partial F}{\partial x_\alpha}\right|$. The norm of the gradient of *F* has the dimension of the inverse of a length, and also σ_i^s (it is a surface per unit of volume), and this length, which must be a local characteristic of *F*, can only be proportional to the curvature radius of the interface at the point being considered. We can see as well that $n_i^\alpha(\left|\frac{\partial F}{\partial x_\alpha}\right|)_{F=0}=-(\frac{\partial F}{\partial x_\alpha})_{F=0}$, or $n_i^\alpha\left|\frac{\partial F}{\partial x_\alpha}\right|\delta(F)=-\frac{\partial F}{\partial x_\alpha}\delta(F)$, and therefore $n_i^\alpha\sigma_i^s=-\frac{\partial \Phi_i}{\partial x_\alpha}$.

We will now show that [2.2] represents, in the whole space and at the interface, the balance of the mass of phase *i*. We can rewrite this equation as:

$$\Phi_i(\frac{\partial}{\partial t}(\rho_i)+\frac{\partial}{\partial x_\alpha}(\rho_i v_i^\alpha))=-\rho_{is}\frac{\partial}{\partial t}(\Phi_i)-\rho_{is}v_{is}^\alpha\frac{\partial}{\partial x_\alpha}(\Phi_i)-\rho_{is}(v_{is}^\alpha-v_s^\alpha)n_i^\alpha\sigma_i^s$$

where the left-hand term no longer contains derivatives of discontinuous functions. In accordance with the rules of calculating distributions, the terms $\frac{\partial \Phi_i}{\partial t},\frac{\partial \Phi_i}{\partial x_\alpha}$ have a Dirac peak structure, non-nil only at the interface, as already shown for σ_i^s. In these conditions, this equation is $\frac{\partial}{\partial t}(\rho_i)+\frac{\partial}{\partial x_\alpha}(\rho_i v_i^\alpha)=0$ in phase *i*, as it should be. In the other phases, it is reduced to 0 = 0, as it should be. At the interface between phase *i* and the others, it gives us: $0=-\rho_{is}\frac{\partial}{\partial t}(\Phi_i)-\rho_{is}v_{is}^\alpha\frac{\partial}{\partial x_\alpha}(\Phi_i)-\rho_{is}(v_{is}^\alpha-v_s^\alpha)n_i^\alpha\sigma_i^s$. The geometry of the interface in motion showing that $\frac{\partial \Phi_i}{\partial t}=-v_s^\alpha\frac{\partial \Phi_i}{\partial x_\alpha}$ it gives $\rho_{is}(v_{is}^\alpha-v_s^\alpha)\frac{\partial}{\partial x_\alpha}(\Phi_i)=-\rho_{is}(v_{is}^\alpha-v_s^\alpha)n_i^\alpha\sigma_i^s$, which is true since $-n_i^\alpha\sigma_i^s=\frac{\partial \Phi_i}{\partial x_\alpha}$.

We can present the phenomenon of phase change as a phenomenon of production (or disappearance) of mass in phase i at the interface, which has a zero volumetric rate everywhere except at the interface, and is therefore written as:

$$-\rho_{is}(v_{is}^{\alpha}-v_{s}^{\alpha})n_{i}^{\alpha}\sigma_{i}^{s}=m_{i}\sigma_{i}^{s} \qquad [2.2]$$

This relationship can be seen as an interface balance for the mass of phase i, with the same general form [2.1b] as in section 2.1, since there is no mass of i in phase i'. The rate causes a loss of mass if it is negative, i.e when the relative velocity of phase i is in the direction emerging from i. There will be a gain in the opposite case and a value of zero when the relative velocity is zero, i.e. the interface velocity is the same as that of phase i at the level of this interface. This loss or gain of mass can represent vaporization or condensation when phase i and the phase in contact with it by this portion of the interface are the liquid and gas phases of the same substance; for example, we can also consider a chemical reaction at the interface, which adsorbs phase i or produces it from the other phase. The mass flow rate per unit of surface m_i must be specified in each case by a formula that represents the implied physical phenomenon.

Note that this assumes that the whole interface has only one normal direction, the two normal vectors exiting from each of the phases present at an interface being equal and opposite. This is always the case if there are only two fluid phases, but with three fluid phases or two fluids and a solid, there are certain lines where three interfaces join together. In the case of contact between two solid phases, or between two grains of a single solid phase, there can also be points or lines of the interface with several normals, at the angular points of the grains surface. We will suppose here that these local situations are very rare, assuming that they disturb only one portion of the interface that is negligible compared to the whole.

When there are only two phases, a single instantaneous interface density suffices, which can also be defined by the two gradients of the phase indices. For cases of three or more phases, it is necessary to define as many local interface densities as phase pairs: the value σ_i^s, defined as above, is therefore the interface density of i with any other phase. The interface density between i and i' $\sigma_{i,i'}^s$ is identical to σ_i^s or $\sigma_{i'}^s$, where these quantities are both non-zero and equal, and zero elsewhere.

It is also interesting to calculate the overall balance of the mass in the medium, irrespective of the phase, and to do this we must add up all of the mass balance equations for all of the phases. We define $\rho=\sum_i \rho_i \Phi_i$, meaning that ρ has the value of ρ_i where phase i is present, that of ρ_j where phase j is present, etc., and $\rho v^\alpha = \sum_i \rho_i \Phi_i v_i^\alpha$, which results in $v^\alpha = \sum_i \Phi_i v_i^\alpha$. We obtain from [2.1c] simply:

$$\frac{\partial}{\partial t}(\rho) + \frac{\partial}{\partial x_\alpha}(\rho v^\alpha) = 0 \qquad [2.3]$$

As the mass that disappears from phase i via exchange with another phase necessarily appears in this other phase (or vice versa), it is clear that the sum in all phases of the mass exchange terms is zero everywhere: $m_i + m_{i'} = 0$ and $\sum_i m_i \sigma_i^s = 0$. The interface, if it is capable of removing mass from phase i, transfers this mass instantaneously to the adjacent phase.

From [2.4], it can be observed that the classic continuity equation is still valid in this piecewise continuous medium, provided that we consider it in the sense of distributions, since the variables involved can be discontinuous.

In each phase i, equation [2.4] is precisely written in a classic manner, and in the classic sense:

$$\frac{\partial}{\partial t}(\rho_i) + \frac{\partial}{\partial x_\alpha}(\rho_i v_i^\alpha) = 0$$

Equation [2.4] also has significance for the interface between phases. We can develop this equation by introducing in it $\rho = \sum_i \rho_i \Phi_i$ and $\rho v^\alpha = \sum_i \rho_i \Phi_i v_i^\alpha$ and removing the phase indices from the derivatives, as done previously in [2.2] with the Kataoka method. At the end of the calculation, we find simply the result of the interface mass balance found in section 2.1:

$$\rho_{is}(v_{is}^\alpha - v_s^\alpha) n_i^\alpha + \rho_{i's}(v_{i's}^\alpha - v_s^\alpha) n_{i'}^\alpha = 0, \text{ or } m_i + m_{i'} = 0 \qquad [2.4]$$

Multiplied by σ_i^s, this relationship corresponds to $\sum_i m_i \sigma_i^s = 0$ in the whole medium.

2.3. Momentum balances

The instantaneous balance equation of the momentum in phase i is written, using the same approach, as:

$$\frac{\partial}{\partial t}(\rho_i \Phi_i v_i^\beta) + \frac{\partial}{\partial x_\alpha}(\rho_i \Phi_i v_i^\alpha v_i^\beta) = \frac{\partial(\Phi_i \sigma_i^{\alpha\beta})}{\partial x_\alpha} + \rho_i \Phi_i g_\beta + m_i \sigma_i^s v_{is}^\beta + f_{i'i,s}^\beta \sigma_i^s \qquad [2.5]$$

Here, the first term on the right-hand side represents the effect of contact forces in phase i only. The force of gravity, which is exerted in the volume of the phase, is a volumetric production rate proportional to the volume fraction of i, the second term on the right-hand side. The effect of the force of contact between phase i and the adjacent phase is found in the last term on the right-hand side, correctly proportional to the area of the interface σ_i^s : $f_{i'i,s}^{\beta}$ is the force, per surface unit, resulting from the action of phase i' in contact with i at this point. The second to last term, proportional to m_i, is the contribution to the momentum balance due to the mass exchange between phases. We could have written the momentum exchange between phases more generally as $\rho_{is}\dot{w}_i^{s\beta} = (m_i v_{is}^{\beta} + f_{i'i,s}^{\beta})\sigma_i^s$, but it would then have become clear that this exchange includes a contribution due to the exchange of mass, if it exists, and contact force, and is of course proportional to σ_i^s.

The Cauchy stress tensor in phase i, $\sigma_i^{\alpha\beta}$, can be written, with defining the pressure as one-third of the opposite of the reversible stress tensor trace, as: $\sigma_i^{\alpha\beta} = -p_i\,\delta_{\alpha\beta} + \tau_i^{\alpha\beta}$. This stress tensor and its decomposition can be defined in all types of phases: when the medium is fluid, the tensor $\tau_i^{\alpha\beta}$ is simply the viscous stress tensor, but if it is solid, it also includes the reversible Cauchy tensor deviator. We will retain this unusual style of writing in the following for the sake of simplicity.

In any case, whether the medium i is fluid or solid, these quantities must be known so that [2.6] can be used. This is indeed the case because for each material, there are *constitutive laws* that enable us to express them in terms of the other specific extensive variables of the medium, either simple equations of state if the medium is supposed to be in reversible evolution, or including a part due to an irreversible behavior. These are classic relationships. The only difficulty that can occur here is when a solid phase is composed of several grains in a simply connected zone where $\Phi_l = 1$. In this case, inside the solid phase itself, sliding can occur between grains, and in these conditions, the Cauchy tensor of the medium (which is no longer really solid) is not well defined. In most of this book, we will consider only situations in which contacts between solid particles are very brief, without lasting sliding between one another, so as to be able to consider that the solid medium is well known. This requires simultaneously sufficient agitation of the grains and a suitably small volume fraction of the solid. It is difficult to estimate the limit immediately, but it must not in any case exceed the random loose packing value (which depends on the distributions of the sizes and shapes of the grains; it is 0.55 for spheres of the same diameter), perhaps one half. This is not the case in the so-called granular media, except in rare cases of granular gases, and we will study these media particularly in Part 3 of this book.

Again, we will consider equation [2.6] and define more precisely the term of contact force between phases. As we saw earlier for the mass balance equation, we see that [2.6] in phase i is reduced to the classic equation, and to 0 = 0 in the other phases. At the interface, using Kataoka's reasoning from section 2.2, this gives:

$$v_{is}^{\beta}(\rho_{is}\frac{\partial}{\partial t}(\Phi_i)-\rho_{is}v_{is}^{\alpha}\frac{\partial}{\partial x_{\alpha}}(\Phi_i))=\sigma_{is}^{\alpha\beta}\frac{\partial}{\partial x_{\alpha}}(\Phi_i)+m_i\sigma_i^s v_{is}^{\beta}+f_{i'i,s}^{\beta}\sigma_i^s$$

or: $v_{is}^{\beta}m_i\sigma_i^s=-\sigma_{is}^{\alpha\beta}n_i^{\alpha}\sigma_i^s+m_i\sigma_i^s v_{is}^{\beta}+f_{i'i,s}^{\beta}\sigma_i^s$

We can see the significance of already defining the mass exchange contribution. Finally:

$$\sigma_{is}^{\alpha\beta}n_i^{\alpha}\sigma_i^s=f_{i'i,s}^{\beta}\sigma_i^s \quad [2.6]$$

This means that the contact force per surface unit exerted on phase i is indeed equal and opposite to the force exerted by phase i at the interface, even when mass exchange is occurring.

When we disregard any surface phenomenon and exchange of mass between phases, the force exerted on phase i is exerted by the other phase in contact with it at this point, and this means that: $f_{i'i,s}^{\beta}+f_{ii',s}^{\beta}=0$, and therefore $\sigma_{is}^{\alpha\beta}n_i^{\alpha}+\sigma_{i's}^{\alpha\beta}n_{i'}^{\alpha}=0$ at any point of the common interface between i and i'. Without surface tension, but with exchange of mass, it is the exchange of momentum between i and i' that is equal and opposite to that between i' and i, since the system of two phases in contact preserves its momentum, and thus at any point of the common interface, $m_i\sigma_{i,i'}^s v_{is}^{\beta}+m_{i'}\sigma_{i,i'}^s v_{i's}^{\beta}+\sigma_{is}^{\alpha\beta}n_i^{\alpha}\sigma_{i,i'}^s+\sigma_{i's}^{\alpha\beta}n_{i'}^{\alpha}\sigma_{i,i'}^s=0$. In the general case, in which we consider surface tension, we must consider that it is the interface that acts on phase i as well as on the phase in contact on the other side, and therefore the interface balance of momentum, at any point of the common interface between i and i', is written as:

$$m_i\sigma_{i,i'}^s v_{is}^{\beta}+m_{i'}\sigma_{i,i'}^s v_{i's}^{\beta}+\sigma_{is}^{\alpha\beta}n_i^{\alpha}\sigma_{i,i'}^s+\sigma_{i's}^{\alpha\beta}n_{i'}^{\alpha}\sigma_{i,i'}^s=F_{capi,i'}^{\beta}\sigma_{i,i'}^s \quad [2.7]$$

We need to know how to express the capillary force per surface unit, which depends, of course, on the two phases in contact and necessitates an additional appropriate formula, for example the Laplace formula $F_{capi,i'}^{\beta}=a_{i,i'}\kappa^s n_i^{\beta}$, where κ is the mean curvature of the interface and a is the surface tension constant (not always constant, incidentally).

The instantaneous momentum equation of the medium, including all phases, is again the sum of equations [2.6] for all phases (even taking surface tension into account, the interface has neither mass nor momentum).

$$\frac{\partial}{\partial t}\left(\rho v^{\beta}\right)+\frac{\partial}{\partial x_{\alpha}}(\rho v^{\alpha} v^{\beta})=\frac{\partial \sigma^{\alpha\beta}}{\partial x_{\alpha}}+\rho g_{\beta}+F_{cap}^{\beta}\sigma^{s} \qquad [2.8]$$

Here again, ρv^{β} is the momentum of phase i if we are in this phase, or of another phase if we are in that phase. Since phases do not overlap, it is easy to verify that: $\sum_i \rho_i \Phi_i v_i^{\alpha} v_i^{\beta} = \rho v^{\alpha} v^{\beta}$.

The value $F_{cap}^{\beta}\sigma^{s}$ concerns all of the interfaces, meaning that $F_{cap}^{\beta}\sigma^{s} = \sum_{i,i'} F_{capi,i'}^{\beta}\sigma_{i,i'}^{s}$. Since the portions of interface i,i' do not overlap, we can also say that $\sigma^{s} = \sum_{i,i'} \sigma_{i,i'}^{s}$ and that F_{cap}^{β} is defined for the set of interfaces in a discontinuous manner, with $F_{capi,i'}^{\beta}$ following the portion of interface i,i' it involves.

The Cauchy stress tensor of the medium is: $\sigma^{\alpha\beta} = \sum_i \Phi_i \sigma_i^{\alpha\beta}$; in general, we can expect that each component of this tensor will be discontinuous at the interfaces when looking from one phase to another. However, in all cases, relationship [2.8] shows that at these interfaces, there are three relations binding the six components of the tensor on one side to the six components on the other side. With no exchange of mass between the phases, the physics indicates that the velocities of each phase are identical at the interface (disregarding, as earlier explained, interfaces between two solid phases or two grains of the same solid phase, which can slide). With an exchange of mass, only the tangential velocities at the interface remain identical, and the non-nil normal velocities are related by the mass balance of the interface. Of course, in each phase, this tensor satisfies the constitutive laws of that phase, which link it to other characteristics of the phase in question.

Equation [2.9] reduces in each phase to the classic momentum balance equation (without a surface tension term). It also has a meaning on the interface that we can now specify using the integral balance at the interface, and we again obtain [2.8].

2.4. Energy balances

The instantaneous energy balance equation for phase i can be written more easily with the total energy per unit of mass, $e_{ti} = e_i + \frac{1}{2} v_i^\beta v_i^\beta$, i.e. the internal energy plus the kinetic energy, and we get the equation:

$$\frac{\partial}{\partial t}(\rho_i \Phi_i e_{ti}) + \frac{\partial}{\partial x_\alpha}(\rho_i \Phi_i v_i^\alpha e_{ti}) = \frac{\partial}{\partial x_\alpha}(\Phi_i(-j_i^{Q\alpha} + \sigma_i^{\alpha\beta} v_i^\beta)) + \rho_i \Phi_i v_i^\beta g_\beta$$

$$+ m_i \sigma_i^s e_{tis} + f_{i'i,s}^\beta \sigma_i^s v_{is}^\beta + q_{is} \sigma_i^s \qquad [2.9]$$

The first term on the right side represents the exchange of energy by the fluid particle of phase i with the same phase i around it, either by heat exchange (conduction) or by the power of contact forces, and the second term on the right-hand side is the power of remote forces, in the volume of the phase, that is due to the gravity. The heat flux via conduction in the phase $j_i^{Q\alpha}$ must be given by a law of irreversible phenomenon appropriate for the phase in which it is occurring, and since this law involves the temperature, we must also have an equation of state to calculate the temperature as a function of internal energy, and possibly of mass by unit of volume.

The last term is the heat exchange at the interface, contributed to phase i by the other phase in contact. The total exchange of energy between phases in contact also includes the power of contact force, the second last term, and the contribution due to mass exchange is the term before it.

We prove this equation in the same way as earlier, by showing that it gives the classic balance equation again in phase i and 0 = 0 in the other phases. At the interface, still with the same reasoning, we again observe, given [2.7], that the heat contributed to phase i is the same as the heat that enters this phase via conduction, or:

$$q_{is} \sigma_i^s = -j_{is}^{Q\alpha} n_i^\alpha \sigma_i^s \qquad [2.10]$$

In the absence of surface phenomena, the sum of all of the energy balance equations gives the balance equation for the total energy of the medium, in which the interphase exchange terms are correctly eliminated:

$$\frac{\partial}{\partial t}(\rho e_t) + \frac{\partial}{\partial x_\alpha}(\rho v^\alpha e_t) = \frac{\partial}{\partial x_\alpha}(-j_\alpha^Q + \sigma^{\alpha\beta} v^\beta) + \rho v^\beta g_\beta \qquad [2.11a]$$

The total energy of the medium is again $\sum_i \rho_i \Phi_i e_{ti} = \rho e_t$, i.e. that of the phase we are examining. The interface balance for energy is written, for every small portion of interface *i,i′*, as:

$$m_i \sigma^s_{i,i'} e_{tis} + \sigma^{\alpha\beta}_{is} n^\alpha_i v^\beta_{is} \sigma^s_{i,i'} + q_{is} \sigma^s_{i,i'} + m_{i'} \sigma^s_{i,i'} e_{ti's} + \sigma^{\alpha\beta}_{i's} n^\alpha_{i'} v^\beta_{i's} \sigma^s_{i,i'} + q_{i's} \sigma^s_{i,i'} = 0 \quad [2.11b]$$

However, taking capillarity into account, we must consider that the interfaces also have an amount of energy per surface unit, say e_s, and the total energy balance is:

$$\frac{\partial}{\partial t}\left(\rho e_t + e_s \sigma^s\right) + \frac{\partial}{\partial x_\alpha}(\rho v^\alpha e_t) + v^\alpha_s \frac{\partial}{\partial x_\alpha}(e_s \sigma^s) = \frac{\partial}{\partial x_\alpha}(-j^Q_\alpha + \sigma^{\alpha\beta} v^\beta)$$
$$+ \rho v^\beta g_\beta + (F^\beta_{cap} v^\beta_s + \Gamma_s e_s)\sigma^s$$

The interface energy per unit of volume (which is the interface energy per surface unit multiplied by the interface area per unit of volume) is carried with the velocity of the interface, and is produced (or destroyed) both by the power of surface tension force and the possible creation or destruction of the interface. As above, we have written σ^s as the sum of all of the $\sigma^s_{i,i'}$, and $e_s, v^\beta_s, F^\beta_{cap}, \Gamma_s$ depend on the pair of phases separated by the portion of interface being considered. This is the theory of capillarity, which gives e_s and $F^\beta{}_{cap}$ both in accordance. The interface balance for energy is therefore no longer written in an algebraic form as in [2.11b]; only at the interface, the total energy balance above gives:

$$\frac{\partial}{\partial t}(e_s) + v^\alpha_s \frac{\partial}{\partial x_\alpha}(e_s) = \sum_{i,i'} (\rho_{is} e_{tis} m_i + q_{is} + \sigma^{\alpha\beta}_{is} n^\alpha_i v^\beta_{is}) + F^\beta_{cap} v^\beta_s$$

It is not easy to specify the terms of these equations, but in many problems the energy involved in the capillary phenomena can be completely disregarded as compared to the variations contributed by other phenomena, and we will therefore limit ourselves to balance equation [2.11a].

The energy balance is often used in terms of enthalpy for fluid media. By definition, the enthalpy in phase *i* is: $h_i = e_i + p_i / \rho_i$. The total enthalpy is the sum of enthalpy and kinetic energy. Remember that the pressure is defined as the opposite of one-third of the reversible Cauchy tensor trace, and that $\sigma^{\alpha\beta}_i = -p_i \delta^{\alpha\beta} + \tau^{\alpha\beta}_i$. We can deduce from [2.9], with [2.10], [2.2] and [2.6]:

$$\frac{\partial}{\partial t}(\rho_i \Phi_i h_{ti}) + \frac{\partial}{\partial x_\alpha}(\rho_i \Phi_i v^\alpha_i h_{ti}) = \frac{\partial}{\partial x_\alpha}(\Phi_i(-j^{Q\alpha}_i + \tau^{\alpha\beta}_i v^\beta_i)) + \frac{\partial}{\partial t}(\Phi_i p_i)$$

$$+ \rho_i \Phi_i v^\alpha_i g^\alpha + (m_i + \sigma^{\alpha\beta}_{is} v^\beta_{is} - j^{Q\alpha}_{is}) n^\alpha_i \sigma^s_i$$

Or $\frac{\partial}{\partial t}(\Phi_i p_i) = \Phi_i \frac{\partial}{\partial t} p_i + p_i \frac{\partial}{\partial t}\Phi_i = \Phi_i \frac{\partial}{\partial t} p_i - p_i v_s^\alpha \frac{\partial}{\partial x_\alpha}\Phi_i = \Phi_i \frac{\partial}{\partial t} p_i + p_i v_s^\alpha n_k^\alpha \sigma_i^s$

and

$$\frac{\partial}{\partial t}(\rho_i \Phi_i h_{ti}) + \frac{\partial}{\partial x_\alpha}(\rho_i \Phi_i v_i^\alpha h_{ti}) = \frac{\partial}{\partial x_\alpha}(\Phi_i(-j_i^{Q\alpha} + \tau_i^{\alpha\beta} v_i^\beta)) + \Phi_i \frac{\partial}{\partial t}(p_i)$$

$$+\rho_i \Phi_i v_i^\alpha g^\alpha + m_i \sigma_i^s h_{tis} - \tau_{is}^{\alpha\beta} v_{is}^\alpha n_i^\beta \sigma_i^s - j_{is}^{Q\alpha} n_i^\alpha \sigma_i^s \qquad [2.12a]$$

Without capillarity, at any point of an interface between phases i and i', the interface balance relationship becomes:

$$\left(h_{ti} - h_{ti'}\right) m_i \sigma_{i,i'}^s - \left(\tau_i^{ml} v_i^l - \tau_{i'}^{ml} v_{i'}^l\right) n_i^m \sigma_{i,i'}^s + \left(j_i^{Qm} - j_{i'}^{Qm}\right) n_i^m \sigma_{i,i'}^s + (p_i - p_{i'}) v_s^m n_i^m \sigma_{i,i'}^s = 0 \quad [2.12b]$$

The last term is nil if we disregard the effects of mass exchange, because the momentum interface balance is reduced to the equality of pressures of each phase at the interface. As a general rule, it remains very weak.

2.5. Position and interface area balance equations

By their mathematical definition, phase indices simply follow the movement of interfaces, and can thus be calculated by a simple convection equation:

$$\frac{\partial}{\partial t}\Phi_i + v_s^\alpha \frac{\partial}{\partial x_\alpha}\Phi_i = 0 \qquad [2.13]$$

The velocity v_s^α is the velocity of the interface of phase i, irrespective of the phases i' in contact with it. In principle, for v_s^α to be valid everywhere, we must extend its definition (continuously) into the phases, but each term of [2.13] is zero except at the interface.

If there is no exchange of mass between the phases, then the interface velocity is always the common velocity of the phases in contact. If mass exchange occurs via vaporization, condensation, sublimation, adsorption, etc., the relative velocity $v_{ris}^\alpha = v_{is}^\alpha - v_s^\alpha$ can be calculated according to the law representing the phenomenon taking place [2.2]. Therefore, in every case, we must couple equation [2.13] with the equation that gives the velocity field of the medium [2.8].

We can see that if the phase in question has the constant density ρ_i, then [2.13] results in: $\frac{\partial}{\partial t}\rho_i\Phi_i+\rho_i v_i^\alpha\frac{\partial}{\partial x_\alpha}\Phi_i-\rho_i v_{ri}^\alpha\frac{\partial}{\partial x_\alpha}\Phi_i=0$, or $\frac{\partial}{\partial t}\rho_i\Phi_i+\rho_i v_i^\alpha\frac{\partial}{\partial x_\alpha}\Phi_i=$ $-\rho_i v_{ri}^\alpha n_i^\alpha\sigma_i^s$.

In addition, for constant ρ_i, $\Phi_i\frac{\partial v_i^\alpha}{\partial x_\alpha}=0$, and also $\Phi_i\frac{\partial \rho_i v_i^\alpha}{\partial x_\alpha}=0$, we obtain the mass balance for phase *i*: $\frac{\partial}{\partial t}\rho_i\Phi_i+\frac{\partial}{\partial x_\alpha}\rho_i v_i^\alpha\Phi_i=m_i\sigma_i^s$.

It is also important to note that for giving sense to this equation, first the recognition of the presence of interfaces is required. Beginning from an initial situation where the medium is known at all points, and therefore with the initial position of the interfaces, this is obvious if any interfacial interaction never exits. But these interactions will necessarily appear if there are coalescences or breakages of parcels of the same phase, something that is desirable to be studied. In these conditions, it is necessary that during these interactions we are able to determine the instant at which a portion of the interface disappears, and also to determine which portion it is. This is not a phenomenon of continuous media and is not included in the equations written until now. It is also necessary that the law governing this interaction is known in terms of the variables of the medium in order to be able to use [2.13] fully in these situations.

The distributions σ_i^s are directly related to Φ_i. We have seen that $-n_i^\alpha\sigma_i^s=\frac{\partial\Phi_i}{\partial x_\alpha}$ everywhere. The calculation of the field of Φ_i in the flow by [2.13], at each instant, allows us to deduce σ_i^s. However, it is necessary to identify the function *F*(*x*,*t*) that represents the interface in order to use the formula $\sigma_i^s=\delta(F)\left|\frac{\partial F}{\partial x_\alpha}\right|$, and also, as we have seen, to determine how two interfaces can disappear when they are too close to each other, a physical phenomenon that occurs at a lower scale than that of continuous media. This requires specific treatment of the results of calculation [2.13]. This is also necessary in order to find the production rate of interface of *i*, which is useful in the total energy balance equation above, which is defined by:

$$\frac{\partial}{\partial t}\sigma_i^s+\frac{\partial}{\partial x_\alpha}v_s^\alpha\sigma_i^s=\Gamma_{si}\sigma_i^s \qquad [2.14]$$

It would be interesting, using [2.13], to find an equation of this type that would allow us to find σ_i^s directly. This has been studied, and we refer readers to the study by Morel [MOR 07], who considered the situations without exchange of mass between phases and provided various forms for this equation:

$$\frac{\partial}{\partial t}\sigma_i^s + \frac{\partial}{\partial x_\alpha}(\sigma_i^s v_s^\beta n_i^\beta n_i^\alpha) = \sigma_i^s v_s^\beta n_i^\beta \frac{\partial n_i^\alpha}{\partial x_\alpha}$$

$$\text{or: } \frac{\partial}{\partial t}\sigma_i^s + \frac{\partial}{\partial x_\alpha}(\sigma_i^s v_s^\alpha) = \sigma_i^s(\delta_{\alpha\beta} - n_i^\alpha n_i^\beta)\frac{\partial v_s^\alpha}{\partial x_\beta} \qquad [2.15a]$$

However, these forms cannot be used independently of [2.13], since the normal at the interface is always involved. It is also possible to use [2.13] to establish an equation for the normal, which is written as:

$$\frac{\partial}{\partial t}n_i^\beta + v_s^\alpha \frac{\partial}{\partial x_\alpha}(n_i^\beta) = -\frac{\partial v_s^\alpha}{\partial x_\beta}n_i^\alpha + \frac{\partial v_s^\alpha}{\partial x_\gamma}n_i^\alpha n_i^\gamma n_i^\beta \qquad [2.15b]$$

However, as we have mentioned, equation [2.13] and the following equations do not involve the physical phenomenon of surface tension, since they are only geometric. The interfacial phenomena that can lead to the agglomeration of two solid particles are not represented either. Surface tension is involved in the equation used to calculate the velocity field, [2.8], but this is not enough to take into account all the phenomena that occur at the interfaces. The surface tension force in [2.8] modifies the velocity field so as to prevent it from obtaining interface curvature radii that are too small, but in cases where two quasi-flat parallel interfaces are approaching one another, it cannot prevent the distance between these two interfaces from decreasing below the physical thickness of an interface. We must therefore reconsider equations [2.15a] and [2.15b] to represent an interface that is not only geometric but also physical, and at least add a term (or a modification) to them, which makes it possible to cause the mutual annihilation of two parallel interfaces.

2.6. Extension for a fluid phase that is a mixture

The balance equations above can be generalized when the continuous phase i is itself a mixture of several fluid species that might possibly react with each other (this is the case, for example, when it is a mixture of several reactive gases), and each species may enter into or exit from the other phase in a different manner (this might be the case if another phase is made up of drops of a liquid mixture whose certain components vaporize). It is sufficient then to consider, in addition, the balance of different masses per unit of volume of the various constituents of the phase in question. If we write ρ_{ji} as the local and instantaneous volumetric mass of substance *j* in phase *i*, its instantaneous balance equation is written as:

$$\frac{\partial}{\partial t}\left(\rho_{ji}\Phi_i\right)+\frac{\partial}{\partial x_\alpha}(\rho_{ji}\Phi_i v_i^\alpha)=\frac{\partial}{\partial x_\alpha}(-\Phi_i j_\alpha^{Dj})+\rho_i\dot{w}_{ji}^{sm}+\rho_i\Phi_i\dot{w}_j^{ch} \quad [2.16]$$

The first term on the right side corresponds to the molecular diffusion flux of substance *j* in phase *i*, and it can exist in a solid or liquid phase as well as a gas phase, although with much lower intensity. The second term is the production term (positive or negative) of substance *j* in this phase at the interface with the other phases in contact, when there is a physicochemical process that justifies it. This can be due to vaporization or condensation or due to a surface chemical reaction, adsorption or desorption, and each physical phenomenon enables us to obtain a law for this exchange of mass, taking into account $\sum_j \rho_i\dot{w}_{ji}^{sm}=\rho_i\dot{w}_i^{sm}$. The last term is the rate of volumetric production due to chemical processes that can occur in the bulk of phase *i*.

The interface balance relationship now takes the general form [2.1b] as in section 2.1, with a process of production or disappearance at the interface, and now includes, on the side of phase *i*, the convection and diffusion fluxes:

$$\Gamma_{ji}^{sm}\sigma_i^s=-(\rho_{jis}(v_{is}^\alpha-v_s^\alpha)+(j_{ji}^{D\alpha})_s)n_i^\alpha\sigma_i^s \quad [2.17]$$

We can also use the mass fraction of substance *j* in phase *i*, Y_{ji} such that $\rho_{ij}=\rho_i Y_{ji}$. The laws of balance of momentum and energy of phase *i* remain unchanged, but the expression of the heat flux in phase *i* must now include the contribution due to the diffusion flux of the substance *j* in this phase.

When we consider a substance *j* that can move from one phase to another, the instantaneous interface mass balance equation, similar to [2.4], can be generally written as:

$$m_i Y_{jis}-(j_{ji}^{D\alpha})_s n_i^\alpha+m_{i'}Y_{ji's}-(j_{ji'}^{D\alpha})_s n_{i'}^\alpha=0 \quad [2.18a]$$

Because, as it should be, the mass of *j* produced at the interface from the side of *i* is the mass destroyed on the side of *i'*, $\Gamma_{ji}^{sm}=-\Gamma_{ji'}^{sm}$. If phase *i'* is formed entirely of substance *j*, which is the case, for example, when phase *i'* is a pure liquid composed of substance *j*, and substance *j* in *i* is the vapor of this liquid, this balance equation is reduced to:

$$m_i Y_{jis} - (j_{ji}^{D\alpha})_s n_i^\alpha - m_i = 0 \qquad [2.18b]$$

However, very often substance j is present in only one of the phases in contact, or i, and thus the interface balance relationship retains form [2.17], which we can also write as:

$$m_i Y_{jis} - (j_{ji}^{D\alpha})_s n_i^\alpha = \Gamma_{ji}^{sm} \qquad [2.19]$$

The right-hand term, which is the surface production rate of j at the interface, must be given by a law that represents the physicochemical phenomena (called as heterogeneous) involved at the interface in question. The surface rates of the different species reacting at an interface have algebraic relationships between them that represent the types of interfacial chemical or physical reactions involved.

2.7. Completing the description of the medium

The balance equations above, differential balances and interface balances, cannot completely describe the state and evolution of the medium by themselves. They must be completed by the equations of state of the various phases, and by the laws giving the irreversible fluxes of momentum and heat, as well as laws enabling us to calculate the possible exchange of mass between phases, and capillarity. All of these laws are known in classic physics, even if they constitute only an approximation of reality. We stated in section 2.3 that here we will not be considering solid media composed of several particles that can remain in contact for a relatively long time; thus, the constitutive laws of the solid phase(s) are those of the material that composes them. Finally, it is important to specify the conditions of continuity at interfaces between phases; since we are excluding contacts other than short collisions between solid particles, we will exclude any possible sliding between two solid particles, and thus the adherence of the phases at the interface invariably results in a condition in which the tangential velocities of the two phases are identical. Likewise, the perfect thermal contact condition implies an equality of temperatures between the two phases at the interface. It could be replaced in special cases, if required, by a condition that links these two temperatures, according to the heat flux exchanged.

With all these equations and conditions, the problem has now been completely described. All that is still needed is to have the boundary and initial conditions related to a given domain of space, in order to calculate for any point the values defined so far, at least in principle. We can verify that such a problem is solved without much difficulty, for simple situations corresponding to "small" problems with a given geometric interface (flat, for example). We can also see that the solution is unique and stable, meaning that the problem has been "well posed".

Examples of this will be given in Chapter 5. Without predicting the existence of a single solution in every case, the obvious difficulty of calculation for more complicated situations is that certain unknowns have discontinuities at points in space that can shift. However, it is possible to calculate numerical approximations of these using appropriate methods, admitting certain errors that can in principle be made as small as desired by improving the numerical grid of the domain. Some numerical calculations of this type already exist for situations that are not too complicated, with a relatively small number of particles or drops.

However, as stated earlier, the objective of modeling is not to carry out such direct time-consuming calculations, but to provide the definition of new, more macroscopic variables, and their use for easier and faster quantitative descriptions.

Chapter 3

Description of a "Mean Multiphase Medium"

3.1. The need for a mean description

The direct numerical calculation of Chapter 2's equations to study practical devices is a highly laborious task using modern computers, and will be so even with future ones. There are appropriate numerical methods, called volume-of-fluid or level set methods (see, for example, [SCA 99] and [MÉN 07]). But the grid sizes and time steps necessary to represent practical flows with a good enough approximation are extremely small; it is particularly clear that the smallest bubble, drop or solid particle must be represented with much more than one grid point. Although simple piecewise continuous media with a small number of pieces, in a small device, with a low flow velocity and implementing a simple two-phase situation, have already been calculated, "real" problems are too large, their geometry are too complex and their velocities are too large, leading to several dozens or hundreds of millions of grid points and an unimaginable calculation time.

Moreover, in order to be able to undertake a direct calculation, the knowledge of all initial conditions is necessary for all phases and in all parts of the domain of interest (and a phase does not necessarily occupy one simply connected domain), as well as of the boundary conditions for all phases and according to time. But the exact knowledge of conditions of this type that may correspond to a given physical experiment is impossible in practice because the experimentalist cannot control them with sufficient accuracy: the shape, the position or exact initial velocity (including rotation velocity) of each grain or drop in a suspension cannot be controlled. Even in the case of an immobile solid phase of the medium (for example, if the flow occurs across a porous medium), the structure, shape and size of this porous medium are never exactly known in practice.

In addition, an approximate knowledge of these conditions is not generally sufficient because flows (and their equations as well, mainly due to their nonlinearity) are usually unstable; a small difference in the initial or boundary conditions is likely to lead, after a certain time, to large differences in the behavior of the system, in some places or everywhere. Turbulent fluid media with constant density are sites of instabilities; we know this because the flow becomes unstable when it has overly strong velocity gradients, if a certain generally defined Reynolds number is too high. Practical multiphase flows bring into play situations of the same type in the vast majority of cases. There are not only instabilities in one or more fluid phases due to the velocity gradients at the walls, but also in a fluid phase near a solid particle or an interface with another fluid where too great a difference in velocity is found. Other instabilities are due to differences in density for fluids that are heavy or subject to forces of inertia. The surface tension and interactions between drops or bubbles give additional reasons for instability in multiphase flows; these have not all been studied in depth, and in the most common conditions they are frequently mingled in hydrodynamic instability leading to turbulence. Collisions and contacts between particles that are neither solid nor regular in shape are also sources of unstable behavior in flows where they are present. It is impossible to determine in all these situations the threshold conditions for the development of instabilities, and this threshold can be simply the initiation of a macroscopic motion. We note these instabilities during experimentations. Most of the time, if the same experiment is done twice – or rather, what we think is the same experiment – we will not see exactly the same thing in every detail, but merely an overall similarity. It follows that even when we can calculate basic equations, we cannot ensure that they will correspond to the experiment that has actually been done unless the flow is stable, which is neither the most common nor the most interesting case.

Even with this practical inability to control conditions exactly, and with the difficulty of calculating sudden variations, it is still interesting to consider the evolutions of complex multiphase media in a way that eliminates the "details", since experiments have shown that these evolutions tend to resemble one another overall. This overall similarity can be explained by the fact that instabilities enter progressively into a saturation phase once they have begun, and the overall behavior that results no longer depends on the disturbances that caused them. With this in mind, we must define and study new values for each point of the medium that filter out the details and constitute a local average of each of the physical values previously defined. These averages must have two roles: they must cause the sensitivity at the initial and boundary conditions to disappear, or at least reduce this sensitivity if it is overly large; and they must also cause the disappearance of sudden variations depending on space or time that require calculation grids that are too small. The problem here is actually the same that has occurred since the time of

Reynolds in the study of classical turbulent flows, and it is interesting to use the experience acquired in this area to justify the approach we must adopt here.

3.2. How are mean values defined?

It is possible to define averages in four different ways, as has been done since the work of Reynolds.

3.2.1. *Temporal average*

The first way considers a temporal average in a certain interval of time. We define the average of any value g in the flow as:

$$\overline{g}(x,t) = \frac{1}{\Delta t} \int_{t}^{t+\Delta t} g(x,t')dt' \qquad [3.1]$$

To define g at any point x at any time, if g_i is defined only in phase i, we must of course consider $g_i \Phi_i$ and not g_i. The case where g can be discontinuous, or a Dirac function, has no influence on this definition. If Δt is large enough in relation to temporal variations of g that are too sudden, it is clear that the integral will "smooth" them out. We can also expect that spatial variations will be smoothed out as well, since they are linked to the temporal variations by a characteristic velocity, of the motion of multiphase structures or of the flow, but more or less following the value of this velocity. When this velocity can be very low at important places in the device, Δt that is necessary to be effective can be very large.

This averaging operation involves only one experiment, but we can expect that the variations in the initial and boundary conditions, if they are on a small scale of time compared to Δt (the characteristic spatial size of these variations must also correspond to a temporal scale via an adapted characteristic velocity, but one that is also not well known), will also be smoothed out. In addition, it is necessary for Δt to remain small in relation to the characteristic time of the overall temporal variations of g, for example those due to deliberate variations in the power supply conditions of the device being considered, if we want to continue to display them accurately. Therefore, the choice of Δt is not fixed in a precise manner. The preceding conditions only give some vague indications, and we cannot show them to be invariably compatible. We can assume that if the frequencies of fluctuations to be smoothed out are much larger than those of the overall device, which we do not want to smooth out, we can find a convenient middle value of Δt, and that the average is not too sensitive to its precise choice. If the purpose is an experiment with a permanent regime, without any characteristic frequency of the overall device, as was often the case in the early studies of turbulent flows, it is enough to take a very large value of Δt. If this is not the case, and if the existing frequencies are not well separated, the dependence on Δt of the mean value remains clear, and the use of this average will include a certain amount of arbitrariness.

The use of mean values for theoretical predictions requires being able to build balance equations that these mean values must satisfy, using the basic equations of a piecewise continuous medium. To do this, certain equalities are very useful, which are called Reynolds relationships. If we consider any two characteristic values of the medium, g and h, and one constant λ, it is useful to be able to show that

$$\overline{(\lambda g)} = \lambda \overline{g}\,,\quad \overline{g+h} = \overline{g} + \overline{h}\,,\quad \overline{\left(\frac{\partial g}{\partial x_\alpha}\right)} = \left(\frac{\partial \overline{g}}{\partial x_\alpha}\right),\quad \overline{\left(\frac{\partial g}{\partial t}\right)} = \left(\frac{\partial \overline{g}}{\partial t}\right),\ \text{and}\ \overline{\left(\overline{g}\right)} = \overline{g},$$

which with the second, implies that $\overline{(gh)} = \overline{g}\overline{h} + \overline{(g-\overline{g})(h-\overline{h})}$. Of course, if the variables can be discontinuous, it is now necessary to understand the derivatives in the sense of distributions.

It is clear that the temporal average verifies the first three, but the latter two are more difficult to justify. In reality, it is possible to demonstrate the third, which will be explained later, but the fourth cannot be exactly shown no matter the value of Δt. We can see, however, that it will be verified if the variations of g have a slow component and a rapid one, if these two components are well separated and if Δt is located in the middle of the separation interval, which can sometimes be a realistic situation, as mentioned above.

3.2.2. *Volumetric average*

The second way of defining an average considers a small volume around a point being considered, which is called the representative elementary volume. Thus, for example, we define the mean value of the index of phase i as the volume fraction of this phase:

$$\overline{\Phi}_i = \frac{1}{Vol}\iiint_{Vol} \Phi_i dv \qquad [3.2]$$

where the integration volume Vol is small but not too small. This quantity can take continuous values between 0 and 1, whereas Φ_i takes only the values of 0 or 1.

The mass fraction of phase i will be an average weighted by the volumetric mass:

$$\tilde{\Phi}_i = \iiint_{Vol} \rho_i \Phi_i dv / \iiint_{Vol} (\sum_i \rho_i \Phi_i) dv$$

For the momentum (and the internal energy), it is generally preferable to also use an average weighted by the volumetric mass:

$$\tilde{v}_i^\alpha = \iiint_{Vol} \rho_i v_i^\alpha \Phi_i dv / \iiint_{Vol} \rho_i \Phi_i dv \qquad [3.3]$$

It is clear that this definition of the average, weighted or not, depends in principle on the size of the volume being considered. Here again, this size is not well defined; it must be large enough to smooth out the fluctuations of the values depending on time or space, and so it must be larger than the size of the largest parcels in the phase; and additionally, it must include parcels that are all the more numerous because they are small, so that the average does not vary suddenly if a parcel exits the volume. But it cannot be too large in relation to the size of the overall flow in order to enable the representation of the variations that occur in it, at its scale, and this is quite difficult if a bubble is almost of the same size as the conduit that contains the flow. In addition, this size must also depend on the spatial and temporal scales of possible fluctuations of initial and boundary conditions, and must be large enough to be able to smooth out the influence of fluctuations.

Here again, the problem can only be easily solved if the mean medium is homogeneous because we can extend the integration volume to infinity. But any practical device has finite boundaries, and thus the choice of a representative elementary volume that will not affect the result can be impossible if the scales of spatial variation are not separate enough. Note that we can give a certain physical interpretation of this volumetric average if we have a measurement device that is not extremely small and that, specifically, gives as a result the integral over its measurement volume of the value being considered. In this case, the size of the volume is fixed by the measuring probe used, and not by the flow.

With regard to Reynolds relationships, for averages that are not weighted by mass, the situation is similar to the case of temporal averages: the first two are obvious, and the other two can be demonstrated, as we will see, but the relationship $\overline{(\overline{g})} = \overline{g}$ and consequently $\overline{(gh)} = \overline{g}\overline{h} + \overline{(g-\overline{g})(h-\overline{h})}$ can only be approximately valid for cases in which the spatial scales of the variations have two widely separated ranges and if the integration volume is well chosen between the two. For weighted averages, it is clear that only the first two are valid, but we can take this point into account in theoretical developments.

3.2.3. *Statistical average*

Another way of defining an average is to consider a statistical average, at one point, with a large number of independent experiments that are similar (or independent and similar calculations) but have certain uncontrollable fluctuations in the initial and boundary conditions. This point of view was introduced for turbulent flows by Kampé de Fériè around 1920. Thus, averages are defined simply as an arithmetical average using values at a single point and a single instant after the beginning of the experiment corresponding to each experiment n ($n < N$). For example:

$$\bar{\Phi}_i = \lim(N \to \infty) \frac{1}{N} \sum_{n=1}^{N} \Phi_i^n \qquad [3.4]$$

This quantity is the "presence rate" (or probability of presence) of phase i, and not exactly the volume fraction of this phase, defined by [3.2]. These two quantities, when the second one can be defined independently of the representative elementary volume under the conditions indicated above, can be identified following the ergodic assumption. We will, therefore, refer later on to the presence rate as the volume fraction.

In the same way, we define the mean volumetric mass of phase i by: $\overline{\rho_i \Phi_i} = \lim(N \to \infty) \frac{1}{N} \sum_{n=1}^{N} \Phi_i^n \rho_i^n$ because $\rho_i \Phi_i$ is well defined at every point, and the mean volumetric mass of the medium by: $\bar{\rho} = \lim(N \to \infty) \frac{1}{N} \sum_{n=1}^{N} (\sum_i \rho_i^n \Phi_i^n) = \sum_i \overline{\rho_i \Phi_i}$.

Weighted averages by the volumetric mass can also be defined, for example:

$$\tilde{\Phi}_i = \lim(N \to \infty) (\sum_{n=1}^{N} \rho_i^n \Phi_i^n / \sum_{n=1}^{N} \sum_i \rho_i^n \Phi_i^n) \qquad [3.5]$$

In the same manner, we often refer to this quantity as the mass fraction of phase i. The mean velocity and energy of phase i, weighted by the volumetric mass, are:

$$\tilde{v}_i^{\alpha} = \lim(N \to \infty) (\sum_{n=1}^{N} \rho_i^n \Phi_i^n v_i^{\alpha n} / \sum_{n=1}^{N} \rho_i^n \Phi_i^n) = \overline{\rho_i \Phi_i v_i} / \overline{\rho_i \Phi_i} \qquad [3.6]$$

$$\tilde{e}_i = \lim(N \to \infty) (\sum_{n=1}^{N} \rho_i^n \Phi_i^n e_i^n / \sum_{n=1}^{N} \rho_i^n \Phi_i^n) \qquad [3.7]$$

This new definition of the average is completely different from the preceding definitions. However, in a statistically homogeneous medium of infinite extent in which the representative volume can be infinite, practice has shown that these two ways of calculating averages give practically identical values, and this has enabled us to claim the ergodic assumption, which states simply that they are equivalent. Likewise, for a medium in a statistically stationary system, it considers the temporal averages with Δt having an infinite value as equivalent to the statistical mean.

We can see that in all cases, the statistical average provides values that are no longer dependent on fluctuations in the initial and boundary conditions. The temporal and spatial variations of the averages will be necessarily smoothed out in a non-random manner by a procedure that depends neither on an integration volume nor on an interval of time Δt.

Its practical use, which requires obtaining a large (theoretically infinite) number of independent values representative of the same quantities, is not very easy, but we will not linger on this point here as it is well known by statisticians. Its theoretical use benefits from the fact that all of the Reynolds relationships are precisely verified. Because establishing an average consists of making additions, it is clear that the average of a sum will be the sum of averages, and the average of a derivative will be the derivative of the average. Moreover, since a mean value is no longer random, it is clear that its average will be itself, and the relationships $\overline{(\bar{g})} = \bar{g}$ and $\overline{(gh)} = \bar{g}\bar{h} + \overline{(g-\bar{g})(h-\bar{h})}$ will, therefore, always be precisely verified. For averages weighted by volumetric mass, we can easily show the associated relationships:

$$\overline{(\rho g)} = \bar{\rho}\tilde{g}\,,\ \overline{(\tilde{g})} = \tilde{g}\,,\ \overline{(\rho\tilde{g})} = \bar{\rho}\tilde{g}\,,\ \overline{(\rho g h)} = \bar{\rho}\tilde{g}\tilde{h} + \overline{\rho(g-\tilde{g})(h-\tilde{h})}$$

3.2.4. *Filtered average*

The fourth way of defining averages is the way used in what is called large eddy simulations (LES) for classical turbulent flows, but this was also proposed as early as the 1960s and 1970s for porous environments ([MAR 67], cited in Chapter 1). It is based on the definition of a filter kernel function, or $f(t,t'-t,x,x'-x)$ (x and x' being three-dimensional vectors), of which the integral in the space of t' and x' is 1, whatever the values of x and t. The domain of definition of $t'-t$ and $x'-x$ is usually finite and f is zero at the boundaries (it may be infinite with a convenient function f).

Thus, the filtered average of any function, for example g_i, is defined by:

$$\bar{g}_i(t,x) = \iiint\int_{-\infty}^{\infty} g_i(x',t')\Phi_i(x',t')f(t,t'-t,x,x'-x)dt'dx' \qquad [3.8]$$

It is also possible with the same filter to define averages weighted by volumetric mass:

$$\overline{\rho_i \Phi_i \tilde{g}_i}(t,x) = \iiint \int_{-\infty}^{\infty} \rho_i g_i(x',t') \Phi_i(x',t') f(t,t'-t,x,x'-x) dt' dx'$$

Because function f has a unitary integral, it follows that its dimension is the inverse of the product of a time and a volume. In fact, we can define a normalized kernel function from a dimensionless function $f^*((t'-t)/\theta,(x'-x)/\Delta)$ with the same properties. The positive $\theta(x,t)$ and $\Delta(x,t)$ are directly related to the domain of space and period of time implied in the filter, and can be chosen by the user.

If we now define the kernel as $f(t,x,t'-t,x'-x) = \dfrac{1}{\theta \Delta^3} f^*((t'-t)/\theta,(x'-x)/\Delta)$, f is normalized and the values θ, Δ are characteristic scales in time and space of the filter. If the function f is not a function of x and t other than in $x'-x$ and $t'-t$, these scales are constants in the entire domain being considered.

The volumetric average corresponds, as we can easily see, to a kernel constant in a given volume *Vol*, and zero elsewhere, and applied at each instant; therefore, it is proportional to $\delta(t-t')$. This is a specific case of filtering, with a zero scale in time, and the space scale of which is the cubic root of *Vol*. Likewise, the temporal average corresponds to a kernel proportional to $\delta(x'-x)$, and is, therefore, without characteristic spatial size, and it is constant between $t' = t$ and $t' = t+\Delta t$ and zero elsewhere, of characteristic temporal scale Δt.

This definition generalizes the ideas of volumetric averages and temporal averages presented above. We can see that the shape of function f can select an area of time and space around t,x, to which the average will be particularly related, if f is zero elsewhere in this area. In fact, the order of magnitude of the area in question is nothing but $\theta\Delta^3$, but the weighting exercised by each part of the area on the calculation of the average is directly dependent on the shape of the filter. It is even possible for f to be negative for certain values of $x'-x$ or $t'-t$, provided that its integral remains positive.

However, this definition still does not provide any support for chosing the characteristic scales of time and space of the filter, and it even introduces a degree of arbitrariness via the choice of the filter shape. It is again clear that the averages are dependent on the characteristic scales of the filters, even though this dependence is increasingly reduced when these sizes are diminished. Neither does it link the filter scales to the scales of fluctuation at the initial and boundary conditions so that it is not certain that these fluctuations will be smoothed out, or that the instabilities they cause will be eliminated for the filtered equations. On the other hand, this is obviously the case with statistical averages, which can no longer be unstable (but not necessarily stationary) if all possible fluctuations of the initial or boundary conditions are included.

With regard to Reynolds relationships, this type of average has roughly the same properties as temporal and volumetric averages. The first two relationships result from the linearity of integrals. The two others can be demonstrated precisely, provided that the characteristic scales θ, Δ are constants. For example:

$$\overline{\frac{\partial g}{\partial x}} = \iiint\int_0^\infty \frac{\partial g}{\partial x'}(x',t')f(t'-t,x'-x)dt'dx' =$$

$$\iiint\int_0^\infty \frac{\partial}{\partial x'}(g(x',t')f(t'-t,x'-x))dt'dx' - \iiint\int_0^\infty g(x',t')\frac{\partial}{\partial x'}f(t'-t,x'-x)dt'dx' =$$

$$0 + \iiint\int_0^\infty g(x',t')\frac{\partial}{\partial x}f(t'-t,x'-x)dt'dx' = \frac{\partial}{\partial x}(\iiint\int_0^\infty g(x',t')f(t'-t,x'-x)dt'dx') = \frac{\partial \overline{g}}{\partial x}$$

In the above equation, the first integral on the second line is zero because the filter is zero at the borders of its volume, and the derivative of f in relation to x' is simply the opposite of its derivative in relation to x, since it is only a function of $x'-x$. If the filter is an additional function of x and t, there is an additional term to add.

The relationships $\overline{(\overline{g})} = \overline{g}$ and $\overline{(gh)} = \overline{g}\overline{h} + \overline{(g-\overline{g})(h-\overline{h})}$, conversely, can only remain approximate, and in a manner that depends on the scales of the filter as much as on its exact shape.

3.3. Which average to choose, according to their advantages and disadvantages?

We have seen that the statistical average is more strictly defined than the temporal average, independent of any volume or time interval of definition, and that it enables us to prove the Reynolds relationships precisely. It is thus perfectly adapted to a situation in which various scales are interwoven in a multiphase medium. Moreover, it is, by nature, independent of the random fluctuations at the initial and boundary conditions, between various realizations. Therefore, it is of great importance for defining mean quantities with smoothed-out variations.

However, it is clear that the physical meaning of these mean quantities is somewhat abstract in the sense that they concern a set within an infinity of similar experiments rather than a given experiment that the experimentalist has just conducted, or wants to conduct, desiring that each experiment can in principle give a result for the mean value. This situation may seem unfortunate or impractical, however the statistician can propose improvements to the limited knowledge that averages contain; it is also possible to define standard deviations around these means, and even probability density functions or joint probability density functions, which enable us to better understand the situation by taking into account all of the risks. We will see in the following how we can obtain equations to calculate averages, and we will also indicate how, even if it is more difficult, we can find equations for standard deviations and probability densities. It is also possible to define several classes of initial or boundary disturbances and to characterize these different classes by different averages and/or probability densities.

A practical flaw of the statistical approach is that it eliminates knowledge of instabilities as real motions and, thus, does not allow us to know more about the shapes, structures and accelerations these motions possess, even at large scales in which we are interested. In addition, although it can address fluctuations through standard deviations, it can in no way distinguish the spatial scale of these fluctuations. Although the smallest structures and briefest accelerations pose problems for calculation, the largest structures and slowest movements can more easily be calculated. Thus, if we want to examine these large-scale movements and structures, the LES approach, based on a filter that only eliminates or smooths out fluctuations on the smallest scale, is of great importance. We may also note that the overlapping of the scales of structures poses no particular problem for this approach, similarly to the statistical approach. We have seen that it is dependent on the sizes of the filter, and the modeling associated with it will have to be explicitly dependent on this size. This is not a flaw as long as we do not want to use sizes that are too large; rather it can provide different points of view for the study of flows and, thus, can be transformed into an advantage, even though this will of course require additional work.

We will use the statistical approach in the following and show the developments and models that it requires. It is also possible to choose the filtered LES approach, following the same route as the statistical approach. The path is more complicated but relatively similar in the end, and we will pay special attention to it in Chapter 8.

Chapter 4

Equations for the Mean Continuous Medium

4.1. Global balance equations for the mean medium

By averaging instantaneous equations [2.3], [2.8] and [2.11a] or [2.11b], we can obtain the mean balances of global mass, global momentum and global energy of the mean medium, that is within the medium for any phase. As decided, the averages here are statistical averages, and thus the Reynolds relationships are precisely verified and we can use them without restriction. Later in this book, we will explore the application of this method with filtered averages for large eddy simulation (LES) modeling.

4.1.1. *Total mass*

First, we obtain immediately the total mass balance:

$$\frac{\partial}{\partial t}(\bar{\rho})+\frac{\partial}{\partial x_\alpha}(\bar{\rho}\tilde{v}^\alpha)=0 \qquad [4.1]$$

Here, we have generally written $\overline{\sum_i \rho_i\Phi_i}=\bar{\rho}$. Because of the use of the mean velocity weighted by the volumetric mass, we have $\overline{\sum_i \rho_i v_i^\alpha\Phi_i}=\bar{\rho}\tilde{v}^\alpha$, and the averaged continuity equation has the same form as the instantaneous equation. Unlike instantaneous values, mean values are continuous in space and time.

4.1.2. *Total momentum*

The mean balance of momentum is easily obtained as:

$$\frac{\partial}{\partial t}\left(\bar{\rho}\tilde{v}^{\beta}\right)+\frac{\partial}{\partial x_{\alpha}}(\bar{\rho}\tilde{v}^{\alpha}\tilde{v}^{\beta})=\frac{\partial}{\partial x_{\alpha}}(\bar{\sigma}^{\alpha\beta}-\overline{\rho v'^{\alpha}v'^{\beta}})+\bar{\rho}g_{\beta}+\overline{F^{\beta}_{cap}\sigma^{s}} \qquad [4.2]$$

An additional term appears on the right-hand side here because the average of the product $\overline{\rho v^{\alpha}v^{\beta}}$ is broken down into $\overline{\rho v^{\alpha}v^{\beta}}=\bar{\rho}\tilde{v}^{\alpha}\tilde{v}^{\beta}+\overline{\rho(v^{\alpha}-\tilde{v}^{\alpha})(v^{\beta}-\tilde{v}^{\beta})}$ (using a consequence of the Reynolds relationships). For the sake of simplicity, we have used the notation $v'^{\alpha}=v^{\alpha}-\tilde{v}^{\alpha}$. The fluctuation of velocity v' depends on the experiment considered; it is the difference between the velocity measured at this point in this experiment and the mean velocity, weighted by the volumetric mass, calculated with N experiments. This definition does not take into account the phase that is present at the point being considered in the different experiments; there is no need to do so for the moment. This new term is quite similar to a Reynolds tensor, which is defined in classic turbulent flows. We will call it dynamic or even turbulent, extending the classical use. This was not the preference of the first researchers to study two-phase turbulent flows; for example, Lance and Bataille [LAN 91], for a liquid flow with bubbles, referred to the turbulent fluctuations in liquid medium (around its mean velocity), as well as in gas medium (around its mean velocity), but referred to as pseudo-turbulence fluctuations relative to the mean velocity of the global medium due to the fact that the mean velocities of the gas and liquid were not generally equal. However, the range of situations termed "turbulent" has widened to include numerous types of flows since then, and we will use the adjective in a broader sense here. This will not prevent us from characterizing fluctuations in as specific and detailed a manner as the particularities of the situations being studied require. We will see in Chapter 9 that the thermodynamics of irreversible processes implies particular shapes for the terms to be modeled, to be adapted further to the precise physical phenomena involved.

Like the standard Reynolds tensor, this term is an additional flux of mean momentum produced by velocity fluctuations: its trace carries a contribution which is added to the classic pressure, and the rest produces a dispersion of this mean momentum around the mean trajectories defined by the mean velocity field $\tilde{v}^{\alpha}$. This effect is added to the mean friction contained in $\bar{\sigma}^{\alpha\beta}$ and is thus represented by additional friction. We can in principle calculate this new term using non-averaged equations, which we would solve many times using slightly different initial and boundary conditions in such a way as to then calculate the term according to its basic definition via statistical average. However, this task would take several centuries to accomplish using current numerical methods. It is the empirical-theoretical modeling of this term that will allow us, as for classic turbulent flows, to "close" this equation and to make it usable.

We can define a total Cauchy tensor, a continuous function of time and space, by:

$$\sigma_t^{\alpha\beta} = \overline{\sigma}^{\alpha\beta} - \overline{\rho v'^{\alpha} v'^{\beta}}$$

The capillarity term in [4.2] is also a continuous function that must be modeled and can also be written as the product of the mean area of interface per unit volume and the mean capillary stress per unit interface area:

$$\overline{F_{cap}^{\beta}\sigma^s} = \overline{F_{cap}^{\beta}}^s \, \overline{\sigma}^s$$

4.1.3. *Total energy*

The balance equation of the mean total energy is:

$$\begin{aligned} &\frac{\partial}{\partial t}\left(\overline{\rho}\tilde{e}_t + \overline{e_s\sigma^s}\right) + \frac{\partial}{\partial x_\alpha}(\overline{\rho}\tilde{v}^\alpha \tilde{e}_t) + \frac{\partial}{\partial x_\alpha}(\overline{e_s\sigma^s}\tilde{v}_s^\alpha) \\ &= \frac{\partial}{\partial x_\alpha}(-\overline{j}_\alpha^Q + \overline{\sigma^{\alpha\beta}v^\beta} - \overline{\rho v'^\alpha e_t} - \overline{e_s\sigma^s v'^\alpha_s}) + \\ &\overline{\rho}\tilde{v}^\beta g_\beta + \overline{e_s\sigma^s \frac{\partial v_s^\alpha}{\partial x_\alpha}} + \overline{(F_{cap}^\beta v_s^\beta + \Gamma_s e_s)\sigma^s} \end{aligned} \qquad [4.3a]$$

The mean total energy is: $\tilde{e}_t = \tilde{e} + \frac{1}{2}\tilde{v}^\alpha \tilde{v}^\alpha + \frac{1}{2}\frac{\overline{\rho v'^\alpha v'^\alpha}}{\overline{\rho}}$, meaning that it includes the mean internal energy of the medium, the kinetic energy of the mean momentum and a third contribution, which is the kinetic energy of velocity fluctuations, which as we have said also contains a contribution due to the fact that the different phases do not have the same mean velocities. This kinetic energy is often negligible and disregarded with respect to the first term of the sum, but if we take it into account as we will see later on for the modeling of the additional Cauchy tensor, we can also take it into account here.

We can also see in the first term on the right in [4.3a], two additional terms of total energy dispersion flux due to velocity fluctuations: the first is a turbulent dispersion flux of the mean total energy and the second is a dispersion flux of surface energy. We can see that the first is also written showing only the total energy fluctuation: $\overline{\rho v'^{\alpha} e_t} = \overline{\rho v'^{\alpha}(\tilde{e}_t + e_t')} = \overline{\rho v'^{\alpha} e_t'}$. All the terms corresponding to the average of a product are still unknown here and must be modeled in the same way as the generalized Reynolds tensor, but are continuous functions of time and space.

Several terms correspond to the contribution of capillarity, and if we overlook them, we obtain the simplified form, which possesses two more terms of total energy flux than the instantaneous equation:

$$\frac{\partial}{\partial t}(\bar{\rho}\tilde{e}_t) + \frac{\partial}{\partial x_\alpha}(\bar{\rho}\tilde{v}^\alpha \tilde{e}_t) = \frac{\partial}{\partial x_\alpha}(-\bar{j}_\alpha^Q + \bar{\sigma}^{\alpha\beta}\tilde{v}^\beta + \overline{\sigma^{\alpha\beta} v'^{\beta}} - \overline{\rho v'^{\alpha} e_t}) + \bar{\rho} g_\beta \tilde{v}^\beta \quad [4.3b]$$

We can define the total enthalpy by: $\rho h_t = \rho e_t + p$, where p is the pressure of the global medium given by $p = \sum_i \Phi_i p_i$, and thus $\bar{\rho}\tilde{h}_t = \bar{\rho}\tilde{e}_t + \bar{p}$ and $\rho h'_t = \rho e'_t + p'$.

Finally, we obtain:

$$\frac{\partial}{\partial t}(\bar{\rho}\tilde{e}_t) + \frac{\partial}{\partial x_\alpha}(\bar{\rho}\tilde{v}^\alpha \tilde{h}_t) = \frac{\partial}{\partial x_\alpha}(-\bar{j}_\alpha^Q + \overline{\tau^{\alpha\beta}}\tilde{v}^\beta + \overline{\tau^{\alpha\beta} v'^{\beta}} - \overline{\rho v'^{\alpha} h_t}) + \bar{\rho} g_\beta \tilde{v}^\beta \quad [4.3c]$$

Here, again, we can define a total dispersion flux of mean total enthalpy:

$$j_\alpha^{Ht} = -\bar{j}_\alpha^Q + \overline{\tau^{\alpha\beta}}\tilde{v}^\beta + \overline{\tau^{\alpha\beta} v'^{\beta}} - \overline{\rho v'^{\alpha} h_t}$$

The last contribution of which is a turbulent dispersion flux of enthalpy.

If we are not interested in studying how the phases are distributed in the medium, the preceding mean equations are sufficient. We can, however, see that they bring additional terms into play compared to the primitive equations, which will have to be modeled.

4.2. Balance equations for the phases of a mean medium

In general, we wish to describe each of the phases and not only the global medium, although we can obtain only a rough understanding in the statistical sense. We can do this using the instantaneous phase balance equations in Chapter 2 and calculating averages for each phase, i.e. using the equations in which the phase index is involved for each phase (we also refer to these averages as being conditioned by the phase index).

4.2.1. *Phase mass*

The mean balance equation of the mass fraction of phase i is written using [2.1] as:

$$\frac{\partial}{\partial t}\left(\bar{\rho}_i^*\bar{\Phi}_i\right)+\frac{\partial}{\partial x_\alpha}(\bar{\rho}_i^*\tilde{v}_i^\alpha\bar{\Phi}_i)=\overline{m_i\sigma_i^s} \qquad [4.4a]$$

Here, we have defined $\overline{\rho_i\Phi_i}=\bar{\rho}_i^*\bar{\Phi}_i$. The value $\overline{\rho_i\Phi_i}$ is the mean volumetric mass of phase i in the mixture, at a single point, taking into account not only eventualities where phase i is present, but also those in which it is not present (and for these, the value of the volumetric mass is zero since $\Phi_i=0$). The value $\bar{\rho}_i^*$ is the average of the volumetric mass only for eventualities where phase i is present, and we can call this the "intrinsic mean volumetric mass of phase i". Moreover, we can also use the mean mass fraction of phase i defined in [3.5] and such that $\bar{\rho}\tilde{\Phi}_i=\overline{\rho_i\Phi_i}=\bar{\rho}_i^*\bar{\Phi}_i$. Using $\bar{\rho}\tilde{\Phi}_i=\bar{\rho}_i^*\bar{\Phi}_i$, and noting that $\overline{\rho_i v_i^\alpha\Phi_i}=\bar{\rho}_i\tilde{v}_i^\alpha=\bar{\rho}\tilde{v}_i^\alpha\tilde{\Phi}_i=\bar{\rho}_i^*\tilde{v}_i^\alpha\bar{\Phi}_i$, we can obtain with these different notations different forms of [4.4a].

But we can also write $\bar{\rho}_i^*\tilde{v}_i^\alpha\bar{\Phi}_i=\overline{\rho_i v_i^\alpha\Phi_i}=\bar{\rho}\tilde{v}^\alpha\tilde{\Phi}_i+\overline{\rho v'^\alpha\,\Phi_i}$, where v'^α is the velocity fluctuation relative to the mean velocity of the mixture, whatever the phase (as in section 4.1) because $v'^\alpha\,\Phi_i$ is the velocity fluctuation of phase i relative to the mean velocity of all of the phases (and not to the mean velocity of phase i itself), which represents a dispersion of phase i around the average of all of the phases. Thus, there appears in equation [4.4a] a term representing the apparent dispersion flux of phase i mass:

$$\frac{\partial}{\partial t}\left(\bar{\rho}\tilde{\Phi}_i\right)+\frac{\partial}{\partial x_\alpha}(\bar{\rho}\tilde{v}^\alpha\tilde{\Phi}_i)=\frac{\partial}{\partial x_\alpha}(-\overline{\rho v'^\alpha\,\Phi_i})+\overline{m_i\sigma_i^s} \qquad [4.4b]$$

Note also that the dispersion term can be written as well with the index i for the volumetric mass because in any case, $\rho\Phi_i=\rho_i\Phi_i$.

In addition, the last term is the mean mass exchange between phase i and the others that are likely to be in contact with it at this point. The factor m_i can also have a different value or even be zero depending on the phase in contact, and we can write for more detail, $\overline{m_i\sigma_i^s}=\sum_{i'}\overline{m_i\sigma_{i,i'}^s}$. We can also write this term in a condensed manner or by showing the mean interfacial area per unit volume, multiplied by the mean mass flow rate of phase i exchanged per interface area unit: $\overline{m_i\sigma_i^s}=\overline{m}_i^s\,\bar{\sigma}_i^s$.

Moreover, we must remember that for each point of interface *i,i'*, we have: $m_i + m_{i'} = 0$.

4.2.2. *Phase momentum*

The mean momentum of phase *i* satisfies the balance equation drawn from [2.5] as below:

$$\frac{\partial}{\partial t}\left(\bar{\rho}_i^* \bar{\Phi}_i \tilde{v}_i^\beta\right) + \frac{\partial}{\partial x_\alpha}(\bar{\rho}_i^* \bar{\Phi}_i \tilde{v}_i^\alpha \tilde{v}_i^\beta) = \frac{\partial}{\partial x_\alpha}(\bar{\Phi}_i \bar{\sigma}_i^{*\alpha\beta} - \overline{\rho_i \Phi_i v'^\alpha_i v'^\beta_i}) + \bar{\rho}_i^* \bar{\Phi}_i g_\beta + \overline{m_i \sigma_i^s v_i^\beta} + \overline{f_{i,s}^\beta \sigma_i^s} \qquad [4.5]$$

The quantity $\bar{\sigma}_i^{*\alpha\beta}$ is the intrinsic average of the Cauchy tensor of phase *i*, or $\bar{\sigma}_i^{*\alpha\beta} = -\bar{p}_i^* \delta_{\alpha\beta} + \bar{\tau}_i^{*\alpha\beta}$, and we see again in this equation an additional flux of momentum, which takes into account fluctuations in velocity around the mean velocity of the phase. Fluctuations are defined for one phase in an identical manner as in a turbulent single-phase flow around the mean velocity of the phase in question. They are both due to instabilities (with nonlinear saturation phenomena) in a fluid medium subject to a velocity gradient, and possible gradients of temperature and remote forces, but in addition, there are physical phenomena induced by the presence of other phases. Remember that we have defined $v'^\alpha_i = v_i^\alpha - \tilde{v}_i^\alpha$, while for [4.4b], it was $v'^\alpha = v_i^\alpha - \tilde{v}^\alpha$.

The terms of exchange between phases can be written again in terms of the mean interfacial area, $\overline{m_i \sigma_i^s v_i^\beta} = \overline{m_i v_i^\beta}^s \,\overline{\sigma_i^s} = \bar{m}_i^s \bar{\sigma}_i^s \bar{v}_i^{s\beta}$, but unknown correlations are brought into play here again as well. The final factor here is an averaged velocity (of phase *i*) at the interface, which is not the mean velocity of phase *i* at this point, independent of the position of the interface. In addition, when there is no exchange of mass between the two phases in contact, their velocities at the interface are identical; so, their mean velocities at the interface will be identical as well, while the mean velocities within the two phases generally are not.

The term of mean contact force $\overline{f_{i,s}^\beta \sigma_i^s}$, with the other phases *i'*, whatever the value of *i'* (other than *i*, of course), needs some discussion.

Because of [2.6], it is defined simply by $\overline{f_{i,s}^\beta \sigma_i^s} = \overline{\sigma_i^{\alpha\beta} n_i^\alpha \sigma_i^s}$. Surprisingly, the force exerted by the other phases on *i* can be expressed only in terms of phase *i*, but it has to be kept in mind that contact modifies the characteristics of phase i itself, at the interface in particular. In fact, the interface balance relationship [2.7], which is rewritten here:

$$m_i\sigma^s_{i,i'}v^\beta_{is} + m_{i'}\sigma^s_{i,i'}v^\beta_{i's} + f^\beta_{i'i,s}\sigma^s_{i,i'} + f^\beta_{ii',s}\sigma^s_{i,i'} = F^\beta_{capi,i'}\sigma^s_{i,i'},$$

also allows us to write this term in terms of the characteristics of the phases in contact with i: $\overline{m_i\sigma^s_i v^\beta_{is}} + \overline{f^\beta_{i,s}\sigma^s_i} = -\sum_{i'\neq i}\overline{(m_{i'}\sigma^s_{i,i'}v^\beta_{i's} + \sigma^{\alpha\beta}_{i's}n^\alpha_{i'}\sigma^s_{i,i'})} + \sum_{i'\neq i}\overline{F^\beta_{capi,i'}\sigma^s_{i,i'}}$, which is reduced, if we disregard capillary effects and the effect of interphase mass exchange, simply as: $\overline{f^\beta_{i,s}\sigma^s_i} = -\sum_{i'\neq i}\overline{\sigma^{\alpha\beta}_{i'}n^\alpha_{i'}\sigma^s_{i,i'}} = \sum_{i'\neq i}\overline{\sigma^{\alpha\beta}_i n^\alpha_i\sigma^s_{i,i'}}$. Then, this force is also expressed in terms of the characteristics of the phases i' in contact at the point being considered in different eventualities.

We can split the Cauchy tensor like that of a fluid by defining the pressure as the opposite of one-third of the trace of the reversible Cauchy tensor:

$$\overline{f^\beta_{i,s}\sigma^s_i} = \overline{\sigma^{\alpha\beta}_i n^\alpha_i\sigma^s_i} = \overline{(-p_i\delta^{\alpha\beta} + \tau^{\alpha\beta}_i)n^\alpha_i\sigma^s_i}$$

If phase i is fluid, then we can see that the mean contact force is, on the one hand, due to the pressure at the interface and, on the other hand, due to the viscous friction at the interface. If phase i is solid, we can use the same writing as stated above, but the tensor $\tau^{\alpha\beta}_i$ will also contain a reversible part without trace. We can thus write the contact force on phase i globally as the sum of these two contributions:

$$\overline{f^\beta_{i,s}\sigma^s_i} = \overline{F}^\beta_{pi} + \overline{F}^\beta_{\tau i} \qquad [4.6a]$$

However, it is often useful to consider, in the pressure contribution, the effect of the mean pressure at the interface, $\overline{p}^s_i = \overline{p_i\sigma^s_i}/\overline{\sigma}^s_i$, which is not the averaged pressure in phase i (at this same point). Thus, we can write:

$$\overline{f^\beta_{i,s}\sigma^s_i} = -\overline{p}^s_i\overline{n^\beta_i\sigma^s_i} + \overline{(\tau^{\alpha\beta}_i - p"_i\delta^{\alpha\beta})n^\alpha_i\sigma^s_i} = -\overline{p}^s_i\overline{n^\beta_i\sigma^s_i} + \overline{\tau^{\times\alpha\beta}_i n^\alpha_i\sigma^s_i}$$

where $p"_i = p_i - \overline{p}^s_i$, to obtain:

$$\overline{f^\beta_{i,s}\sigma^s_i} = -\overline{p}^s_i\overline{n^\beta_i\sigma^s_i} + \overline{\tau^{\times\alpha\beta}_i n^\alpha_i\sigma^s_i} = \overline{p}^s_i\frac{\partial\Phi_i}{\partial x_\beta} + \overline{F}^{\times\beta}_{is} \qquad [4.6b]$$

The interface relationship is: $m_i v_i^\beta \sigma_{i,i'}^s + m_{i'} v_{i'}^\beta \sigma_{i,i'}^s + f_{i'i,s}^\beta \sigma_{i,i'}^s + f_{ii',s}^\beta \sigma_{i,i'}^s = F_{capi,i'}^\beta \sigma_{i,i'}^s$, or:

$$m_i \sigma_{i,i'}^s (v_i^\beta - v_{i'}^\beta) - \overline{p}_i^s n_i^\beta \sigma_{i,i'}^s + \tau_i^{\times\alpha\beta} n_i^\alpha \sigma_{i,i'}^s - \overline{p}_{i'}^s n_{i'}^\beta \sigma_{i,i'}^s + \tau_{i'}^{\times\alpha\beta} n_{i'}^\alpha \sigma_{i,i'}^s = F_{capi,i'}^n n_i^\beta \sigma_{i,i'}^s$$

Because at the interface, the phase velocities always have equal tangential components (there is at least one fluid phase and thus an adherence condition), we also have:

$$m_i \sigma_{i,i'}^s (v_i^n - v_{i'}^n) n_i^\beta - \overline{p}_i^s n_i^\beta \sigma_{i,i'}^s + \tau_i^{\times\alpha\beta} n_i^\alpha \sigma_{i,i'}^s - \overline{p}_{i'}^s n_{i'}^\beta \sigma_{i,i'}^s + \tau_{i'}^{\times\alpha\beta} n_{i'}^\alpha \sigma_{i,i'}^s = F_{capi,i'}^n n_i^\beta \sigma_{i,i'}^s .$$

In addition, at the interface, $v_i^\alpha - v_s^\alpha = v_{ri}^\alpha$, and likewise for i', which gives: $v_i^n - v_s^n = v_{ri}^\alpha n_i^\alpha = -m_i / \rho_i$ and $v_{i'}^n - v_s^n = v_{ri'}^\alpha n_i^\alpha = -v_{ri'}^\alpha n_{i'}^\alpha = m_i / \rho_{i'}$; thus, $v_i^n - v_{i'}^n = -m_i (1/\rho_i - 1/\rho_{i'})$ and also:

$$-(m_i)^2 (1/\rho_i - 1/\rho_{i'}) n_i^\beta \sigma_{i,i'}^s - \overline{p}_i^s n_i^\beta \sigma_{i,i'}^s + \tau_i^{\times\alpha\beta} n_i^\alpha \sigma_{i,i'}^s - \overline{p}_{i'}^s n_{i'}^\beta \sigma_{i,i'}^s + \tau_{i'}^{\times\alpha\beta} n_{i'}^\alpha \sigma_{i,i'}^s = F_{capi,i'}^n n_i^\beta \sigma_{i,i'}^s$$

On average, at i,i' interfaces, we can thus write:

$$\overline{-(m_i)^2 (1/\rho_i - 1/\rho_{i'}) n_i^\beta \sigma_{i,i'}^s} + \overline{p}_i^s \frac{\partial \overline{\Phi}_i}{\partial x_\beta} + \overline{F}_{is}^{\times\beta} - \overline{p}_{i'}^s \frac{\partial \overline{\Phi}_i}{\partial x_\beta} + \overline{F}_{i's}^{\times\beta} = \overline{F_{capi,i'}^n n_i^\beta \sigma_{i,i'}^s}$$

On the other hand, the instantaneous relationship has a normal component at the interface, which gives: $-(m_i)^2 (1/\rho_i - 1/\rho_{i'}) + (-p_i + p_{i'}) = F_{capi,i'}^n$, disregarding the viscosity in volume. Averaging the interfaces of this type, we obtain:

$$\overline{-(m_i)^2 (1/\rho_i - 1/\rho_{i'})}^s + (-\overline{p}_i^s + \overline{p}_{i'}^s) = \overline{F_{capi,i'}^n}^s$$

Defining $(m_i)^2 (1/\rho_i - 1/\rho_{i'}) = \overline{(m_i)^2 (1/\rho_i - 1/\rho_{i'})}^s + ((m_i)^2 (1/\rho_i - 1/\rho_{i'}))''$ and $F_{capi,i'}^n = \overline{F_{capi,i'}^n}^s + (F_{capi,i'}^n)''$ in the mean interface relationship, it can be simplified as:

$$\overline{-((m_i)^2 (1/\rho_i - 1/\rho_{i'}))'' n_i^\beta \sigma_{i,i'}^s} + \overline{F}_{is}^{\times\beta} + \overline{F}_{i's}^{\times\beta} = \overline{F_{capi,i'}^n{}'' n_i^\beta \sigma_{i,i'}^s}$$

The fluctuations that remain here in the capillarity and mass exchange terms are only due to the fluctuating orientations of the interfaces. These phenomena occur at the scale of the thickness of the interface, and we may think that they are only very slightly sensitive to orientation, contrary to the pressure, which depends on the environment in the phases. In these conditions, we see that the forces $\overline{F}_{is}^{\times\beta}$

compensate for the phases in contact: for interface *i,i'*, we have $\overline{F}_{is}^{\times\beta} + \overline{F}_{i's}^{\times\beta} = 0$, and for all possible interfaces: $\sum_i \overline{F}_{is}^{\times\beta} = 0$. The mean contact force $\overline{F}_{is}^{\times\beta}$ by the other phases is not due only to friction, but also to fluctuations in pressure around the mean pressure at the interface; we will see in Chapter 5 that this is the classic contact force minus the generalized Archimedes force.

Remember, to conclude, that if the calculation of the momentum and of mass of phase i is done with [4.4a] and [4.5], this replaces the use of [4.4b] and of an empirical approximate formula for the generalized diffusion flux of mass that appears in it.

4.2.3. *Energies of each phase*

The mean energy of phase i in the medium is given by averaging the energy equation of phase i, taken from [2.9]:

$$\frac{\partial}{\partial t}\left(\overline{\rho}_i^*\overline{\Phi}_i\tilde{e}_{ti}\right)+\frac{\partial}{\partial x_\alpha}(\overline{\rho}_i^*\overline{\Phi}_i\tilde{v}_i^\alpha\tilde{e}_{ti})=\frac{\partial}{\partial x_\alpha}(\overline{\Phi}_i(-\overline{j_i^{Q\alpha}}^* +\overline{\sigma_i^{\alpha\beta}}^*\tilde{v}_i^\beta +\overline{\sigma'^{\alpha\beta}_i v'^{\alpha}_i})-\overline{\rho_i\Phi_i v'^{\alpha}_i e_{ti}})$$
$$+\overline{\rho}_i^*\overline{\Phi}_i\tilde{v}_i^\beta g_\beta+\overline{m_i\sigma_i^s e_{ti}}+\overline{f_{i,s}^\beta\sigma_i^s v_i^\beta}+\overline{q_{is}\sigma_i^s} \quad [4.7a]$$

Like the others, this equation includes terms of turbulent energy dispersion flux, due to velocity fluctuations of the phase, with this fluctuation occurring around the mean velocity of phase i. We can also write this equation considering dispersion fluxes due to fluctuations of velocity around the mean velocity of the medium, and thus, we obtain:

$$\frac{\partial}{\partial t}\left(\overline{\rho}_i^*\overline{\Phi}_i\tilde{e}_{ti}\right)+\frac{\partial}{\partial x_\alpha}(\overline{\rho}_i^*\overline{\Phi}_i\tilde{v}^\alpha\tilde{e}_{ti})=\frac{\partial}{\partial x_\alpha}(\overline{\Phi}_i(-\overline{j_i^{Q\alpha}}^* +\overline{\sigma_i^{\alpha\beta}}^*\tilde{v}_i^\beta +\overline{\sigma'^{\alpha\beta}_i v'^{\alpha}_i})-\overline{\rho_i\Phi_i v'^{\alpha} e_{ti}})$$
$$+\overline{\rho}_i^*\overline{\Phi}_i\tilde{v}_i^\beta g_\beta+\overline{m_i\sigma_i^s e_{ti}}+\overline{f_{i,s}^\beta\sigma_i^s v_i^\beta}+\overline{q_{is}\sigma_i^s} \quad [4.7b]$$

Note that formulas [4.7a] and [4.7b] seem to be the same: only an i index is lacking in each term of the second. But this is not a mistake; this index changes the meaning of the two terms concerned. We can also use in these equations the relationships: $\overline{\rho}_i = \overline{\rho}_i^*\overline{\Phi}_i = \overline{\rho}\tilde{\Phi}_i$, and they have slightly different forms.

The mean total energy of phase i is defined by $\tilde{e}_{ti} = \tilde{e}_i + \frac{1}{2}\tilde{v}_i^\alpha\tilde{v}_i^\alpha + \frac{1}{2}\frac{\overline{\rho v'^{\alpha}_i v'^{\alpha}_i}}{\overline{\rho}}$. The last term of this sum is simply the kinetic energy of the turbulence in phase i, which is generally negligible compared to the mean internal energy of the phase, but which can also be taken into account by a model.

The mean exchange terms of energy linked to mass exchange, momentum and heat exchanges, which are at the end of the right-hand side of these equations, are still unknown terms, even though they no longer have peak structures. We can write them differently as:

$$\overline{m_i \sigma_i^s e_{ti}} = \overline{m_i}^s \bar{\sigma}_i^s \overline{e}_{ti}^s \,,\; \overline{f_{i,s}^{\beta} \sigma_i^s v_i^{\beta}} = \overline{\sigma_i^{\alpha\beta} v_i^{\beta} n_i^{\alpha}}^s \bar{\sigma}_i^s \,,\; \overline{q_{is} \sigma_i^s} = -\overline{j_i^{Q\alpha} n_i^{\alpha}}^s \bar{\sigma}_i^s$$

We can see that they are proportional to the mean area of interface per unit volume, but the factors of mean total energy at the interface, mean contact power at the interface and mean normal heat flux are still unknown. Like the additional flux terms, they could be calculated precisely by direct numerical simulations of the instantaneous equations in Chapter 2, followed by adequate calculations of averages. Because this would be far too time-consuming a task, they must also be modeled.

We must also remember that at each point of the interface, an energy interface balance is verified, which here in the absence of surface tension phenomena (see Chapter 2) is:

$$m_i \sigma_{i,i'}^s e_{ti} + f_{i'i,s}^{\beta} v_i^{\beta} \sigma_{i,i'}^s + q_{is} \sigma_{i,i'}^s + m_{i'} \sigma_{i,i'}^s e_{ti'} + f_{ii',s}^{\beta} v_{i'}^{\beta} \sigma_{i,i'}^s + q_{i's} \sigma_{i,i'}^s = 0$$

On average, taking into account the fact that several *i'* phases can be in contact with phase *i*, we have:

$$\overline{m_i e_{ti} \sigma_i^s} + \overline{f_{i,s}^{\beta} v_i^{\beta} \sigma_i^s} + \overline{q_{is} \sigma_i^s} = -\sum_{i' \neq i} (\overline{m_{i'} \sigma_{i,i'}^s e_{ti'}} + \overline{f_{i',s}^{\beta} v_{i'}^{\beta} \sigma_{i,i'}^s} + \overline{q_{i's} \sigma_{i,i'}^s})$$

The power of contact forces is often disregarded in this term of interphase exchange. Mass exchanges between phases are often of great importance with regard to energy exchanges, and their effects must be carefully considered.

We would have been able to write the mean energy balance in terms of enthalpy starting from [2.12]. This would have given us:

$$\begin{aligned} &\frac{\partial}{\partial t}\left(\bar{\rho}_i^* \bar{\Phi}_i \tilde{h}_{ti}\right) + \frac{\partial}{\partial x_\alpha}(\bar{\rho}_i^* \bar{\Phi}_k \tilde{v}_i^{\alpha} \tilde{h}_{tk}) = \frac{\partial}{\partial x_\alpha}(\bar{\Phi}_i(-\overline{j_{Qi}^{\alpha}}^* + \overline{\tau_i^{\alpha\beta}}^* \tilde{v}_i^{\beta} + \overline{\tau'^{\alpha\beta}_i v'^{\beta}_i}^*) \\ &-\overline{\rho_i \Phi_i v'^{\alpha}_i h_{ti}}) + \bar{\rho}_i^* \bar{\Phi}_i \tilde{v}_i^{\beta} g_\beta + (\overline{m_i \sigma_i^s h_{ti}} + \overline{\tau_i^{\alpha\beta} n_i^{\alpha} v_i^{\beta} \sigma_i^s} + \overline{q_{is} \sigma_i^s}) + \bar{\Phi}_i \frac{\partial \bar{p}_i^*}{\partial t} \end{aligned} \qquad [4.8]$$

4.2.4. *Phase volume*

There is another mean balance equation, which is the average of the balance of the indicative phase function [2.13], which gives a probability of presence (or volume fraction) of the phase:

$$\frac{\partial}{\partial t}\bar{\Phi}_i + \overline{v_s^\alpha \frac{\partial}{\partial x_\alpha}\Phi_i} = 0 \qquad [4.9a]$$

This equation does not have a classic form and is not often used. We can rewrite it in the form of a mean balance equation, which is then a phase volume fraction balance equation, because $\bar{\Phi}_i$ can be assimilated to the volume fraction of phase i, with the three terms on the right to be modeled:

$$\frac{\partial}{\partial t}\left(\bar{\Phi}_i\right)+\frac{\partial}{\partial x_\alpha}(\tilde{v}_i^\alpha \bar{\Phi}_i)=\frac{\partial}{\partial x_\alpha}\left(-\overline{\Phi_i v'^\alpha_i}\right)+\overline{\Phi_i \frac{\partial v_i^\alpha}{\partial x_\alpha}}+\overline{\left(v_i^\alpha - v_s^\alpha\right)\frac{\partial \Phi_i}{\partial x_\alpha}} \qquad [4.9b]$$

We have seen in Chapter 2, section 2.5, that if phase i has a constant volumetric mass, the instantaneous phase volume balance is equivalent to the instantaneous mass balance of this phase; thus, the same is true for the mean balance equation.

4.3. Complete representation of the mean medium

With the definition of mean values we have discussed, and the mean balance equations, is it possible to describe completely a multiphase medium and its evolutions?

It is clear that it will be necessary in any case to link these balance equations to three other types of laws:

– First, it will be necessary to find the equations of state for the mean values for each phase, using instantaneous equations of state.

– Second, we must model the mean terms of exchange between phases.

– Finally, it will be necessary to find additional laws for the additional dispersion fluxes (which we will often call turbulent) which we have seen appear in mean equations.

We will discuss mean equations of state later, and the next two chapters will be dedicated to turbulent dispersion fluxes and interphase exchanges terms. First, we will show how it is possible to build two types of models, each one representing the mean multiphase medium by a group of different variables, and including, of course, as many variables as there are equations to calculate. To do this, we will have to discuss the problem of the link between the intrinsic mean pressure and mean interfacial pressures for the phases.

4.3.1. *Global representation*

The simplest way of representing a multiphase mean medium consists of considering it simply as a fluid mixture; we then examine only the global mean medium, defining at each point a mean volumetric mass of the medium, a mean velocity and a mean mass fraction for each phase in the medium. It is possible in this context to focus only on the mean internal energy and temperature of the medium, as well as to consider the mean internal energy and temperature of each phase. The variables that characterize the medium, in cases where we are interested only in the mean temperature of the medium, are $\overline{\rho}, \tilde{v}, \overline{p}, \tilde{e}, \tilde{T}, \tilde{\Phi}_i$, with for this last variable as many values of i as the number of phases minus one, since the sum of the mass fractions is obviously 1.

We can use the mean balance equations [4.1], [4.2] and [4.3b] (for the sake of simplicity, we will disregard the effects of surface tension here) as well as [4.4b]. These must be completed, as is usual for a mixture of fluids, by two mean equations of state, one linking the mean pressure (still defined as $\overline{p} = \sum_i \overline{\Phi_i p_i}$ since we have seen that $\overline{\sigma}^{\alpha\beta} = \sum_i \overline{\Phi_i \sigma_i^{\alpha\beta}}$) to the mean volumetric mass and mean internal energy, and the other enabling us to calculate the temperature (mean again, and this time weighted by the volumetric mass) using the internal energy and possibly the pressure or volumetric mass. In this way, the six variables defined above will be accompanied by six equations: four partial differential equations and two algebraic equations. We need to note here that the (mean) balance equations are in the same form (exactly the same if we disregard surface effects) as those used for a single-phase fluid. The multiphase specificity of the fluid is thus modeled only in the equations of state, and the choice of their forms is thus a critical part of the modeling approach.

A second part of this modeling consists of finding an expression of the turbulent mass dispersion flux term for phase i (relative to the mean velocity), as it appears in [4.4b], and of the total Cauchy tensor found in [4.3b] in terms of the mean variables we have chosen to characterize the medium. This essential task must be examined and discussed in detail by bringing into play all the physical phenomena of the situation being considered. The modeling of the mass dispersion flux term replaces the use of the mean momentum equation for phase i, and thus eliminates the need to seek a model for the interphase momentum exchange term. Conversely, the interphase mass exchange terms must be modeled (if they exist), and this constitutes the third part of the modeling process.

This global representation can be slightly enriched if we are interested in the mean temperature of each phase, and not only in the mean temperature of the mixture. In this case, without changing what was said above, we can thus consider the balance equations [4.6b] instead of [4.3b], and so, we are able to establish an equation of state for each phase linking the mean temperature of the phase to the mean internal energy of this phase. In this case, we must also have a model for the interphase heat exchange terms.

4.3.2. *Multifluid representation*

The second approach consists of representing the mixture as a set of fluids, with one fluid corresponding to each phase. We will consider, in the following, only two phases for the sake of simplicity. The mixture is thus represented by 11 mean quantities, $\overline{\rho}_1^*$, $\overline{\rho}_2^*$, $\tilde{v}_1$, $\tilde{v}_2$, $\overline{p}_1^*$, $\overline{p}_2^*$, $\tilde{e}_1$, $\tilde{T}_1$, $\tilde{e}_2$, $\tilde{T}_2$, and $\overline{\Phi}_1$. We can, instead of the volume fraction (or probability of presence), use the mass fraction because $\overline{\rho}_i^*\overline{\Phi}_i = \overline{\rho}\tilde{\Phi}_i$ and $\overline{\rho} = \sum_i \overline{\rho}_i^*\overline{\Phi}_k$.

– In the general case in which the two phases are *two compressible media*, there are 11 unknowns to calculate. If the first phase is a compressible gas and the second phase is a compressible liquid or solid, we have two equations of state, which give $\tilde{T}_1(\tilde{e}_1, \overline{\rho}_1^*)$ and $\overline{p}_1^*(\overline{\rho}_1^*, \tilde{T}_1)$, and two equations of state giving $\overline{p}_2^*(\overline{\rho}_2^*)$ and $\tilde{e}_2(\tilde{T}_2)$: four in all. These equations of state are found from the instantaneous equations of state of the media being considered, without too much difficulty, but still require some discussions; we will return to this subject in the next paragraph. The mass balances, momentum balances and energy balances [4.4a], [4.5] and [4.7a] for $i = 1$ and 2, constitute six equations; so, we have 10 equations for 11 unknowns. We lack only the balance equation of $\overline{\Phi}_1$, or [4.9a], conveniently modeled.

We must model the balance equation of the volume fraction of the phases in order to obtain a complete model of the medium. However, this equation is not generally studied in detail on the basis of equation [4.9b] and is even often avoided and replaced by $\overline{p}_1^* = \overline{p}_2^*$, a relationship that seems arbitrary, but we will see its explanation in the discussions below. Baer and Nunziato's model seems to have been the first one proposed for modeling [4.9a] in 1986 [BAE 86]. For a two-phase medium without surface tension and with no interphase mass exchange, it is written as:

$$\frac{\partial}{\partial t}\overline{\Phi}_1 + \tilde{v}_1^\alpha \frac{\partial \overline{\Phi}_1}{\partial x_\alpha} = \frac{\overline{\Phi}_1(1-\overline{\Phi}_1)}{\mu_{BN}}(\overline{p}_1^* - \overline{p}_2^*) \qquad [4.10]$$

The difference in mean pressure between the phases is assumed to be the direct reason for the variation in the volume fraction, with a coefficient that contains a characteristic response time of this process. The original model by Baer and Nunziato deals with a granular medium in which a shock wave, and even a detonation, can occur. There are also intergranular pressure and reactive phenomena that consume the solid phase and are represented by an additional reaction term in [4.10], but we do not consider these effects in the preceding formula. We will see in Chapter 9 that formula [4.10] is in agreement with the thermodynamics of irreversible processes.

More recent models have been studied by many researchers. We can find examples by Gavriliuk and Saurel [GAV 02] and by Coquel *et al.*[COQ 02].

– If the first of the two phases has *constant volumetric mass*, there is one unknown less, as well as one (mean) equation of state less. There are now 10 unknowns: $\bar{\rho}_1^*, \tilde{v}_1, \tilde{v}_2, \bar{p}_1^*, \bar{p}_2^*, \tilde{e}_1, \tilde{T}_1, \tilde{e}_2, \tilde{T}_2, \bar{\Phi}_1$. There are only three equations of state and, again, six mass, momentum and energy balance equations; we are still lacking one equation. Because the first phase has constant volumetric mass, its velocity field is at zero divergence at all times, which gives the mean relationship $\overline{\Phi_1 \frac{\partial v_1^\alpha}{\partial x_\alpha}} = 0$, and this constraint should in principle allow us to calculate the then unknowns. But this relationship does not directly bring into play the unknowns listed above. We can also write it as: $\overline{\frac{\partial \Phi_1 v_1^\alpha}{\partial x_\alpha}} = \overline{v_1^\alpha \frac{\partial \Phi_1}{\partial x_\alpha}}$ or $\frac{\partial}{\partial x_\alpha} \bar{\Phi}_1 \tilde{v}_1^\alpha = -\overline{v_1^\alpha n_1^\alpha \sigma_{12}^s}$ (note that if the medium i has constant volumetric mass, the mean velocity weighted by the volumetric mass and the unweighted one are identical, $\tilde{v}_i^\alpha = \overline{v_i^\alpha}$, although this is not the case for the mean velocities of the global medium) and undertake a modeling of the right-hand side. This remains to be done, however. We cannot write a new balance equation for the volume fraction for the phase with constant volumetric mass because this equation is already written in the form of a mass balance; we can use a Baer and Nunziato equation for the other phase. Here again, the hypothesis of equality of the mean phase pressures, which remains to be proved, solves the problem.

– In the very common situation in which *two phases have constant volumetric mass*, there are only nine independent variables instead of 11. There are six classic balance equations available to us, and we have one equation of state for each phase linking the temperature and internal energy; thus, we have eight equations in total. We are lacking one relationship, and here, there are two incompressibility relationships: $\overline{\Phi_1 \frac{\partial v_1^\alpha}{\partial x_\alpha}} = 0$ and $\overline{\Phi_2 \frac{\partial v_2^\alpha}{\partial x_\alpha}} = 0$. Fortunately, we can show that these two relationships are not independent of each other due to the two mass balances already used. For the sake of simplicity, we will take a case in which there is no interphase mass exchange. The mass balances are written here as $\frac{\partial \overline{\Phi}_i}{\partial t}+\frac{\partial}{\partial x_\alpha}\left(\overline{\Phi}_i \tilde{v}_i^\alpha\right)=0$, for $i = 1$ and $i = 2$, and $\frac{\partial}{\partial x_\alpha}\left(\overline{\Phi}_1 \tilde{v}_1^\alpha\right)+\frac{\partial}{\partial x_\alpha}\left(\overline{\Phi}_2 \tilde{v}_2^\alpha\right)=0$. On the other hand, if we take into account the fact that, at the interface, the normal velocities of the two phases are identical, the result is $\overline{v_2^\alpha n_2^\alpha \sigma_{12}^s}=-\overline{v_1^\alpha n_1^\alpha \sigma_{12}^s}$ (remember that the interface normal vectors of the two phases are opposite), and thus the two incompressibility relations, such as $\frac{\partial \overline{\Phi_i v_i^\alpha}}{\partial x_\alpha}=-\overline{v_i^\alpha n_i^\alpha \sigma_{12}^s}$, result in $\frac{\partial}{\partial x_\alpha}\overline{\Phi}_1 \tilde{v}_1^\alpha=-\frac{\partial}{\partial x_\alpha}\overline{\Phi}_2 \tilde{v}_2^\alpha$, the same relationship. Thus, we must use one incompressibility relationship, and one is sufficient, to complete the representation of the mean medium, but it still remains to be modeled. The use of a Baer and Nunziato relationship is no longer justified because the two mass balance equations already contain volumetric fraction balance equations. Here again, the hypothesis of equality of mean phase pressures allows us to obtain the necessary additional relationship, and this warrants further discussion.

Is it possible to invoke a local mean pressure equilibrium that can be established very rapidly, thus resulting in $\overline{p}_1^* = \overline{p}_2^*$?

We can already say from the interface balance relationships that at the interface, insofar as capillarity and mass exchange are negligible, at each instant, $p_1 = p_2$. We can thus deduce, if we take a conditioned statistical average at the interface, that $\overline{p}_1^s = \overline{p}_2^s$. It remains to justify $\overline{p}_i^s = \overline{p}_i^*$. On the one hand, in a small volume limited by interfaces or walls, the pressure equilibrium is reached by the multiple propagation and reflection of waves (acoustic or resulting from the vibration of the medium), and this is done in an even shorter time if they propagate rapidly in comparison to the characteristic velocity in which we are interested. On the other hand, this local equilibrium is certainly influenced by outside conditions that could be overly non-homogeneous (statistically of course). In a small volume near the interface, locally, much smaller than the mean size of the device being studied, this non-homogeneity is not necessarily felt. Thus, for media where mean pressure is slowly variable in space, with low velocities compared to the velocity of sound, and in which we are not interested in acoustic phenomena, it seems justified to assume that $\overline{p}_1^* = \overline{p}_1^s, \overline{p}_2^* = \overline{p}_2^s$.

This reasoning can even be applied in the presence of capillarity or mass exchange. We saw earlier (in section 4.2 the mean interface balance relationship: $-\overline{(m_1)^2(1/\rho_1 - 1/\rho_2)}^s + (-\overline{p}_1^s + \overline{p}_2^s) = \overline{F_{cap1,2}^n}^s$). It results in a link between the two mean pressures at the interface, even though this is not simple equality.

When this approximation is doubtful due to the fact that we are interested in rapid phenomena, we can still use a generalized form of [4.10] written as:

$$\frac{\partial}{\partial t}\overline{\Phi}_i + \overline{v}_s^\alpha \frac{\partial \overline{\Phi}_i}{\partial x_\alpha} = \frac{\overline{\Phi}_i(1-\overline{\Phi}_i)}{\mu_{BNi}}(\overline{p}_i^* - \overline{p}_i^s)\text{, for } i = 1.2 \quad [4.11]$$

With the preceding interface relationship related to $\overline{p}_i^s$, and provided that $\overline{v}_s^\alpha$ is known, these two new equations provide us with the 13 equations necessary for the 13 unknowns. The instantaneous equilibrium $\overline{p}_i^* = \overline{p}_i^s$ corresponds to the boundary situation in which the characteristic response time is very small, meaning that the constants μ_{BNi} are very small. The value $\overline{v}_s^\alpha$ is none other than the mean value at the interface of the interface velocity itself, such that $\overline{v}_s^\alpha \overline{\sigma}_i^s = \overline{v_s^\alpha \sigma_i^s}$, and we can have a model of this when studying the terms of energy exchange between phases, which we will discuss in the next chapters.

When this approximation is doubtful due to excessive non-homogeneity, of mean pressure for example, we can propose a mean pressure equilibrium relationship taking into account the spatial pressure gradient in a linear form. This has not been studied before now, but we can seek it with the general form:

$$\overline{p}_i^s - \overline{p}_i^* = \frac{C_{Pi}}{\overline{\sigma}_i^s}\left(\frac{\partial \overline{p}_i^*}{\partial x_\beta}\frac{\partial \overline{p}_i^*}{\partial x_\beta}\right)^{1/2}, \ i = 1.2 \quad [4.12]$$

This type of relationship, and the dimensionless coefficient C_{Pi}, could be studied beneficially using direct numerical simulation of the two-phase medium because real measurements of pressure are very difficult. The more the interface in the zone being considered possesses folds and wrinkles (and thus the stronger $\overline{\sigma}_i^s$ is), the more pressure waves can rapidly equalize the two mean pressures. A difference in behavior of one phase compared to another can also be explained by a difference in the velocity of sound, perhaps justifying different values of μ_{BNi}, C_{Pi}.

Remember that we are not considering granular media here that are too dense with solid particles. We have excluded this situation so that lasting contact between particles does not require us to reconsider the equations of state of the phase composed of these particles. In addition, we can also see that if the grains are too numerous in such a way that multiple contacts can connect long chains of particles, possibly touching the boundaries of the medium, the stresses inside the grains phase will no longer be very directly linked with those at the interface with the fluid phase. The condition of pressure equality of the two phases at each point of the interface will remain valid (disregarding surface effects and the effect of mass exchange), but

$\overline{p}_p^* \neq \overline{p}_p^s$ for the solid particles will prevent us from assuming the equality of the mean pressures of the two phases. If we preserve $\overline{p}_f^* = \overline{p}_f^s$, it will result that $\overline{p}_p^s = \overline{p}_f^*$.

4.4. Mean equations of state

Equations of state, for a classic compressible fluid, relate thermodynamic pressure and temperature to volumetric mass and internal energy per unit of mass. For a deformable solid, they usually give the temperature and all of the components of the elastic stress tensor in terms of the strain tensor and mass internal energy. Except for the part dedicated to granular media, we are interested only in the variation of volumetric mass of the solid in terms of pressure (pressure is the opposite of one-third of the elastic stress tensor trace), and a limit case could be when the volumetric mass of the solid is constant. For our multiphase medium, then, we know the instantaneous equations of state for the phases, but we still require the mean equations of state.

In *multifluid representations*, we must be able to obtain the equations of state that will enable us to relate the mean pressures of each phase to the mean volumetric masses and mean internal energies per unit mass. Likewise, we must be able to define the mean temperature in each phase. This is not too difficult if the instantaneous equations of state are simple themselves. To further clarify these ideas, we will give an example below, disregarding capillarity for the sake of simplicity.

If we consider that the internal energy of each phase is a linear function of temperature (as for a perfect gas or an incompressible liquid), it is easy to average this function and find:

$$\overline{\rho_i \Phi_i e_i} = C_{vi} \overline{\rho_i T_i \Phi_i} + e_i^0 \overline{\rho_i \Phi_i},$$

thus: $\overline{\rho}_i \overline{\Phi}_i \tilde{e}_i = C_{vi} \overline{\rho}_i \overline{\Phi}_i \tilde{T}_i + e_i^0 \overline{\rho}_i \overline{\Phi}_i$,

and finally: $\tilde{e}_k = C_{vk} \tilde{T}_k + e_k^0$.

These equations directly define the mean temperature $\tilde{T}_k$ and give their expressions in terms of the variables $\tilde{e}_k$, which must satisfy their balance equations.

With regard to pressure, if we consider, for example, that phase *i* is an ideal gas, by averaging the equation of state that defines the pressure in this phase, we have:

$$\overline{p_1 \Phi_1} = R_1 \overline{\rho_1 T_1 \Phi_1},$$

and thus: $\overline{p}_1^* = \overline{\rho}_1^* R_1 \tilde{T}_1$.

It is not realistic to consider that other phases are still ideal gases because all the gases would be mixed. If we assume very simply that the other phases are liquids or deformable solids with linear compressibility: $p_i = K_i(\rho_i - \rho_{i0})$, then the average immediately gives:

$$\overline{p}_i^* = K_i(\overline{\rho}_i^* - \rho_{i0})$$

If certain phases can be considered to have constant volumetric mass, which is common, their pressures are not then given by an equation of state, but are the results of calculations coupled with balance equations, which must include as many volume fraction balances as numbers of phases, minus one. As we have seen, if we replace these equations by the equality of mean phase's pressures, they are all equal to $\overline{p}_1^*$ and the mean balance equations of the volume fractions are not useful.

It is important to note, however, that the equations of state we have used above for each phase are very simple. If, for example, the first gas phase has a calorific capacity that is variable with temperature, and we have written this in a linear manner, we will have found:

$$\overline{\rho_1 \Phi_1 e_1} = C_{v1} \overline{\rho_1 T_1 \Phi_1} + \frac{1}{2} C'_{v1} \overline{\rho_1 T_1^2 \Phi_1} + e_1^0 \overline{\rho_1 \Phi_1},$$

In this case, it is necessary to know the new variable $\overline{\rho_1 \Phi_1 T_1^2} \neq \overline{\rho}_1^* \overline{\Phi}_1 (\tilde{T}_1)^2$. Likewise, a nonlinear law for the definition of pressure in a compressible fluid or a deformable solid will require one or more additional variables related to fluctuations of pressure and volumetric mass. In these conditions, the mean equations of state will be more complicated, bringing in variables other than the averages of pressure, temperature and internal energy, but will pose no difficulty in writing. To calculate these new variables, in principle, we can again find balance equations using instantaneous balance equations, but this will pose additional problems of modeling. Thus, it is only worth doing this if the situation being studied really requires it, that is, if it is highly probable that the additional variables compared to the classic variables play an important role.

With a *global modeling* of a multiphase medium taking into account a single mean velocity, we need in the momentum balance, as an equation of state, only the reversible part of the mean global Cauchy tensor, which is related to the mean pressure of the medium (we are not working with granular media). Moreover, we need another equation of state to link the mean temperature to the mean internal energy. As has been stated, finding these two equations must take completely into account the specificities of the medium, and it is a crucial stage. We must start from the basis of the equations of state of each constituent medium, average them, bring into play the mean values defined above and justify the approximation necessary for other independent mean variables to be discarded. When all of the phases are fluids, the hypothesis of equality of mean phase pressure has often also been used to simplify the work. It makes it very easy to find the equation of state for the global mean pressure in terms of the global volumetric mass, the mass or volume fractions of the phases, and the mean temperature(s). However, in any case, the choice of global equations of state is much more difficult to justify than the choice of mean equations for a multifluid model.

As an example, we may consider a two-phase medium with a gas phase containing drops or, symmetrically, a liquid phase containing bubbles. We will propose a model of equations of state with the hypothesis that the bubbles or drops are small enough for the mean values of the temperature of the liquid and the gas to be very close and also for the mean pressures of the gas and the liquid to be practically equal. If the gas phase is represented by an ideal gas, we have $p_g = \rho_g RT_g$, and, on average, we obtain: $\overline{p_g \Phi_g} = \overline{\rho_g \Phi_g RT_g}$, from which $\bar{p}_g^* \bar{\Phi}_g = \overline{\rho_g \Phi_g} R\tilde{T}_g = \bar{\rho}_g^* \bar{\Phi}_g R\tilde{T}$. Additionally, we have: $\bar{\rho} = \bar{\rho}_l^* \bar{\Phi}_l + \bar{\rho}_g^* \bar{\Phi}_g$. By identifying the mean pressure of the medium to the mean pressure in the gas, we have: $\bar{p} = \dfrac{(\bar{\rho} - \bar{\rho}_l^* \bar{\Phi}_l)}{1 - \bar{\Phi}_l} R\tilde{T}$, and if we assume that the liquid has constant volumetric mass, the equation of state of the medium giving the pressure is $\bar{p} = \dfrac{(\bar{\rho} - \rho_l \bar{\Phi}_l)}{1 - \bar{\Phi}_l} R\tilde{T}$.

Another form is also: $\bar{p} = \dfrac{(1 - \tilde{\Phi}_l)}{(1/\bar{\rho} - \tilde{\Phi}_l / \rho_l} R\tilde{T}$, because $\bar{\rho}\tilde{\Phi}_l = \rho_l \bar{\Phi}_l$.

Likewise, starting from $\rho_g e_g = \rho_g C_{vg} T_g + \rho_g q_g^0$ and $\rho_l e_l = \rho_l C_{vl} T_l + \rho_l q_l^0$ (for the liquid, in this case, we are discarding the contribution of pressure), we can find $\bar{\rho}\tilde{e} = \overline{\rho_g e_g} + \overline{\rho_l e_l} = \bar{\rho}(C_{vg} \tilde{\Phi}_g + C_{vl} \tilde{\Phi}_l)\tilde{T} + \bar{\rho}(1 - \tilde{\Phi}_l) q_g^0 + \bar{\rho}\tilde{\Phi}_l q_l^0$, and finally, the second equation of state is written as: $\tilde{e} = (C_{vl} \tilde{\Phi}_l + C_{vg}(1 - \tilde{\Phi}_g))\tilde{T} + \tilde{\Phi}_l (q_l^0 - q_g^0) + q_g^0$.

When we study cavitation in a liquid flow, we are interested in a fluid containing bubbles of its own vapor. The two phases are in vaporization–condensation equilibrium at each instant of their evolution, but this is at the level of their interfaces and not with regard to their mean pressure and temperature. In some models, it is directly given as an equation of state $\bar{\rho} = \bar{\rho}(\bar{p}, \tilde{T}, \bar{\Phi}_l)$, following an empirical formula. It follows that in this situation, the two variables $\bar{\rho}, \bar{\Phi}_l$ are not independent, contrary to our work context in section 4.3. The variable $\bar{\Phi}_l$ is no longer a base variable in the problem, and there is thus no longer any need for the mean equation of mass balance for the liquid phase, [4.4b], and therefore no longer a need to model the dispersion flux that figures in it. In fact, in this case, this balance equation just states that the two terms of phase change that feature in it, that of vaporization and that of condensation, are very large and compensate mutually for one another, erasing the influence of dispersion and convection.

In the case of granular media that are dense and not too rapid, the non-diagonal part of $\bar{\sigma}^{\alpha\beta}$ can no longer be disregarded and it poses additional problems. We will discuss on this further in the chapters dedicated to granular media.

4.5. Extensions

4.5.1. *Extension when a fluid phase is a mixture*

We saw in Chapter 2 that ρ_{ji}, the local and instantaneous volumetric mass of substance *j* in phase *i*, satisfies the instantaneous balance equation:

$$\frac{\partial}{\partial t}\left(\rho_{ji}\Phi_i\right)+\frac{\partial}{\partial x_\alpha}(\rho_{ji}\Phi_i v_i^\alpha)=\frac{\partial}{\partial x_\alpha}(-\Phi_i j_\alpha^{Dj})+\rho_i \dot{w}_{ji}^{sm}+\rho_i\Phi_i \dot{w}_j^{ch}$$

The volumetric rate of interfacial reaction is, of course, linked to the instantaneous area of interface per unit volume, $\rho_i \dot{w}_{ji}^{sm}=\Gamma_j^s \sigma_i^s$, and the surface rate of reaction Γ_j^s must be given according to the physical–chemical phenomena involved at the interface being considered. The last term of the differential balance equation corresponds to the chemical processes occurring in the volume of phase *i*.

Taking the average of this balance equation, it is clear that we obtain:

$$\frac{\partial}{\partial t}\left(\bar{\rho}_{ji}^*\bar{\Phi}_i\right)+\frac{\partial}{\partial x_\alpha}(\bar{\rho}_{ji}^*\bar{\Phi}_i\tilde{v}_i^\alpha)=\frac{\partial}{\partial x_\alpha}(-\bar{\Phi}_i\overline{j_\alpha^{Dj}}^*-\overline{\rho_{ji}\Phi_i v_i'^\alpha})-\overline{\Gamma_j^s\sigma_i^s}+\bar{\rho}_i^*\bar{\Phi}_i\tilde{\dot{w}}_j^{ch} \quad [4.13]$$

We have written: $\overline{\rho_{ji}\Phi_i}=\bar{\rho}_{ji}^*\bar{\Phi}_i$; we can also use $\tilde{Y}_{ji}$, the mean mass fraction of substance *j* in phase *i*, such that $\overline{\rho_{ji}\Phi_i}=\bar{\rho}_i^*\tilde{Y}_{ji}\bar{\Phi}_i$.

In a classic manner, we see the appearance here of an additional term corresponding to the diffusion flux, due to the turbulence in phase *i*, of substance *j* in phase *i*. This turbulent diffusion term of *j* is dependent on the correlation of fluctuations in velocity and concentration of *j*, in phase *i*.

The interfacial reaction and volume chemical reaction terms give rise to mean surface and volume rates of mass production of *j*, due to these two kinds of phenomena.

The laws pertaining to balance of momentum and of energy in phase i are unchanged, but the expression of the heat flux in phase i must now include the contribution due to the mass flux of substances j in this phase.

4.5.2. *Extension for dispersed media*

Another interesting generalization concerns dispersed flows in which one or more phases are dispersed in the form of a large number of solid or liquid particles in a fluid continuous phase. This is the case in an aerosol, for example. In general, the medium is diluted concerning particles, or is not very dense, but collisions can occur between these particles. Often, in these media, the forms of the particles do not vary greatly and have little effect on their behavior, but this is not necessarily the case concerning their size.

It is common to define for these media a number of particles per unit volume. This definition is a volumetric average and can only be clear and independent of the measurement volume if we consider particles of which the maximal volume is much smaller than the measurement volume. In addition, this number can be subject to temporal fluctuations if it is too low, unless a statistical mean number of particles, as described above, is considered in addition to the volumetric average.

Particles can also have different natures, and this nature can be significant. In this case, we will define several numbers of particles per unit volume, one per type of particles. An even more interesting natural extension consists of considering a continuous set of types of particles: this is the case, for example, when we must consider particles of the same nature that have various volumes (or various diameters when they are quasi-spherical), which can vary without discontinuity in a certain range, while of course remaining limited to a maximal value that is much lower than the size of the device.

The value of interest in representing this situation is the number of particles $\overline{\gamma}^N\left(\vec{M},t,\upsilon\right)d\upsilon$ of volume included between υ and $\upsilon+d\upsilon$, per unit volume of the medium, around a point $\vec{M}$. The quantity $\overline{\gamma}^N$ is called the numerical density of particles, and the particles of which the sizes are similar to the size υ, at near $d\upsilon/2$, constitute a category of sizes. The number of particles per unit volume of the medium, with all categories mingled, or overall numerical concentration C^N, is obtained by adding together the number of particles $\overline{\gamma}^N(\upsilon)d\upsilon$ that are in each category or: $C^N\left(\vec{M},t\right)=\int_\upsilon \overline{\gamma}^N(\upsilon,\vec{M},t)d\upsilon$.

When the particles present are all of the same volume, and thus of the same radius (or equivalent radius if they are only quasi-spherical), the only C^N, the number of particles per unit volume, often written simply as n, is sufficient to represent the dispersed medium. Insofar as the size of the particles is very small in relation to a length scale of the variation of the global properties of the flow (for example, the diameter of the tube in which it occurs), the probability of the presence of the particle phase $\overline{\Phi}_p$ at a point of flow is a true local volumetric average, and thus, $\overline{\Phi}_p = \frac{4}{3}\pi r^3 C^N$, where r is the common radius of the particles. If this radius is constant and known, the knowledge of $\overline{\Phi}_p$ leads directly to that of C^N.

It is possible to generalize this notion of number per unit volume, with the statistical average and the filtered average, using the indicative phase functions and avoiding any discussion concerning the respective sizes of the particles and of the volume being considered. We must first define, in the manner most useful to us (but clearly), the different categories of particles we wish to consider, including the possibility of recognizing these categories by their volume in a continuous manner, and perhaps also by other characteristics that could be useful. The volumetric average of Φ_i gives the volume fraction of the phase in question. Then, let us now call this category a phase. When we have identified the phase by the volume of particles that compose it, it is clear that the volume fraction of this phase is simply the product of the number of particles per unit volume and the volume chosen to characterize the phase. When we use the statistical average, the average of the phase-indicative function defines the probability of presence of the phase, identifiable as the volume fraction if the volume of particles in question is small compared to a volume in which the flow being considered can be assumed to be homogeneous. This condition is nearly always fulfilled for solid particles, but not always for liquid parcels or gas bubbles. With a phase defined by the volume of particles, we can then call the ratio of $\overline{\Phi}_i$ and of the typical volume of the phase, the number per unit volume of particles of the type being considered, and to find the numerical concentration density we must again divide by the $d\upsilon$ being considered. With the filtered average, a similar definition is still possible, even if the filter is not exactly a volume filter.

When we define different categories of particles because they have different volumes, and not only because they are of different types and thus of different physical phases, we must note that the instantaneous balance equation of the indicative function associated with each category can be more complicated than that defined in Chapter 2. If we define the categories of particles as having volume υ (at $\delta\upsilon/2$ approximately, of course) *at an initial instant,* and leaving the volumes to evolve subsequently, the indicative function of each category will follow a transport equation identical to [2.13], but the volume υ is then liable to vary afterward if the physical phenomena that cause it are present. It is usual for the choice of categories to be based on the volumes at each instant, and so particles can change category if their volume changes. In these conditions, if the physical phenomena of vaporization-condensation, fragmentation, and agglomeration are possible, the balance equation of the category indicative function is no longer simply a transport like [2.13]; it must have additional terms of production and destruction, which take into account these changes of category, and if they still have a Dirac peak structure they are not easy to write.

Whatever the case, in the same way as we have obtained [4.9b] from [2.13], we can in the most general case obtain a balance equation of numerical density, which can be written formally as:

$$\frac{\partial}{\partial t}\overline{\gamma}^N + \frac{\partial}{\partial x^\alpha}\left(\overline{\gamma}^N \tilde{v}_\upsilon^\alpha\right) = \overline{\Gamma}^N_{vap-con} + \overline{\Gamma}^N_{Frag} + \overline{\Gamma}^N_{Agglo} \qquad [4.14]$$

In this equation, $\overline{\gamma}^N$ has been obtained by a statistical average, the velocity $\tilde{v}_\upsilon^\alpha$ is the mean velocity of the particles in the category in question; that is, of volume υ (only and not particles of all sizes). The terms (mean as well) of number production in the right-hand part take into account not only the mass exchange due to changes in physical phase, vaporization–condensation, for example (which was already part of the volume fraction equation), but also the phenomena of possible fragmentations and agglomerations. These phenomena must then be studied in detail to provide explicit expressions of these terms. We have noted again the mean velocity of the class of particles $\tilde{v}_\upsilon^\alpha$ and not $\overline{v}_\upsilon^\alpha$; if the volumetric mass of the category is constant, which is generally considered, these two velocities coincide.

The mean velocity of the category of particles, using the hypothesis of constancy of its volumetric mass, can be calculated by a balance equation of momentum similar to [4.5], but in which the mass exchange term is replaced by the three intercategory exchange terms of numbers, the averages of which are used in the equation of $\overline{\gamma}^N$ above.

4.6. Boundary conditions

The modeling we seek is thus based on the definition of a certain number of variables, to be calculated by the same number of equations, most of which are partial differential equations. To be able to make the calculation, in a certain domain of space, a number of conditions at the boundaries of this domain must also be given, as well as initial conditions if we are not seeking only stationary solutions. On the one hand, current mathematical analysis techniques allow us to predict the properties of these conditions, but it is the physics, on the other hand, that we must use to find them in the end. This is the final part of the task of modeling.

With regard to the mathematical aspect, we know how to discuss the number of these conditions to find a single solution to the problem, which will be stable relative to the disturbances of these conditions at the initial time or at the boundaries. The response to these questions is not always known, of course, but it is in some simple cases, and we also know that it is not possible without certain properties of these conditions. We will not go into further detail in this discussion, but simply reiterate that each partial differential equation is solved with boundary conditions that can be of three types: the Dirichlet's condition prescribes the value of the variable at the boundaries; the Neumann's condition prescribes its normal derivative at the boundaries, and we can also use a mixed condition giving a linear combination of the first two types at each point of the boundaries. This condition is often called a Robin's condition. We can also give one or another of these three types for portions of the boundaries. With regard to initial conditions for non-stationary problems, Dirichlet's conditions are generally used for all of the variables.

With regard to the physical aspect, the first question raised is the choice of position of the boundaries of the domain. Certain boundaries are directly indicated by the physical properties of the problem; these are solid walls that contain the medium being studied. Others are necessary in order to avoid having to consider an overly large domain of calculation; these are boundaries where we believe we have a good understanding of the medium, no matter what the phenomena that can occur in the domain we are studying. It is at the boundaries where walls exist that a problem of modeling is still posed, and the first issue is the definition of "wall" itself.

The wall is often considered ideally as a smooth wall with a perfectly defined position, meaning that random imperfections are at a much smaller scale than the small scales of variation of the flow, and in this case smaller than the particles or parcels of the phases in the flow. But common practical situations include rough walls, due to construction or because they become rough through use, especially if the flow includes particles. In addition, if we are interested in thermal or chemical phenomena, the presence of impurities (of little-known size and type) on the wall can also cause locally erratic behavior that can have consequences at the macroscopic level. With our statistical modeling approach, these random fluctuations at the wall can also be averaged, without raising more problems in principle than those existing within the medium, but this can cause certain practical problems nonetheless.

If the geometric extension of rough patches is important, our average approach leads us to define a mean wall zone to which these rough patches are confined, instead of a simple wall surface. In this zone, the effect of the wall is not translated into boundary conditions to be imposed, but it appears in averaged balance equations as additional mean exchange terms with the "wall phase", which are non-zero only in this rough wall zone. The knowledge in this zone of the mean wall area per unit volume $\bar{\sigma}_w^s$ (which characterizes roughness) is a quantity necessary in order to evaluate these terms. The modeling of these exchange terms according to the principles described in the next chapter will then allow us to represent the effect of the mean wall in a continuous manner in the wall zone. However, if this wall zone is rather thin, particularly because the wall is effectively very smooth, this thinness makes it difficult to calculate it in a continuous manner. We will, therefore, consider in this zone a "median wall surface", on which the effect of the wall will be represented by specific conditions. It is clear that these boundary conditions will also be averages: if the wall is smooth, the median wall is the real wall, and they will also be statistical averages at the point of the wall. If the wall is rough, even slightly, they will not only be statistical averages at a given point, but also averages in the (small) thickness of the wall zone at this point. The form of these conditions can thus, in principle, be different from the boundary conditions for single-phase flows.

For a single-phase fluid medium (and an impermeable wall), we set the condition that at the wall, the velocity of the fluid is equal to that of the wall, which is justified by experimentation and by the kinetic theory of gases in the conditions of a continuous medium. This adherence condition is valid whether the wall is rough or smooth. For a solid medium, on the other hand, which will move against a wall, experimentation shows that there is a possibility of sliding between the two solids determined by a law often modeled like Coulcmb's law of friction. This law is often presented as taking into account the microscopic rough patches of two walls in contact. For a multiphase medium in flow, can we still justify an adherence condition, especially if one of the phases is composed of solid particles? Is it necessary to consider a slip velocity, and how do we determine this? One might think that for a medium composed of solid particles, a condition of Coulomb's friction must also be used, with a friction coefficient different from that of the solid, however. Nevertheless, if the size of the wall rough patches is of the same order of magnitude as the size of the solid particles, or larger, we might also think that the situation is similar to that of a fluid, with the wall able to retain particles, and experiments show that the condition of non-slip of the mean velocity at the wall is still valid. More detailed studies of this question have been done, which require taking into account a certain roughness of the wall, for situations called granular gas situations (see, for example, [CAO 96]). In general, it is not surprising that the result is that the most general Robin-type condition is the most realistic: the tangential velocity of the particle fluid (or the global fluid) at the wall is not that of the wall, but its difference with the (tangential) velocity of the wall is related to the tangential velocity normal derivative at the wall:

$$(\overline{u}_w^t - U_w^t) = a \left.\frac{\partial \overline{u}^t}{\partial n}\right|_w$$

The coefficient a depends on the roughness of the wall and on the volume fraction of particles; it can be found out via theoretical calculations but only for perfectly known wall configurations. It tends toward zero for a very rough wall, thus recovering a classic adherence condition. For the common case of a wall of which the configuration is poorly understood, it is logical to consider again a condition of this type, but with a coefficient a which must thus be empirical and such that we can again find the adherence condition for very rough walls.

This condition does not replace the interface balance relationship of momentum, which states that (with an impermeable wall) the contact forces exerted by the fluid on the wall are equal and opposite to the forces that the wall exerts on the fluid. Both relationships are necessary; if the wall is at rest, or has a given displacement velocity, we can find the force that must be exerted on the wall for this, and if the force is known, we can deduce whether the wall will move or not. If the link between the fluid velocity at the wall and the wall velocity is not known, this is impossible. We may think that this type of condition is not necessary for a phase composed of solid particles and that fluid phases preserve an adherence condition in any case. This is clearly true for a smooth wall, but it is not immediately demonstrable for a rough wall because the condition is an average in the wall zone and not only an average of the whole. In any case, the question is always raised when we are representing a medium in a global manner.

This condition of possible slip of the velocity field has a counterpart concerning temperature fields. The balance equations of (mean) internal energy or enthalpy allow us to calculate these mean temperature fields only if a thermal condition is given at the boundaries of the calculation domain. In a single-phase flow, a condition of equality between the temperature of flow at the wall and the known temperature of the surface of this wall (at each point of it) is often used. Another situation is to consider, as being known, the heat flux entering (or exiting) the fluid at the level of the wall, which, for heat transfer by conduction, consists of prescribing the temperature gradient normally at the wall. These two different conditions correspond to differently controlled physical situations; in the first case, control of the wall temperature is used to calculate the heat flux exchanged between the fluid and the wall; in the second, we deduce it from controlling the temperature flux of the fluid at the wall, and that of the wall, assumed to be identical at the same point. For the case of a solid-particle fluid flowing near a wall, equating the value of the temperature of this fluid near a point of the wall (even if this is a mean temperature) to the temperature of this wall at this point is not a straightforward hypothesis. It is understandable for very small particles likely to remain near the wall for some time, perhaps, but this is probably not generally the case, and it is also dependent on the roughness of this wall. Here again, a condition enabling a difference in (mean) temperature between the wall and the particle fluid near this wall is possible, but this difference cannot be left free. A Robin condition is also a very realistic approximation here, with the coefficient b to be found:

$$(\tilde{T}_w - \overline{T}^p) = b \left.\frac{\partial \tilde{T}}{\partial n}\right|_w$$

As above, it is mainly for the mean temperature of the particle's medium that this condition appears realistic, but it can also be so for a global representation.

We will see later on the interest of the calculation of the kinetic energy of velocity fluctuations for the modeling of turbulent flows, whether they are single-phase or multiphase. Here again, we face the problem of finding a condition at the wall for this kinetic energy, and the granular gas situation has also shown the justification for a Robin condition, but one in this case that is nonlinear and coupled with the slip velocity, of the type:

$$(c'k_w(\overline{u}^t_w - U^t_w) - k_w^{3/2}) = ck_w \left.\frac{\partial k^{1/2}}{\partial n}\right|_w$$

Additionally, when we are interested in the dispersion of a phase in a flow between solid walls, we must write a boundary condition at these walls, which represents the influence of the wall on this phase. The wall can retain certain solid particles, at least for some time, and certain fluid phases can be adsorbed into it. The boundary condition at the wall is none other than a mass interface balance for this phase, in which the phase emission or retention rate at the wall is modeled. In cases where this rate is negligible, the mass flux of the phase being considered, perpendicularly to the wall, must be zero at the wall. If we use a global modeling process that links this flow to the gradient of the volumetric fraction of the phase, this is equivalent to giving a condition that the normal gradient for $\overline{\Phi}_i$ is nil. With a multifluid model, this indicates that the normal velocity of this phase at the wall is zero, while its velocity tangential to the wall follows the discussions in the preceding paragraph.

If the physical phenomena involved are such that they lead to a strong variation of volume fractions in a thin layer near the wall, it is useful to include this layer in the averaged wall. This done, we must define the adapted boundary conditions for the volume fraction as well as for the other variables involved. For example, the phenomenon of a boiling liquid near a wall is typical of two-phase liquid–vapor flows, often called a boiling crisis. The problem is thus one of modeling the relationships between heat flux, volume fractions and temperature for the phases and the solid wall. Numerous experimental researches have been developed vis-à-vis these conditions; we will return to this topic in Chapter 11, section 11.3. It is clear that these conditions of heat transfer at the wall can cause critical problems.

We will conclude by reiterating that this problem of boundary conditions near walls is an important part of modeling, which must be examined for each specific multiphase flow.

PART 2

Modeling: A Single Approach Adaptable to Multiple Applications

Part 2 discusses the principles of modeling the effect of *irreversible processes* occurring in the flows. There are, on the one hand, irreversible processes linked to exchanges between phases, masses, and momentum or energy, as well as the interest of defining and studying exchanges in the form of exchange fluxes per unit of interface area. Just as importantly, there are also additional "dispersion fluxes" inside a phase, for momentum or energy, that are due to fluctuations, often turbulent in nature, the modeling of which will be discussed. We will see that they may be modeled by generalizing approaches that are usual in single phase turbulent flows, but there remains at least one major problem to be studied: the influence of the presence of phases on the characteristic length scales. We will then study the "mean interface area per unit of volume" quantity, which plays a critical role in all of these questions related to exchanges between phases, and, finally, the generalization of the approach for filtered averages such as the *large eddy simulation* will be discussed. We will then reconsider most of these questions from the general point of view of the thermodynamics of irreversible processes, which will require a generalization of the classic thermodynamic approach. Approximations that have been carried out classically will be reinforced for the most part, but there will also be interesting openings. The final two chapters of this part center on experimental data, which are essential for the development of realistic hypotheses within the framework of physical modeling through the use of balance equations, as well as for their validation. Examples will be given of recent detailed studies of multiphase situations that are still relatively unknown, after a summary of current and new experimental methods. Even though many experimental results have been obtained over the past 50 years, it has been preferred to concentrate here on examples of recent experimental studies which focus on especially delicate points.

Chapter 5

The Modeling of Interphase Exchanges

5.1. General methodology

In Chapter 4, we precisely defined the terms due to the interphase exchanges in mean balance equations. All of them can also be expressed as the product of the mean interface area per unit of volume $\overline{\sigma}_i^s$ and a mean flux of exchange per surface unit (the dimension of which depends on the quantity exchanged). We will see that this appears most of the time in the formulas studied, but sometimes it will remain implicit.

The mean force of interphase contact per unit of volume has been defined as $\overline{f_{i,s}^{\beta}\sigma_i^s} = \overline{F}_{pi}^{\beta} + \overline{F}_{\tau i}^{\beta} = -\overline{p_i\sigma_i^s n_i^{\beta}} + \overline{\tau_i^{\alpha\beta}\sigma_i^s n_i^{\alpha}}$, but another form is also useful:

$$\overline{f_{i,s}^{\beta}\sigma_i^s} = -\overline{\overline{p}_i^s n_i^{\beta}\sigma_i^s} + \overline{\tau_i^{\times\alpha\beta} n_i^{\alpha}\sigma_i^s} = \overline{p}_i^s \frac{\partial \overline{\Phi}_i}{\partial x_{\beta}} + \overline{F}_{is}^{\times\beta}.$$

Except for the first term of this second expression, the proportionality to $\overline{\sigma}_i^s$ can be shown.

The energy exchange terms are primarily due to the exchange of heat and the power of contact stresses. We can also write them per unit of interface area:

$$\overline{q_{is}\sigma_i^s} = -\overline{j_i^{Q\alpha} n_i^{\alpha}}^s \overline{\sigma}_i^s , \quad \overline{f_{i,s}^{\beta} v_i^{\beta}\sigma_i^s} = \overline{\sigma_i^{\alpha\beta} v_i^{\beta} n_i^{\alpha}}^s \overline{\sigma}_i^s ,$$

The exchange of mass between phases, if it exists, also results in an exchange of momentum and energy. We have:

$$\overline{m_i\sigma_i^s} = \overline{m_i}^s \overline{\sigma}_i^s , \quad \overline{m_i v_i^{\beta}\sigma_i^s} = \overline{m_i}^s \overline{\sigma}_i^s \overline{v}_i^{s\beta} , \quad \overline{m_i h_{ti}\sigma_i^s} = -\overline{m_i}^s \overline{\sigma}_i^s \overline{h}_{ti}^s .$$

In the moving medium, the interfaces themselves have fluctuating positions, orientations and curvatures. The local and instant values of p, u, τ, etc., at one interface depend on the local and instant state of the two adjacent media in a certain volume near the interface, and also on its curvature (much more than the orientation). The correlations of fluctuations of these p, u and τ at the interface must therefore take this neighboring volume into account. So, it is impossible to calculate all of these terms exactly if the instant flow field is not known, and we must seek approximate models that can be calculated based solely on the knowledge of mean local quantities. So, a physical methodology for the modeling of these exchange terms can only be based on the study of "local exchange configurations" attached to an interface element. Numerical simulations based on the equations in Chapter 2 would be very useful in the study of interface elements where the exchanges occur in different representative situations. The updating of new numerical methods will make this increasingly possible. However, there are many kinds of representative configurations and many possible temperature and flow fields near the interfaces. These are dependent first on the type of multiphase flow being studied, and also on the mean conditions of the medium in the zone considered of this particular type of flow.

Until now, the modeling of these terms has been studied separately for three major types of multiphase flows: flows in which solid particles are dispersed in a fluid carrier phase, flows in which a liquid contains gas bubbles and flows in which liquid drops are dispersed in a gas or non-miscible liquid phase. In these three cases, the parcels of phases are well dispersed (with the exception of the carrier phase) and have almost always been assumed to be quasi-spherical, as well as rather small in size, if not very small in comparison to the scales of variation of the flow itself. Because of this, the pertinent local exchange configuration is a quasi-spherical particle, relatively isolated from the others in a continuous carrier phase, and the analytical or experimental results concerning this exchange configuration, which have been obtained separately, have been used for modeling. Once this local exchange configuration has been chosen, the physically reasonable methodology of modeling includes three stages:

– The first stage consists of studying, combining theory and experimentation if necessary, the exchanges around these solid, gas or liquid parcels, in terms of the conditions at the boundaries of the local domain being considered, in order to deduce exchange flux formulations.

– The second stage consists of relating the exchange, for instance, contact force calculated for each configuration to a mean exchange per unit of volume, for instance the mean force that is exerted on the phase of the mean continuous medium (we will see for instance that it is not sufficient simply to multiply the force undergone by one particle by the number of particles per unit of volume).

– The third stage in obtaining the desired models requires the establishment of a link between the conditions at the boundaries of the local domain studied and the averaged quantities to which the balance equations relate. Here again, some of the most crucial hypotheses and approximations must be made.

Multiphase flows in which the phase parcels can be of similar size as a scale of variation of the flow also have a practical importance, and we have emphasized our desire to study these. This situation occurs for instance in the case of bubbly flows in which vaporization forms large bubbles that can even display almost the same diameter as the tube in which the flow is occurring, or during the formation of drops in a gas medium from a round jet of liquid. These situations are more rarely studied, but they can be examined using the same methodology, provided that we define a different local exchange configuration in the vicinity of almost flat elements of interface. We will study this new exchange configuration in some detail here. It will be seen that this study, even with the same methodology, requires certain specific adaptations compared to the one of spherical parcels, but shares many similarities with it. In particular, the mean interface area per unit of volume $\overline{\sigma}_i^s$ will show its great usefulness, while the numerical densities of particles or parcels of different types are used more in the classic case.

We will consider successively the exchanges of momentum, energy and mass, with the last two coupled. As the preceding formulas show, we are placing ourselves in the context of the definition of statistical averages. We will consider the transposition of modeling for filtered average, thus enabling a large eddy simulation (LES) approach, in Chapter 8.

5.2. Interface between phases and its mean area per unit of volume

Before describing in detail the local exchange configurations and the modeling associated with it, we will look more closely at the method of calculating the mean interface area per unit of volume $\overline{\sigma}_i^s$.

5.2.1. *Case of a suspension of liquid or solid particles*

If the multiphase medium is composed of liquid or solid spherical or quasi-spherical particles in a continuous phase, the mean interface area per unit of volume is obviously related to the number of these particles per unit of volume.

If all of the particles do have the same known and constant radius, we have $\overline{\Phi}_p = \frac{4}{3}\pi r^3 C^N, \overline{\sigma}_p^s = 4\pi r^2 C^N$, thus $\overline{\sigma}_p^s = 3\overline{\Phi}_p / r$. The knowledge of the volume fraction of particles directly gives the mean interface area per unit of volume, at known and constant radius r, there is no additional problem to solve.

If the particles have a wide range of radii and thus different volumes, we have seen that the numerical density of particles of a certain volume also follows a balance equation, the form of which is similar to the others:

$$\frac{\partial}{\partial t}(\overline{\gamma}^N) + \frac{\partial}{\partial x_\alpha}(\overline{\gamma}^N \tilde{v}^\alpha) = \frac{\partial}{\partial x_\alpha}(-\overline{\rho v_p^{"\alpha} \gamma^N}) + \overline{\Gamma}^N$$

We will rewrite it here with a dispersion term relative to the mean velocity of the medium, and we have seen that there can be sources or sinks of numbers of particles following their coalescence or fragmentation. The terms representing these phenomena must be modeled, if the multiphase medium studied is actually subject to these phenomena.

The total number of particles per unit of volume is $C^N = \int \overline{\gamma}^N (v)dv$. To find the mean volume area, we must then calculate the integral: $\overline{\sigma}_p^s = \int_0^\infty 4\pi r(v)^2 \overline{\gamma}^N (v)dv$.

We can thus calculate the interface area per unit of volume directly according to the knowledge of the numerical particle densities of volume v per unit of volume. But the balance equations for these quantities can be numerous, and their modeling can be rather complicated.

5.2.2. *Case of a medium containing parcels of variable shapes and sizes*

If the multiphase medium contains parcels of variable shapes, with some of them being as large as the flow itself, for example a gas–liquid medium containing bubbles in some places and drops in others (may be also small bubbles in the drops and small drops in the bubbles), and consequently with a very extended interface that touches the walls between which the medium flows, we can no longer define precise categories of particles nor the number of particles in a category. The notion of a number of particles does not enable us to describe the distribution of phases in the medium. However, the mean interface area per unit of volume $\overline{\sigma}_{ii'}^s$, between phases i and i'", or between phase i and any other phase, $\overline{\sigma}_i^s$, remains completely defined. Their usefulness for the mean calculation of all exchange phenomena between phases is very interesting, insofar as we can define and calculate the exchange terms, per surface unit, which can take into account in a realistic manner, on average, the characteristics of the different interfaces to be considered.

We have already mentioned (Chapter 2, section 2.5) that a balance equation for σ_i^s can of course be deduced from the transport equation of the corresponding phase index. Morel [MOR 07] has discussed different forms of this equation and shown that it can be written as:

$$\frac{\partial \sigma_i^s}{\partial t} + \frac{\partial}{\partial x^\alpha}(\sigma_i^s v_s^\alpha) = \sigma_i^s (\delta^{\alpha\beta} - n_i^\alpha n_i^\beta) \frac{\partial v_s^\alpha}{\partial x^\beta}$$

However, we have emphasized that this form does not take into account the destruction of interfaces by mutual interaction, which occurs at a scale that this piecewise continuous medium equation cannot represent.

Whatever the case, we must take the statistical average of this equation in order to find a balance equation for the mean interfacial area per unit of volume. As we might expect, this averaged equation is not directly in a closed form, meaning that it includes new correlation terms of which an approximation must be found. It is easy to find the form of this equation from [2.15a], reiterated above, we obtain:

$$\frac{\partial \bar{\sigma}_i^s}{\partial t} + \frac{\partial}{\partial x^\alpha}(\bar{\sigma}_i^s \bar{v}_s^\alpha) = \overline{+\sigma_i^s (\delta^{\alpha\beta} - n_i^\alpha ni) \frac{\partial v_s^\alpha}{\partial x^\beta}}$$

The mean velocity of the interface, a statistical average without weighting but conditioned by the fact that the interface is present at the point being considered, is involved here, although this does not mean that its value is easy to measure. The last term represents the evolution rate of the surface due to the stretching induced by its velocity field. This is the simultaneous consequence of the velocity fields of the two adjacent media (and very often at least one of these fields is turbulent) and surface tension stresses. This rate of evolution, depending on circumstances and points in the mean medium, can lead to the (mean) growth or decrease of the interface. We will see in Chapter 7 how these effects can be studied and modeled, and how we can add the phenomenon of interactions of interfaces.

We will, in a very general manner, use this mean interface area per unit of volume, or mean interface density, to give expressions of the exchange terms between phases. In models of multiphase flows, these terms are more often given using numbers of particles per unit of volume. It is mostly because interface density is more generally usable, not only for a completely dispersed phase, which we are using here. In addition, we will see that it is possible to model situations where coalescences and fragmentations of drops occur in the medium. These models will be discussed in Chapter 7.

Nevertheless, though the interface exchanges increase with the interface area, we will see that they are not simply linear functions of this area, and therefore the knowledge of the mean area is not sufficient to model mean exchanges. Because of this, the simple knowledge of the mean interface area per unit of volume does not perfectly solve the modeling of exchanges; we must also know the fluctuations of the interface area, formally equivalent to the distribution of sizes of particles in a dispersed medium.

5.2.3. *Case of a suspension of particles of constant and known sizes*

If we look again at the case of solid dispersed particles, it is also possible to find an equation for the mean interfacial area per unit of volume, and this equation is quite a bit simpler than the equations of $\overline{\gamma}^N$ if these particles cannot agglomerate or fragment, or lose or gain mass and volume. The mean interface area per unit of volume then follows a simple conservation equation, the same as the preceding equation with the last term being zero: $\frac{\partial \overline{\sigma}_i^s}{\partial t} + \frac{\partial}{\partial x^\alpha}(\overline{\sigma}_i^s \overline{v}_s^\alpha) = 0$. The problem is thus simply knowing the mean velocity of the interface.

If the solid particles are spherical and all of the same size and material, we have seen that $\overline{\sigma}_p^s = \frac{3}{r}\overline{\Phi}_p$, and the balance equation for the area per unit of volume must therefore be the same as that of the volume fraction and also of the mass fraction, since all the particles have the same volumetric mass. The mean interface velocity is thus exactly the mean velocity of the mass of the particles (weighted or not by the volumetric mass of the particles, which is the same), even if all the particles do not move in the same manner, we can write:

$$\frac{\partial \overline{\sigma}_p^s}{\partial t} + \frac{\partial}{\partial x^\alpha}(\overline{\sigma}_p^s \overline{v}_p^\alpha) = 0$$

When all the solid particles do not have the same radius, the mean velocity of the interface area is no longer the mean velocity of the mass, since the displacement of each particle depends on its radius. Thus, we can write:

$$\frac{\partial \overline{\sigma}_p^s}{\partial t} + \frac{\partial}{\partial x^\alpha}(\overline{\sigma}_p^s \overline{v}_p^\alpha) = \frac{\partial}{\partial x^\alpha}(-\overline{\sigma}_p^s(\overline{v}_s^\alpha - \overline{v}_p^\alpha))$$

and $(\overline{v}_s^\alpha - \overline{v}_p^\alpha)$ is a dispersion velocity of the particle interface relative to the particle mass velocity. To model this dispersion phenomenon due to the fluctuations in velocity of the particles (whatever the cause) and depending on the difference in size between the particles, a gradient model is often used, which defines a diffusion coefficient, and the previous equation is written as:

$$\frac{\partial \bar{\sigma}_p^s}{\partial t} + \frac{\partial}{\partial x^\alpha}(\bar{\sigma}_p^s \bar{v}_p^\alpha) = \frac{\partial}{\partial x^\alpha}(D_{sp}\frac{\partial \bar{\sigma}_p^s}{\partial x^\alpha})$$

This equation can therefore replace all of the balance equations for $\bar{\gamma}^N$, mentioned in section 5.2. But the diffusion coefficient D_{sp} is a function of the velocity fluctuations of particles, and is also dependent on the distribution of particle sizes. If this distribution varies from one point to other point in the flow, it must still be calculated in terms of $\bar{\gamma}^N$, which does not simplify anything. However, if this distribution has a similar shape at any point, which is expressed in terms of the values $\bar{\Phi}_p$ and $\bar{\sigma}_p^s$, which we can calculate, we can then model the dispersion with the form $D_{sp} = D_{sp}(\bar{\Phi}_p, \bar{\sigma}_p^s)$. This dependence in terms of the characteristics of the fluctuations is similar to that of the other diffusion coefficients, which will be discussed later in Chapter 5, but the exact form of this dependence in terms of $\bar{\Phi}_p$ and $\bar{\sigma}_p^s$ remains to be found; it has never been studied .

We can also write this mean interface area balance equation by bringing in a dispersion term relative to the mean velocity of the global medium:

$$\frac{\partial \bar{\sigma}_p^s}{\partial t} + \frac{\partial}{\partial x^\alpha}(\bar{\sigma}_p^s \tilde{v}^\alpha) = \frac{\partial}{\partial x^\alpha}(-\overline{\sigma_p^s v'_s})$$

We are also led to model this dispersion term with a diffusion coefficient, but here it has to be written in terms of the kinetic energy of the fluctuations in the medium and of their length scale. It also depends, of course, on the size of the particles, meaning on $\bar{\Phi}_p$ and $\bar{\sigma}_p^s$, but in a different manner that remains to be modeled.

5.3. Forces of contact and friction between phases

We are interested here in the modeling of contact forces between phases. We will first consider the classic case of the local exchange configuration of a quasi-spherical particle embedded in a carrier fluid. The radius of these particles is assumed to be known and constant, we will consider later the case of drops or bubbles in evaporation–condensation. This radius has always been assumed to be small in comparison to the scale of variation of the flow itself. To study situations in which phase parcels, drops or bubbles, can be large in size, in section 5.3.3, we will consider an exchange configuration at an interface that is very slightly curved. We have split the contact force showing a contribution due to pressure and due to viscous friction, $\overline{F}_{pi}^{\beta} + \overline{F}_{\tau i}^{\beta} = -\overline{p_i \sigma_i^s n_i^{\beta}} + \overline{\tau_i^{\alpha\beta} \sigma_i^s n_i^{\alpha}}$; we will consider these successively. For some cases, we prefer to use a slightly different splitting $\overline{F}_{pi}^{\beta} + \overline{F}_{\tau i}^{\beta} = \overline{p}_i^s \frac{\partial \overline{\Phi}_i}{\partial x_{\beta}} + \overline{F}_{is}^{\times\beta}$ as recalled in 5.1.

5.3.1. *Pressure forces on spherical particles in a non-viscous flow*

5.3.1.1. *Single spherical solid particle in a given non-viscous flow*

As mentioned in Chapter 4, force is exerted on particles by the surrounding fluid by means of the pressure field, if this is not uniform. We know, for example, that gravity causes a vertical pressure gradient in a fluid and in consequence produces Archimedes force on a particle in it, and a similar force also exists if the flow conditions themselves can cause a large-scale pressure gradient around the particle, independently of gravity. We must address these two effects at the same time, referring to a "generalized Archimedes force". But this is not all: if the flow is not stationary (relative to the particle), the fluid flow around the particle deforms when it accelerates or slows down, applying forces of its own. This effect is called the added mass effect, because it can be represented by inertia that is added to the particle (in addition to that due to its own mass), as the particle entrains some surrounding fluid. These effects have been studied in detail for almost 100 years for an isolated sphere in motion in a given flow and also for other bodies.

The general study by Auton *et al.* [AUT 88] , for a weightless and non-viscous flow, concludes that the stress exerted on a spherical particle by a given fluid flow is:

$$F_{p\oplus}^{\alpha} = \frac{4\pi r_p^3}{3}(\rho_f \frac{d_f V_f^{\alpha}}{dt} + \rho_f C_M (\frac{d_f V_f^{\alpha}}{dt} - \frac{d_p V_p^{\alpha}}{dt}) + \rho_f C_L ((V_f - V_p) \times curl V_f)^{\alpha})$$

where V_f is the velocity of the fluid at the point where the particle is located and V_p is the velocity of the center of the particle (which does not rotate). The added mass term is the second term and $C_M = 1/2$. The third term is added in case of a rotational flow; it gives a force perpendicular to the relative velocity between the fluid and the particle, and is therefore a "lift" (and $C_L = 1/2$ for a sphere). The formula is obtained at the limit case where the flow does not vary too rapidly and the $curl V_f$ is weak.

We can extend this result to include the effect of gravity by writing simply:

$$F_{p\oplus}^{\alpha} = \frac{4\pi r_p^3}{3}(\rho_f (\frac{d_f V_f^{\alpha}}{dt} - g^{\alpha}) + C_M \rho_f (\frac{d_f V_f^{\alpha}}{dt} - \frac{d_p V_p^{\alpha}}{dt}) + \rho_f C_L ((V_f - V_p) \times curl V_f)^{\alpha})$$

Remember that in any case gravity attracts particles as well as fluid, which adds the force $+\rho_p \frac{4\pi r_p^3}{3} g^\alpha$ into the momentum equation for the particle, and then the difference in volumetric mass between the particle and the fluid will appear in the momentum equation. Also, the momentum of the non-viscous fluid satisfies $\rho_f \frac{d_f V_f^\alpha}{dt} - \rho_f g^\alpha = -\frac{\partial p_f}{\partial x_\alpha}$, which enables us to bring in the pressure gradient of the fluid (at the position of the particle):

$$F_{p\oplus}^\alpha = \frac{4\pi r_p^3}{3}\left(-\frac{\partial p_f}{\partial x_\alpha} + C_M\left(-\frac{\partial p_f}{\partial x_\alpha} - \rho_f\left(\frac{d_p V_p^\alpha}{dt} - g^\alpha\right)\right) + \rho_f C_L((V_f - V_p)\times curl V_f)^\alpha\right) \quad [5.1a]$$

A more precise study also shows that a memory effect (called Basset force) exists, containing the effects of past evolutions of the fluid around the particle, which give an even more complicated additional term (see [GAT 83] and [NIG 91]).

For the added mass term to be important, the characteristic times of relative accelerations must be smaller than the Stokes time related to friction at the interface, which we will define precisely later on. This occurs only for particles that are very light in comparison to the fluid, and this term is not usually retained except for bubbles. The historical Basset term is also kept only very rarely. The contribution of lift, due to the *curl* in the fluid's velocity, has often been disregarded as well. This is what we will do next, keeping only the first term of [5.1a], but this should be revisited for certain specific situations.

5.3.1.2. *How to calculate the mean force on the phase of particles, knowing the force exerted on one single particle?*

When the medium is composed of C^N particles per unit of volume in a fluid, we may conclude that the force exerted by the fluid on the "particles' phase" is simply C^N times the force exerted on the particle. This is true for a globally homogeneous medium (that is medium in which the particles are distributed in a statistically uniform manner in space), but in a general non-homogeneous case, there is also an additional contribution related to gradients of the mean properties of the medium.

This problem has been discussed in depth by Nigmatulin [NIG 91]. It is explained that if we calculate the average in a given volume, certain particles in this volume are necessarily "riding" on the surface that defines this volume. If we split the studied mean force in $\overline{F}_{pP}^\beta = C^N F_{p\oplus}^\beta + \overline{F}_{sP}^\beta$, it has been found that it is realistic to write $\overline{F}_{sP}^\beta = \frac{\partial \overline{\sigma}_{sPf}^{\alpha\beta}}{\partial x_\alpha}$, where the new tensor $\overline{\sigma}_{sPf}^{\alpha\beta}$ is symmetrical. This contribution is related to normal forces on portions of surfaces, within the considered volume, belonging to particles "riding" on the boundary of this volume. The contributions of these riding particles are modified, and these modifications do compensate for each other only in a globally homogeneous medium, without gradients of mean quantities, and thus of tensor $\overline{\sigma}_{sPf}^{\alpha\beta}$.

Because of its form, this contribution can be grouped with the term $\frac{\partial(\overline{\Phi}_p \overline{\sigma}_p^{*\alpha\beta})}{\partial x_\alpha}$ of the mean momentum balance of the phase of particles, although it is actually due to the effect of the fluid. The sum $\overline{\sigma}_{\mathrm{Pr}ed}^{\alpha\beta} = \overline{\Phi}_p \overline{\sigma}_p^{*\alpha\beta} + \overline{\sigma}_{sPf}^{\alpha\beta}$ is, therefore, called the particulate phase reduced stress tensor, and this gives the impression that the force of contact between phases is C^N times the contact force on a particle, but this is only a definition (see, for example, [JAC 00]). This effect of non-homogeneity is also discussed more generally in the recent article by Lhuillier [LHU 09], where we see that we can also consider other gradients to intervene. We will return to this subject later on when we discuss the forces of friction between particles.

In any case, we must still model $\overline{\sigma}_{sPf}^{\alpha\beta}$ or $\overline{\sigma}_{\mathrm{Pr}ed}^{\alpha\beta}$. If the volume fraction of particles (non-deformable solids) in the medium is such that contacts between particles are very rare, the reduced stress tensor must be negligible, since the particulate phase does not have any more internal contact forces. We only consider pressure forces here, and therefore only the diagonal part of the tensor, and in this medium without contacts $\overline{\sigma}_{sPf}^{\alpha\beta} - \overline{\Phi}_p \overline{p}_p^* \delta_{\alpha\beta} = 0$. In situations where we can assume $\overline{p}_p^* = \overline{p}_p^s$ and $\overline{p}_f^* = \overline{p}_f^s$ (without the effects of surface tension and without mass exchange), we have $\overline{p}_p^* = \overline{p}_f^*$ and, therefore, we can write: $\overline{\sigma}_{sPf}^{\alpha\beta} = \overline{\Phi}_p \overline{p}_f^* \delta_{\alpha\beta}$. Symmetrically, the action of the particulate phase on the fluid phase, $\overline{F}_{pP}^{\beta} = -\overline{F}_{pf}^{\beta}$, includes, in addition to the opposite of the sum of the fluid forces on all particles, the divergence of a tensor that is equal and opposite to the first, $\overline{\sigma}_{sfP}^{\alpha\beta} = -\overline{\sigma}_{sPf}^{\alpha\beta} = -\overline{\Phi}_p \overline{p}_f^* \delta_{\alpha\beta}$. It also allows us to define for the fluid a reduced tensor: $\overline{\sigma}_{fred}^{\alpha\beta} = \overline{\Phi}_f \overline{\sigma}_f^{*\alpha\beta} - \overline{\Phi}_p \overline{p}_f^* \delta_{\alpha\beta}$. When the particles are submitted to many collisions and even rub together, we can generalize the previous writing as $\overline{\sigma}_{sPf}^{\alpha\beta} = \overline{\Phi}_p \overline{p}_p^s \delta_{\alpha\beta}$, as it is the matter of contact force at the fluid–particles interface, and without surface tension or mass exchange, $\overline{\sigma}_{sPf}^{\alpha\beta} = \overline{\Phi}_p \overline{p}_f^s \delta_{\alpha\beta}$. It follows that if the hypothesis $\overline{p}_f^* = \overline{p}_f^s$ remains valid (for the fluid only), we can preserve the relationship $\overline{\sigma}_{sPf}^{\alpha\beta} = \overline{\Phi}_p \overline{p}_f^* \delta_{\alpha\beta}$. However, the reduced tensor of the particulate phase, due to interactions between particles, $\overline{\sigma}_{\mathrm{Pr}ed}^{\alpha\beta} = \overline{\Phi}_p \overline{\sigma}_p^{*\alpha\beta} + \overline{\Phi}_p \overline{p}_f^* \delta_{\alpha\beta}$ will no longer be zero.

A priori models of a two-phase medium, which do not use the fact that it is the average of a piecewise continuous medium (for example [JAC 00]), are interested only in $\overline{\sigma}^{\alpha\beta}_{\Pr ed}$ and $\overline{\sigma}^{\alpha\beta}_{fred}$, and do not utilize the mean pressures at the interface. When particles collide with each other or rub together, the problem of the modeling of these tensors is combined with the issues posed by the other phenomena discussed here. Conversely, the approach we are using here enables us to justify $\overline{\sigma}^{\alpha\beta}_{sPf} = \overline{\Phi}_p \overline{p}^*_f \delta_{\alpha\beta}$ directly (this seems to have been proposed first by Nigmatulin [NIG 91]), and to better separate the different phenomena involved in the model.

However, what we have said so far is not enough to obtain a usable formulation of the contribution of Archimedes force to the mean contact force. Formula [5.1a] takes into account the pressure in the surrounding fluid without the particle, at the position of the particle, a fictional quantity for the real two-phase medium. The mean equations do not give us access to this pressure, they give access only to a mean pressure. If, in the last stage of the approach, we identify this fictional pressure to intrinsic mean pressure, it is then that we disregard the effects of fluctuations around this mean pressure, or take them into account later in $\overline{F}^{\times\beta}_P$ instead of $\overline{F}^{\beta}_{\tau P}$.

By identifying this pressure with the mean pressure of the fluid at the point being considered, we can now write the contribution of the Archimedes force to the contact stress. If we use the reduced stress tensor at the same time, the force can be summarized as:

$$\overline{F}^{\beta}_{pP} = -\overline{\Phi}_p \frac{\partial \overline{p}^*_f}{\partial x_\beta} \qquad [5.1b]$$

and without using it, but keeping our definition of contact force, we have:

$$\overline{F}^{\beta}_{pP} = -\overline{\Phi}_p \frac{\partial \overline{p}^*_f}{\partial x_\beta} + \frac{\partial}{\partial x_\beta}(\overline{\Phi}_p \overline{p}^*_f) = \overline{p}^*_f \frac{\partial \overline{\Phi}_p}{\partial x_\beta} \qquad [5.1c]$$

In these formulas, we have used the relationship $\frac{4\pi r_p^3}{3} C^N = \overline{\Phi}_p$, which is still only an approximation, disregarding the volume fraction of particles gradient, which is all the more valid because the particle radius is small compared to the characteristic length scale of this gradient.

Concerning the effects of added mass and *curlV*$_f$, the effects of mean accelerations and mean *curlV*$_f$ could be also taken into account, but their turbulent fluctuations have to be included in $\overline{F}^{\times\beta}_P$. Note, finally, that the statistic averages here have been identified to volumetric averages, since in the case of particles it is always assumed (as in [NIG 91]) that the radius of these is always small compared to the scales of the variations of flow.

5.3.2. *Friction on solid particles in steady flow*

5.3.2.1. *The Stokes theory*

A uniform viscous flow around a solid particle, without the influence of gravity, is another interesting local configuration that is relatively frequently considered. The force exerted by the fluid on the particle is therefore due solely to friction, meaning that the contribution of pressure is zero. This has been studied often, in a theoretical manner with certain approximations, as well as through experimentation.

The *classic Stokes theory* considers a fluid with constant physical properties and assumed to be weightless, and is applicable to cases where the relative velocity between the spherical particle and the fluid is low. It allows us to find the force applied to the particle:

$$F_{\tau p}^{\alpha} = 6\pi\rho_f \nu_f r_p (v_f^{\alpha} - v_p^{\alpha}) \qquad [5.2]$$

The index f indicates the fluid surrounding the particle (of radius r_p), and the difference of velocity in the parentheses is that of the velocity of the fluid far away from the particle compared to that of the particle (see, for example, [LAN 87]). We can also say that, as previously, the velocity of the fluid is that of a flow field that is uniform here, existing independently of the particle, taken at the position of the center of the particle. In the calculations, the Reynolds number of the particle $R_p = \frac{2r_p}{\nu_f}\left|v_p - v_f\right|$ is assumed to be very small and only the first-order terms in relation to this parameter are retained. Since the force per unit of volume has the same dimension as a momentum per unit of volume and time, this formula shows first that we can define a characteristic time of particle friction as $\tau_{Sp} = \frac{2}{9}\frac{\rho_p r_p^2}{\rho_f \nu_f}$, which is called the Stokes time of the particle. The time necessary for a particle to slow down by friction with a fluid to divide by two (for example) its difference in velocity relative to the surrounding fluid is proportional to the Stokes time, all the smaller when the particle is small, but also as large as it is heavy in comparison to the fluid.

As previously, we must now deduce from this formula the term of friction force of the fluid on the particulate phase, per unit of volume, $\overline{F}_{\tau p}^{\beta}$.

5.3.2.2. *Mean force in an homogeneous medium*

It is sufficient to multiply the force acting on a particle by the number of particles per unit of volume. In addition, the linear form with respect to the velocities is averaged by causing the mean velocities to appear, and assimilating the mean velocity of the fluid far away from the particles at the mean velocity of the flow in this area seems valid if the particles are not too large relative to the length scale of the variations of the flow (remember here that the densities of the fluid and of the particles are constant, and the averages weighted by the densities are thus equal to classic averages). This gives:

$$\overline{F}^{\beta}_{\tau p} = C^N 6\pi\rho_f v_f r_p \overline{(v^{\beta}_f - v^{\beta}_p)} = \overline{\Phi}_p \frac{9}{2r_p^2}\rho_f v_f (\tilde{v}^{\beta}_f - \tilde{v}^{\beta}_p) \quad [5.3a]$$

We can also write this formula, remembering that all particles are identical, as:

$$\overline{F}^{\beta}_{\tau p} = \frac{1}{2}\rho_f v_f \frac{\overline{\sigma}^s_{pf}}{\overline{\Phi}_p}(\tilde{v}^{\beta}_f - \tilde{v}^{\beta}_p)\overline{\sigma}^s_{pf} \quad [5.3b]$$

Here, we have used the relationships $\frac{4\pi r_p^3}{3}C^N = \overline{\Phi}_p$ and $4\pi r_p^2 C^N = \overline{\sigma}^s_{pf}$, valid approximations for small particles and small volume fraction gradients.

The situation obviously becomes more complicated if the distribution of particle sizes varies in the medium, due to the physical phenomena involved. The complications are also extended if the statistical distribution of the particle velocities at a given point is dependent on the distribution of the radii, which is completely understandable physically. In these conditions, it is clear that we must calculate the distribution $\gamma^N(v)$ in order to be able to better calculate the friction force, and all of the exchange terms. We will discuss these situations in Part 4 of this book.

5.3.2.3. *Mean force in a non-homogeneous medium*

In the case of the pressure force in section 5.3.1, it is in principle necessary to add a term, which can be written as $\frac{\partial}{\partial x_\alpha}(\overline{\tau}^{\alpha\beta}_{sPf})$. With the same reasoning as in section 5.3.1, applied to the Cauchy tensor deviator, we write $\overline{\tau}^{\alpha\beta}_{sPf} = \overline{\Phi}_p \overline{\tau}^{s\alpha\beta}_p$. As mentioned in section 5.3.1, instead of taking it into account as a force between phases, we can add it to the stress tensor deviator of the medium, which then becomes the reduced stress tensor for the dispersed phase, the one that is simply due to contact between particles. However, this additional term is often disregarded in the momentum balance equation of the particular phase, in relation to the term corresponding to the fluid pressure, as discussed in section 5.3.1.

However, as discussed by [NOZ 87] and [LHU 09], influences of other gradients can also be expected. If we write Γ as a measure of the mean velocity gradient U of the medium, Lhuillier proposes:

$$\overline{F}_{\tau p}^{\beta} = -\frac{1}{2}\rho_f \nu_f \frac{\overline{\sigma}_{pf}^{s2}}{\overline{\Phi}_p}(\tilde{v}_p^{\beta} - \tilde{v}_f^{\beta}) - \rho_f \nu_f \Gamma D_{\beta\alpha}^{P} \frac{\partial \overline{\Phi}_p}{\partial x_\alpha} + \rho_f \nu_f B_{\beta\alpha\gamma\delta}^{P} \frac{\partial^2 U^\delta}{\partial x_\alpha \partial x_\gamma}$$

The last term contains the Faxen force, shown when $\dfrac{\partial^2 U^\delta}{\partial x_\alpha \partial x_\alpha} \neq 0$. This was discussed for non-turbulent media, and the tensorial coefficient $D_{\alpha\beta}^{P}$ is dimensionless, but the formula has its origin in phenomenological reasoning, and we can extend it to a turbulent version. In this turbulent context, it has been suggested independently to add to the difference in mean velocities a so-called "drift velocity" related to the volume fraction gradient [SIM 00]. Thus, it is written:

$$\overline{F}_{\tau p}^{\beta} = -\frac{1}{2}\rho_f \nu_f \frac{\overline{\sigma}_{pf}^{s2}}{\overline{\Phi}_p}(\tilde{v}_p^{\beta} - \tilde{v}_f^{\beta} - v_{Dp}^{\beta}) \qquad [5.4]$$

The drift velocity is given as $v_{Dp}^{\beta} = -\dfrac{D_{tpf}}{\overline{\Phi}_p \overline{\Phi}_f} \dfrac{\partial \overline{\Phi}_p}{\partial x_\beta}$, where D_{tpf} is a (scalar) turbulent diffusion coefficient (in square meters per second). It must be related to the characteristics of velocity fluctuations of the phases, as well as to the molecular viscosity of the fluid phase and to the size of the particles concerned, in a manner to be studied and which we must discuss again later.

5.3.2.4. *In a turbulent flow, the Stokes formula must be "adjusted"*

We already know that when *the particulate Reynolds number is not low*, the Stokes formula must be modified. The physical justification for this modification is that, on the one hand, we must depart from the linear approximation that constitutes the Stokes theory and, on the other hand, the high local velocity gradients in the vicinity of the particle can create turbulence. These two things are translated in terms of a multiplicative coefficient, dependent on the Reynolds number of the particle, and experiments enable us to give empirical formulations of this coefficient. In all situations where turbulence is created, the mean result is the creation of an "eddy viscosity" that is added to the molecular viscosity, and this turbulence always makes it possible, at the limit of a very large Reynolds number, to render the global properties, and thus the mean drag, insensitive to molecular viscosity and to the Reynolds number. In the case we are examining here, we must thus find that this coefficient, a function of the Reynolds number, tends when this number is large to be proportional to this number, such that the molecular viscosity of the fluid is no longer involved in the calculation of the mean force exerted.

A classic form [NIG 91] of the force on the particle is:

$$F_{\tau p}^{\alpha} = 6\pi \nu_f r_p \rho_f (v_f^{\alpha} - v_p^{\alpha})(1 + \frac{\left(R_p\right)^{1/2}}{6} + \frac{1,2R_p}{72}) \text{, where } R_p = \frac{2r_p}{\nu_f}\left|v_p - v_f\right|.$$

We can see that if $R_{\rm p}$ is high, this formula becomes independent of ν_f, and is simplified to:

$$F_{\tau p}^{\alpha} = 0,2\pi r_p^2 \rho_f (v_f^{\alpha} - v_p^{\alpha})\left|v_f^{\alpha} - v_p^{\alpha}\right|$$

With the hypothesis of the homogeneity of the two-phase medium, and by identifying the velocities with the mean velocities of the medium, the expression for the force of contact between the fluid and particles, per unit of volume of the medium, is thus:

$$\overline{F}_{\tau p}^{\beta} = -\frac{1}{2}\rho_f \nu_f \frac{\overline{\sigma}_{pf}^{s2}}{\overline{\Phi}_p}(\tilde{v}_p^{\beta} - \tilde{v}_f^{\beta})(1 + \frac{1}{\sqrt{6}}\left(\frac{\left|\tilde{v}_p - \tilde{v}_f\right|\overline{\Phi}_p}{\nu_f \overline{\sigma}_{pf}^{s}}\right)^{1/2} + 0,1\frac{\left|\tilde{v}_p - \tilde{v}_f\right|\overline{\Phi}_p}{\nu_f \overline{\sigma}_{pf}^{s}}) \qquad [5.5]$$

which gives, at the limit of large difference in velocities:

$$\overline{F}_{\tau p}^{\beta} = -0,05\rho_f (\tilde{v}_p^{\beta} - \tilde{v}_f^{\beta})\left|\tilde{v}_p - \tilde{v}_f\right|\overline{\sigma}_{pf}^{s}$$

To take into account the gradients of the volume fraction of the particles, we can, of course, add once again to these formulas a drift velocity.

However, these formulas do not take into account the *pre-existing turbulence* in the fluid phase, in comparison to the difference in mean velocities, nor the fluctuations in particles velocity (passing at this point) around their mean velocity. Pre-existing turbulence also accompanies some pressure fluctuations, which make it so that the friction force is then more what we have called $\overline{F}_p^{\times\beta}$ than $\overline{F}_{\tau p}^{\beta}$. We must take velocity fluctuations into account if the fluctuations of the differences between the fluid velocity and the particle velocity are not negligible in comparison to the difference of mean velocities, but then the characteristic length scale of fluid turbulence, compared to the radius of particles, affects the method of doing this. In fact, turbulent fluctuations of scales larger than the diameter of the particle will generate fluctuations in the fluid–particle velocity difference, while fluctuations of scales smaller than the diameter of the particle will create additional eddy viscosity. The generalization in this sense of formula [5.5] cannot be done easily, and should be studied more closely.

If we assume *relatively large particles*, these cannot closely follow the fluctuations in velocity of the fluid (their Stokes time is too long). In addition, some of the fluid's velocity fluctuations occur at scales smaller than their size. We can, therefore, represent these fluctuations by an eddy viscosity coefficient that is proportional to the square root of the kinetic energy of these fluctuations and to the radius of the particle (which marks the largest scale of this part of the fluctuations). We can suggest the reapplication of the Stokes formula with the mean velocities, which also assumes that the difference in velocities between fluid and particle never becomes too large:

$$\overline{F}_p^{\times\beta} = -C\overline{\Phi}_p\rho_f \frac{(k_{fp})^{1/2} r_p}{r_p^2}(\tilde{v}_p^\beta - \tilde{v}_f^\beta)$$

where C is a constant and k_{pf} is the part of the kinetic energy of the fluctuations at a scale lower than the radius. The estimation of the relationship between k_{pf} and k_f thus requires knowing the form of the velocity spectrum, and depends, of course, on the ratio between the radius of the particles and the integral length scale of the fluctuations. We are finally led to:

$$\overline{F}_p^{\times\beta} = -C_{tp}(l_t / r_p)\overline{\Phi}_p\rho_f(\tilde{v}_p^\beta - \tilde{v}_f^\beta)k_f^{1/2} / r_p \qquad [5.6]$$

The coefficient C_{tp} must take into account the proportion of turbulence occurring on a small scale, and must be determined through experimentation if a more precise theoretical study is not carried out. The integral scale of turbulence must be estimated through the turbulence model, which we will discuss in Chapter 6.

However, the situation is more delicate for *relatively small particles*, with shorter Stokes time, in a flow that is already turbulent, since we do not know, in particular, how to distinguish between $\overline{F}_p^{\times\beta}$ and $\overline{F}_{\tau p}^{\beta}$. This problem is still largely unsolved. We will see in Chapter 6, particularly in section 6.1.5, that the modeling of $\overline{F}_p^{\times\beta}$ is crucial in order to determine the manner in which particles disperse in a turbulent medium, and this will lead us back to discussing this point when proposing formula [6.17].

5.3.2.5. *Wall effects*

One spherical particle in a viscous sheared flow, that is subject to a transverse velocity gradient, in a flow domain of size large compared to the diameter of the particle, is another interesting local configuration. Such a flow is rotational, and we saw in section 5.3.1 that this leads to the appearance of a lift force in a non-viscous medium, that is for an infinite particulate Reynolds number. Lift is also present when viscosity is in play, i.e. for a lower Reynolds number.

Saffman (see, e.g., [GUA 12]) has studied this type of situation. If the coordinate x is the direction of the velocity of the fluid relative to the particle and y is the direction of the velocity gradient, he has shown that the particle is subject to a lift force in the direction y of value:

$$F_{LSp} = 6\pi\mu_f r_p K(\frac{\rho_f |\gamma| r_p^p}{\mu_f})^{1/2} \frac{\gamma}{|\gamma|}(V_f - V_p)$$

with $K \approx 0.343$, where $\gamma = \dfrac{\partial V_f}{\partial y}$.

This situation is interesting when we are examining a flow carrying solid particles near a wall. The lift force can explain the migration of the particles noted in experiments (in various manners depending on the case). However, it appears that there is also a direct effect of the wall, which pushes away the particles if they are too near, since this increases the fluid pressure between the wall and the particle approaching it, as shown in the review paper [MAT 04]. Taking into account these two effects jointly in the general modeling of multiphase flows, for any velocity field, has never been attempted, to our knowledge. Generally speaking, the study of particles in flows near walls is a subject that needs to be studied more closely.

5.3.2.6. *When the medium becomes dense in particles*

It is clear that the neighboring particles influence increasingly the fluid flow around a particle.

We can already note in formulas [5.3b] to [5.6] that the appearance of the mean velocity of the fluid is already capable of taking into account some effect of other particles, since the mean contact force due to all of the particles is taken into account in the fluid mean momentum balance equation for calculating this mean velocity. We can also note the same thing concerning the fluid mean pressure for calculating the generalized Archimedes force in [5.1b]. In many situations of this type in statistical physics, this "mean field" hypothesis is considered as a good first approximation to take into account all the collective effects. Nevertheless, modifications of the velocity field in the vicinity of a particle due to the presence of the closest neighbors are obviously not represented by the fluid mean velocity. To do this, it is usual to introduce a multiplicative corrective factor, as a function of the volume fraction of particles to formulae like [5.3a] and following ones. In [NIG 91], we can find expressions of this factor based on studies of periodic or random configurations. A corrective multiplicative factor in [5.3a] or [5.3b] is calculated numerically as $\Psi = (1-\frac{3}{2}\overline{\Phi}_p^{1/2})^{-1}$, for the first case, $\Psi = (1-\overline{\Phi}_p)^{-5}$, for the second case. There are numerous formulas proposed (including the popular Ergun equation), as there have been many different experiments in various ranges of $\overline{\Phi}_p$.

As a limiting case, in a very dense situation close to a granular medium, a local exchange configuration in which the fluid opens a trail between numerous particles that are almost in contact, like within a porous medium, is much more realistic than that of a flow around a single sphere. Moving even more at the limit, the case of a flow established in a cylindrical channel is a rough caricature of this, but it is the only one (with the plane channel we will discuss in 5.3.3) able to give explicit formulas.

Calculating the friction exerted by the walls of a very long cylindrical tube on the laminar flow it contains is a classic problem in fluid mechanics. We can show that the force exerted on the fluid per unit of length is proportional to the outgoing velocity of the flow (see [LAN 87]):

$$F_f = -4\pi\rho_f \nu_f V_f \qquad [5.7a]$$

If the volume fraction of the solid in the medium, assumed to be globally homogeneous, is $\overline{\Phi}_p$, the unit of length of a tube of radius r_f is contained in a solid cylinder of total radius r_p, such that $1-\overline{\Phi}_p = \frac{4\pi r_f^2}{4\pi r_p^2}$. Moreover, the interface area per unit of volume is $\overline{\sigma}_{pf}^s = \frac{2\pi r_f}{4\pi r_p^2}$. As a result, if we identify the outgoing velocity of the fluid with the relative mean velocity between the fluid and the solid, and if we assume that the orientation probability of the channel in the local exchange configuration is homogeneous in space, we can deduce from this the force per unit of volume exerted by the fluid on the solid, with a constant close to ½ to be determined by experimentation:

$$\overline{F}_{\tau p}^{\beta} = -C'_D \rho_f \nu_f (\tilde{v}_p^{\beta} - \tilde{v}_f^{\beta}) \frac{\overline{\sigma}_{pf}^{s2}}{1-\overline{\Phi}_p} \qquad [5.7b]$$

The resemblance with the formula corresponding to a local configuration of a solid sphere [5.3b] is striking, as is the respective role of the volume fraction of the solid phase. The correcting function here is $\Psi = (C_{st}/2C'_D)\bar{\Phi}_p/(1-\bar{\Phi}_p)$. We find the form of the Carman–Kozeni correction for porous media, in which the proportionality constant is around 10. The local configuration of the isolated sphere assumes that $\bar{\Phi}_p$ is rather small, while here we can assume $1-\bar{\Phi}_p$ to be rather small. In fact, for the case of a medium highly loaded with solid particles, we have said that we cannot go too close to the random loose packing boundary, where $\bar{\Phi}_{p\max} = 0.55$ for spherical particles of the same size.

To clarify further the phenomena that can occur in situations of suspensions with high solid particle density, it is useful to study configurations of pairs of particles in motion. Some studies have already been done; we may refer to [GUA 12]. Interactions between particles occur even if they remain separate, and modify the "pair velocity" in given media. When two particles approach each other, we note a "lubricating" effect of contact: the pressure between particles increases and gas escapes laterally, which slows down their approach and can even prevent their contact, if no force is permanently applied. The movements of three particles (or more) in interaction are unstable.

5.3.3. *Slightly curved liquid–gas interfaces*

We will look now at multiphase media that are not simply very dispersed media, but show in some places very large phase parcels occupying a large part of the flow. In liquid–gas plug flows, large bubbles that are highly non-spherical in shape and occupy all or nearly all of the conduit section maintain interfaces that are only slightly curved and greatly extended. Similarly, when a liquid jet fragments into smaller or larger drops, an extended, stretched interface that is more or less conical, in fluctuating motion, separates the essentially liquid jet from a dense spray. The exchanges that occur in the vicinity of these slightly curved interfaces, and thus different from drops or bubbles surfaces, warrant particular study.

The pertinent local exchange configuration to be studied is that of a quasi-flat piece of interface through which, at least, two close fluid media exchange momentum. The expression of the mean contact force per unit of volume for the global medium is still the same, $\overline{f_{i,s}^{\beta}\sigma_i^s} = \overline{\left(\tau_i^{\alpha\beta} - p_i\delta^{\alpha\beta}\right)\sigma_i^s n_i^{\alpha}}$, but it is no longer useful to bring in the contribution of some particle; we must study it in a different way.

5.3.3.1. *The effect of mean pressure*

The effect of mean pressure on the mean contact force at a slightly curved interface is taken into account directly by using the mean pressure at the interface, $\overline{p}_i^s = \overline{p_i \sigma_i^s} / \overline{\sigma}_i^s$. We have already mentioned this in Chapter 4, section 4.2.2. Thus, we can write:

$$\overline{f_{i,s}^{\beta} \sigma_i^s} = -\overline{p}_i^s \overline{n_i^{\beta} \sigma_i^s} + \overline{(\tau_i^{\alpha\beta} - p"_i \, \delta^{\alpha\beta}) n_i^{\alpha} \sigma_i^s} = -\overline{p}_i^s \overline{n_i^{\beta} \sigma_i^s} + \overline{\tau_i^{\times\alpha\beta} n_i^{\alpha} \sigma_i^s}$$

where $p"_i = p_i - \overline{p}_i^s$, and therefore:

$$\overline{f_{i,s}^{\beta} \sigma_i^s} = \overline{p}_i^s \frac{\partial \overline{\Phi}_i}{\partial x_{\beta}} + \overline{F}_{is}^{\times\beta} \qquad [5.8]$$

Using the instant interface balance relationship, we have shown that:

$$-\overline{(m_i)^2 (1/\rho_i - 1/\rho_{i'})}^s + (-\overline{p}_i^s + \overline{p}_{i'}^s) = \overline{F_{capi,i'}^n}^s$$

and also that the forces $\overline{F}_{is}^{\times\beta}$ must compensate themselves for phases in contact $\sum_i \overline{F}_{is}^{\times\beta} = 0$.

The term $\overline{p}_i^s \frac{\partial \overline{\Phi}_i}{\partial x_{\beta}}$ in [5.8] is precisely the effect of the mean pressure on the mean contact force for the interfaces that may be present there.

In section 5.3.1, we already saw a similar term when studying the generalized Archimedes force for the case of a medium with discrete particles. If we consider formula [5.1b], which takes account completely of the Archimedes force on the medium, we see that the approximation $\overline{p}_i^s = \overline{p}_i^*$ allows us to identify it with [5.8]. Thus, we recover the same thing; the term $\overline{p}_i^s \frac{\partial \overline{\Phi}_i}{\partial x_{\beta}}$ is none other than the generalized Archimedes effect for any portion of interface at this point. In $\overline{F}_{is}^{\times\beta}$, we can again find the effects of pressure fluctuations relative to the mean pressure on this portion of the interface, which can contain the effects of added mass or the effects of a velocity curl during fluctuations, but the effect of the mean pressure field is no longer there.

5.3.3.2. *A flat interface between two fluids rubbing together*

A simple interesting local exchange configuration is shown by a piece of almost flat interface between two fluids rubbing together. Then, we can study the term $\overline{\tau_i^{\alpha\beta}\sigma_i^s n_i^\alpha}$ of the mean contact force in this configuration. It is interesting to see that we can solve exactly the small problem of scheme 5.1.

The flat interface, infinitely extended, separates a fluid (*g*) in motion with a velocity different from that of another fluid (*l*). The velocity of these two fluids is assumed to be fixed at a given distance from the interface, *d* for the gas and *d'* for the liquid. These velocities are tangential to the interface, constant and $V_g(d) > V_l(-d')$, and we are seeking to calculate the friction force of one fluid on the other at the interface, in a steady regime. The two fluids are assumed to be of constant volumetric mass and Newtonian viscosity, as well as weightless. There is no mass exchange between the two phases.

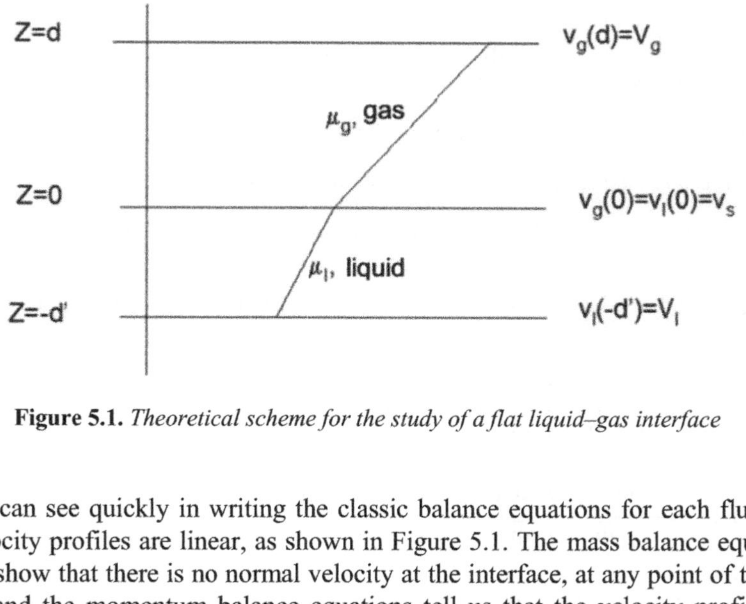

Figure 5.1. *Theoretical scheme for the study of a flat liquid–gas interface*

We can see quickly in writing the classic balance equations for each fluid that the velocity profiles are linear, as shown in Figure 5.1. The mass balance equations simply show that there is no normal velocity at the interface, at any point of the two fluids, and the momentum balance equations tell us that the velocity profiles are linear:

$$\mu_g \frac{dv_g}{dz} = cst = \mu_l \frac{dv_l}{dz}.$$

The constant is none other than the friction force per unit of interface area of the liquid phase on the gas phase, and we can deduce from this:

$$\mu_g \frac{V_g - v_s}{d} = F_l = \mu_l \frac{v_s - V_l}{d'} \text{, from which } v_s = \frac{\mu_g d' V_g + \mu_l d V_l}{\mu_l d + \mu_g d'} .$$

The friction force exerted on the liquid in the common direction of V_l and V_g tangential to the interface, per unit of volume of the two-phase medium, is:

$$F_l^f = \frac{\mu_l \mu_g}{(\mu_l d + \mu_g d')(d + d')}(V_g - V_l)$$

Let us consider now that the turbulent two-phase medium is composed of fluctuating identical flow elements displaying the geometric structure, as shown in Figure 5.1, a piece of interface and two fluid layers of thicknesses *d* and *d'*. We look at a given position where the volume fraction of the gas is $\bar{\Phi}_g = \frac{d}{d + d'}$ and the interface area per unit of volume is $\bar{\sigma}^s = \frac{1}{d + d'}$. We can, therefore, calculate the lengths *d* and *d'* as $d = \frac{\bar{\Phi}_g}{\bar{\sigma}^s}$ and $d' = \frac{\bar{\Phi}_l}{\bar{\sigma}^s}$. In addition, if we assume that the velocities V_l and V_g fluctuate, as does the position of the interface relative to the point being considered, the mean velocities of the phases are $\tilde{v}_i = \frac{1}{2}(\bar{V}_i + \bar{v}_s)$ (since the volumetric masses of the phases are constant, $\tilde{v}_l = \bar{v}_l, \tilde{v}_g = \bar{v}_g$).

In a medium where these interfaces are distributed homogeneously, it simply gives:

$$\bar{F}_{ls}^{f\beta} = -\frac{2\mu_g \mu_l}{(\mu_g \bar{\Phi}_l + \mu_l \bar{\Phi}_g)}(\tilde{v}_l^\beta - \tilde{v}_g^\beta)\bar{\sigma}^{s2} \qquad [5.9a]$$

and also $\bar{v}_s^\beta = \dfrac{\mu_g \bar{\Phi}_l \tilde{v}_g^\beta + \mu_l \bar{\Phi}_g \tilde{v}_l^\beta}{\mu_l \bar{\Phi}_g + \mu_g \bar{\Phi}_l}$

In a medium that includes a phase volume fraction gradient, as for the particle configuration, the addition of a drift velocity can be justified as well.

We can generalize this reasoning somewhat considering that the medium is composed of elements of this type, but *d* and *d'* have also fluctuating values. When we can assume that the joint probability density of these quantities, for all elements liable to be found at a given point in the medium, has a self-similar shape depending only on the two values $\bar{\Phi}_g, \bar{\sigma}^s$ at this point, then we recover a similar formula (for reasons of dimensions) but with a multiplicative constant (of the order of unity) that is a function of $\overline{\Phi}_g$:

$$\overline{F}_{ls}^{f\beta} = -\frac{2C_D \mu_g \mu_l}{(\mu_g \overline{\Phi}_l + \mu_l \overline{\Phi}_g)}(\tilde{v}_g^\beta - \tilde{v}_l^\beta)\overline{\sigma}^{s2}$$

This assumption of similitude of probability density of the sizes has the advantage of simplicity; if we want to avoid it, however, things become more complicated.

In numerous practical cases, the liquid is such that $\mu_l \gg \mu_g$, and thus we find the simplified form:

$$\overline{F}_{ls}^{f\beta} = -\frac{2C_D \mu_g}{\overline{\Phi}_g}(\tilde{v}_g^\beta - \tilde{v}_l^\beta)\overline{\sigma}^{s2} \qquad [5.9b]$$

The resemblance of this formula to the Stokes law for the drag of a solid sphere [5.4] is striking, although the geometry is completely different.

5.3.3.3. *For a turbulent medium*

Where the turbulence includes smaller scales than $d + d'$, these small scale fluctuations modify the friction force. For a first approximation, the effect of these can be represented as creating, on average for these small scales, an additional eddy viscosity. Then, we can replace the molecular coefficient μ_i with $\mu_i + C\rho_i k_{id}^{1/2} l_i$, where k_{id} is the kinetic energy of this portion of the turbulence, of a scale smaller than l_i, in phase i. This is the size $d + d'$ that selects this scale, and thus in fact $l_i = 1/\overline{\sigma}^s$, independent of the phase. Thus, we find:

$$\overline{v}_s^\beta = \frac{(\mu_g + C\rho_g k_{gd}^{1/2}/\overline{\sigma}^s)\overline{\Phi}_l \tilde{v}_g^\beta + (\mu_l + C\rho_l k_{ld}^{1/2}/\overline{\sigma}^s)\overline{\Phi}_g \tilde{v}_l^\beta}{(\mu_l + C\rho_l k_{ld}^{1/2}/\overline{\sigma}^s)\overline{\Phi}_g + (\mu_g + C\rho_g k_{gd}^{1/2}/\overline{\sigma}^s)\overline{\Phi}_l} \quad \text{and}$$

$$\overline{F}_{sl}^{f\beta} = -\frac{2C_D(\mu_g + C\rho_g k_{gd}^{1/2}/\overline{\sigma}^s)(\mu_l + C\rho_l k_{ld}^{1/2}/\overline{\sigma}^s)}{(\mu_l + C\rho_l k_{ld}^{1/2}/\overline{\sigma}^s)\overline{\Phi}_g + (\mu_g + C\rho_g k_{gd}^{1/2}/\overline{\sigma}^s)\overline{\Phi}_l}(\tilde{v}_l^\beta - \tilde{v}_g^\beta)\overline{\sigma}^{s2}$$

The estimation of k_{id} requires knowledge of the integral scale of the turbulence in the medium i as well as at least the approximate knowledge of the form of the spectrum of this turbulence, meaning the distribution of the total energy for the different scales, between (roughly) the largest, the integral scale, and the smallest, the Kolmogorov scale. Again, at least temporarily, we can represent this estimation using k_i itself by rendering the coefficient C as a function of the ratio of the turbulence scale to the interface folding scale or $l_t\bar{\sigma}^s$.

A situation in which the liquid is much more "abrasive" than the gas, meaning that $\mu_l + C\rho_l k_{ld}^{1/2} \gg \mu_g + C\rho_g k_{gd}^{1/2}$, again gives:

$$\bar{F}_{sl}^{f\beta} = -\frac{2C_D(\mu_g\bar{\sigma}^s + C\rho_g k_g^{1/2})}{\bar{\Phi}_g}(\tilde{v}_l^\beta - \tilde{v}_g^\beta)\bar{\sigma}^s,$$

and at the limit where the turbulence is dominant over the molecular viscosity:

$$\bar{F}_{sl}^{f\beta} = -\frac{2C_D C\rho_g k_{gd}^{1/2}}{\bar{\Phi}_g}(\tilde{v}_l^\beta - \tilde{v}_g^\beta)\bar{\sigma}^s$$

We again have:

$$\bar{F}_{sl}^{f\beta} = -\frac{2C_D C(l_t\bar{\sigma}^s)\rho_g k_g^{1/2}}{\bar{\Phi}_g}(\tilde{v}_l^\beta - \tilde{v}_g^\beta)\bar{\sigma}^s \qquad [5.9c]$$

In the above example, we have not considered the possibility of vaporization or condensation. If this is the case, this implies that the velocities of the two phases also have a component perpendicular to the interface, toward the gas or the liquid. This does not change anything in the preceding calculation, except that the velocities to be considered are tangential velocities $v_i^{t\beta} = v_i^\beta - v_i^\alpha n_i^\alpha n_l^\beta$, and the normal velocities $v_i^{n\beta} = v_i^\alpha n_i^\alpha n_l^\beta$ are constant in each of the phases. Since the normal velocity, due solely to the change of state, is (except in extraordinary cases) very small in comparison to the difference in the mean velocities of the phases, we can disregard this fact.

5.3.3.4. *The contribution of pressure fluctuations*

In the situations featuring large bubbles or large non-spherical drops, for the force $\bar{F}_{is}^{\times\beta} = \overline{(-p_i"\delta_{\alpha\beta} + \tau_i^{\alpha\beta})\sigma_i^s n_i^\alpha}$ between phases, the contribution of pressure fluctuations is also of significance. This contribution $\overline{(-p_i"\sigma_i^s n_i^\beta)}$ only exists, of course, for curved instant interfaces, which is a situation that is much more likely than the flat interface in Figure 5.1, and of which the average (not necessarily curved) retains the trace of instant asymmetries.

Without trying to study these curved interfaces precisely, we can say that the mean result of fluctuating pressure forces on phase i will again be to bring closer the mean velocities of the two phases in contact, but to do this without simultaneously reducing the fluctuations in each of these velocities, which conversely is caused by friction between the phases. We can thus also suggest, at least in a linear approximation, a formula linking the mean force to the difference of the mean velocities of the phases, but without any link to the viscosity, as a consequence of an "isotropization" of the fluctuations. For example, we can use a formula similar to [5.9c]:

$$\overline{(-p_l{}''\sigma_l^s n_l^\beta)} = -C_p(\frac{\rho_l}{\rho_g},\bar{\Phi}_g)\bar{\rho}k_*^{1/2}(\tilde{v}_l^\beta - \tilde{v}_g^\beta)\bar{\sigma}^s \qquad [5.10]$$

where the velocity scale $k_*^{1/2}$ is linked only to the kinetic energies of the velocity fluctuations in the two phases in contact, in a way that experiments or numerical simulations can define precisely. These are not fluctuations on a smaller scale than the size $d + d'$ of the local exchange component in play, but those of all sizes, which contribute to the complete folding of the interface, and this does not require us to bring in the ratio of the scales $l_t\bar{\sigma}^s$. In the first approximation, we may consider using the global kinetic energy k, without particular attention to the phase being considered, that we will study in Chapter 6. Since [5.10] is based simply on dimensional reasoning, we cannot say that the proportionality coefficient C_p is not dependent on the volume fraction of the gas phase. Moreover, this coefficient can also be a function of the ratio of the volumetric masses of the two phases involved, in this case liquid and gas. Note that the mean volumetric mass of the medium appears, and is thus dependent on this volume fraction and the two volumetric masses, but we could also have used the volumetric mass of the gas phase, and then the coefficient would have been different. Finally, we can see that the addition of formulas [5.9b] and [5.10] will give, for this situation of slightly curved interfaces in fluid media, a result relatively similar to that of formula [5.6] for spherical particles, with both in a high Reynolds number case.

5.3.4. *Drops or bubbles*

If we are interested in dispersed multiphase flows in which the parcels are drops or bubbles, we are led to define configurations of momentum exchange relatively similar to those of a solid spherical particle. First, this requires that we disregard mass exchange between phases; second, the deformation of drops or bubbles from the spherical shape; and third, the fact that the interior of the bubbles or drops is a fluid, which is thus mobile due to the motion of the carrier fluid along the interface. If we accept these approximations, the same local exchange configuration as for a solid particle is sufficient, and we can transpose it for the results given above, and thus formulas [5.1a]–[5.7a], for all of the aspects we have discussed. However, we must discuss the disturbance effects mentioned.

With regard to *drops*, motion of the fluid inside the drops appears when they are subjected to an external flow, and this produces two internal contra-rotating swirls. Generally speaking, we can study the influence of the movement of the fluid in a drop or bubble, in the same conditions as Stokes (Hadamard studied this problem in 1911), see [GUA 12]. Retaining this spherical form, we obtain a drag force that is expressed as:

$$F_{\tau p}^{\alpha} = 6\pi\rho_f \nu_f r_p (v_f^{\alpha} - v_p^{\alpha})(1+\frac{2\mu_f}{\mu_i})/(1+\frac{\mu_f}{\mu_i})$$

We find again [5.1a] when the internal viscosity is very strong in comparison to that of the surrounding fluid.

A second more important effect is that vaporization at the interface can profoundly change the drag force of a drop; this force can be very significantly diminished, as shown clearly by numerical studies [SIR 99]. To take this into account, it has been proposed based on these studies to weight the formulas without vaporization by a multiplicative coefficient $(1+B)^{-0,2}$. B is the Spalding transfer parameter, due to the vaporization of the drop, the definition of which is given in section 5.6.2. A typical value for a drop in a flame is 8, which results in dividing the drag force of the drop by around 1.7 in comparison to that of a solid sphere of the same radius.

With regard to *bubbles*, the situation is more complicated; on the one hand, added mass terms are essential, and on the other hand, the big bubbles lose their shape easily, with or without an exchange of mass. The density of the gas in a bubble is much lower than that of the surrounding liquid, and we can see that the inertia of the external liquid entrained by the bubble in its motion must therefore play a much stronger role than the inertia of the bubble itself. A discussion of the problems of shape can be found in [JAK 08]. A curious aspect of the change in shape is that this can amplify the lift force, causing the movement of the bubbles to be rather unstable depending on conditions. Current knowledge has not yet enabled us to take this effect into account in a general model, but very big bubbles can be modeled as many elements of slightly curved interfaces, as considered in section 5.3.3.

The forces applied to a vapor bubble by a given flow of corresponding liquid, when this bubble grows through the boiling of hot liquid or when it collapses by condensation of cold liquid, have been studied in detail recently, through numerical simulation, by Magnaudet and colleagues [LEG 98a], and also Merle *et al.* [MER 05]. These studies confirm that the importance of generalized Archimedes forces, lift and added mass is as great as that of the friction term, and enables us to specify the values of the constants C_M and C_L for different bubble Reynolds numbers.

The contribution of added mass is affected by the volume fraction of bubbles, this has been studied, in particular, by Zhang and Prosperetti [ZHA 94], with a theoretical model. For small volume fractions, they suggest $C_M = \frac{1}{2}(1+2,76\overline{\Phi}_p + O(\overline{\Phi}_p^2))$, but it appears that the characteristics of the non-stationary velocity field also play a role to some extent in the numerical coefficients, as well as the ratio of volumetric masses between the carrier fluid and the particles. The added mass contribution also requires taking into account the variation in volume of the bubble. Therefore, we have to add to the force with the same value of the constant:

$$F^{\alpha}_{vap\oplus} = 4\pi r_p^2 C_M \rho_f \frac{dr_p}{dt}(v_f^{\alpha} - v_p^{\alpha}).$$

The momentum equation for a single bubble, with the added mass term, thus finally becomes:

$$\rho_p \frac{4\pi r_p^3}{3}\frac{dV_p^{\alpha}}{dt} = \rho_p \frac{4\pi r_p^3}{3} g^{\alpha} + \frac{4\pi r_p^3}{3} F_B^{\alpha} + (\frac{\rho_p}{\rho_p + \rho_f C_M}) F^{\alpha}_{vap\oplus},$$

with $F_B^{\alpha} = (\frac{\rho_p}{\rho_p + \rho_f C_M})(-(1+C_M)\frac{\partial p_f}{\partial x_{\alpha}} + F^{\alpha}_{\tau p} + \rho_f C_L((V_f - V_p) \times rot V_f)^{\alpha})$

To find the mean force exerted on the phase of the bubbles, as discussed in section 5.3.1, we multiply first by the volume fraction of the bubbles and equate the pressure of the carrier fluid at the center of the bubble with the mean pressure of the fluid. The last term is often disregarded, and we will disregard it in this case as well. We find:

$$\overline{F}_B^{\alpha} = \overline{\Phi}_p (\frac{\rho_p}{\rho_p + \rho_f C_M})(-(1+C_M)\frac{\partial \overline{p}_f^*}{\partial x_{\alpha}} + \overline{F}^{\alpha}_{\tau p})$$

This expression can be directly used if we use the reduced stress tensor in the phase of bubbles, and this is zero if we disregard the contacts between bubbles. And the reduced stress tensor of the fluid carrier phase must consider the opposite of the additional contact term caused by the non-uniformity of the global medium (see section 5.3.1).

This additional contact stress term is $+\frac{\partial}{\partial x_\beta}(\overline{\Phi}_p \overline{p}^*_p)$, neglecting viscous effects, and since the surface tension at the interface cannot be disregarded here, we must rewrite it as $\frac{\partial}{\partial x_\beta}(\overline{\Phi}_p(\overline{p}^*_f - 2a/r_p))$, where a is the surface tension constant involved, or in terms of mean interface area per unit of volume, $\frac{\partial}{\partial x_\beta}(\overline{\Phi}_p(\overline{p}^*_f - 2a\overline{\sigma}^s_{pf}/3\overline{\Phi}_p))$.

The contribution due to vaporization requires, to calculate the variation of the (mean) radius, knowledge of the mean mass flow rate exchanged during the growth or collapse of bubbles, which we will study later (section 5.5.2). If we use $\overline{\Gamma}^s_{Mb}$ to refer to the mass of vapor released per unit of time and volume of the medium, we have $\rho_p \frac{dr_p}{dt}\overline{\sigma}^s_p = \overline{\Gamma}^s_{Mb}$ and thus:

$$\left(\frac{\rho_p}{\rho_p + \rho_f C_M}\right)\overline{F}^\alpha_{vap\oplus} = C_M\left(\frac{\rho_p}{\rho_p + \rho_f C_M}\right)\frac{\rho_f}{\rho_p}\overline{\Gamma}^s_{Mb}(\overline{v}^\alpha_f - \overline{v}^\alpha_p) \approx \overline{\Gamma}^s_{Mb}(\overline{v}^\alpha_f - \overline{v}^\alpha_p)$$

Situations in which bubbles are liable to rotate on themselves in the liquid, where several bubbles interact in their movements, are fairly common, without speaking of their coalescence. Can they have significant consequences, both for the stresses applied and for mass exchanges at the interface with the liquid phase? We do not know today how to answer this question. These situations have started being studied in detail by experimental studies with new optical methods and even through direct numerical simulation. We will give examples of this in Chapter 11, section 11.2.

5.4. Heat transfers at the surface of a particle, without mass exchange

Energy exchanges between phases, either in the balance equation for the total energy of a phase or in that for the total enthalpy, include a contribution due to mass exchange, one due to viscous friction, in addition to heat transfer through conduction or radiation. We are only studying situations here that do not involve mass exchanges. The power of the friction force is negligible when the velocities of the flows always remain lower than approximately 20% of the speed of sound in the medium, which is very generally the case.

This is the last term we will study here: $\overline{q_{is}\sigma_i^s} = -\overline{j_i^{Q\alpha} n_i^\alpha}^s \overline{\sigma_i}^s$, for the case in which it is due to conduction. An interesting local exchange configuration for this thermal problem is again that of a solid spherical particle that is subject, as previously, to a uniform fluid flow, neglecting gravity. This flow is also uniform in temperature, which is, for example, higher than that of the particle, whose temperature is assumed also uniform at a certain initial instant. The theoretical study of this situation is not possible in an analytical manner, even if we assume a fluid with constant physical properties and a low relative particle velocity, as for the Stokes theory. However, this is a classic problem in thermal sciences; numerical and experimental studies have been performed, and we can discuss here the form that the results take.

What interests us particularly here is the global heat flux that enters the sphere through conduction, and the volume averaged temperature in the whole sphere. It is clear that the problem is not stationary; after the initial instant, the averaged temperature of the sphere will grow continuously, toward the value of the temperature (assumed to be constant at some distance from the sphere) of the surrounding fluid flow, to which it will become closer asymptotically. Likewise, the global heat flux absorbed by the sphere will evolve with time, decreasing to zero when the temperatures of the sphere and the surrounding fluid will be equal. Since these two evolutions are monotonic, it is possible to eliminate the time and to plot a law linking flux to mean temperature, the parameters of which will be the physical characteristics of the materials being considered and the initial and boundary conditions chosen and measured. We can write formally:

$$\varphi_Q = \varphi_Q(T_p, V, r_p, T_F, T_{p0}, \lambda_F, C_{pF}, \nu_F, \rho_F, \lambda_p, C_p, \rho_p)$$

The values T_p, T_{p0}, T_F and V are, respectively, the mean (volumetric) temperature of the sphere and its initial value, and the temperature and velocity of the fluid relatively far from the surface of the sphere (with respect to the sphere radius), with the two latter assumed to be constant. The other values are constant parameters of the fluid and the sphere. If we look more closely at the equations governing the problem, we can see, on the one hand, that the differences in temperature, and not their absolute values, are involved here, and on the other hand, the parameters work in groups. Thus, we actually have:

$$\varphi_Q = \varphi_Q(T_F - T_p, V, r_p, T_F - T_{p0}, \lambda_F / \rho_F C_{pF}, \nu_F, \lambda_p / \rho_p C_p)$$

Moreover, dimensional analysis allows us to write this relationship in a simpler non-dimensional form as:

$$\frac{\varphi_Q}{4\pi r_p(T_F - T_p)\lambda_F} = \phi(\frac{Vr_p}{\nu_F}, \frac{T_F - T_p}{T_F - T_{p0}}, \frac{\lambda_F/\rho_F C_{pF}}{\lambda_p/\rho_p C_p}) \qquad [5.11a]$$

On the left-hand side, we can see the Nusselt number of the particle, and the first parameter on the right-hand side is the Reynolds number of this particle. When the evolution of the particle's temperature is advanced enough, near equilibrium, the difference $T_F - T_p$ is low and it is sufficient to write: $\frac{\varphi_Q}{4\pi r_p(T_F - T_p)\lambda_F} = \Psi(\frac{Vr_p}{\nu_F}, \frac{\lambda_F/\rho_F C_{pF}}{\lambda_p/\rho_p C_p})$. Several books give examples of such a formula, often in the form of curves. As for the case of friction, if the Reynolds number becomes large, we must expect that the heat flux itself will no longer depend on the molecular characteristics of the fluid, thus, in this case, on the viscosity and heat conduction coefficients. Thus, the function Ψ becomes proportional to the Reynolds number, and the ratio $\frac{\nu_f}{\rho_f \lambda_f / C_{pf}} = \Pr_f$, the Prandtl number, appears.

In the same way as for the momentum exchange in section 5.3.1, the heat flux exchanged by the particulate phase with the fluid phase will be C^N (the number of particles per unit of volume) times the flux exchanged by a particle, plus a divergence term that can be identified with the opposite of the divergence of the conduction heat flux inside the particles, if these particles have little or no contact between them. Thus, for particles of identical radius, we obtain:

$$\overline{q_{ps}\sigma_p^s} = \frac{1}{3}\frac{\bar{\sigma}_p^s}{\bar{\Phi}_p}\lambda_F(\tilde{T}_F - \tilde{T}_p)\Psi - \frac{\partial}{\partial x_\alpha}(\bar{\Phi}_p \overline{j_p^{Q\alpha}}^*) \qquad [5.11b]$$

Then, this last divergence term will cancel out itself with the term of the heat flux inside the particles, when the particles (practically) never touch. As previously for the velocities in section 5.3.1, we have considered that we can in the first approximation identify the characteristics of the fluid far from the particle, T_F and T_p, with mean characteristics. The average of the mean particle temperatures takes into account the possible fact that all of the particles were not at the same initial temperature. Ψ is the function of formula [5.11] but expressed in terms of the Reynolds number, which is calculated with the mean velocity of the fluid relative to that of the particles, and of the thermal diffusivities' ratio.

We could again discuss the intervention of turbulence in this formula, as well as the difference in mean velocities involved in the Reynolds number on which the function Ψ is dependent. Discussions similar to those of section 5.3.2 can be developed; certain empirical formulas exist. Generalizations for the situation in which the medium is composed of particles of different radii and/or for a relatively dense medium can be discussed as well, in the same manner as for friction.

5.5. Heat and mass transfers during boiling

When a liquid–vapor interface in a multiphase flow is the site for boiling phenomenon, fluxes of mass and heat exchanged per unit of area are coupled, and follow specific laws. This situation is very common in water–vapor flows, and we must study it more closely. The mean energy exchange term includes in principle three contributions. When we disregard the power of friction force, which goes along with neglecting the kinetic energy compared to internal energy, only two contributions remain. These are, in equation [4.8] for enthalpy from Chapter 4: $\overline{m_i\sigma_i^s h_i} + \overline{q_{is}\sigma_i^s} = \overline{m}_i^s\overline{\sigma}_i^s\overline{h}_i^s + \overline{q_{is}\sigma_i^s}$. We must therefore study these two terms, separating the factors of the first, since the exchange of mass itself must be used for mass exchange in the phase mass balance equation.

5.5.1. *Slightly curved liquid–gas interfaces*

The simplest local exchange configuration that allows us to begin modeling is that of a flat double layer of a liquid and its vapor, separated by an interface that is also flat. We will consider in this first configuration that the two fluids have only one velocity normal to the interface: the liquid moves toward the interface, at which it transforms into vapor, and this then moves away from the interface. However, we could consider various fluid velocity fields relative to the interface, which of course would make the situation more complicated. We can see fairly easily that the consideration of a linear profile of velocity tangential to the interface, in each phase, is possible without changing the calculations concerning heat and mass exchanges. The basic situation is shown in Figure 5.2.

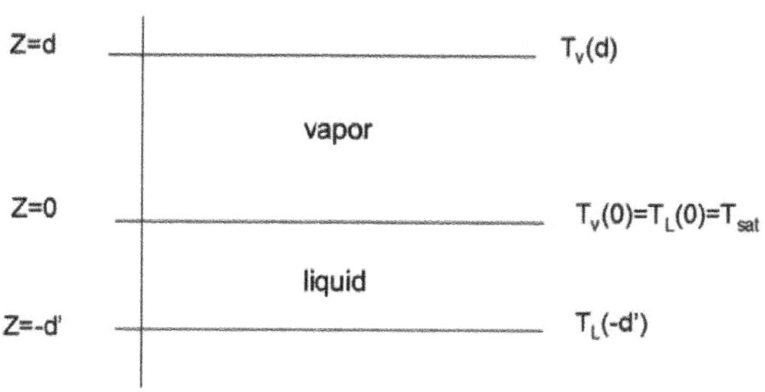

Figure 5.2. *Theoretical scheme of the study of a flat liquid–vapor interface*

To ensure a steady regime, we must assume that the thickness of the liquid, d', is known (and finite), as well as that of the gas, d; and that a certain mass flow per unit of area enters in liquid phase at $z = -d'$, where the temperature is known and constant, and the same mass flow rate exits from gas phase at $z = d$, where the temperature is also known and constant.

Of course, for the phase change to occur, the temperature of the liquid at $-d'$ must be lower than the saturation temperature T_{sat} of the liquid–vapor couple being considered (depending on the pressure, which is uniform in both layers in the absence of velocity gradients in them), while the temperature of the gas at $z = d$ is higher than it. The thermodynamic equilibrium of vaporization–condensation at the interface tells us that the interface temperature, the same for liquid and gas, is equal to the saturation temperature.

It is clear that this local exchange configuration is a very simple example of the situations that can be seen at elementary interface portions in a real multiphase flow; even if the interface is not curved, the liquid and the gas can have various motions relative to the interface, even preventing any steady regime to be maintained (but if the conditions at the boundaries are not varying too rapidly, a quasi-steady state can often be invoked). However, this configuration can be approached in certain conditions, and in any case serves to show how the phenomena of vaporization and heat exchange become intermingled with one another.

The physical problem above can be translated into equations, with balance equations for each of the phases given in Chapter 2, and it can be shown that the configuration remains stationary only if the mass flow rate that is injected into the liquid in $z = -d'$ has a specific value, dependent on the temperatures at $z = -d'$ and $z = d$, and also on the physical characteristics of the liquid and the gas. We will summarize a simplified version of this demonstration below.

In the liquid, disregarding the kinetic energy relative to the internal energy as well as the gravity forces (at this length scale the Froude number is very small), the energy balance equation (we are in steady regime and the quantities vary only as function of the normal coordinate at interface z) is written as:

$$\rho_L u_L \frac{dh_L}{dz} - \frac{d}{dz}\left(\frac{\lambda_L}{C_L}\frac{dh_L}{dz}\right) = 0$$

The mass balance gives simply $\rho_L u_L = \dot{m}$, where the constant $\dot{m}$ is the exact mass flow rate per unit of area injected into $z = -d'$, which it will be necessary to extract in $z = d$. The momentum balance in the liquid phase is replaced, because the velocities involved are low, by the constancy of static pressure, which is therefore equal to the pressure on the interface, common to the gas and the liquid because the interface is almost flat.

In the gas, with the same hypotheses, we have:

$$\rho_V u_V \frac{dh_V}{dz} - \frac{d}{dz}\left(\frac{\lambda_V}{C_V}\frac{dh_V}{dz}\right) = 0,$$

and also $\rho_V u_V = \dot{m}$. Likewise, the pressure in the gas is constant, and thus equal to that of the liquid.

Enthalpies are related to temperatures and pressure by equations of state. We will simplify these, for example, into $h_V = C_V T_V + L, h_L = C_L T_L$ (the pressure dependence is neglected), where the calorific capacities are constants, as is L. L is the latent vaporization heat at 0 K, related to that at T$_{\text{sat}}$ by $h_V(T_{sat}) - h_L(T_{sat}) = L_{Vsat} = (C_V - C_L)T_{sat} + L$.

To be able to solve the differential equations above, we must give the boundary conditions as well as those at the interface. At the two liquid and vapor boundaries, we have assumed that the temperatures were known, $h_V(d) = C_V T_V(d) + L, h_L(-d') = C_L T_L(-d')$. At the interface, we have $T_L(0) = T_V(0) = T_{sat}(p)$. But we must also satisfy the interface balance relationship discussed in Chapter 2. It is written, from [2.12b], if we disregard kinetic energy and the power of viscous stress, with the notations from the balance equations above:

$$\dot{m}h_{Vsat} - \frac{\lambda_V}{C_V}\left(\frac{dh_V}{dz}\right)_0 = \dot{m}h_{Lsat} - \frac{\lambda_L}{C_L}\left(\frac{dh_L}{dz}\right)_0$$

It means that the enthalpy that enters into the gas (through normal convection and conduction) is none other than the enthalpy exiting the liquid (through normal convection and conduction).

We can see already that each second-order differential equation has two conditions at the boundaries of its domain, and that an additional relationship remains to be verified, the interface balance relationship. This relationship imposes a condition that must be satisfied by the quantity $\dot{m}$. If we develop the calculations, we find, defining $\dot{m}h_{Vsat} - \frac{\lambda_V}{C_V}\left(\frac{dh_V}{dz}\right)_0 = \dot{m}h_{Lsat} - \frac{\lambda_L}{C_L}\left(\frac{dh_L}{dz}\right)_0 = K$, that the temperature profiles in the liquid and in the gas are exponential:

$$\frac{T_V - K/\dot{m}C_V + L/C_V}{T_{sat} - K/\dot{m}C_V + L/C_V} = \exp(\dot{m}\frac{C_V}{\lambda_V}z) \text{ and } \frac{T_L - K/\dot{m}C_L}{T_{sat} - K/\dot{m}C_L} = \exp(\dot{m}\frac{C_L}{\lambda_L}z)$$

and the boundary conditions at $z = d$ and $z = -d'$ give the possibility of calculating both K and $\dot{m}$.

We will look first at the vaporized mass flow rate per unit of interface area. Eliminating K from the two preceding relationships, unfortunately, we cannot find an explicit general relationship to give this mass flow rate, but it follows:

$$\frac{h_L(-d') - h_{Lsat}\exp(-\dot{m}d'C_L/\lambda_L)}{1-\exp(-\dot{m}d'C_L/\lambda_L)} = \frac{h_V(d) - h_{Vsat}\exp(\dot{m}dC_V/\lambda_V)}{1-\exp(\dot{m}dC_V/\lambda_V)} \qquad [5.12a]$$

We note immediately that if $h_L(-d') = h_{Lsat}$ and also $h_V(d) = h_{Vsat}$, this relationship is only possible if the vaporized flow rate is zero, which is quite physical: if both temperatures relatively far from the interface are equal, and equal to that of the interface, which is necessarily $T_{sat}(p)$, the system is steady with no mass exchange. This implicit formula can be used only with difficulty in the modeling of mass exchange by boiling.

Nevertheless, we can find explicit approximations of it. The first one is valid if the temperature of the liquid is close to the saturation temperature, $h_L(-d') \approx h_{Lsat}$, meaning that there is not really a temperature gradient in the liquid. Thus, the left-hand side of formula (5.12a) equals 1 and we take from it:

$$\dot{m} = \frac{\lambda_V}{C_V d} Ln(\frac{h_V(d) - h_{Lsat}}{L_{Vsat}}) = \frac{\lambda_V}{C_V d} Ln(\frac{C_V(T_V(d) - T_{sat}) + L_{Vsat}}{L_{Vsat}}) \qquad [5.12b]$$

It is necessary in this case that $h_V(d) > h_{Lsat}$, meaning that $C_V T_V(d) + L > C_L T_{sat}$, which is verified since $T_V(d) > T_{sat}$, and we can see that the vaporized flow rate is indeed positive.

The other approximate formula concerns the case in which d' is large enough for the exponential in which it is involved to be negligible, even though the thermal conductivity of the liquid can be high. Thus:

$$\dot{m} = \frac{\lambda_V}{C_V d} Ln(\frac{h_V(d) - h_L(-d')}{h_{Vsat} - h_L(-d')}) \qquad [5.13]$$

In this case, the logarithm argument must remain positive; since $T_L < T_{sat}$, the denominator is always positive and the numerator must be positive also, then $C_V T_V(d) + L > C_L T_L(-d')$, which is, again, always verified if $T_V(d) > T_{sat}$, and $T_L(-d') < T_{sat}$, and $\dot{m} > 0$.

For the two formulas, the vaporized flow rate is indeed zero if $T_V(d) = T_{sat}$, meaning that $h_V(d) = h_{Vsat}$. We can find two other approximations by using the same hypotheses for the gas part.

In every case, the heat fluxes exchanged in the presence of a phase change are linked by the interface balance relationship written above from [2.12b], which means simply that the heat flux that enters through conduction into the liquid is the one that exits the gas by conduction, reduced by the heat necessary for vaporization, and then can be rewritten:

$$\lambda_L \left(\frac{dT_L}{dz}\right)_0 = \lambda_V \left(\frac{dT_V}{dz}\right)_0 - \dot{m} L_{Vsat}$$

Moreover, we have already defined $\dot{m} h_{Vsat} - \frac{\lambda_V}{C_V}\left(\frac{dh_V}{dz}\right)_0 = \dot{m} h_{Lsat} - \frac{\lambda_L}{C_L}\left(\frac{dh_L}{dz}\right)_0 = K$,

and K can also be calculated as:

$$K = \dot{m}\frac{h_{Lsat}\exp(-\dot{m}d' C_L/\lambda_L) - h_L(-d')}{\exp(-\dot{m}d' C_L/\lambda_L) - 1} = \dot{m}\frac{h_{Vsat}\exp(\dot{m}dC_V/\lambda_V) - h_V(d)}{\exp(\dot{m}dC_V/\lambda_V) - 1} \qquad [5.14]$$

Thus we obtain, as a function of $\dot{m}$, expressions of the heat fluxes entering into the liquid or the gas. For example, we obtain the heat flux entering the liquid by:

$$\lambda_L \left(\frac{dT_L}{dz}\right)_0 = \dot{m}\frac{h_{Lsat} - h_L(-d')}{1 - \exp(-\dot{m}d' C_L/\lambda_L)} \qquad [5.15]$$

We can see that, for the conditions of the first approximation, where $T_L(-d') = T_{sat}$, the conduction flux entering the liquid is zero, but this is not the case in the second approximation, where it is directly proportional to the vaporized mass flow rate $\lambda_L \left(\frac{dT_L}{dz}\right)_0 = \dot{m}(h_{Lsat} - h_L(-d'))$.

As the constant K can be expressed either on the liquid side or on the gas side, as indicated by formula [5.14], the conduction flux entering the liquid can also be expressed in terms of gas enthalpy and gas conduction. It is, therefore, possible to retain the symmetry of expressions.

This local exchange configuration can represent what occurs locally near a piece of interface that is very slightly curved in a turbulent two-phase water–vapor flow. Let us now consider, as in section 5.3.3, that this flow is composed of fluctuating fluid elements displaying this geometrical structure. We must relate d, d', $T_V(d), T_L(-d')$ to the mean temperatures of the liquid and the gas and to the length scales of the two-phase medium. If we assume again, as we did in section 5.3.3, that these scales are controlled by the local geometry of the phases, they must be related to the volume fraction of the liquid and to the interface area per unit of volume, thus by $\bar{\Phi}_L = \dfrac{d'}{d+d'}$, and $\bar{\sigma}^s_{\lg} = \dfrac{1}{d+d'}$. In addition, we must relate the temperatures $T_V(d), T_L(-d')$ to $\tilde{T}_V, \tilde{T}_L$, which can be calculated locally with the mean energy or enthalpy balances of the phases. If we simply identify the mean temperatures and those at d or $-d'$, we obtain, for example, for the case of the second approximation:

$$\dot{m} = C_D \frac{\lambda_V \bar{\sigma}^s_{\lg}}{C_V(1-\bar{\Phi}_L)} Ln(\frac{\tilde{h}_L - \tilde{h}_V}{\tilde{h}_L - h_{Vsat}}) \qquad [5.16]$$

and the mean mass exchange per unit of volume is:

$$\overline{m_l \sigma^s_{\lg}} = -C_D \frac{\lambda_V (\bar{\sigma}^s_{\lg})^2}{C_V(1-\bar{\Phi}_L)} Ln(\frac{\tilde{h}_L - \tilde{h}_V}{\tilde{h}_L - h_{Vsat}}) = -\overline{m_g \sigma^s_{\lg}} \qquad [5.17]$$

The constant C_D again may contain the influence of the distribution of d, d' of the fluctuating flow elements, and will be one if these are all identical.

In all cases, the heat flux given to the liquid by conduction, per unit of volume, is:

$$\overline{q_l \sigma^s_{\lg}} = \overline{m_l \sigma^s_{\lg}} \frac{h_{Lsat} - \tilde{h}_L}{1 - \exp(-\dot{m}\bar{\Phi}_L C_L / (\lambda_L \bar{\sigma}^s_{\lg}))} \qquad [5.18]$$

If we calculate exactly the mean temperatures in the fluids taking into account the variation with z, we obtain only that there is a multiplicative constant that is slightly lower than the unit. It is also possible to adapt these formulas to a case in which turbulence will occur at scales lower than $d + d'$, the scale of the fluid element being considered, using the same reasoning as mentioned in section 5.3.3.

The term for enthalpy exchange by mass exchange also includes the factor $\bar{h}^s_i$, which is the average at the interface of the enthalpy of the phase being considered (and subject to mass transfer). The local study we have just conducted gives us its value directly; this is none other than $h_{isat} = h_i(T_{sat})$, on the liquid or vapor side.

5.5.2. *Bubbles*

Other local exchange configurations are obviously of interest in multiphase flows, which are more or less probable and more or less simple. For example, the equivalent of the preceding problem for a spherical vapor bubble is much more adapted to the case of highly curved interfaces.

The *growth of a bubble* of vapor from a liquid, within this liquid at rest, heated above the saturation temperature, has been studied, for example [PLE 54]. Studies through numerical simulations of this situation, also including the influence of a flow of liquid relative to the bubble, have been conducted recently by Magnaudet and colleagues ([LEG 98a, MER 05]).

For a bubble at rest in a liquid, the result is that, after a certain time, such that the initial conditions lose their influence, the radius of the bubble increases, in a quasi-steady regime, with a rate proportional to the inverse of the square root of the time:

$$\frac{d(r_b/r_0)}{dt} = \left(\frac{4J_a^2 f(J_a)}{\pi r_0^2}\frac{\lambda_L}{\rho_L C_L}\right)^{1/2}\frac{1}{\sqrt{t}}$$

The Jacob number is $J_a = \rho_L C_L |T_\infty - T_{sat}|/(\rho_g L_V)$, $f(J_a)$ is a function that tends toward ¾ if J_a tends toward infinity, attains it practically as soon as J_a is greater than one, and grows toward infinity as $1/J_a$ if J_a tends toward zero.

Moreover, the heat flux by conduction emerging from the liquid toward the bubble is given by:

$$\frac{Q_L}{2\pi\lambda_L r_b(T_\infty - T_{sat})} = \frac{16}{\pi}J_a f(J_a)$$

In this quasi-steady regime, the entire heat flux is used for the change of state, with the temperature in the bubble being uniform and equal to T_{sat}. It is possible to deduce from this the instant mass flow rate of vaporized liquid per unit of area for one bubble:

$$\dot{m} = \frac{8J_a^2}{\pi}\frac{\rho_g \lambda_L}{\rho_L C_L}\frac{f(J_a)}{r_b}$$

To deduce from this the mean mass flow rate of vaporization and the mean heat flux exchanged per unit of volume in a liquid medium with bubbles, the problem is again that it is not enough to simply multiply the flow vaporized in one single bubble by C^N bubbles present per unit of volume. As for the contact force in section 5.3.1.2, the problem of bubbles "riding" on the surface of any given control volume also exist again here, and a compensation for these bubbles occurs only when the bubbles are statistically homogeneously distributed. It is then natural to expect an additional term proportional to the gradient of the mean interface area per unit volume to appear, for the mean vaporized mass flow rate as well as for the mean heat flux between phases. However, to our knowledge, this problem has not been really studied.

In addition, we need again to bring in the mean temperature of the liquid in which the bubbles occur in the place of T_∞, and relate the radius of bubbles that can occur at the point studied in the liquid to $\bar{\Phi}_g, \bar{\sigma}^s_{lg}$. This allows us to deduce the following models, the structure of which is similar to that of formulas [5.17] and [5.18]:

$$\overline{m_b \sigma^s_{\lg}} = \frac{24(J_a(\tilde{T}_L))^2}{\pi} f(J_a(\tilde{T}_L)) \frac{\rho_g \lambda_L}{\rho_L C_L} \frac{(\bar{\sigma}^s_{lg})^2}{\bar{\Phi}_g} \quad [5.19]$$

and $$\overline{q_{sb} \sigma^s_{\lg}} = -\frac{24}{\pi} J_a(\tilde{T}_L) f(J_a(\tilde{T}_L)) \lambda_L (\tilde{T}_L - T_{sat}) \frac{(\bar{\sigma}^s_{lg})^2}{\bar{\Phi}_g} \quad [5.20]$$

Here, we have assumed that the bubbles all have the same radius; if this is not the case we can adapt these formulas, but we must know the distribution of the radii.

The numerical study by Magnaudet and colleagues [LEG 98a] on the influence of a flow relative to the bubble shows a regime in which the vaporized mass flow rate and the heat flux are proportional to the square root of the Péclet number of the bubble, that is the product of the Reynolds number by the Prandtl number. It seems, therefore, that, at least up to a Péclet number of 10^4, we can multiply the preceding formulas by a factor $(1 + KP_e^{1/2})$ (K being a numerical constant), a property that we will also discuss for drops in the next section.

A vapor bubble collapses if, from a given initial size (and a temperature equal to the saturation temperature), it is placed in a liquid medium at a temperature lower than the saturation temperature. Due to the condensation of the vapor, the interface regresses, and its radius decreases until the point of total disappearance. This situation has also been studied, and it appears in this case that the quasi-steady regime seen previously is not valid (for a sufficient duration of the life of the bubble, before its total disappearance), unless the condensation is slow, and thus the Jacob number is low enough. Taking into account the behavior equivalent to $f(J_a)$ at low J_a, we find in this case that the square of the radius of the bubble collapses proportionally to time $r_b^2 = r_0^2 - 2J_a \dfrac{\lambda_L}{\rho_L C_L} t$, and therefore

$$r_b \frac{dr_b}{dt} = -J_a \frac{\lambda_L}{\rho_L C_L} \qquad [5.21]$$

We still have $\dfrac{Q_L}{2\pi\lambda_L r_b (T_\infty - T_{sat})} = \dfrac{2\rho_L C_L}{J_a \lambda_L} r_b \dfrac{dr_b}{dt}$ and from this we can deduce formulas similar to [5.19] and [5.20] for the vaporized mass flow rate (condensed here) and the heat flux supplied (here) to the liquid phase. For higher Jacob numbers, the collapse regime of the bubble is fully transient, numerical simulations enable us to calculate it, and various approximate formulas also exist. We refer the readers to [LEG 98a].

5.6. Mass and heat exchanges by vaporization

Numerous situations of multiphase flow bring into play liquid–gas interfaces with evaporation at the interface, when the gas medium adjacent to the liquid is composed of a mixture of an incondensable gas and the vapor of the liquid in question, and not only of vapor. It is possible to study this phenomenon in the same local exchanges configurations as for boiling.

The composition of the gas medium can thus vary depending on the distance from the interface (and we will see that it must vary). This variation of composition can be assessed as a variation of the mass fraction Y_V, following the notation in Chapter 2, or as a variation of the partial vapor pressure p_V, or as a variation of the vapor mole number per unit of volume of the mixture, N_V. At the interface, the vaporization–condensation equilibrium implies that the partial pressure is the saturating vapor pressure at the temperature of the interface, whose value is common to the liquid and the gas mixture.

5.6.1. *Mass transfer by evaporation at a flat interface*

Let us consider first the local exchange configuration with a flat interface separating a liquid and a gas medium in which this liquid vaporizes, a situation shown in Figure 5.3.

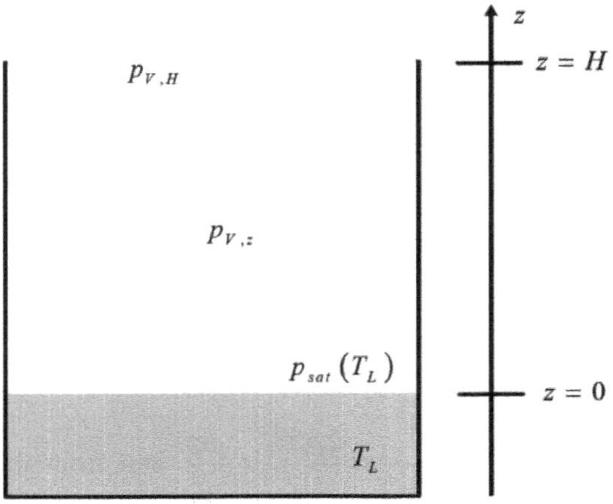

Figure 5.3. *Theoretical scheme for the study of the evaporation of a liquid*

5.6.1.1. *Evaporation of a liquid at given temperature*

If we consider that the liquid is at a known temperature T_L, and that the partial pressure of the vapor is maintained at a constant level (at a value lower than p_{sat}) at a certain distance H from the interface, it is possible to calculate the mass flux that evaporates steadily.

The mass transfer from the liquid phase toward the gas phase involves the combination of three phenomena: the first is a displacement of molecules of water vapor in the gas phase due to the partial pressure gradient; this is a phenomenon of molecular diffusion. The second is a liquid–vapor transformation at the surface to maintain the equilibrium between the water and its vapor, since this equilibrium is disturbed by the motion due to molecular diffusion. The third is the motion of the interface, or equivalently, the liquid is convected toward the interface and the vapor convected away.

The field of vapor partial pressure above the interface, as function of z attached to the interface, can be studied in the continuous gas medium with the balance equations given in Chapter 2, section 2.2.3.

Considering in a classic manner that the gas mixture is a perfect mixture, there is a direct relationship between the partial pressure p_v and the mass fraction of vapor in the mixture, a relationship valid at each point z: $p_V = p\frac{N_V}{N_{mel}} = p\frac{M_{mel}}{M_V}Y_V$. Here, p is the total pressure in the gas, which is uniform in the field as the movements of the fluid are very slow, and can thus be assumed to be known; the *Ms* are the molar masses of the vapor and the mixture, respectively. The vapor mass balance in the gas medium, at each z, can thus be used.

In a steady regime, in the absence of chemical reactions, the classic balance, equation [2.8], enables us to write:

$$\frac{d\,\rho u Y_V}{dz} = \frac{d}{dz}(\rho D_v \frac{dY_V}{dz})$$

Here, on the left, we have the convection flux and, on the right, the diffusion flux. u is the velocity of the gas mixture (for example vapor plus air) in direction z and D_v is the diffusion coefficient of the vapor in the air. We have, therefore, used Fick's law to express the vapor diffusion flux. Obviously, we still have $Y_v + Y_a = 1$, for the whole height, and the total mass balance equation (air plus vapor) gives, of course, $\frac{d}{dz}(\rho u) = 0$, thus $\rho u = cst = \dot{m}$, where $\dot{m}$ is the mass flow rate vaporizing at each instant at $z = 0$ (per unit of area).

We will assume for the sake of simplicity that the product ρD_v is a constant, and so it is simple to integrate the differential equation. First, we obtain:

$$\dot{m}Y_V - \rho D_v \frac{dY_V}{dz} = cst = \dot{m} \quad [5.22]$$

The integration constant represents the total mass flow rate (convection plus diffusion) at the interface from the gas side. Also, the interface balance (see Chapter 2) tells us that the vapor mass flow rate exiting the interface is the same as the mass flow rate of vaporizing liquid, $\dot{m}$. This results in the written equality. It is clear that $\dot{m}$ is also the total mass flow rate (convection plus vapor diffusion) exiting above $z = H$ into the free air.

Integrating once again, we obtain the profile of $Y_v(z)$:

$$Ln\left(\frac{1-Y_V}{1-Y_V(0)}\right) = \frac{\dot{m}}{\rho D_v} z$$

In applying the boundary conditions at H, where $Y_V(H) = Y_{V,H}$, and also at $z = 0$, where $Y_V(0) = Y_{Vsat}(T_L)$, we find the relationship that must satisfy the vaporized mass flow rate so that the steady regime can be maintained:

$$\dot{m} = \frac{\rho D_v}{H} Ln\left(\frac{1-Y_{V,H}}{1-Y_{Vsat}(T_L)}\right) \quad [5.23]$$

We can note that, since $1-Y_V=Y_a$, the profile of Y_a is $Ln\left(\frac{Y_a}{1-Y_{Vsat}}\right)=\frac{\dot{m}}{\rho D_v}z$.

Using the relationship given above between the mass fraction and the partial pressure, and assuming that the vapor is always in a low enough proportion in the mixture for M_v and M_{mel} to be very close, we also obtain:

$$\dot{m}=\frac{\rho D_V}{H}Ln\left(\frac{p-p_{V,H}}{p-p_{sat}(T_L)}\right)$$

This formula is often called Stefan's equation. If we add the two hypotheses stating that the total pressure is clearly higher than $p_{V,H}$ and p_{sat}, both hypotheses being verified when the temperature of the water and the gas are lower than 30°C, we have: $\frac{P-P_{V0}}{P-P_{sat}}\approx 1$. Stefan's equation, therefore, takes the form of Dalton's equation $\dot{m}=\frac{\rho D_V}{H}\left(p_{sat}(T_L)-p_{V,H}\right)$.

Dalton's and Stefan's equations, even though they are only strictly applicable to simple one-dimensional problems of water evaporation in a tube, clearly show the dependence of the evaporation rate on water temperature: when the water temperature increases, the saturation pressure also increases, and so the evaporation rate increases. They also show directly that the evaporation rate is even higher when the partial vapor pressure in H is weak. We can also see that the formula is reversible: there is condensation at the interface and not vaporization, if the partial vapor pressure of H is higher than the saturation pressure at the temperature of the liquid. This is not only simply a matter of vaporization, but indeed of vaporization–condensation equilibrium.

5.6.1.2. *Evaporation of a layer of liquid close to a layer of gas*

To date, the theoretical approach has assumed the temperature of the liquid, and more precisely that of the liquid at the interface, to be known. The theory will be complete only if we now study how to calculate this temperature at the interface (common to the liquid and the gas), in terms of the mean temperature of the liquid and the gas, or more easily in terms of the temperature of these phases at a certain distance, H and H', from the interface. To do this, we must repeat the study of section 5.5.1, with slight necessary modifications.

Only two aspects must be modified. On the one hand, the enthalpy of the gas medium is now dependent on the composition of the medium plus the temperature, and is given by $h_g=Y_Vh_V+Y_ah_a=(C_gY_V+C_gY_a)T_g+LY_V$ (we are still assuming for the sake of simplicity that the calorific capacity of the mixture is constant, but the enthalpy of the formation of the mixture is obviously a linear function of the vapor content; we can consider the air formation enthalpy to be zero). On the other hand, the enthalpy balance in the gas phase is now written as:

$$\rho_g u_g \frac{dh_g}{dz} = \frac{d}{dz}(\lambda_g \frac{dT_g}{dz} + \rho_g D_V h_V \frac{dY_V}{dz} + \rho_g D_V h_a \frac{dY_a}{dz})$$

The heat flux is now composed of the heat flux by conduction plus the heat flux caused by species diffusion. If we assume that the thermal diffusivity of the gas mixture is equal to the mass diffusivity of the vapor, $\lambda_g/(\rho_g C_g) = D_V$ (we say that the Lewis number of the vapor is one), we find the same equation in terms of enthalpy:

$$\rho_g u_g \frac{dh_g}{dz} - \frac{d}{dz}(\frac{\lambda_g}{C_V}\frac{dh_g}{dz}) = 0 \qquad [5.24]$$

As we have said, the condition at the interface ($z = 0$) is no longer only $h_L(0)=h_L(T_{sat}(p))$, but $Y_V(0) = Y_{V,sat}(T_s)$, which is translated in terms of enthalpy by $h_L(0) = h_L(T_s)$ where T_s is given by the inverse of the function Y_{Vsat}: $T_s = Y_{Vsat}^{-1}(Y_V(0))$.

Finally, the energy balance relationship at the interface [2.12a], still disregarding the kinetic energy and viscous friction (and the pressure difference at the interface), is written taking into account the fact that the heat flux from the gas side is due to species diffusion plus conduction. With $\lambda_g/(\rho_g C_g) = D_V$, we obtain:

$$\dot{m}h_{gs} - \frac{\lambda_g}{C_g}\frac{dh_g}{dz})_s = \dot{m}h_{Ls} - \frac{\lambda_L}{C_L}\frac{dh_L}{dz})_s \qquad [5.25]$$

We can see that, giving details of the enthalpies and heat fluxes, we obtain:

$$\dot{m}(C_g T_{gs} + LY_{Vs}) - \lambda_g \frac{dT_g}{dz})_s - L\frac{\lambda_g}{C_g}\frac{dY_V}{dz})_s = \dot{m}C_L T_{Ls} - \lambda_L \frac{dT_L}{dz})_s$$

We have also seen that $\dot{m}Y_{Vs} - \frac{\lambda_g}{C_g}\frac{dY_V}{dz})_s = \dot{m}$, always assuming $\lambda_g/(\rho_g C_g) = D_V$.

This gives $-\lambda_L(\frac{dT_L}{dz})_s + \dot{m}C_L T_{Ls} = -\lambda_g(\frac{dT_g}{dz})_s + \dot{m}C_g T_{gs} + \dot{m}L$, meaning finally that:

$$-\lambda_L(\frac{dT_L}{dz})_s = -\lambda_g(\frac{dT_g}{dz})_s + \dot{m}L_V(T_s)$$

The interface balance relationship thus states simply that the conduction flux exiting the liquid is equal to that entering the gas plus the latent heat of the vaporization of the mass flow at the interface temperature.

With balance equation [5.24] and interface relationship [5.25], we find equations similar to the case of boiling in section 5.5.1 and we are thus led to a similar calculation. The vaporized mass flow rate $\dot{m}$ is given by an implicit formula transposing [5.11a]:

$$\frac{h_L(-H') - h_{Ls}\exp(-\dot{m}d'C_L/\lambda_L)}{1-\exp(-\dot{m}H'C_L/\lambda_L)} = \frac{h_g(H) - h_{gs}\exp(\dot{m}HC_g/\lambda_g)}{1-\exp(\dot{m}HC_g/\lambda_g)}$$

If we use, for example, the approximation of [5.12a], with $h_L(-H') \approx h_{Ls}$, taking the changes into account, we obtain:

$$\dot{m} = \frac{\rho_g D_V}{H} Ln(\frac{h_g(H) - h_L(T_s)}{L_{Vs}}) \qquad [5.26]$$

We concluded earlier in [5.23] that $\dot{m} = \frac{\rho_g D_V}{H} Ln\left(\frac{1-Y_{V,H}}{1-Y_{Vsat}(T_s)}\right)$, and because of these two relationships, we can find both the temperature of the liquid at the interface and the vaporized mass flow rate. The interface temperature is given by the solution of the implicit equation obtained by equating the two equations:

$$\frac{1-Y_{V,H}}{1-Y_{Vsat}(T_s)} = \frac{h_g(H) - h_L(T_s)}{L_{Vs}} \qquad [5.27a]$$

If we use the second approximation, where $\dot{m}H'C_L/\lambda_L \gg 1$, we have $\dot{m} = \frac{\rho_g D_V}{H} Ln(\frac{h_g(H) - h_L(-H')}{h_g(T_s) - h_L(-H')})$, which allows us to calculate the surface temperature by:

$$\frac{1-Y_{V,H}}{1-Y_{Vsat}(T_s)} = \frac{h_g(H)-h_L(-H')}{h_g(T_s)-h_L(-H')} \qquad [5.27b]$$

In all the cases, the vaporized mass flow rate is again given by formula [5.23], where T_L is indeed T_s.

Here again, the use of these results for the expression of the mean interphase mass transfer term requires the identification of the values in H and H'. The discussion to be given on this subject is the same as in section 5.5.1. Beginning with [5.23], by linking the scales H and H' to the values $\bar{\Phi}_g, \bar{\sigma}^s_{lg}$, as previously with d and d', we obtain:

$$\overline{m_g \sigma^s_{lg}} = C_D \frac{\rho_g D_V}{\bar{\Phi}_g} (\bar{\sigma}^s_{lg})^2 Ln\left(\frac{1-\tilde{Y}_V}{1-Y_{Vsat}(T_s)}\right) \qquad [5.28]$$

and the surface temperature is given by introducing into [5.27a] or [5.27b] the mean enthalpy and mass fractions in the place of these same quantities in H and H'.

The heat flux that enters through conduction into the liquid phase is now, again using [5.15] but with $h_L(T_s)$ instead of h_{Lsat}:

$$\lambda_L(\frac{dT_L}{dz})_0 = \dot{m}\frac{h_L(T_s)-\tilde{h}_L}{1-\exp(-\dot{m}\bar{\Phi}_l C_L/(\lambda_L \bar{\sigma}^s_{lg}))}$$

which gives:

$$\overline{q_{ls}\sigma^s_{lg}} = C_D \frac{\rho_g D_V}{\bar{\Phi}_g} (\bar{\sigma}^s_{lg})^2 Ln\left(\frac{1-\tilde{Y}_V}{1-Y_{Vsat}(T_s)}\right) \frac{h_L(T_s)-\tilde{h}_L}{1-\exp(-\dot{m}\bar{\Phi}_l C_L/(\lambda_L \bar{\sigma}^s_{lg}))} \qquad [5.29]$$

These formulas can also be expressed with the characteristics of the vapor, as mentioned above.

5.6.2. *Evaporation of a drop*

Another interesting local exchange configuration is that of the vaporization of a drop in a gas medium. This situation has been studied both theoretically and experimentally for applications to engines, mostly using liquid fuels. We may refer to, for example, [SIR 99] or [BOR 00].

The theoretical study of the evaporation of a drop can be done relatively easily in the context of quasi-steady approximation, that is assuming a steady "virtual" mass supply at the center of the drop in such a way to ensure the constancy of the diameter of this drop during its evaporation. The drop is assumed to be isolated and spherical in shape, without motion relative to the air surrounding it, and in the absence of gravity, such that the geometry of the velocity and temperature fields remains spherical. We will use the results of this analysis here without redoing them, and comment on the results.

With the same simplifying hypotheses as used previously, meaning that $D_V = cste$ and the Lewis number of the vapor is equal to one, it is notable that we find that the mass flow rate vaporized from a drop is $\dot{m}_d = 4\pi r_d \rho_g D_V Ln\left(\frac{1-Y_{V\infty}}{1-Y_{Vsat}(T_s)}\right)$, meaning that the mass flow rate per unit of surface is:

$$\dot{m} = \frac{\rho_g D_V}{r_d} Ln\left(\frac{1-Y_{V\infty}}{1-Y_{Vsat}(T_s)}\right) \qquad [5.30]$$

a formula practically identical to [4.16]: it is the radius of the drop in play instead of H, and the mass fraction of the vapor very far from the drop plays the role of that at distance H.

In addition, the surface temperature of the drop must satisfy the implicit equation:

$$\frac{Y_{Vsat}(T_s)-Y_{V\infty}}{1-Y_{Vsat}(T_s)} = \frac{C_g(T_{g\infty}-T_s)}{L_{Ve}} \qquad [5.31]$$

where

$$L_{Ve} = L_{Vs} + \frac{1}{\dot{m}}(\lambda_L \frac{dT_L}{dr})_s \qquad [5.32]$$

The "Spalding transfer parameter" B is defined as one or another of the sides of equality [5.31]. Here, it replaces the Jacob number used for bubbles.

The calculation of the heat flux that enters by conduction into the drop, $(\lambda_L \frac{dT_L}{dr})_s$, poses a more delicate problem here. If we consider that, in reality, the drop is fed from its center, this heat flux would be exactly what is necessary to elevate the temperature of the liquid above its injection temperature, at the center, to its surface temperature. But in reality, the drop is not fed, and we do not know the temperature at the center of the drop, which moreover increases as time passes from an initial temperature to the surface temperature. We must calculate the heat flux by conduction in the liquid through the solution of the energy balance equation in the liquid sphere, and this is a non-steady problem. However, if we assume that the thermal conductivity of the liquid λ_L is very high, it follows that the drop is practically homogeneous in temperature ($\frac{dT_L}{dr}$ is practically zero, though the flux is not). It suffices, then, to equal the surface temperature with the mean temperature of the drop, which can be calculated through the energy balance equation of the liquid phase $T_s = \tilde{T}_L$, and the two equations [5.31] and [5.32] can thus give the entering heat flux. We have:

$$\left(\lambda_L \frac{dT_L}{dr}\right)_s = \dot{m}\left(\frac{C_g(T_{g\infty} - T_s)}{Y_{Vsat}(T_s) - Y_{V\infty}}(1 - Y_{Vsat}(T_s)) - L_{Vs}\right)$$

The heat flux exiting the gas phase, taking into account the contribution due to species diffusion, is such that $\dot{m}h_{gs} - \frac{\lambda_g}{C_V}\frac{dh_g}{dr})_s = \dot{m}h_{Ls} - \lambda_L \frac{dT_L}{dr})_s$, therefore:

$$\left(\frac{\lambda_g}{C_V}\frac{dh_g}{dr}\right)_s = \dot{m}\left(\frac{C_g(T_{g\infty} - T_s)}{Y_{Vsat}(T_s) - Y_{V\infty}} - L\right)(1 - Y_{Vsat}(T_s))$$

All of these questions have been studied in detail, in particular, by Sirignano and colleagues [SIR 99]. When the drop is in motion, the calculation of internal thermal conduction is even more delicate, since the friction of exterior gas causes a movement of the liquid in the drop, which tends to homogenize its temperature more rapidly than conduction alone.

The calculation of the mean interphase mass exchange per unit of volume of the two-phase medium can again be done based on formulas concerning a drop. For the same reasons that have been discussed for vaporization of bubbles in section 5.5.2, the exchange in the medium is C^N times (the number of drops per unit of volume) the exchange from one drop, only if the drops are homogeneously distributed and then the gradient of the drops volume fraction is zero. An additional term proportional to this gradient has to be taken into account in principle, but is never studied in practice. To deduce from this local exchange configuration a model usable for exchanges of mass and energy, we must again identify the mass vapor fraction far from the drop with the mean value of this quantity in the gas phase at the point being considered, $\tilde{Y}_V$. It is clear that these two values are linked, but their identification, which is a so-called "mean field" approximation and is often made, includes an error that is difficult to solve. No other hypothesis on this subject has been discussed, however, and to do it we need more complete direct numerical simulations for instant fields of drops and of the gas medium, taking into account interactions between the drops, if possible.

Thus, we can find the mass exchange term in a form very similar to [5.28]:

$$\overline{m_g \sigma_{lg}^s} = \overline{m_g}^s . \bar{\sigma}_{lg}^s = \frac{\rho_g D_V}{3\bar{\Phi}_g} (\overline{\sigma_{lg}^s})^2 Ln\left(\frac{1-\tilde{Y}_V}{1-Y_{Vsat}(\tilde{T}_L)}\right) \quad [5.33]$$

and the heat flux entering the liquid is:

$$\overline{q_{ls}\sigma_{lg}^s} = \overline{m_g \sigma_{lg}^s} \left(\frac{C_p(\tilde{T}_g - \tilde{T}_L)}{Y_{Vsat}(\tilde{T}_L) - \tilde{Y}_V}(1 - Y_{Vsat}(\tilde{T}_L)) - L_{Vs}\right) \quad [5.34]$$

These formulas, as they did previously, assume that all of the drops have the same radius. If this is not the case, and if the distribution of drop sizes can vary from one point to other point in the flow, we must be able to know this distribution of sizes at every point, which can again be achieved with balance equations for the numerical densities $\bar{\gamma}^N(v)$, which we discussed in Chapter 4. This is a considerable task. When we can think (and verify) that this distribution of drops only varies according to two parameters, the knowledge of the volume fraction of drops $\bar{\Phi}_g$ and the mean interfacial area per unit of volume $\bar{\sigma}_{lg}^s$ is sufficient to take its variation into account. Thus, a formula similar to [5.33] can be used, with a multiplicative constant that is solely a function of the volume fraction, the form of which depends on that of the distribution.

Note, here, that these formulas give the vaporization but also the condensation, if needed. The vaporized mass flow rate is positive if $\tilde{Y}_V < Y_{Vsat}(\tilde{T}_L)$, and negative for condensation with the opposite inequality.

The most realistic configuration, however, concerns drops in motion relative to the gas medium surrounding them. As for the study of friction of drops or particles, the formulas here must be modified by a multiplicative factor that brings in the Reynolds number of the drop, and the exact form of which is essentially empirical. An example of this factor is the Ranz–Marshall factor: $(1+0,3(R_{ed})^{1/2}(\text{Pr})^{1/3})$, where $R_{ed} = 2\left|v_d - v_g\right| r_d / v_g$, valid up to a Reynolds number value of approximately 150. Note that this multiplicative factor is very similar to the factor found for bubbles by Magnaudet and colleagues, as mentioned in section 5.5.4.2. To use this factor with formula [5.33], we must express it in terms of mean values: in the first approximation, we can bring in only the mean velocities of the gas and the drops, as we have done for formula [5.4]. The influence of turbulence, in the gas as well as the liquid, remains to be studied, for the case in which the velocity fluctuations are as large as the difference between the mean velocities.

A local exchange configuration that is intermediate between the quasi-flat liquid–gas interface subject to tangential flows (see section 5.6.1) and a spherical drop of liquid subject to a uniform gas flow considered here is that of a quasi-flat interface subject to flows that are perpendicular to it, and thus with a common stagnation point at the interface. This situation has been studied in [BUR 00].

5.6.3. *Combustion of a drop*

When a flame is stabilized in a spray of combustible drops in air (for example), some droplets can be surrounded with a "flamelet" that consumes the vapors escaping from the liquid. The temperature increases sharply on the median surface of this flamelet, and is directly involved in the vaporized mass flow rate. This local exchange situation can also be studied, assuming a spherical geometry of the phenomena around the drop (itself spherical as well), that is disregarding any rapid motion of the drop relative to the fluid medium. We can, again, find details on this situation in [BOR 00] and [SIR 99]. We will summarize the results in the following.

To simplify the representation of the reactive processes, which occur in the gas phase, we assume them to be nearly equivalent to a single, irreversible K+νOx→(1+ν)P reaction, where K is the combustible, which is the vapor of the drop; Ox is the oxidant (the air or the oxygen it contains); and P is the products. We can, therefore, assign to it a stoichiometric constant ν and a heat of reaction, given, q, per mass of combustible. We are defining here the mass stoichiometric coefficient, meaning that the reaction occurs between 1 kg of combustible and ν kg of oxidant. The processes of mass and heat diffusion in the gas medium are again represented by a diffusivity coefficient equal to D_V for all of these processes. In these conditions, we can define the quantity Z such that $Z = Y_K - Y_{Ox} / \nu$, at each point of the gas phase, which satisfies a balance equation similar to that of an inert gas substance. Its value is equal to $Z_s = Y_{Ksat}(T_s)$ at the surface of the drop, since K is the vapor of this drop, and, for fairly rapid combustion, the oxidant of the surrounding air cannot diffuse to the surface of the drop because it is consumed by the reaction before getting there, and its value is equal to $Z_\infty = -Y_{Ox\infty} / \nu$ far from the drop. We can, thus, apply the same theoretical approach as for the vapor mass fraction in the case of the simple vaporization of a drop.

We find that the vaporized mass flow rate from a drop is written as:

$$\dot{m}_d = 4\pi r_d \dot{m} = 4\pi r_d \rho_g D_V Ln\left(\frac{1+Y_{Ox\infty}/\nu}{1-Y_{Ksat}(T_s)}\right) \quad [5.35]$$

In the same way, the surface temperature of the drop must satisfy the implicit equation:

$$\frac{Y_{Ksat}(T_s)-Y_{ox\infty}/\nu}{1-Y_{Ksat}(T_s)} = \frac{C_g(T_{g\infty}-T_s)+qY_{Ox\infty}/\nu}{L_{Ve}} \quad [5.36]$$

with

$$L_{Ve} = L_{Vs} + \frac{1}{\dot{m}}(\lambda_L \frac{dT_L}{dr})_s$$

Supposing again that the thermal conductivity of the liquid is very high, and thus that $T_s = T_L$, these equations can give the heat flux entering the liquid:

$$-(\lambda_L \frac{dT_L}{dr})_s = \dot{m}((1-Y_{Ksat}(T_L))\frac{C_g(T_{g\infty}-T_L)+qY_{Ox\infty}/\nu}{Y_{Ksat}(T_L)-Y_{ox\infty}/\nu} - L_V(T_L))$$

As we did previously, we must identify the characteristics far from the drop with the mean values in the gas phase, and the temperature of the liquid with its mean value, in order finally to obtain a mean mass exchange term between liquid and gas phase in the two-phase medium. Similarly, we find:

$$\overline{m_g}^s.\overline{\sigma}_{lg}^s = \frac{\rho_g D_V}{3\overline{\Phi}_g}(\overline{\sigma_{lg}^s})^2 Ln\left(\frac{1+\tilde{Y}_{Ox}/\nu}{1-Y_{Ksat}(\tilde{T}_L)}\right) \quad [5.37]$$

$$\overline{q_{sl}\sigma_{lg}^s} = \overline{m_g}^s.\overline{\sigma}_{lg}^s((1-Y_{Ksat}(\tilde{T}_L))\frac{C_g(\tilde{T}_g-\tilde{T}_L)+q\tilde{Y}_{Ox}/\nu}{Y_{Ksat}(\tilde{T}_L)-\tilde{Y}_{ox}/\nu} - L_V(\tilde{T}_L)) \quad [5.38]$$

In a two-phase flame, however, each drop does not possess a flamelet around itself. These flamelets also occur around groups of several drops, as discussed in [BOR 00]. The complete model must again decide, in each zone of the two-phase flame, the proportion of drops that burn with a flamelet in their immediate vicinity or that vaporize without a flamelet nearby.

Chapter 6

Modeling Turbulent Dispersion Fluxes

We showed in Chapter 4 that multiphase flows can be represented either globally, as a single flow with a mean velocity, a mean energy and volume fractions of various phases, or as a "multifluid" flow, where each phase has its own mean velocity and mean energy. In both cases, velocity fluctuations exist, defined with respect to either the mean velocity of the flow or the mean velocities of the phases, and these fluctuations produce additional flux terms in the balance equations. Here, we study the modeling of these fluxes for the two kinds of representation.

6.1. Global modeling

6.1.1. *General information*

When we are only interested in the overall flow, the flux terms to be modeled are not just the dispersion fluxes due to fluctuating velocity. It is necessary, in theory, to also model the mean viscous stress tensor: $\bar{\tau}^{\alpha\beta} = \sum_i \overline{\Phi_i \tau_i^{\alpha\beta}}$ (recall that for the solid phase we should also put in it the non-diagonal elements of the Cauchy tensor), and $\bar{j}_\alpha^Q = \sum_i \overline{\Phi_i j_i^{Q\alpha}}$ the mean heat flux in the overall medium. The additional fluxes, that we will call dynamic or turbulent, are $-\overline{\rho v'^\alpha v'^\beta}$ and $+\overline{\tau^{\alpha\beta} v'^\beta} - \overline{\rho v'^\alpha h'_t}$, as well as $-\overline{\rho v'^\alpha \Phi_i}$ necessary for the calculation of mean mass fractions. The importance of these terms is primarily (and exclusively in the first two cases) due to the sufficiently high velocity of the flow, but it is not easy to give a straightforward *a priori* threshold for this situation.

Global modeling of the viscous stress tensor of a multiphase flow fluid poses the traditional problem of the "rheology" of this overall fluid. Even if each constituent is a viscous Newtonian fluid, and thus if the mean viscous stress tensor of each phase is calculated according to the gradient of the mean velocity of this phase, there is no possibility of expressing the mean stress tensor of the entire medium as a function of the mean velocity and the volume fractions of the overall medium, which are the only quantities which appear in the mean equations. It is the same for the heat flux, even when the phases simply follow Fourier's law. Emulsions or flows with bubbles are further complicated by the influence of surface tension. For flows loaded with solid particles, the situation is basically the same, even if the particles do not collide with each other and do not have lasting contact with friction, which of course would again complicate a detailed study of the phenomena. Taking the particles to be rigid, but with some inelasticity in the collisions, is a common way of modeling the situation, which simplifies reality and makes it possible to go ahead with certain calculations, but cannot solve the problem completely.

Friction in a multiphase medium where small particles are dispersed homogeneously in a fluid has been studied extensively. Experiments show that an approximation, which is often suitable, consists of defining an "apparent viscosity coefficient", which increases with the volume fraction of the particles. A theoretical approach for this was proposed by Einstein in 1906, for a low volume fraction of particles $\overline{\Phi}_p$, and by supposing that they have the same density as the fluid. This viscosity coefficient can be written simply as:

$$\frac{\mu}{\mu_f} = (1 + \frac{5}{2}\overline{\Phi}_p) \qquad [6.1a]$$

Experimental results indicate that this coefficient increases even more when there is a high load of particles and seems to tend toward infinity for a maximum load, for example according to the Krieger–Dougherty expression:

$$\frac{\mu}{\mu_f} = (1 - \frac{\overline{\Phi}_p}{\Phi_{\max}})^{-2.5\Phi_{\max}} \qquad [6.1b]$$

We can take $\Phi_{\max} = 0.64$, corresponding to "random close packing" of hard spheres, but, in fact, this value depends on the medium, that is the distribution of sizes and the "preparation" of this medium.

Experiments show, however, that for high particle content, the medium becomes non-Newtonian, the apparent viscosity becomes dependent on shear, then the characteristic times of shear variations have to be known, and there is no commonly accepted method for modeling these phenomena (for more detail on this subject, we can refer to [GUA 12]). We can also apply Einstein's formula for liquid droplets in a highly diluted condition, of course; the generalization to more dense conditions has not been studied, especially as such conditions introduce the problem of coalescence.

The difficulty in global modeling of the viscous stress tensor explains the importance in multifluid modeling, about which we will discuss next, but the difficulties of the latter make the first more practical, primarily in situations where the flow of the medium is rather fast so that $\overline{\tau}^{\alpha\beta}$ and $\overline{j}_{\alpha}^{Q}$ are weak compared to the additional dispersion fluxes, $-\overline{\rho v'^{\alpha} v'^{\beta}}$ and $-\overline{\rho v'^{\alpha} h'_{t}}$. Under these conditions, it is not convenient to neglect the viscous and conductive phenomena completely; but simple approximations of these terms are sufficient. Indeed, they are essential in certain zones of the flow, and the phenomena that they represent also influence the modeling of additional dispersion fluxes. Later, we will discuss how to approach these influences, in the turbulent situations which interest us, without completely dealing with the problem of apparent viscosity and apparent conduction.

Dynamic additional fluxes are formally similar to the Reynolds terms in traditional turbulent flows. However, they are only partially similar physically. On the one hand, the fluctuations of velocities or other quantities around their average are not exclusively due to the well-known traditional turbulence of a fluid – with or without constant density – but are also caused by other types of instabilities due to the presence of other phases. On the other hand, they must take account of the fact that the fluctuations concerning position and velocity of the phases in the initial conditions or at the boundaries can impart particular characteristics to the global velocity field. For example, the presence of two different mean velocities (for a fluid with only two phases) implies that the half trace of $\overline{\rho v'^{\alpha} v'^{\beta}}$, $\overline{\rho v'^{\alpha} v'^{\alpha}}/2=\overline{\rho}\tilde{k}$, which we will name the "kinetic energy of the fluctuations", can be decomposed into three contributions:

$$\overline{\rho}\tilde{k}=\overline{\rho_1\Phi_1 v'^{\alpha}_1 v'^{\alpha}_1}/2+\overline{\rho_2\Phi_1 v'^{\alpha}_2 v'^{\alpha}_2}/2+\frac{\overline{\rho}_1^*\overline{\rho}_2^*}{\overline{\rho}}\overline{\Phi}_1\overline{\Phi}_2(\tilde{v}_1^{\alpha}-\tilde{v}_2^{\alpha})(\tilde{v}_1^{\alpha}-\tilde{v}_2^{\alpha})$$

When the first two contributions are definitely smaller than the third, the overall medium is not really turbulent, but is rather an intermittent ensemble of two flows; this very important contribution can be identified, as "pseudo-turbulent" according to Lance and Bataille [LAN 91]. In this case, modeling with two fluids is more justified. But when the first two contributions – or one of them – are of the same order of magnitude as the third, in such a way that the two probability densities of the velocity fluctuations conditioned on each phase have a notable covering, then we cannot say that the velocity fluctuations of the medium are very different from those of a traditional turbulent medium.

It is by "analogy" with traditional turbulent flows that we approach modeling. On the one hand, the random fluctuations are at least partially due to similar sources of instabilities in a nonlinear mode toward saturation. On the other hand, the thermodynamics of irreversible processes shows us that the terms due to the lack of knowledge of the medium must be modeled in a way that satisfies the "second principle", whatever their precise source, and this explains the use of a "coefficient of turbulent viscosity" whereas this term could seem to be a contradiction. We will show in Chapter 9, by extending the traditional approach of irreversible thermodynamics, how it justifies certain usual elements of the models.

As for traditional turbulent flows, we first try to represent the turbulent dispersed flux terms with a gradient law of the Boussinesq type. Indeed, the phenomenon of dispersion, for the mass, energy or momentum, was very often suitably represented by such a law, which can be precisely explained by the thermodynamics of irreversible processes.

The overall medium is described by equations [4.1], [4.2] and [4.4b], and then we have:

$$-(\overline{\rho v'^{\alpha} v'^{\beta}} - \frac{1}{3}\overline{\rho v'^{\gamma} v'^{\gamma}\delta_{\alpha\beta}}) = \mu_e(\frac{\partial \tilde{v}_\alpha}{\partial x_\beta} + \frac{\partial \tilde{v}_\beta}{\partial x_\alpha} - \frac{2}{3}\frac{\partial \tilde{v}_\gamma}{\partial x_\gamma}\delta_{\alpha\beta}) \quad [6.2]$$

$$+\overline{\sigma_d^{\alpha\beta} v'^{\beta}} - \overline{\rho v'^{\alpha} h'_t} = \frac{\mu_e}{\Pr_e}\frac{\partial \tilde{h}_t}{\partial x_\alpha} \quad [6.3]$$

and $$-\overline{\rho v'^{\alpha}\Phi_i} = \frac{\mu_e}{Sc_e}\frac{\partial \tilde{\Phi}_i}{\partial x_\alpha} \quad [6.4]$$

Of course, we still have to calculate the "effective viscosity coefficient" μ_e and find realistic values for the effective Prandtl and Schmidt numbers that are generally considered constants. When following the traditional approach for single-phase turbulent flows, these coefficients will be sought as a function of two quantities that are considered sufficient to represent the effects of fluctuations: a kinetic energy and a characteristic length or time scale. In this framework, we can use formulations such as the Prandtl mixing length law or a k-ε model. We will study these possibilities in the following.

The new effects that arise here are, first, the influence of the volume fractions and, second, of the ratio of intrinsic densities of the phases. However, when it comes to solid particles or liquid droplets of constant geometry, the influence of their diameter is also a question. Finally, the influence of surface tension forces can be important in certain cases. All these problems have not been completely studied and inevitably cannot be studied in detail with this approach. We will return to this, depending on the case.

Just as is done for turbulence in a single-phase medium, it is also possible to find a model for the "Reynolds tensor" $\overline{\rho v'^{\alpha} v'^{\beta}} / \overline{\rho}$, which avoids [6.2]. The approach of modeling the balance equations of this tensor is similar to that known for traditional turbulence, but it is also necessary to take into account the fact that the flow does not have a constant density and thus that the effect of the pressure is not only a return to isotropy. To avoid [6.3] and [6.4], we can also study the equations for mass dispersion flux and enthalpy dispersion flux. We will discuss the dispersion flux in section 6.1.5.

6.1.2. *Kinetic energy of the "global fluctuations"*

By definition, the global kinetic energy of the fluctuations is:

$$\overline{\rho}k = \overline{\rho v'^{\beta} v'^{\beta}} / 2 = \overline{\rho(v^{\beta} - \tilde{v}^{\beta})(v^{\beta} - \tilde{v}^{\beta})} / 2$$

$$= \overline{\rho v^{\beta} v^{\beta}} / 2 - \overline{\rho}\tilde{v}^{\beta}\tilde{v}^{\beta} / 2 = \sum_i \overline{\rho_i \Phi_i v_i^{\beta} v_i^{\beta}} / 2 - \overline{\rho}\tilde{v}^{\beta}\tilde{v}^{\beta} / 2$$

The last term is simply the kinetic energy (per unit of volume) of the mean movement and can be obtained from the equation of mean global momentum, which we saw in Chapter 4 (equation [4.2]):

$$\frac{\partial}{\partial t}\left(\overline{\rho}\tilde{v}^{\beta}\right) + \frac{\partial}{\partial x_{\alpha}}\left(\overline{\rho}\tilde{v}^{\alpha}\tilde{v}^{\beta}\right) = -\frac{\partial \overline{p}}{\partial x_{\beta}} + \frac{\partial}{\partial x_{\alpha}}\left(\overline{\tau}^{\alpha\beta} - \overline{\rho v'^{\alpha} v'^{\beta}}\right) + \overline{\rho}g_{\beta} + \overline{F_{cap}^{\beta}\sigma^{s}}$$

Except for very particular cases, the energy associated with surface tension phenomena is, in general, very weak (compared to the instantaneous kinetic energy) and we will neglect its effect in the following. We multiply this equation (scalar product) by the mean global velocity, and after some traditional manipulations, we obtain:

$$\frac{\partial}{\partial t}\left(\overline{\rho}\tilde{v}^{\beta}\tilde{v}^{\beta} / 2\right) + \frac{\partial}{\partial x_{\alpha}}\left(\overline{\rho}\tilde{v}^{\beta}\tilde{v}^{\beta}\tilde{v}^{\alpha} / 2\right) = \frac{\partial}{\partial x_{\alpha}}\left(\tilde{v}^{\beta}\overline{\sigma}^{\alpha\beta} - \tilde{v}^{\beta}\overline{\rho v'^{\alpha} v'^{\beta}}\right) + \overline{\rho}g_{\beta}\tilde{v}^{\beta}$$

$$+ \overline{\rho v'^{\alpha} v'^{\beta}}\frac{\partial \tilde{v}^{\beta}}{\partial x_{\alpha}} + \overline{p}\frac{\partial \tilde{v}^{\beta}}{\partial x_{\beta}} - \overline{\tau}^{\alpha\beta}\frac{\partial \tilde{v}^{\beta}}{\partial x_{\alpha}}$$

Regarding the sum in the first term of k, it is necessary to start with each equation for the instantaneous velocity of the phase, which we established in Chapter 2, equation [2.5], to find an equation for $\rho_i \Phi_i v_i^\beta v_i^\beta / 2$, average it and sum over all the phases. Equation [2.5] is:

$$\frac{\partial}{\partial t}\left(\rho_i \Phi_i v_i^\beta\right)+\frac{\partial}{\partial x_\alpha}(\rho_i \Phi_i v_i^\alpha v_i^\beta)=\frac{\partial(\Phi_i \sigma_i^{\alpha\beta})}{\partial x_\alpha}+\rho_i \Phi_i g_\beta + m_i \sigma_i^s v_i^\beta + f_{i'i,s}^\beta \sigma_i^s$$

where by using the balance equation for the mass of phase i we obtain:

$$\rho_i \Phi_i \frac{\partial}{\partial t}\left(v_i^\beta\right)+\rho_i \Phi_i v_i^\alpha \frac{\partial}{\partial x_\alpha}(v_i^\beta)=\frac{\partial(\Phi_i \sigma_i^{\alpha\beta})}{\partial x_\alpha}+\rho_i \Phi_i g_\beta + f_{i'i,s}^\beta \sigma_i^s$$

This last equation is well defined: the left-hand terms and the second term on the right are the traditional terms in phase i and are zero outside that phase; the last term on the right is zero everywhere except at the interface; the first term on the right can be written as: $\Phi_i \dfrac{\partial}{\partial x_\alpha}(\sigma_i^{\alpha\beta})+\sigma_i^{\alpha\beta}\dfrac{\partial}{\partial x_\alpha}(\Phi_i)$ and the first term is the usual one in the phase and zero elsewhere; the second term is zero everywhere except at the interface because $\dfrac{\partial \Phi_i}{\partial x_\alpha}=-n_i^\alpha \sigma_i^s$.

If we multiply it (scalar product) by v_i^β, extended appropriately out of phase i, we obtain, after some calculations:

$$\rho_i \Phi_i \frac{\partial}{\partial t}\left(v_i^\beta v_i^\beta / 2\right)+\rho_i \Phi_i v_i^\alpha \frac{\partial}{\partial x_\alpha}(v_i^\beta v_i^\beta / 2)=\frac{\partial(v_i^\beta \Phi_i \sigma_i^{\alpha\beta})}{\partial x_\alpha}-\Phi_i \sigma_i^{\alpha\beta}\frac{\partial(v_i^\beta)}{\partial x_\alpha}$$
$$+\rho_i \Phi_i v_i^\beta g_\beta + v_{is}^\beta f_{i'i,s}^\beta \sigma_i^s$$

and then:

$$\frac{\partial}{\partial t}\left(\rho_i \Phi_i v_i^\beta v_i^\beta / 2\right)+\frac{\partial}{\partial x_\alpha}(\rho_i \Phi_i v_i^\alpha v_i^\beta v_i^\beta / 2)-m_i \sigma_i^s v_{is}^\beta v_{is}^\beta / 2$$

$$=\frac{\partial(v_i^\beta \Phi_i \sigma_i^{\alpha\beta})}{\partial x_\alpha}-\Phi_i \sigma_i^{\alpha\beta}\frac{\partial(v_i^\beta)}{\partial x_\alpha}+\rho_i \Phi_i v_i^\beta g_\beta + v_{is}^\beta f_{i'i,s}^\beta \sigma_i^s$$

By averaging, we obtain an equation for $\overline{\rho_i \Phi_i v_i^\beta v_i^\beta / 2}$, which is the mean kinetic energy of the phase. We obtain:

$$\frac{\partial}{\partial t}\left(\overline{\rho_i \Phi_i v_i^\beta v_i^\beta / 2}\right)+\frac{\partial}{\partial x_\alpha}(\overline{\rho_i \Phi_i v_i^\beta v_i^\beta v_i^\alpha / 2}) = \frac{\partial}{\partial x_\alpha}(\overline{\sigma_i^{\alpha\beta} \Phi_i v_i^\beta})$$

$$+\overline{\rho_i \Phi_i g_\beta v_i^\beta} + \overline{\Phi_i p_i \frac{\partial v_i^\beta}{\partial x_\beta}} - \overline{\Phi_i \tau_i^{\alpha\beta} \frac{\partial v_i^\beta}{\partial x_\alpha}} + \overline{m_i \sigma_i^s v_{is}^\beta v_{is}^\beta / 2} + \overline{\sigma_{is}^{\alpha\beta} n_i^\alpha v_{is}^\beta \sigma_i^s}$$

We then take the sum over all the phases:

$$\frac{\partial}{\partial t}\left(\sum_i \overline{\rho_i \Phi_i v_i^\beta v_i^\beta / 2}\right)+\frac{\partial}{\partial x_\alpha}(\sum_i \overline{\rho_i \Phi_i v_i^\beta v_i^\beta v_i^\alpha / 2}) = \frac{\partial}{\partial x_\alpha}(\sum_i \overline{\sigma_i^{\alpha\beta} \Phi_i v_i^\beta})$$
$$+\sum_i \overline{\rho_i \Phi_i g_\beta v_i^\beta} + \sum_i \overline{\Phi_i p_i \frac{\partial v_i^\beta}{\partial x_\beta}} - \sum_i \overline{\Phi_i \tau_i^{\alpha\beta} \frac{\partial v_i^\beta}{\partial x_\alpha}}$$

$$+\sum_i \overline{m_i \sigma_i^s v_{is}^\beta v_{is}^\beta / 2} + \sum_i \overline{\sigma_{is}^{\alpha\beta} n_i^\alpha v_{is}^\beta \sigma_i^s}$$

Subtracting the equation for $\overline{\rho}\tilde{v}^\beta \tilde{v}^\beta / 2$ then gives the equation of the kinetic energy of the fluctuations. We have directly:

$$\frac{\partial}{\partial t}(\overline{\rho}k) + \frac{\partial}{\partial x_\alpha}(\overline{\rho}k\tilde{v}^\alpha) =$$

$$\frac{\partial}{\partial x_\alpha}(\sum_i \overline{\rho_i \Phi_i v_i^\beta v_i^\beta (\tilde{v}^\alpha - v_i^\alpha)}/2 + \tilde{v}^\beta \overline{\rho v'^\alpha v'^\beta} + \sum_i \overline{\sigma_i^{\alpha\beta} \Phi_i v_i^\beta} - \tilde{v}^\beta \overline{\sigma}^{\alpha\beta})$$

$$-\overline{\rho v'^\alpha v'^\beta} \frac{\partial \tilde{v}^\beta}{\partial x_\alpha} + \sum_i \overline{\Phi_i p_i \frac{\partial v_i^\beta}{\partial x_\beta}} - \overline{p}\frac{\partial \tilde{v}^\beta}{\partial x_\beta} - \sum_i \overline{\Phi_i \tau_i^{\alpha\beta} \frac{\partial v_i^\beta}{\partial x_\alpha}} + \overline{\tau}^{\alpha\beta} \frac{\partial \tilde{v}^\beta}{\partial x_\alpha}$$

$$+\sum_i \overline{m_i \sigma_i^s v_{is}^\beta v_{is}^\beta / 2} + \sum_i \overline{\sigma_{is}^{\alpha\beta} n_i^\alpha v_{is}^\beta \sigma_i^s}$$

The right-hand term on the second line can also be written with $v'^\alpha_i = v_i^\alpha - \tilde{v}^\alpha$:

$$\frac{\partial}{\partial x_\alpha}(-\sum_i \overline{\rho_i \Phi_i v'^\beta_i v'^\beta_i v'^\alpha_i / 2} - \sum_i \overline{p_i \Phi_i v'^\alpha_i} + \sum_i \overline{\tau_i^{\alpha\beta} \Phi_i v'^\beta_i})$$

or, in a shorter form:

$$\frac{\partial}{\partial x_\alpha}(\overline{-\rho v'^\beta v'^\beta v'^\alpha/2} - \overline{pv'^\alpha} + \overline{v'^\beta \tau^{\alpha\beta}})$$

and on the third line:

$$-\overline{\rho v'^\alpha v'^\beta}\frac{\partial \tilde{v}^\beta}{\partial x_\alpha} + \sum_i \overline{\Phi_i p_i \frac{\partial v'^\beta_i}{\partial x_\beta}} - \sum_i \overline{\Phi_i \tau_i^{\alpha\beta}\frac{\partial v'^\beta_i}{\partial x_\alpha}}$$

$$= -\overline{\rho v'^\alpha v'^\beta}\frac{\partial \tilde{v}^\beta}{\partial x_\alpha} + \overline{p\frac{\partial v'^\alpha}{\partial x_\alpha}} - \overline{\tau^{\alpha\beta}\frac{\partial v'^\beta}{\partial x_\alpha}}$$

Immediately, it appears that the last two terms of the kinetic energy equation are zero if there is no exchange of mass between phases: on the one hand, the first term is zero because m_i is zero, and on the other hand, in this case, the velocity of each phase is the same at the interface $\sum_i \overline{\sigma_{is}^{\alpha\beta} n_i^\alpha v_{is}^\beta \sigma_i^s} = \overline{(\sum_i \sigma_{is}^{\alpha\beta} n_i^\alpha \sigma_i^s) v_s^\beta}$, and as the momentum relation at the interface gives that $\sum_i \sigma_{is}^{\alpha\beta} n_i^\alpha \sigma_i^s = 0$, the second term is, therefore, zero.

If there is an exchange of mass between phases, then, taking into account the equality of the tangential velocities at the interface, there is a term which describes the "source of the kinetic energy of the fluctuations":

$$\bar{\rho}\dot{w}_k^m = \sum_{i,i'} \overline{(-p_{is} v_{is}^n + p_{i's} v_{i's}^n)\sigma_{i,i'}^s} + \sum_{i,i'} \overline{m_i \sigma_{i,i'}^s (v_{is}^{n2} - v_{i's}^{n2})/2}$$

The notation v_{is}^n is the modulus of the normal component of the velocity of phase i (or $v_{i's}^n$ for phase i') at the interface between i and i' in question. This term is non-identically zero because the mean relation of the momentum of the i,i' interface is: $\overline{(-p_{is} + p_{i's})\sigma_{i,i'}^s} + \overline{m_i \sigma_{i,i'}^s (v_{is}^n - v_{i's}^n)} = 0$, but it is generally very weak.

Thus, we finally arrive at:

$$\frac{\partial}{\partial t}(\bar{\rho}k) + \frac{\partial}{\partial x_\alpha}(\bar{\rho}k\tilde{v}^\alpha) = \frac{\partial}{\partial x_\alpha}(\overline{-\rho v'^\beta v'^\beta v'^\alpha/2} - \overline{pv'^\alpha} + \overline{v'^\beta \tau^{\alpha\beta}})$$

$$-\overline{\rho v'^\alpha v'^\beta}\frac{\partial \tilde{v}^\beta}{\partial x_\alpha} + \overline{p\frac{\partial v'^\alpha}{\partial x_\alpha}} - \overline{\tau^{\alpha\beta}\frac{\partial v'^\beta}{\partial x_\alpha}} + \bar{\rho}\dot{w}_k^m \qquad [6.5a]$$

which can also be written as:

$$\frac{\partial}{\partial t}(\bar{\rho}k)+\frac{\partial}{\partial x_\alpha}(\bar{\rho}k\tilde{v}^\alpha)=\frac{\partial}{\partial x_\alpha}(-\overline{\rho v'^\beta v'^\beta v'^\alpha/2}-\overline{p'v'^\alpha}+\overline{v'^\beta\tau^{\alpha\beta}})$$

$$-\overline{\rho v'^\alpha v'^\beta}\frac{\partial\tilde{v}^\beta}{\partial x_\alpha}+\overline{p'\frac{\partial v'^\alpha}{\partial x_\alpha}}-\overline{v'^\alpha}\frac{\partial\bar{p}}{\partial x_\alpha}-\overline{\tau^{\alpha\beta}\frac{\partial v'^\beta}{\partial x_\alpha}}+\bar{\rho}\dot{w}_k^m \qquad [6.5b]$$

In this equation, as in all equations involving averages, no variable is discontinuous. We note that this equation, except for the weak last term due to the exchange of mass, could have been obtained directly from the instantaneous equation [2.8] of Chapter 2, by neglecting the discontinuous nature of certain factors. This last term is the result of these discontinuities, but it is zero when there is no exchange of mass between phases, and is in general very weak; so, we will not consider it hereafter.

The first term on the right-hand side is a dispersion term, with a contribution from the velocity fluctuations themselves, a contribution from the pressure fluctuations and a contribution from the viscosity. The second term is a "production" term, which is due to the gradients of the mean velocity (which is generally positive). The last term is the usual dissipation term, which is always negative. When the kinetic energy is dissipated, it of course does not disappear, but rather turns into a form of internal energy. Indeed, if we infer the mean kinetic energy (established above) from the total energy of the medium (equation [4.3b] from Chapter 4), we obtain the internal energy where this term is seen with an opposite sign. We have:

$$\frac{\partial}{\partial t}(\bar{\rho}\tilde{e}_t)+\frac{\partial}{\partial x_\alpha}(\bar{\rho}\tilde{v}^\alpha\tilde{e}_t)=\frac{\partial}{\partial x_\alpha}(-\bar{j}_\alpha^Q+\bar{\sigma}^{\alpha\beta}\tilde{v}^\beta+\overline{\sigma^{\alpha\beta}v'^\beta}$$

$$-\overline{\rho v'^\alpha e}-\overline{\rho v'^\alpha v^\beta v^\beta}/2)+\bar{\rho}g_\beta\tilde{v}^\beta$$

from which we subtract:

$$\frac{\partial}{\partial t}\left(\sum_i\overline{\rho_i\Phi_i v_i^\beta v_i^\beta}/2\right)+\frac{\partial}{\partial x_\alpha}(\sum_i\overline{\rho_i\Phi_i v_i^\beta v_i^\beta\tilde{v}^\alpha}/2)$$

$$=\frac{\partial}{\partial x_\alpha}(\sum_i\overline{\sigma_i^{\alpha\beta}\Phi_i v_i^\beta}-\sum_i\overline{\rho_i\Phi_i v_i^\beta v_i^\beta v'^\alpha_i}/2)$$

$$+\sum_i\overline{\rho_i\Phi_i g_\beta v_i^\beta}+\sum_i\overline{\Phi_i p_i\frac{\partial v_i^\beta}{\partial x_\beta}}-\sum_i\overline{\Phi_i\tau_i^{\alpha\beta}\frac{\partial v_i^\beta}{\partial x_\alpha}}$$

which gives:

$$\frac{\partial}{\partial t}(\bar{\rho}\tilde{e})+\frac{\partial}{\partial x_\alpha}(\bar{\rho}\tilde{v}^\alpha\tilde{e})=\frac{\partial}{\partial x_\alpha}(-\bar{j}^Q_\alpha-\overline{\rho v'^\alpha e})-\overline{p\frac{\partial v^\beta}{\partial x_\beta}}+\overline{\tau^{\alpha\beta}\frac{\partial v^\beta}{\partial x_\alpha}}$$

or $$\frac{\partial}{\partial t}(\bar{\rho}\tilde{e})+\frac{\partial}{\partial x_\alpha}(\bar{\rho}\tilde{v}^\alpha\tilde{e})=\frac{\partial}{\partial x_\alpha}(-\bar{j}^Q_\alpha-\overline{\rho v'^\alpha e}-\bar{p}\overline{v'^\alpha})-\bar{p}\frac{\partial\tilde{v}^\beta}{\partial x_\beta}$$

$$-\overline{p'\frac{\partial v'^\alpha}{\partial x_\alpha}}+\overline{v'^\alpha}\frac{\partial\bar{p}}{\partial x_\alpha}+\overline{\tau^{\alpha\beta}\frac{\partial v^\beta}{\partial x_\alpha}}$$

The last term of this equation consists of the dissipation rate of k plus the dissipation of the mean movement, which is in general much smaller than the preceding term.

The production term (which can be positive or negative depending on the case) due to the mean pressure of [6.5b] represents an exchange between the kinetic energy of the fluctuations and the mean internal energy. We find it in the equation above with the opposite sign, and it should be stressed that this exchange can be reversible, in contrast to the dissipation. The anterior pressure fluctuation term, as well as other pressure terms, are also found there and result from the total energy and the kinetic energy of the mean movement. The terms $-\overline{p'\frac{\partial v'^\alpha}{\partial x_\alpha}}+\overline{v'^\alpha}\frac{\partial\bar{p}}{\partial x_\alpha}$ disappear if the density is constant because the field of velocity fluctuations has zero divergence at each instant and the unweighted average of the velocity fluctuation, $\overline{v'^\alpha}$, is identically zero. Here, it is the average weighted by the density which is zero and not $\overline{v'^\alpha}$.

Finally, we note that equations [6.5a] and [6.5b] are valid whatever the source of the fluctuations. It is only in the model which we will discuss now that the physical properties of the (extended) turbulence will be involved.

6.1.3. *Modeling the kinetic energy of the fluctuations*

All the right-hand side terms should be modeled for the equation for k to be valid, either in form [6.5a] or in form [6.5b].

Since the first term is a dispersion term, it can be modeled traditionally with a gradient law, as $\frac{\partial}{\partial x_\alpha}((\mu+\mu_e/S_{ck})\frac{\partial k}{\partial x_\alpha})$, in both cases. The coefficient $\mu=\bar{\rho}\nu$ is the global viscosity coefficient of the fluid medium in slow flow, which we discussed above; μ_e is the coefficient that appears in formulas [6.2]–[6.4], and S_{ck} is a Schmidt number related to k.

The production by mean velocity gradient term can be modeled precisely using the Boussinesq approach, by formula [6.2], and a little later, we will discuss the cases where this is not advisable. It should be noted that hidden in this term is a "dilation" contribution: $-\frac{2}{3}\bar{\rho}k\frac{\partial\tilde{v}_\gamma}{\partial x_\gamma}$ which decreases the kinetic energy of the fluctuations if the mean density decreases along a (mean) streamline and increases in the opposite case.

The dissipation term is modeled in a traditional way, by using a length or time scale, as:

$$-\overline{\tau^{\alpha\beta}\frac{\partial v'^{\beta}}{\partial x_\alpha}}=-\bar{\rho}\varepsilon=-\bar{\rho}\frac{k}{\tau_t}=-\bar{\rho}C_D\frac{k^{3/2}}{l_t} \qquad [6.6a]$$

where C_D, dimensionless, is a constant or a function of the important parameters and variables of the problem, which we will discuss again in the following section, and it is necessary to find equations or algebraic formulas for one of the three quantities: the dissipation rate itself ε (per unit of mass), the turbulence characteristic time scale τ_t or the (turbulence) fluctuations integral length scale l_t.

The term related to the mean pressure in [6.5b] is in a form which, in general, does not ask for an additional approximation. We can write it in various ways:

$$-\overline{v'^{\alpha}}\frac{\partial\bar{p}}{\partial x_\alpha}=-(\bar{v}^\alpha-\tilde{v}^\alpha)\frac{\partial\bar{p}}{\partial x_\alpha}=\frac{\overline{\rho' v^\alpha}}{\bar{\rho}}\frac{\partial\bar{p}}{\partial x_\alpha}=\frac{\overline{\rho' v''^\alpha}}{\bar{\rho}}\frac{\partial\bar{p}}{\partial x_\alpha} \text{ or } v''^\alpha=v^\alpha-\bar{v}^\alpha .$$

Indeed, in this global model, the instantaneous density of the medium is a function, $\rho=\sum_{i=1,N}\Phi_i\rho_i$, that can often be reduced to $\rho=\rho_1(1-\Phi_2)+\rho_2\Phi_2$, where

ρ_1 and ρ_2 are constants. Then, we can show that $-\overline{v'^{\alpha}} = \dfrac{\overline{\rho' v^{\alpha}}}{\bar{\rho}} = \dfrac{\rho_2 - \rho_1}{\rho_1} \overline{v'^{\alpha} \Phi_2}$, and modeling the flux $\overline{v'^{\alpha} \Phi_2}$, allows us to calculate $\overline{v'^{\alpha}}$. Inappropriate modeling of this flux, however, can introduce miscalculations of this term. We see that if the mean pressure gradient is low, then the influence of this term in the kinetic energy can also be low.

The pressure fluctuation term requires modeling, and the underlying physical problems remain to be studied in detail. There again, direct numerical simulations with the equations from Chapter 2 will be very advantageous. However, until now, existing studies did not show this term to have a crucial role, and neglecting it has not produced any major problems (see [DEM 07]). Insofar as, in the averaged equations of momentum, we can suppose that the mean pressures in each phase are identical, we can consider the pressure fluctuations that intervene in this term are not exactly related to the density fluctuations between phases, and thus, this term can be neglected. When the surface tension is important (many small bubbles in a liquid medium) or if the mass exchange between phases is very significant, this term could be very important. In the case of turbulent gas flows with a variable density, this term also exists and has always been modeled as being proportional to the Mach number of the turbulence, which would seem to indicate that it is negligible except for the case of very rapid flows [CHA 02].

With the modeling of the equation of *k*, the expression for the effective viscosity coefficient can be obtained by simply using dimensional bases. Indeed, this coefficient has the dimension of a density multiplied by an effective diffusivity, and diffusivity is in meters squared per second. Then, because we suppose that two base quantities are sufficient to express it, we have:

$$\mu_e = \bar{\rho} C_\mu \frac{k^2}{\varepsilon} = \bar{\rho} C_\mu k \tau_t = \bar{\rho} \frac{C_\mu}{C_D} k^{1/2} l_t \qquad [6.6b]$$

C_μ is dimensionless and is constant or is a function of important parameters and quantities. In the case of a monophasic medium with high turbulence, i.e. when $-\overline{\rho v'^{\alpha} v'^{\beta}} \gg \overline{\tau}^{\alpha\beta}$, the approach known as "traditional k-ε" takes C_μ to be constant, determined by experimentation, but approaches known as "RNG k-epsilon" or "realizable k-ε" calculate it as a function of the quantity $\dfrac{Sk}{\varepsilon}$, where $S = (\frac{1}{2}(\frac{\partial \tilde{v}^{\alpha}}{\partial x_\beta} + \frac{\partial \tilde{v}^{\beta}}{\partial x_\alpha})(\frac{\partial \tilde{v}^{\beta}}{\partial x_\alpha} + \frac{\partial \tilde{v}^{\alpha}}{\partial x_\beta}))^{1/2}$ (for the case of a fluid with constant density).

But if $-\overline{\rho v'^{\alpha} v'^{\beta}}$ is not prevalent over $\overline{\tau}^{\alpha\beta}$, then experiments (reinforced by turbulence theory) show that C_μ and C_D must be functions of an appropriate Reynolds number called the "turbulence Reynolds number": $\text{Re}_t = \dfrac{\overline{\rho}k^{1/2}l_t}{\mu}$ (which becomes large precisely if $-\overline{\rho v'^{\alpha} v'^{\beta}} \gg \overline{\tau}^{\alpha\beta}$). In these formulas, we find the length l_t and time τ_t scales that were already introduced in our discussion of the dissipation rate. We will discuss these coefficients and how to calculate the length and time scales in the following section.

For a turbulent flow with several phases, we must think that the coefficients C_μ and C_D can still depend on the difference of the densities of the phases, for example on the ratio between the densities of the liquid and the gas in a medium with droplets or bubbles and even locally on the volume fraction. Experiments are necessary to specify this.

The model known as the "Prandtl mixing length formula", in the case where the mean velocity field is simple shearing, corresponds to the assumption that the equation for k can be replaced by the single equality of its mean velocity gradient production rate and its dissipation rate. The effect of pressure, mean or fluctuating, is then neglected, and if we refer to the preceding discussion, there are many situations where this is suitable. The question should be re-examined, however, if the flow has a notable mean pressure gradient, which can be due to gravity in particular, or to flow in a cyclone. Under conditions where it is valid, we can write the following instead of [6.5a]:

$-\overline{\rho v'^{\alpha} v'^{\beta}}\dfrac{\partial \tilde{v}^{\beta}}{\partial x_{\alpha}} = \overline{\tau'^{\alpha\beta}\dfrac{\partial v^{\beta}}{\partial x_{\alpha}}} = \overline{\rho}C_D\dfrac{k^{3/2}}{l_t}$, or rather $-\overline{\rho v'_x v'_y}\dfrac{\partial \tilde{v}_x}{\partial y} = \overline{\rho}C_D\dfrac{k^{3/2}}{l_t}$. But as $-\overline{\rho v'_x v'_y} = \mu_e\dfrac{\partial \tilde{v}_x}{\partial y} = \overline{\rho}C_\mu k^{1/2}l_t\dfrac{\partial \tilde{v}_x}{\partial y}$, it follows that $(C_\mu)^{1/2}l_t\left|\dfrac{\partial \tilde{v}_x}{\partial y}\right| = (C_D)^{1/2}k^{1/2}$, and finally, we obtain:

$$\mu_e = \overline{\rho}\frac{C_\mu^{\;3/2}}{C_D^{\;1/2}}l_t^2\left|\frac{\partial \tilde{v}_x}{\partial y}\right| \qquad [6.7]$$

The "mixing length", introduced by Prandtl around 1920, is the length $l_M = \dfrac{C_\mu^{3/4}}{C_D^{1/4}}l_t$, and it is proposed to calculate it directly, algebraically, as a function of the geometry of the flow, i.e. the thickness of the velocity gradient layer considered and also the proximity of the walls, which experiments confirm for a turbulent single-phase medium. This quantity is enough then, via [6.7], for calculating the effective viscosity $-\rho\overline{v'_x v'_y}$, which is the important component of the additional friction tensor.

6.1.4. *Length scales for fluctuations and time scale for the dissipation of kinetic energy of fluctuations*

To complete the preceding models, we must now be able to calculate the length scale associated with the velocity fluctuations l_t (or l_M) or ε the dissipation rate of *k*, or the time scale τ_t, that are related by [6.6a]. In fact, a length scale is defined in the kinetic energy dissipation rate [6.6a] and another is defined in the coefficient of turbulent viscosity [6.6b] for turbulent friction, and it is assumed that they are related via dimensionless coefficients C_μ and C_D which are not necessarily constant.

6.1.4.1. *The problem*

In a turbulent single-phase flow, velocity fluctuations length scales are studied through the correlation of these fluctuations at two points separated by a certain distance, around each given point of the flow. This correlation is, therefore, inherently a non-local value, even if it is defined at each point. This study shows that there is a range of lengths over which a "spectrum" of kinetic energy density is distributed, the latter being larger for the greatest lengths and lower for the smallest, close to zero (we can see the exact definition of this spectral energy distribution in classic books, for example [CHA 02]). In order to get a "mean" value of this length scale, we calculate the integral of the energy density distribution and obtain the "integral scale", denoted l_t, which is a length on the side of large scales.

Experiments show that the largest length scales are directly influenced by the "geometry" of the flow, whereas the smallest length scales are related to the viscosity of the fluid being considered, the Kolmogorov scale, practically the lowest limit of the scales range, being $\eta \propto l_t(\mathrm{Re}_t)^{-3/4}$. In fact, it is the unstable flow phenomena that give rise to these large scales in a way that cannot be explained here due to space constraints. Moreover, the experiments also show that the shape of the spectrum, at different points of the same flow and in many situations, is sufficiently similar to be able to be represented by a family of curves, mainly functions of the turbulence Reynolds number, once we know *k* and l_t. So, the length scales for turbulent friction and dissipation that correspond to differently weighted integrals are algebraically related to l_t in a way that, in the most general case, depends on Sk/ε and Re_t and so on the viscosity of the fluid being considered. Finally, experiments also show that the dissipation scale becomes independent of the Reynolds number if this is large, as well as large scales and the integral scale (but the Kolmogorov scale still depends on it, decreasing more and more). This is explained figuratively (and approximately) by a permanent "cascade of kinetic energy" from the large scales sustained by instabilities to the smaller scales where viscosity dissipates the kinetic energy into heat.

How can this situation be modified by the presence of two or several phases in the medium, whether fluid phases or solid particles are dispersed in a fluid? The physical characteristics that will be most important here are the density, which locally modifies the inertia of the medium, and the viscosity or whatever takes this role. Insofar as the dispersed solid particles can be considered to be causing an "apparent viscosity", which can be calculated, the situation is similar to that of a traditional viscous fluid. Does this represent reality sufficiently well? The surface tension in media with several fluids will play a significant part only in a medium where the kinetic energy is low, for moving bubbles in a medium at rest, for example.

6.1.4.2. *Reflections on the phenomena*

Due to continuity, especially when we are interested in averages (and the energy spectrum is defined by a statistical average), we are led to believe that the situation changes only gradually when we gradually increase the mass fraction of the phases other than the carrying fluid phase. Thus, we must still find, in any turbulent multiphase flow, a whole range of length scales and a kinetic energy density spectrum for the fluctuating movements associated with these scales.

The large scales of these movements must still be influenced by the general geometry of the flow. In fact, we could say that this is the mark of a "turbulent situation". This is translated, on the one hand, by the appearance of large scales due to the instabilities of this flow and the kinetic energy of the fluctuations is produced in this zone of large scales. But on the other hand, for zones close to the walls, the geometry on the contrary limits the growing of large scales.

For a multiphase medium consisting of several fluids with different viscosities, it is clear that it is still viscous friction that will ensure dissipation and will modify small scales. If the medium is composed of small drops dispersed in a fluid, with weak concentration, then we can use the apparent viscosity to quantify the phenomena more precisely. If the particles of the phase prove to be larger or more abundant, with different viscosities, then the situation will not be different from that of the same fluid with a variable viscosity (as a function of the non-uniform temperature, for example). The largest viscosity will intervene to modify the largest of the small scales, and we will surely still be able to define a "global Kolmogorov scale" with this largest viscosity (if the proportion of the corresponding fluid is not very small), which will make it possible to pinpoint the area where the viscous dissipation will begin.

We could also think that, as a result of a certain balance of all these effects, the spectrum can still have a typical form, which is found in different places of the same flow and in different types of flows, even if this form can be different and can depend on other parameters. The experimental studies of Lance and Bataille [LAN 91] for flows with bubbles considered these questions for the first time. These questions remain open, the phenomena that occur are complex and varied, they are still being studied with great experimental difficulties, and we will speak about them again in Chapter 11, section 11.2. In particular, we can ask ourselves whether the sizes of the parcels $\bar{\Phi}_i / \bar{\sigma}_i^s$ and the distance between the parcels $(\bar{\Phi}_i)^{2/3} / \bar{\sigma}_i^s$ will have a strong influence on this spectrum. Of course in a fluid medium, these sizes are as greatly influenced by the velocity fluctuations as the other way round, and in a way that depends on the viscosity and the capillarity, two things already taken into account: the first can play through the Reynolds number and the second can play through the Weber number but is only important for low velocities.

Under these conditions, we will still be able to define an integral scale and represent the range of scales with the contribution of a Reynolds number to be defined with the viscosity serving to define what we called the global Kolmogorov scale. The fact that the dissipation scale is no longer dependent on the (turbulence) Reynolds number if this is large will be preserved, but the definition of this Reynolds number can lead probably more easily to not very high values, and then the length scale related to the viscosity, and therefore the Reynolds number, should be taken into consideration.

For a medium containing dispersed solid particles, how is energy dissipated? Of course, in the bulk of the fluid it is dissipated by viscosity due to the velocity gradients that the turbulence causes locally, but it is also dissipated close to the particles insofar as their relative movement still creates velocity gradients close to the interface. Collisions between particles are also accompanied by even stronger gradients in the fluid, and when the particles finally collide and are compressed, viscous dissipation in the solid occurs. The friction between the particles rubbing against each other also leads to dissipation, but we will neglect this friction here; it will be important for the dense granular media in Part 3 of this book. The collisions are only frequent if the number of particles is high and the dissipation which they involve will probably be minor in the many cases which interest us here. Can all the dissipation in the fluid be accounted for by some apparent viscosity, as discussed in section 6.1.1, taking into account the particles of the medium, by locating the dissipation at scales close to an apparent Kolmogorov scale? If the particles are not too small compared to the (large) scales where the energy is produced, we can wonder whether the scale where dissipation is done cannot be fixed directly, without the explicit intervention of the fluid viscosity, dependent on the diameter of the particles or the mean distance between them. This would suppose that the viscosity that acts at these scales is the eddy viscosity due to a smaller scale turbulence. Nevertheless, this will not eliminate the cascade of energy because this scale linked to the particles will be again smaller than the scales at which the energy is produced; this will only change the identification of the scale where dissipation occurs, which will become independent of the molecular viscosity of the fluid. This length scale is also undoubtedly relevant for dissipation during collisions as long as it can be represented, as for "granular gases", by a coefficient of restitution of kinetic energy. However, we do not know the detailed answer to the preceding question and the way in which the shape of the kinetic energy spectrum (related to the correlation of the velocity fluctuations between two close points) is modified compared to that of a turbulent single-phase medium, as we consider a medium loaded with particles, remains unknown.

In any event, all this does not seem to prevent us from considering a spectrum with a typical general shape nor the definition of an integral scale. If the apparent viscosity can be useful here, the shapes of the spectrum should be similar to those of a traditional turbulent spectrum, and we should be able to calculate the scales useful for modeling by using an (apparent) Reynolds number for turbulence. Moreover, if the flow is a less unstable because of the strong apparent viscosity due to the particles, of course, the large scales of the spectrum will be close to the small scales where dissipation occurs, and will remain directly dependent on them. If the apparent viscosity represents phenomena imperfectly, the preceding descriptions would tend to show that the shape of the velocity spectrum could be modified at a length scale linked to the diameter or the mean distance between particles. Instead of a turbulence Reynolds number for representing its shape, it would be necessary to use the ratio between the integral scale and this new particle length scale.

6.1.4.3. *Proposals*

Following the preceding discussions, we will consider the cases of media without solid particles and media that carry them separately.

In media without solid particles, the approaches for modeling turbulent single-phase flows seem to be able to be applied globally, particularly models of the k-ε type, with minor additions as regards principles, provided it is not forgotten that the density is variable and fluctuating. We are again concerned with the equation for k written in [6.5b], with models that use the dissipation rate, ε, and we seek a partial derivative equation for this quantity. The equation for ε is traditional and is written as:

$$\frac{\partial}{\partial t}(\bar{\rho}\varepsilon)+\frac{\partial}{\partial x_\alpha}(\bar{\rho}\varepsilon\tilde{v}^\alpha)=\frac{\partial}{\partial x_\alpha}(\frac{\mu_t}{S_{c\varepsilon}}\frac{\partial\varepsilon}{\partial x_\alpha})-C_1\frac{\varepsilon}{k}(\overline{\rho v'^\alpha v'^\beta}-\frac{2}{3}\bar{\rho}k\delta_{\alpha\beta})\frac{\partial\tilde{v}^\beta}{\partial x_\alpha}-C'_1\frac{2\varepsilon}{3}\bar{\rho}\frac{\partial\tilde{v}^\alpha}{\partial x_\alpha}$$

$$+C_3\frac{\varepsilon}{k}\overline{p'\frac{\partial v'^\alpha}{\partial x_\alpha}}-C_4\frac{\varepsilon}{k}\overline{v'^\alpha}\frac{\partial\bar{p}}{\partial x_\alpha}-C_2\frac{\varepsilon}{k}\overline{\tau'^{\alpha\beta}\frac{\partial v^\beta}{\partial x_\alpha}} \qquad [6.8]$$

The model coefficients, C_1, C'_1, C_2, C_3 and C_4, are formally the same as those used for a traditional turbulent flow. We mentioned above that the term due to pressure fluctuations could definitely be neglected for flows at low Mach number, in which case C_3 is no longer of interest. They will now be able to depend, however, on a ratio of densities, in addition to eventual variation (according to the model variants) with an important other parameter, in particular the turbulence Reynolds number (defined on the basis of a viscosity that is representative of the most viscous fluid). We still need to check that the scales of the phase parcels are not significant.

If we can suppose that the experiments are conducted under conditions with high Reynolds numbers (for gas as well as liquid) and with high Weber numbers, then we can expect that the values of the coefficients defined above no longer depend on these numbers because in this case the parcels' sizes are very small and the shape of the spectrum no longer depends on them, and so the presence of two phases simply produces an effect of density differences.

There still remains a lot of work to be done to validate such an equation for ε in a global turbulent multiphase flow. Such an approach has rarely been used for media with two fluid phases (a liquid and a gas phase), either for media with a very low density of very small drops ($\overline{\Phi}_p$ of the order of a percent) or for dense media with "atomization" of the liquid [VAL 01, DEM 07].

In the case of a medium containing dispersed solid particles that are not so numerous, fairly small particles can be modeled as fluids by using a Reynolds number based on the apparent viscosity defined above. For larger particles, in accordance with preceding discussions, we can expect notable modifications to the shape of the spectrum depending on the ratio l_t / r_p. Nevertheless, the calculation of the integral scale l_t may still be modeled by an equation for the dissipation rate ε, as previously, of a form similar to [6.8], but the constants therein should be re-studied as a function of the ratio l_t / r_p and perhaps also of a Reynolds number where the viscosity of the carrying fluid intervenes. We will also have to expect the explicit intervention in this of the ratio of the density of the solid of the particles and that of the carrying fluid. Such a global model has never been studied, to our knowledge. It should also be stressed that the coefficient C_μ will also be very probably dependent on this ratio. This can be also studied within the framework of a "mixing length" model. For this model, we algebraically connect this length, proportional to the integral length, to the geometric dimensions of the flow, in particular the proximity of the walls. It will be necessary to find out how the ratio l_t / r_p can be taken into account there, and perhaps also the fluid viscosity, decreasing the influence of the geometry of the flow as this ratio becomes less big.

6.1.5. *Further studies on the dispersion flux of a phase*

Modeling the dispersion of a phase by a simple dispersion coefficient, as proposed by formula [6.4], only takes into account the effects of the different intrinsic densities of the phases, and of the volume fraction of these, via the mean density. The effect of the diameter of the droplets, particles or bubbles is not considered. We can try to take it into account, just as an effect that is additional to the intrinsic densities and volume fractions, via the Schmidt number, but further studies have to be done both theoretically and empirically.

It should be mentioned initially that the calculations of Demoulin *et al.* [DEM 07] for a liquid jet dispersing in a calm gas atmosphere gave very realistic results by using $-\overline{\rho v'^{\alpha} \Phi_i} = \frac{\mu_e}{Sc_e} \frac{\partial \tilde{\Phi}_i}{\partial x_\alpha}$, with Sc_e=0.9 (see also [LEB 09]). On the other hand, for the case of a liquid jet assisted by a coaxial gas jet with high velocity, it is proved that the prediction given by the calculation gives a liquid jet that is too long. A modification of the dispersion coefficient has been proposed, where the ratio of the densities of the liquid and the gas used explicitly plays a part, which could be physically justified by the intervention of Rayleigh–Taylor instabilities of the liquid filaments during atomization (see [DEM 07]). However, some more recent results from [STÉ 12] complicate the situation, which will be described in Chapter 11, section 11.1. It should not be forgotten that in any event, the presence of parcels of phase can also directly modify the kinetic energy of global fluctuations, and this should be taken into account in the calculation of this quantity, which naturally has an effect on the dispersion coefficient.

We should be able to study this dispersion more precisely via an equation for the dispersion flux itself, which has also been attempted in the cited works. We note, moreover, that a dissymmetrical effect of the density ratio between the jet and the exterior flow is already known for gas jets (see [CHA 02]), and seems to be able to be taken into account with an equation for the turbulent diffusion flux. It is, therefore, interesting to discuss this aspect now. We will do it in two ways.

6.1.5.1. *From the balance equation for the mean momentum of phase i*

This momentum was studied in Chapter 4, and here we rewrite the obtained equation:

$$\frac{\partial}{\partial t}\left(\overline{\rho}_i^*\overline{\Phi}_i\tilde{v}_i^\beta\right)+\frac{\partial}{\partial x_\alpha}(\overline{\rho}_i^*\overline{\Phi}_i\tilde{v}_i^\alpha\tilde{v}_i^\beta)=\frac{\partial}{\partial x_\alpha}(\overline{\Phi}_i\overline{\sigma}_i^{*\alpha\beta}-\overline{\rho_i\Phi_i v'^\alpha_i v'^\beta_i})+\overline{\rho}_i^*\overline{\Phi}_i g_\beta$$
$$+\overline{\rho_i\dot{w}_i^{\sec h}v_{is}^\beta}+\overline{f_{i,s}^\beta\sigma_i^s} \qquad [6.9]$$

The last term of contact with the other phases can be broken up, as shown in Chapter 5, section 5.3.3, as: $\overline{f_{i,s}^\beta\sigma_i^s}=\overline{p}_i^s\dfrac{\partial\overline{\Phi}_i}{\partial x_\beta}+\overline{F}_{si}^{\times\beta}$, and then:

$$\frac{\partial}{\partial t}\left(\overline{\rho}_i^*\overline{\Phi}_i\tilde{v}_i^\beta\right)+\frac{\partial}{\partial x_\alpha}(\overline{\rho}_i^*\overline{\Phi}_i\tilde{v}_i^\alpha\tilde{v}_i^\beta)=\frac{\partial}{\partial x_\alpha}(\overline{\Phi}_i\overline{\tau}_i^{*\alpha\beta}-\overline{\rho_i\Phi_i v'^\alpha_i v'^\beta_i})-\overline{\Phi}_i\frac{\partial\overline{p}_i^*}{\partial x_\beta}$$
$$-(\overline{p}_i^*-\overline{p}_i^s)\frac{\partial\overline{\Phi}_i}{\partial x_\beta}+\overline{\rho_i\dot{w}_i^{echs}v_{is}^\beta}+\overline{\rho}_i^*\overline{\Phi}_i g_\beta+\overline{F}_{si}^{\times\beta} \qquad [6.10]$$

We will suppose in the following that $\overline{p}_i^*=\overline{p}_i^s$, according to the discussion of Chapter 4, and if we neglect the effect of the exchange of mass between phases and the surface tension forces, the relation at the interface simply gives us $\overline{p}_i^s=\overline{p}$.

If we assume that the balance equation can be reduced to a local law, i.e. not differential, because almost everywhere, or in any case in the zones of interest, the last three terms are definitely prevalent, then we obtain an approximate form:

$$-\overline{\Phi}_i\frac{\partial\overline{p}}{\partial x_\beta}+\overline{\rho}_i^*\overline{\Phi}_i g_\beta+\overline{F}_{si}^{\times\beta}=0 \qquad [6.11]$$

This simplification is often made for slow moving media, and this makes it possible to neglect the nonlinear velocity terms.

The modeling of the last term, the contact force between the phase considered and the carrying fluid, is the most important remaining problem, and it was studied in Chapter 5. In the case where phase *i* is in the form of quasi-spherical dispersed parcels, a model that generalizes Stokes law, [5.8], with a "drift velocity" is most often used:

$$\overline{F}_{si}^{\times\alpha} = -\frac{9}{2}\rho_f \nu_f \frac{\overline{\Phi}_i}{r_i^2}(\tilde{v}_i^{\alpha} - \tilde{v}_f^{\alpha} - v_{Di}^{\alpha}) \text{, with } v_{Di}^{\alpha} = -\frac{D_{tif}}{\overline{\Phi}_i(1-\overline{\Phi}_i)}\frac{\partial \overline{\Phi}_i}{\partial x_{\alpha}}$$

The approximate form is then written as:

$$-\overline{\Phi}_i \frac{\partial \overline{p}}{\partial x_{\beta}} + \overline{\rho}_i^* \overline{\Phi}_i g_{\beta} - \frac{9}{2}\rho_f \nu_f \frac{\overline{\Phi}_i}{r_i^2}(\tilde{v}_i^{\beta} - \tilde{v}_f^{\beta} + \frac{D_{tif}}{\overline{\Phi}_i(1-\overline{\Phi}_i)}\frac{\partial \overline{\Phi}_i}{\partial x_{\beta}}) = 0$$

and finally:

$$\tilde{v}_i^{\beta} - \tilde{v}_f^{\beta} = \frac{r_i^2}{\frac{9}{2}\rho_f \nu_f}(-\frac{\partial \overline{p}}{\partial x_{\beta}} + \overline{\rho}_i^* g_{\beta}) - \frac{D_{tif}}{\overline{\Phi}_i(1-\overline{\Phi}_i)}\frac{\partial \overline{\Phi}_i}{\partial x_{\beta}} \qquad [6.12]$$

The dispersion flux is $\overline{\rho v'^{\alpha}\Phi_i} = \overline{\rho}\tilde{v}_i^{\alpha}\tilde{\Phi}_i - \overline{\rho}\tilde{v}^{\alpha}\tilde{\Phi}_i = \overline{\rho}_i^*\overline{\Phi}_i(\tilde{v}_i^{\alpha} - \tilde{v}^{\alpha})$. For a medium with two phases with constant densities, $\overline{\rho}\tilde{v}^{\alpha} = \rho_i\overline{\Phi}_i\tilde{v}_i^{\alpha} + \rho_f(1-\overline{\Phi}_i)\tilde{v}_f^{\alpha}$; therefore, $\tilde{v}_i^{\alpha} - \tilde{v}^{\alpha} = \tilde{v}_i^{\alpha}(1-\rho_i\overline{\Phi}_i/\overline{\rho}) - \rho_f(1-\overline{\Phi}_i)\tilde{v}_f^{\alpha}/\overline{\rho} = (\tilde{v}_i^{\alpha} - \tilde{v}_f^{\alpha})\rho_f(1-\overline{\Phi}_i)/\overline{\rho}$; it follows that $\overline{\rho v'^{\alpha}\Phi_i} = \rho_i\rho_f\overline{\Phi}_i(1-\overline{\Phi}_i)(\tilde{v}_i^{\alpha} - \tilde{v}_f^{\alpha})/\overline{\rho}$. Relation [6.12], therefore, gives:

$$-\overline{\rho v'^{\beta}\Phi_i} = \frac{2\rho_i r_i^2}{9\overline{\rho}\nu_f}\overline{\Phi}_i(1-\overline{\Phi}_i)(\frac{\partial \overline{p}}{\partial x_{\beta}} - \rho_i g_{\beta}) + \frac{\rho_i\rho_f D_{tif}}{\overline{\rho}}\frac{\partial \overline{\Phi}_i}{\partial x_{\beta}} \qquad [6.13]$$

There we obtain the generalization of the usual gradient law, which takes into account not only the gradient of the volume fraction but also that of the pressure and even the effect of gravity.

We see in formula [6.12], first, that when gravity and the mean pressure gradient are negligible, the diffusion flux can be approximated by a gradient formula. It should be noted that it is only with the addition of the drift velocity that the gradient appears, this can be an *a posteriori* justification of it. The length scale which should play a part in the diffusion coefficient D_{tif}, in principle, is the integral scale of global velocity fluctuations: $D_{tif} = C_{di}k\tau_t = C'_{di}k^{1/2}l_t$, where C and C' are constants to be determined. Via these constants, the diffusion coefficient can depend on the phase, in particular the size of the particles or drops. However, the theoretical developments that we have described here do not provide a way of explicitly calculating the dependence of the coefficient on this size; it should incorporate empirical data.

Next, we see that the effects of the mean pressure gradient and gravity can still be maintained at a high Reynolds number if we adapt the formula for $\overline{F}_{si}^{\times\alpha}$ for this case. We finally see that if the particles are very small, in fact if their *Stokes time* $\tau_{Si} = \frac{2}{9}\frac{\rho_i r_i^2}{\rho_f \nu_f}$ is very small, then we have the case of turbulent diffusion of fluid particles, and the term due to the drift velocity precisely takes into account this turbulent dispersion of fluid particles. Modeling it with a simple gradient of volume fraction is a representation, which is often highly acceptable.

6.1.5.2. *Balance equation for the dispersion flux*

We can also obtain an equation for the dispersion flux itself of phase i relative to the medium, and this is more symmetrical.

Indeed, we saw that $\overline{\rho v'^{\alpha}\Phi_i} = \overline{\rho}_i^* \tilde{v}_i^{\alpha}\overline{\Phi}_i - \overline{\rho}_i^* \tilde{v}^{\alpha}\overline{\Phi}_i = \overline{\rho}_i^* \tilde{v}_i^{\alpha}\overline{\Phi}_i - \overline{\rho}\tilde{v}^{\alpha}\tilde{\Phi}_i$. The first term of this expression is none other than the total flux of the mass of i, already calculated via:

$$\frac{\partial}{\partial t}\left(\overline{\rho}_i^*\overline{\Phi}_i\tilde{v}_i^{\beta}\right) + \frac{\partial}{\partial x_\alpha}(\overline{\rho}_i^*\overline{\Phi}_i\tilde{v}_i^{\alpha}\tilde{v}_i^{\beta}) = \frac{\partial}{\partial x_\alpha}(\overline{\Phi}_i\overline{\tau}_i^{*\alpha\beta} - \overline{\rho_i\Phi_i v'^{\alpha}_i v'^{\beta}_i}) - \overline{\Phi}_i\frac{\partial\overline{p}_i^*}{\partial x_\beta}$$
$$-(\overline{p}_i^* - \overline{p}_i^s)\frac{\partial\overline{\Phi}_i}{\partial x_\beta} + \overline{\rho_i\dot{w}_i^{echs}v_{is}^{\beta}} + \overline{\rho}_i^*\overline{\Phi}_i g_\beta + \overline{F}_{si}^{\times\beta}$$

The equation of the second term $\overline{\rho\Phi_i\tilde{v}^{\beta}} = \overline{\rho}\tilde{\Phi}_i\tilde{v}^{\beta}$ is obtained by multiplying [4.2], taking into account [4.1], by $\tilde{\Phi}_i$:

$$\tilde{\Phi}_i\overline{\rho}\frac{\partial}{\partial t}\left(\tilde{v}^{\beta}\right) + \tilde{\Phi}_i\overline{\rho}\tilde{v}^{\alpha}\frac{\partial}{\partial x_\alpha}(\tilde{v}^{\beta}) = -\tilde{\Phi}_i\frac{\partial\overline{p}}{\partial x_\beta} + \tilde{\Phi}_i\frac{\partial}{\partial x_\alpha}(\overline{\tau}^{\alpha\beta} - \overline{\rho v'^{\alpha}v'^{\beta}}) + \overline{\rho}\tilde{\Phi}_i g_\beta + \tilde{\Phi}_i\overline{F_{cap}^{\beta}\sigma^s}$$

Then from [4.4b], again taking into account [4.1] multiplied by $\tilde{v}^{\beta}$, we have:

$$\tilde{v}^{\beta}\overline{\rho}\frac{\partial}{\partial t}\left(\tilde{\Phi}_i\right) + \tilde{v}^{\beta}\overline{\rho}\tilde{v}^{\alpha}\frac{\partial}{\partial x_\alpha}(\tilde{\Phi}_i) = \tilde{v}^{\beta}\frac{\partial}{\partial x_\alpha}(-\overline{\rho v'^{\alpha}\Phi_i}) + \tilde{v}_\beta\overline{\rho_i\dot{w}_i^{echs}}$$

Their summation gives:

$$\bar{\rho}\frac{\partial}{\partial t}\left(\tilde{\Phi}_i\tilde{v}^\beta\right)+\bar{\rho}\tilde{v}^\alpha\frac{\partial}{\partial x_\alpha}(\tilde{\Phi}_i\tilde{v}^\beta)=-\tilde{\Phi}_i\frac{\partial\bar{p}}{\partial x_\beta}+\tilde{\Phi}_i\frac{\partial}{\partial x_\alpha}(\bar{\tau}^{\alpha\beta}-\overline{\rho v'^\alpha v'^\beta})+\bar{\rho}\tilde{\Phi}_i g_\beta+\tilde{\Phi}_i\overline{F^\beta_{cap}\sigma^s}$$
$$+\tilde{v}^\beta\frac{\partial}{\partial x_\alpha}(-\overline{\rho\Phi_i v'^\alpha})+\tilde{v}_\beta\overline{\rho_i\dot{w}_i^{echs}}$$

where

$$\frac{\partial}{\partial t}\left(\bar{\rho}\tilde{\Phi}_i\tilde{v}^\beta\right)+\frac{\partial}{\partial x_\alpha}(\bar{\rho}\tilde{v}^\alpha\tilde{\Phi}_i\tilde{v}^\beta)=-\tilde{\Phi}_i\frac{\partial\bar{p}}{\partial x_\beta}+\frac{\partial}{\partial x_\alpha}(\tilde{\Phi}_i\bar{\tau}^{\alpha\beta}-\tilde{\Phi}_i\overline{\rho v'^\alpha v'^\beta}-\overline{\rho\Phi_i v'^\alpha}\tilde{v}^\beta)$$
$$+\overline{\rho\Phi_i v'^\alpha}\frac{\partial}{\partial x_\alpha}\tilde{v}^\beta-(\bar{\tau}^{\alpha\beta}-\overline{\rho v'^\alpha v'^\beta})\frac{\partial}{\partial x_\alpha}\tilde{\Phi}_i+\bar{\rho}\tilde{\Phi}_i g_\beta+\tilde{\Phi}_i\overline{F^\beta_{cap}\sigma^s}+\tilde{v}_\beta\overline{\rho_i\dot{w}_i^{echs}}$$

Taking the difference between the two equations finally gives the equation for $\overline{\rho v'^\beta\Phi_i}$:

$$\frac{\partial}{\partial t}\left(\overline{\rho v'^\beta\Phi_i}\right)+\frac{\partial}{\partial x_\alpha}(\overline{\rho v'^\beta\Phi_i}\tilde{v}^\alpha)=\frac{\partial}{\partial x_\alpha}(-\tilde{\Phi}_i\bar{\tau}^{\alpha\beta}+\tilde{\Phi}_i\overline{\rho v'^\alpha v'^\beta}+\overline{\rho\Phi_i v'^\alpha}\tilde{v}^\beta+\bar{\Phi}_i\bar{\tau}_i^{*\alpha\beta}-\overline{\rho_i\Phi_i v'^\alpha_i v'^\beta_i})$$
$$+\tilde{\Phi}_i\frac{\partial\bar{p}}{\partial x_\beta}-\bar{\Phi}_i\frac{\partial\bar{p}_i^*}{\partial x_\beta}-\overline{\rho v'^\alpha\Phi_i}\frac{\partial\tilde{v}^\beta}{\partial x_\alpha}-\overline{\rho v'^\alpha v'^\beta}\frac{\partial\tilde{\Phi}_i}{\partial x_\alpha}+\bar{\tau}^{\alpha\beta}\frac{\partial\tilde{\Phi}_i}{\partial x_\alpha}-(\bar{p}_i^*-\bar{p}_i^s)\frac{\partial\bar{\Phi}_i}{\partial x_\beta}$$
$$+\bar{F}_{si}^{\times\beta}-\tilde{\Phi}_i\overline{F^\beta_{cap}\sigma^s}+\overline{\rho_i\dot{w}_i^{echs}(v^\beta_{is}-\tilde{v}^\beta)} \qquad [6.14]$$

This equation is complicated, but has the usual structure of an equation for turbulent diffusion flux; so, we will model its terms, such as we know it for a gas medium. In fact, before modeling, this equation is very general, whether or not the fluctuations are due to turbulence.

On the right-hand side, there is a term for the diffusion flux of the dispersion flux, to model, then the "production" terms, by mean pressure gradient, by mean velocity gradient and by mean mass fraction gradient, which are not to be modeled. When the dispersion flux in a certain direction is itself positive, the mass fraction or volume fraction gradient in this direction is usually negative and we see, for example, that $-\overline{\rho v'^\alpha v'^\beta}\frac{\partial\tilde{\Phi}_i}{\partial x_\alpha}$ is positive, thus able to further increase $\left(\overline{\rho v'^\beta\Phi_i}\right)$.

Finally, $\bar{F}_{si}^{\times\beta}$ must be a term of flux reduction, and therefore of decorrelation of fluctuations of velocity and volume fraction of phase by irreversible but also reversible phenomena (the pressure fluctuations that contribute to $\bar{F}_{si}^{\times\beta}$). It is indeed the only one able to play this part in a simple situation of homogeneous pressure and mean velocity. The surface tension term is, except in a special case, negligible, and this is also the case, if turbulence is well established, for the term comprising the mean tensor of viscous stresses. The term for the mass exchange between phases is, except in a very exceptional case, negligible.

We can model this equation, and thus propose models for the unknown terms which contribute to it, with the same point of view as usual for turbulent diffusion in a single-phase fluid. The diffusion term is habitually represented by a gradient model. The production terms of the second line do not require modeling, if we also use the approximation identifying all the pressures. We see that the model of the destruction term is in fact that of the term for the contact force between phases and that it must necessarily have a negative sign compared to that of the dispersion flux itself.

It is traditional to also seek a local approximation of this equation, only keeping the local production and destruction terms. The justification is not then the weakness of the velocity, but a local quasi-equilibrium between production and dissipation, when strong mean gradients lead to a high production, a situation that is often invoked, at least in the center of turbulent zones (we saw, for example, that the mixing length theory is based on this approximation). Neglecting the first term of the second part, which is a gradient of flux terms, is all the more justifiable as the turbulent velocity fluctuations are homogeneous in this zone. This gives:

$$+(\tilde{\Phi}_i - \bar{\Phi}_i)\frac{\partial \bar{p}}{\partial x_\beta} - \overline{\rho v'^\alpha \Phi_i}\frac{\partial \tilde{v}^\beta}{\partial x_\alpha} - \overline{\rho v'^\alpha v'^\beta}\frac{\partial \tilde{\Phi}_i}{\partial x_\alpha} + \bar{\tau}^{\alpha\beta}\frac{\partial \tilde{\Phi}_i}{\partial x_\alpha} + \bar{F}_{si}^{\times\beta} = 0 \qquad [6.15]$$

This equation is different from [6.11], even if the latter seems to convey the same information. Indeed, here we have worked on the velocity of phase *i* compared with the velocity of the global medium, and [6.15] thus directly comprises the influence of this global medium, which only appears in [6.11] via the mean pressure gradient. In fact, the approximations made in order to obtain [6.11] are stronger, being concerned with the movement of the particulate phase, they also have an effect on the global medium if the particulate phase is not infinitely diluted, whereas the approximations made here do not influence the movement of the global medium. In addition to the fact that the pressure gradient acts differently (but that translates in a different way the effect of gravity combined with the difference in densities of the phase *i* and of the medium), we explicitly see the mass fraction gradient of *i* appear (or the volume fraction gradient) and the gradient of the mean velocity. These two gradients, it was seen, are production terms of the dispersion flux.

It is necessary to now introduce a model of the contact force between phases. If we still use: $\overline{F}_{si}^{\times\beta} = -\frac{9}{2}\rho_f \nu_f \frac{\overline{\Phi}_i}{r_i^2}(\tilde{v}_i^\beta - \tilde{v}_f^\beta - v_{Di}^\beta)$, which we can write in terms of the Stokes time of the particles $\overline{F}_{si}^{\times\beta} = -\rho_i \frac{\overline{\Phi}_i}{\tau_{Si}}(\tilde{v}_i^\beta - \tilde{v}_f^\beta - v_{Di}^\beta)$, we obtain instead of [6.12]:

$$\tilde{v}_i^\beta - \tilde{v}_f^\beta = \frac{\tau_{Si}}{\rho_i \overline{\Phi}_i}((\tilde{\Phi}_i - \overline{\Phi}_i)\frac{\partial \overline{p}}{\partial x_\beta} - \overline{\rho v'^\alpha \Phi_i}\frac{\partial \tilde{v}^\beta}{\partial x_\alpha} + (\overline{\tau}^{\alpha\beta} - \overline{\rho v'^\alpha v'^\beta})\frac{\partial \tilde{\Phi}_i}{\partial x_\alpha}) - \frac{C_{di} k \tau_t}{\overline{\Phi}_i(1-\overline{\Phi}_i)}\frac{\partial \overline{\Phi}_i}{\partial x_\beta}$$

As $\overline{\rho v'^\alpha \Phi_i} = \rho_i \rho_f \overline{\Phi}_i (1-\overline{\Phi}_i)(\tilde{v}_i^\alpha - \tilde{v}_f^\alpha)/\overline{\rho}$, then we obtain instead of [6.13] the relation:

$$-\overline{\rho v'^\alpha \Phi_i}(\delta_{\alpha\beta} + \frac{\tau_{Si}\rho_f(1-\overline{\Phi}_i)}{\overline{\rho}}\frac{\partial \tilde{v}^\beta}{\partial x_\alpha}) =$$

$$\frac{\tau_{Si}\rho_f(1-\overline{\Phi}_i)}{\overline{\rho}}((\overline{\Phi}_i - \tilde{\Phi}_i)\frac{\partial \overline{p}}{\partial x_\beta} - (\overline{\tau}^{\alpha\beta} - \overline{\rho v'^\alpha v'^\beta})\frac{\rho_i \rho_f}{\overline{\rho}^2}\frac{\partial \overline{\Phi}_i}{\partial x_\alpha}) + C_{di}\frac{\rho_i \rho_f}{\overline{\rho}} k \tau_t \frac{\partial \overline{\Phi}_i}{\partial x_\beta} \qquad [6.16]$$

Recall that $\overline{\rho}\tilde{\Phi}_i = \rho_i \overline{\Phi}_i$; thus, $\tilde{\Phi}_i - \overline{\Phi}_i = (\frac{\rho_i}{\overline{\rho}} - 1)\overline{\Phi}_i$, with $\overline{\rho} = \rho_i \overline{\Phi}_i + \rho_f(1-\overline{\Phi}_i)$, and therefore, $\frac{\partial \tilde{\Phi}_i}{\partial x_\beta} = \frac{\rho_i \rho_f}{\overline{\rho}^2}\frac{\partial \overline{\Phi}_i}{\partial x_\beta}$.

In a homogeneous and isotropic turbulent medium with high turbulence Reynolds number, where $\overline{\tau}^{\alpha\beta} \ll -\overline{\rho v'^\alpha v'^\beta}$, this relation reduces to:

$$-\overline{\rho v'^\beta \Phi_i} = \frac{\tau_{Si}\rho_f(1-\overline{\Phi}_i)}{\overline{\rho}}(\overline{\Phi}_i - \tilde{\Phi}_i)\frac{\partial \overline{p}}{\partial x_\beta} + \frac{\rho_i \rho_f}{\overline{\rho}}(\frac{2\tau_{Si}\rho_f(1-\overline{\Phi}_i)}{\overline{\rho}} + C_{di}\tau_t)k\frac{\partial \overline{\Phi}_i}{\partial x_\beta}$$

We find again the dispersed flux goes in the opposite direction of the volume fraction gradient, with an additional baro-dispersion term.

We can think that the global effect that Nozières [NOZ 87] and Lhuillier [LHU 09] put forward is directly taken into account in [6.15], at least partially, whereas it was neglected in [6.11], at the same time as the velocity terms. We would be tempted to also say that we can do without "drift velocity" because a volume fraction gradient is already present in the equation. But this term, which results in the last term on the right-hand side of [6.16], is necessary in order for us to find the turbulent diffusion for very small-sized particles, which is identical to that of the fluid particles, using time τ_{dt} and not the Stokes time of the particles, which becomes zero. Indeed, all the preceding terms, including the one for the volume fraction gradient, are canceled in this situation. Therefore, we must discuss this question again close to this limit.

To this end, we come back to the first definition $\overline{F}_{is}^{\times\beta} = \overline{(-p_i"\delta_{\alpha\beta} + \tau_i^{\alpha\beta})\sigma_i^s n_i^\alpha}$, which can also be written as $\overline{F}_{is}^{\times\beta} = \overline{(-p_i"\delta_{\alpha\beta} + \tau_i^{\alpha\beta})\frac{\partial \Phi_i}{\partial x_\alpha}}$. In this term, we find, almost identically, the usual expression for the "destruction rate" of the turbulent diffusion flux that takes place in the equation of this flux for a turbulent single-phase flow transporting a fluid species. It is generally supposed that $\overline{p}^* = \overline{p}^s$, the fluctuation $p" = p - \overline{p}^s$ is thus the same as $p' = p - \overline{p}^*$, the usual fluctuation. The difference is that here the index of the phase replaces the mass fraction of the fluid which diffuses, and which is not then discontinuous (see, for example, [CHA 02]).

For large Reynolds numbers, it is the part due to the pressure fluctuations that is supposed to be prevalent in this turbulent model, and it is usually modeled as:

$$-\overline{p_i"\frac{\partial \Phi_i}{\partial x_\beta}} = -\frac{2}{C_d}\frac{\overline{\rho v'^\beta \Phi_i}}{\tau_t} - C\overline{\rho v'^\alpha \Phi_i}\frac{\partial \tilde{v}_\beta}{\partial x_\alpha}$$

where τ_t is the characteristic integral time of the turbulence of the global medium, we defined above that $\tau_t = k/\varepsilon = l_t / C_D k^{1/2}$, C_D, C_d and C are numerical coefficients (of order 1). In fact, the second term is not essential, and only allows the effect of the mean velocity gradient already present in the formula to be corrected a little.

By analogy with the approach for modeling turbulent diffusion fluxes that we must find for very small particles or drops, we are then tempted to propose an analogous formula without using the notion of the drift velocity. The usual form proportional to the difference in mean velocities can be preserved because this difference is proportional to the dispersion flux and because the effect of the two components of the term should be to reduce this difference and this flux to zero. But it is necessary to introduce a correcting function, which would give:

$$\overline{F}_{si}^{\times\beta} = -\rho_i \frac{\overline{\Phi}_i}{\tau_{Si}}(\tilde{v}_i^\beta - \tilde{v}_f^\beta)\phi(\tau_{Si}/\tau_t) \qquad [6.17a]$$

The correcting function ϕ must be close to unity when its argument is large, but should become equivalent to $\frac{\tau_{Si}}{\tau_t}\frac{2\rho_f}{C_d\bar{\rho}}(1-\bar{\Phi}_i)$ if the argument tends toward zero – as an example, this could be something like $\phi = 1/(1+\bar{\rho}\tau_t/(C_d\rho_f(1-\bar{\Phi}_i)\tau_{Si}))$, but it is this function which describes the way the particles diffuse according to their size, therefore it must be found from experiment. Because we know that $\overline{\rho v'^{\alpha}\Phi_i} = \rho_i\rho_f\bar{\Phi}_i(1-\bar{\Phi}_i)(\tilde{v}_i^{\alpha}-\tilde{v}_f^{\alpha})/\bar{\rho}$, the limit of $\bar{F}_{si}^{\times\alpha}$ when the radius of involved particles (and their Stokes time) tends to zero would be quite identical to the usual model for the turbulent diffusion of the fluid particles.

We would then obtain instead of [6.16]:

$$-\overline{\rho v'^{\beta}\Phi_i} - \overline{\rho v'^{\alpha}\Phi_i}\tau_{Si}\left(1-\bar{\Phi}_i\right)\rho_f\frac{\partial\tilde{v}^{\beta}}{\partial x_{\alpha}}/\left(\bar{\rho}\phi\left(\frac{\tau_{Si}}{\tau_t}\right)\right) =$$

$$\frac{\rho_f(1-\bar{\Phi}_i)\tau_{Si}}{\bar{\rho}\phi(\frac{\tau_{Si}}{\tau_t})}\left(\left(\bar{\Phi}_i-\tilde{\Phi}_i\right)\frac{\partial\bar{p}}{\partial x_{\beta}} - \left(\bar{\tau}^{\alpha\beta}-\overline{\rho v'^{\alpha}v'^{\beta}}\right)\frac{\rho_i\rho_f}{\bar{\rho}^2}\frac{\partial\bar{\Phi}_i}{\partial x_{\alpha}}\right) \quad [6.17b]$$

Formulas [6.16] and [6.17b] are, in fact, very similar; the only difference is the intervention of the turbulence time also in the term of baro-dispersion.

For an isotropic homogeneous turbulent medium with a large Reynolds number, we would have:

$$-\overline{\rho v'^{\beta}\Phi_i} = \frac{\rho_f(1-\bar{\Phi}_i)\tau_{Si}}{\bar{\rho}\phi(\frac{\tau_{Si}}{\tau_t})}\left(\left(\bar{\Phi}_i-\tilde{\Phi}_i\right)\frac{\partial\bar{p}}{\partial x_{\beta}} + 2k\frac{\rho_i\rho_f}{\bar{\rho}}\frac{\partial\bar{\Phi}_i}{\partial x_{\beta}}\right)$$

The function $\phi(\frac{\tau_{Si}}{\tau_t})$ is essential for regulating the way in which the size of the particles plays a part. When this tends toward zero, the "Stokes number" $(\frac{\tau_{Si}}{\tau_t})$ tends toward zero and the diffusion flux becomes controlled by the turbulence time, as it should be.

For the atomization and the dispersion of the liquid droplets, it is in the term $\bar{F}_{si}^{\times\beta}$, which does not relate only to well-formed drops but also to parcels and even to "strips of liquid" (as discussed in section 5.3.3), that Demoulin *et al.* [DEM 07] have proposed to add a contribution of Rayleigh–Taylor instabilities at the level of liquid–gas interfaces, of the form: $C_Y\bar{\rho}\bar{\Phi}_i(1-\bar{\Phi}_i)(\frac{1}{\rho_f}-\frac{1}{\rho_i})k\frac{\partial\bar{\Phi}_i}{\partial x_\beta}$. This formulation, after a wise choice of the constant C_y, has made it possible to reproduce the experimental results in a wider range of conditions.

6.1.5.3. *Conclusion*

We must first compare the study of turbulent dispersion by equation [6.11] from section 6.1.5.1 or by equation [6.15] from section 6.1.5.2. These two equations use different approximations, and the approximations implied in the second equation are less restrictive than those of the first equation because they allow the global influence of the other phase to persist via the mean flow. This second equation should, therefore, be preferred, and it can be justified in a larger domain.

Next, by modeling the dispersion of particles, or drops, in a turbulent two-phase flow, we have in fact used two different methods to model the contact force $\bar{F}_{si}^{\times\beta}$ between the particulate phase and the continuous phase. The first is based on the usual approach with a "drift velocity", but is transposed into the framework of a global model. The second is suggested by continuity with the usual model of the turbulent diffusion of species in a single-phase medium. The delicate problem that arises is that of the representation of what occurs when the size of the particles tends toward zero. If we make the radius of particles, or their Stokes time, directly tend to zero in the formula for the drag of a particle, by using the adapted Stokes formula which thus neglects the effect of turbulence pressure fluctuations, that is to say: $\bar{F}_{si}^{\times\alpha}=-\frac{9}{2}\rho_f v_f\frac{\bar{\Phi}_i}{r_i^2}(\tilde{v}_i^\alpha-\tilde{v}_f^\alpha)=\rho_i\frac{\bar{\Phi}_i}{\tau_{Si}}(\tilde{v}_i^\alpha-\tilde{v}_f^\alpha)$, this force becomes infinite, except if $\tilde{v}_i^\alpha-\tilde{v}_f^\alpha$ becomes zero and then has an indeterminacy of the zero multiplied by infinity type. But then the particles, which become fluid particles, do not diffuse, and this does not conform to reality.

By using a drift velocity that is proportional to the gradient of the volume fraction, with a diffusion coefficient directly related to the global turbulence of the medium, we obtain a realistic law within this limit for the usual turbulent diffusion: by using [6.15], we obtain [6.16]. We can note here that the coefficient D_{tif}, and therefore the drift velocity, should depend on the size of the considered particles in a way that is not specified at this level. With this drift velocity, the expression for the force in the limit gives an infinity minus infinity indeterminacy because $\bar{F}_{si}^{\times\alpha}=\rho_i\frac{\bar{\Phi}_i}{\tau_{Si}}(\tilde{v}_i^\alpha-\tilde{v}_f^\alpha)-\rho_i\frac{\bar{\Phi}_i}{\tau_{Si}}v_{Di}^\alpha$, but the use of [6.15] allows it to be removed.

By using the analogy with the balance equation for turbulent diffusion flux of a single-phase medium, we are led to model the contact force using formula [6.17a] in a way that directly removes the indeterminacy, and all the terms of [6.15] then have the same order in the limit.

We can note in any event that, if finally $\tilde{v}_i^\alpha - \tilde{v}_f^\alpha \approx v_{Di}^\alpha \approx -D_{tif} \frac{\partial \bar{\Phi}_i}{\partial x_\alpha}$, the factors $\tilde{v}_i^\alpha - \tilde{v}_f^\alpha - v_{Di}^\alpha$ or $\tilde{v}_i^\alpha - \tilde{v}_f^\alpha$ are not fundamentally different.

The physical justification of the correcting function is the fact that the influence of the viscosity of the fluid should disappear when the flow is very turbulent, with the small scales at which this viscosity acts remaining at the level of what is known as the Kolmogorov scale. This was noted very clearly for single-phase media, and must be found again for multiphase media when the particle size tends to zero, and gradually fades away for larger particles. The physical justification of the drift velocity was originally the intervention of the non-homogeneity of the medium, according to Nozières and Lhuillier, but it also makes it possible to deal with the question of small scales, and the works of Simonin and his colleagues [SIM 00] have indeed shown it, within the framework of a two-fluid approach, about which we will speak further.

In both cases, the obtained formulas [6.16] and [6.17b], which are very close, are not yet explicit: they still comprise either a constant (C_{di}) or a function of the ratio of the Stokes time to the integral time of the turbulence, which must be found by experiment. The variation of the dispersion coefficient according to the size of the particles or the drops depends on these sizes; preceding developments did not allow this point to be clarified. This must take into account the phenomenon of velocity response time of a particle with velocity fluctuation from the continuous phase, that of the selection of spatial scales of the particle fluctuations and that of the influence of pressure fluctuations. At this stage, we can only plan to introduce experimental data to resolve this question, but a two-fluid approach can make it possible to discuss these phenomena in more detail as we will see in the following section.

6.2. "Multifluid" modeling

When we are interested in knowing the different phases more precisely, it is necessary to be able to calculate the mean velocity of each phase, i, and this requires a model for the two tensors $-\overline{\rho_i \Phi_i v'^{\alpha}_i v'^{\beta}_i}$. We do not see the phase dispersion flux appear in this case because the mean velocity of each phase is known from a balance equation. In the same way, if we are interested in the mean energy of a phase, the term $-\overline{\rho_i \Phi_i v'^{\alpha}_i e_{ti}}$ represents an additional dispersion flux of (total) energy in the space which contains this phase, whether this is between the different particles or within a particle if internal movements take place.

The first models for two-phase flows simply neglected these additional flux terms, judging that the intraphase friction term would be prevalent compared to them. If we use this approximation, we also suppose that, in general, the terms due to the viscosity and the thermal conduction are negligible. This type of representation of two-phase flows consists of considering that the two phases each have a different mean velocity, but neglects the fact that "fluid particles" of each phase do not all have the same velocity at a given point. We use [4.4], [4.5] and [4.7] then for the two phases, neglecting the additional and molecular fluxes terms. Within this framework, the flow is not reversible, but this is only due to the friction of one phase on the other or to the exchange of heat between the phases. We can point out that such a hypothesis has the provisional drawback of generating a system of non-stationary equations which is elliptic, whereas it should be hyperbolic or parabolic.

To consider each phase as a turbulent fluid in itself seems to be simply a small deviation from the usual situation only if this phase is fluid, if it is the largest part of the flow and if the other phases are only locally present in the form of small inclusions. In these conditions, it is well understood that we can represent the fluctuations and turbulence in this phase by equations which are completely similar to those used when this phase is alone. But for highly dispersed phases, that seems much less natural, and besides, that explains why the approaches where they are represented in the "Lagrangian" way have been greatly developed. We want to treat the general situation here where the phases can all be in appreciable proportion or where the same phase is not the majority at all points of the flow. We have seen that, formally, it is not a problem to establish Eulerian equations for the mean characteristics of a phase in the medium, whatever the manner in which this phase arises. We will see now that we can also establish equations for quantities that characterize velocity fluctuations in a single phase. Of course, the velocity fluctuations are now partly due to the phenomena internal to a particle of phase (a drop, bubble or solid particle) and partly due to the differences between parcels of the same phase, in variable proportions. In this way, we are led to adapt the model used in the single-phase turbulent flows for each fluid phase, whether quasi-continuous or made up of fluids or dispersed solids, with possibly some reexamination of the approximations.

Initially, we will plan to use the Boussinesq relation for each phase, defining an "effective viscosity coefficient":

$$-(\overline{\rho_i \Phi_i v'^{\alpha}_i v'^{\beta}_i} - \frac{1}{3}\overline{\rho_i \Phi_i v'^{\gamma}_i v'^{\gamma}_i}\delta_{\alpha\beta}) = \mu_{ei}(\frac{\partial \tilde{v}^{\alpha}_i}{\partial x_\beta} + \frac{\partial \tilde{v}^{\beta}_i}{\partial x_\alpha} - \frac{2}{3}\frac{\partial \tilde{v}^{\gamma}_i}{\partial x_\gamma}\delta_{\alpha\beta})$$

and analogous gradient laws for the additional heat flux in the phases. In order to calculate the effective viscosity coefficient, we will call on the definition, and calculation, of a kinetic energy of fluctuations, in each phase, and then on a characteristic length of the fluctuations. As usual, we will write: $\mu_{ei} = \bar{\rho}_i^* \bar{\Phi}_i \frac{C_{\mu i}}{C_{Di}} k_i^{1/2} l_{ti}$, with $\bar{\rho}_i^* \bar{\Phi}_i k_i = \frac{1}{2} \overline{\rho_i \Phi_i v'^{\gamma}_i v'^{\gamma}_i}$, and it will then be necessary to study how to calculate the kinetic energy of the fluctuations in each phase, which will be discussed in the following sections.

We will need, however, to reconsider whether these Boussinesq relations are still justified, or even how to do without them. This will be discussed in the last section of this chapter.

6.2.1. *The kinetic energy of the fluctuations in each phase*

To study the balance equation for the kinetic energy of the fluctuations (or of the "turbulence") of each phase, we start with the balance equations of $\rho_i \Phi_i v_i^\alpha$ and of $\overline{\rho_i \Phi_i v_i^\alpha} = \bar{\rho}_i^* \bar{\Phi}_i \tilde{v}_i^\alpha$. The instantaneous equation is:

$$\frac{\partial}{\partial t}\left(\rho_i \Phi_i v_i^\beta\right) + \frac{\partial}{\partial x_\alpha}(\rho_i \Phi_i v_i^\alpha v_i^\beta) = \frac{\partial(\Phi_i \sigma_i^{\alpha\beta})}{\partial x_\alpha} + \rho_i \Phi_i g_\beta + m_i \sigma_i^s v_i^\beta + f_{i'i,s}^\beta \sigma_i^s,$$

or using the balance equation for mass *i*:

$$\rho_i \Phi_i \frac{\partial}{\partial t}\left(v_i^\beta\right) + \rho_i \Phi_i v_i^\alpha \frac{\partial}{\partial x_\alpha}(v_i^\beta) = \frac{\partial(\Phi_i \sigma_i^{\alpha\beta})}{\partial x_\alpha} + \rho_i \Phi_i g_\beta + f_{i'i,s}^\beta \sigma_i^s$$

This last equation is well defined: the left-hand terms and the second term on the right-hand side are the usual terms in phase *i* and are zero outside this phase; the final term on the right-hand side is zero everywhere except at the interface; the first term on the right-hand side is written: $\Phi_i \frac{\partial}{\partial x_\alpha}(\sigma_i^{\alpha\beta}) + \sigma_i^{\alpha\beta} \frac{\partial}{\partial x_\alpha}(\Phi_i)$; the first part is usual in the phase and zero elsewhere, and the second part is zero everywhere except at the interface, since $\frac{\partial \Phi_i}{\partial x_\alpha} = -n_i^\alpha \sigma_i^s$.

If we multiply it (scalar product) by v_i^β, extended appropriately out of phase i, we obtain, after some calculations:

$$\rho_i \Phi_i \frac{\partial}{\partial t}\left(v_i^\beta v_i^\beta / 2\right) + \rho_i \Phi_i v_i^\alpha \frac{\partial}{\partial x_\alpha}(v_i^\beta v_i^\beta / 2) = \frac{\partial (v_i^\beta \Phi_i \sigma_i^{\alpha\beta})}{\partial x_\alpha}$$

$$-\Phi_i \sigma_i^{\alpha\beta} \frac{\partial (v_i^\beta)}{\partial x_\alpha} + \rho_i \Phi_i v_i^\beta g_\beta + v_{is}^\beta f_{i'i,s}^\beta \sigma_i^s$$

and then:

$$\frac{\partial}{\partial t}\left(\rho_i \Phi_i v_i^\beta v_i^\beta / 2\right) + \frac{\partial}{\partial x_\alpha}(\rho_i \Phi_i v_i^\alpha v_i^\beta v_i^\beta / 2) - m_i \sigma_i^s v_{is}^\beta v_{is}^\beta / 2$$

$$= \frac{\partial (v_i^\beta \Phi_i \sigma_i^{\alpha\beta})}{\partial x_\alpha} - \Phi_i \sigma_i^{\alpha\beta} \frac{\partial (v_i^\beta)}{\partial x_\alpha} + \rho_i \Phi_i v_i^\beta g_\beta + v_{is}^\beta f_{i'i,s}^\beta \sigma_i^s$$

Through averaging, we obtain a balance equation for $\overline{\rho_i \Phi_i v_i^\beta v_i^\beta / 2}$, which is the mean kinetic energy of the phase. We obtain:

$$\frac{\partial}{\partial t}\left(\overline{\rho_i \Phi_i v_i^\beta v_i^\beta} / 2\right) + \frac{\partial}{\partial x_\alpha}(\overline{\rho_i \Phi_i v_i^\beta v_i^\beta} \tilde{v}_i^\alpha / 2) = -\frac{\partial \overline{p_i \Phi_i v_i^\beta}}{\partial x_\beta} + \frac{\partial}{\partial x_\alpha}(\overline{\tau_i^{\alpha\beta} \Phi_i v_i^\alpha} - \overline{\rho_i \Phi_i v'^\alpha_i v_i^\beta v_i^\beta} / 2)$$

$$+\overline{\rho_i \Phi_i g_\beta v_i^\beta} + \overline{\Phi_i p_i \frac{\partial v_i^\beta}{\partial x_\beta}} - \overline{\Phi_i \tau_i^{\alpha\beta} \frac{\partial v_i^\beta}{\partial x_\alpha}} + \overline{m_i \sigma_i^s v_{is}^\beta v_{is}^\beta} / 2 + \overline{\sigma_{is}^{\alpha\beta} n_i^\alpha v_{is}^\beta \sigma_i^s} \qquad [6.18a]$$

In addition, the balance equation for the mean velocity in i is:

$$\frac{\partial}{\partial t}\left(\bar{\rho}_i^* \bar{\Phi}_i \tilde{v}_i^\beta\right) + \frac{\partial}{\partial x_\alpha}(\bar{\rho}_i^* \bar{\Phi}_i \tilde{v}_i^\alpha \tilde{v}_i^\beta) = \frac{\partial}{\partial x_\alpha}(\bar{\Phi}_i \bar{\sigma}_i^{*\alpha\beta} - \overline{\rho_i \Phi_i v'^\alpha_i v'^\beta_i}) + \bar{\rho}_i^* \bar{\Phi}_i g_\beta$$

$$+\overline{m_i \sigma_i^s v_{is}^\beta} + \overline{f_{i'i,s}^\beta \sigma_i^s},$$

or by using the mean balance equation for the mass:

$$\bar{\rho}_i^* \bar{\Phi}_i \frac{\partial}{\partial t}\left(\tilde{v}_i^\beta\right) + \bar{\rho}_i^* \bar{\Phi}_i \tilde{v}_i^\alpha \frac{\partial}{\partial x_\alpha}(\tilde{v}_i^\beta) = \frac{\partial}{\partial x_\alpha}(\bar{\Phi}_i \bar{\sigma}_i^{*\alpha\beta} - \overline{\rho_i \Phi_i v'^\alpha_i v'^\beta_i}) + \bar{\rho}_i^* \bar{\Phi}_i g_\beta$$

$$+\overline{m_i \sigma_i^s v'^\beta_{is}} + \overline{f_{i'i,s}^\beta \sigma_i^s}$$

If we multiply it by $\tilde{v}_i^\beta$, we will find a balance equation for $\bar{\rho}_i^* \bar{\Phi}_i \tilde{v}_i^\beta \tilde{v}_i^\beta / 2$, it is the balance equation of the kinetic energy of the mean movement.

$$\frac{\partial}{\partial t}(\bar{\rho}_i^* \bar{\Phi}_i \tilde{v}_i^\beta \tilde{v}_i^\beta / 2) + \frac{\partial}{\partial x_\alpha}(\bar{\rho}_i^* \bar{\Phi}_i \tilde{v}_i^\beta \tilde{v}_i^\beta \tilde{v}_i^\alpha / 2) = \frac{\partial}{\partial x_\beta}(-\bar{\Phi}_i \tilde{v}_i^\beta \bar{p}_i^* + \bar{\Phi}_i \tilde{v}_i^\beta \bar{\tau}_i^{\alpha\beta} - \tilde{v}_i^\beta \overline{\rho_i \Phi_i v'^\alpha_i v'^\beta_i})$$

$$+\bar{\rho}_i^* \bar{\Phi}_i g_\beta \tilde{v}_i^\beta + \overline{\rho_i \Phi_i v'^\alpha_i v'^\beta_i} \frac{\partial \tilde{v}_i^\beta}{\partial x_\alpha} + \bar{\Phi}_i \bar{p}_i^* \frac{\partial \tilde{v}_i^\beta}{\partial x_\beta} - \bar{\Phi}_i \bar{\tau}_i^{\alpha\beta *} \frac{\partial \tilde{v}_i^\beta}{\partial x_\alpha} + \overline{m_i \sigma_i^s} \tilde{v}_i^\beta \tilde{v}_i^\beta / 2 + \tilde{v}_i^\beta \overline{\sigma_{is}^{\alpha\beta} n_i^\alpha \sigma_i^s} \quad [6.18b]$$

The difference between these two equations, [6.18a] and [6.18b], will give us an equation for:

$$\bar{\rho}_i^* \bar{\Phi}_i k_i = \overline{\rho_i \Phi_i v'^\beta_i v'^\beta_i / 2} = \overline{\rho_i \Phi_i v_i^\beta v_i^\beta / 2} - \bar{\rho}_i^* \bar{\Phi}_i \tilde{v}_i^\beta \tilde{v}_i^\beta / 2$$

The result, after some manipulations, is:

$$\frac{\partial}{\partial t}(\bar{\rho}_i^* \bar{\Phi}_i k_i) + \frac{\partial}{\partial x_\alpha}(\bar{\rho}_i^* \bar{\Phi}_i k_i \tilde{v}_i^\alpha) = \frac{\partial}{\partial x_\alpha}(-\overline{\rho_i \Phi_i v'^\beta_i v'^\beta_i v'^\alpha_i / 2} - \overline{p_i \Phi_i v'^\alpha_i} + \overline{\Phi_i v'^\beta_i \tau_i^{\alpha\beta}}) - \overline{\rho_i \Phi_i v'^\alpha_i v'^\beta_i} \frac{\partial \tilde{v}_i^\beta}{\partial x_\alpha}$$

$$+\overline{\Phi_i p_i \frac{\partial v'^\beta_i}{\partial x_\beta}} - \bar{\rho}_i^* \bar{\Phi}_i \varepsilon_i + \overline{m_i v_{is}^\alpha v_{is}^\alpha \sigma_i^s / 2} - \overline{m_i \sigma_i^s} \tilde{v}_i^\alpha \tilde{v}_i^\alpha / 2 + \overline{\sigma_{is}^{\alpha\beta} n_i^\beta v'^\alpha_{is} \sigma_i^s} \quad [6.19a]$$

where we have noted $\bar{\rho}_i^* \bar{\Phi}_i \varepsilon_i = \overline{\Phi_i \tau_i^{\alpha\beta} \frac{\partial v'^\beta_i}{\partial x_\alpha}}$.

If we define, as in Chapter 5, section 5.3.3, $\sigma_i^{\alpha\beta} = -p_i \delta^{\alpha\beta} + \tau_i^{\alpha\beta} = -\bar{p}_i^s \delta^{\alpha\beta} + \tau_i^{\times\alpha\beta}$, therefore $\tau_i^{\times\alpha\beta} = -(p_i - \bar{p}_i^s)\delta^{\alpha\beta} + \tau_i^{\alpha\beta}$, we have:

$$\overline{\sigma_{is}^{\alpha\beta} n_i^\beta v'^\alpha_{is} \sigma_i^s} = -\bar{p}_i^s \overline{n_i^\alpha v'^\alpha_{is} \sigma_i^s} + \overline{\tau_{is}^{\times\alpha\beta} n_i^\beta v'^\alpha_{is} \sigma_i^s}.$$

Then $-\bar{p}_i^s \overline{n_i^\alpha v'^\alpha_i \sigma_i^s} = \bar{p}_i^s \overline{v'^\alpha_i \frac{\partial \Phi_i}{\partial x_\alpha}} = \bar{p}_i^s (\frac{\partial \overline{\Phi_i v'^\alpha_i}}{\partial x_\alpha} - \overline{\Phi_i \frac{\partial v'^\alpha_i}{\partial x_\alpha}})$ and $\overline{p_i \Phi_i \frac{\partial v'^\alpha_i}{\partial x_\alpha}} = \bar{p}_i^s \overline{\Phi_i \frac{\partial v'^\alpha_i}{\partial x_\alpha}} + \overline{p''_i \Phi_i \frac{\partial v'^\alpha_i}{\partial x_\alpha}}$, where $p''_i = p_i - \bar{p}_i^s$. Finally, we obtain:

$$\frac{\partial}{\partial t}(\overline{\rho}_i^*\overline{\Phi}_i k_i)+\frac{\partial}{\partial x_\alpha}(\overline{\rho}_i^*\overline{\Phi}_i k_i \tilde{v}_i^\alpha)=\frac{\partial}{\partial x_\alpha}(-\overline{\rho_i \Phi_i v'^\beta_i v'^\beta_i v'^\alpha/2}-\overline{p"_i \Phi_i v'^\alpha_i}+\overline{\Phi_i v'^\beta_i \tau_i^{\alpha\beta}})$$

$$-\overline{\rho_i \Phi_i v'^\alpha_i v'^\beta_i}\frac{\partial \tilde{v}_i^\beta}{\partial x_\alpha}+\overline{p"_i \Phi_i \frac{\partial v'^\alpha_i}{\partial x_\alpha}}-\overline{\Phi_i v'^\alpha_i}\frac{\partial \overline{p}_i^s}{\partial x_\alpha}-\overline{\rho}_i^*\overline{\Phi}_i\varepsilon_i$$

$$+\overline{m_i v_{is}^\alpha v_{is}^\alpha \sigma_i^s}/2-\overline{m_i \sigma_i^s}\tilde{v}_i^\alpha \tilde{v}_i^\alpha/2+\overline{\tau_{is}^{\times\alpha\beta} n_i^\beta v'^\alpha_{is}\sigma_i^s} \qquad [6.19b]$$

We recognize an equation whose structure is completely standard. The kinetic energy of turbulence of the phase is convected with the mean velocity of the phase, it is dispersed by the first term of the right-hand side, produced by the mean velocity gradients of the second term on the right-hand side, and dissipated by the viscous effects contained in ε_i. The production terms where the pressure intervenes are zero if the phase in question has a constant density.

What is new is the existence of a term due to the exchange of mass between phases, if the latter exists, and above all a term for the exchange of kinetic energy between phases due to the contact force between them, which is the last term. This uses the tensor $\tau_i^{\times\alpha\beta}$, which comprises not only the effects of the viscosity but also those of the pressure fluctuations around the mean value of the phase at the interface. We have already noted that $\overline{\tau_{is}^{\times\alpha\beta} n_i^\alpha \sigma_i^s}=\overline{F}_{is}^{\times\beta}$, if we note $\tau_{is}^{\times\alpha\beta} n_i^\alpha \sigma_i^s = F_{is}^{\times\beta}$ it follows that $\overline{\tau_{is}^{\times\alpha\beta} n_i^\beta v'^\alpha_{is}\sigma_i^s}=\overline{F_{is}^{\times\alpha} v_{is}^\alpha}-\overline{F}_{is}^{\times\alpha}\tilde{v}_i^\alpha$.

It should be noted that this equation is completely general: it is valid for a multiphase medium with formation of drops, bubbles or particles of all sizes, deformations and breakups in the particles and coalescences. It is just as valid for the carrying phase of a flow with dispersed phases as it is for the phase of dispersed particles, particularly solid ones, which move and knock each other all the time. It is valid whatever the sources of the fluctuations, turbulence or not. We have only assumed that the contacts do not last.

6.2.2. *Modeling the balance equations of the kinetic energies of turbulence*

6.2.2.1. *Traditional terms*

The traditional terms of diffusion, production and dissipation can be approached in the usual way, at least at first approximation, with the usual approach in turbulence.

It is necessary to model the diffusion flux of k_i, $-\overline{\rho_i \Phi_i v'^{\beta}_i v'^{\beta}_i v'^{\alpha}/2}$ $-\overline{p''_i \Phi_i v'^{\alpha}_i} + \overline{\Phi_i v'^{\beta}_i \tau^{\alpha\beta}_i}$, and a gradient law is no less advisable than usual, with an effective diffusion coefficient that is directly analogous to that of momentum.

The term for production due to mean velocity gradient can still be expressed in a usual way with the Boussinesq law given above. The terms that include the pressure, $+\overline{p''_i \Phi_i \dfrac{\partial v'^{\alpha}_i}{\partial x_\alpha}} - \overline{\Phi_i v'^{\alpha}_i} \dfrac{\partial \overline{p}^s_i}{\partial x_\alpha}$, do not pose any problem if all the phases have a constant density (since they are, by definition, zero), but do evidently pose the same problem as for the global representation if this is not the case.

The dissipation term, $\overline{\rho}^*_i \overline{\Phi}_i \varepsilon_i$, can also be related to the characteristic length scale by $\varepsilon_i = C_{Di} k_i^{3/2} / l_{ti}$, this length being given algebraically, as this is done in a mixing length model (but without assuming here that the velocity fluctuations are directly proportional to the mean velocity gradient). In a more general way, the dissipation can also be calculated by its own balance equation, which makes it possible at the same time to have access to the characteristic length scale of the fluctuations. In the following section, we will discuss this calculation for the two different cases with or without solid particles.

Finally, it should be mentioned that a Boussinesq relation of type [6.6b] must be used, for each phase, and the coefficient $C_{\mu i}$ that appears there can, moreover, now depend on the characteristic parameters of the interaction with the other phase. For the case of a flow loaded with solid particles, we will immediately see that a characteristic time of this interaction is the Stokes time, and the coefficients $C_{\mu i}$ can thus depend on the ratio $\tau_{Sp} \varepsilon_i / k_i$, at least.

6.2.2.2. *The terms due to the exchanges between phases*

We can expect that the effect of the contact forces between phases, $\overline{F^{\times\alpha}_{si} v'^{\alpha}_i} = \overline{F^{\times\alpha}_{si} v^{\alpha}_i} - \overline{F}^{\times\alpha}_{si} \tilde{v}^{\alpha}_i$, will be more important for a very dispersed phase, because the contact surface is very large. This term is composed of two contributions: the second is given directly by knowledge of $\overline{F}^{\times\beta}_{si}$, which is discussed in Chapter 5, and the first requires a new model. The mass exchange term is still negligible, except in some special cases when it is not zero.

If we take the example of a flow with two phases, there are two k_i, being k_1 and k_2, and $\overline{F}^{\times\beta}_{s2} = -\overline{F}^{\times\beta}_{s1}$, because the balance relationship at the interface shows that $F^{\times\beta}_{s2} = -F^{\times\beta}_{s1}$ at every instant. The effect of the first term $\overline{F^{\times\alpha}_{si} v^{\alpha}_i}$ on the sum

$\rho_1\overline{\Phi}_1 k_1 + \rho_2\overline{\Phi}_2 k_2$ is exactly zero because the value of the velocity to take into account is that on the interface, and the velocities of the two phases there are always identical.

By neglecting the effect of mass exchange, we can write: $\overline{F_{si}^{\times\alpha} v_i^\alpha} = \overline{F}_{si}^{\times\alpha}\overline{v}_{sf}^\alpha$ where $\overline{v}_{sf}^\alpha$ is an average (weighted by the force of friction) of the velocity on the interface, which is independent of the phase, and we have $\overline{F_{s2}^{\times\alpha} v_2^\alpha} = -\overline{F_{s1}^{\times\alpha} v_1^\alpha} = -\overline{F}_{s1}^{\times\alpha}\overline{v}_{sf}^\alpha$ because $\overline{F}_{s2}^{\times\beta} = -\overline{F}_{s1}^{\times\beta}$. This term is thus a simple exchange, perfectly reversible between k_1 and k_2, in one direction or the opposite at each instant according to the respective directions of the vectors $F^\times$ and v.

On the other hand, the sum of the two other terms $-\overline{F}_{si}^{\times\alpha}\tilde{v}_i^\alpha$ gives $-\overline{F}_{s1}^{\times\alpha}\tilde{v}_1^\alpha - \overline{F}_{s2}^{\times\alpha}\tilde{v}_2^\alpha = -\overline{F}_{s1}^{\times\alpha}(\tilde{v}_1^\alpha - \tilde{v}_2^\alpha)$ and is not therefore zero. With a model of $\overline{F}_{s1}^{\times\beta}$ that is proportional to $-(\tilde{v}_1^\alpha - \tilde{v}_2^\alpha)$, without drift velocity, this sum is positive, which indicates that this is a source of common turbulence (for $\rho_1\overline{\Phi}_1 k_1 + \rho_2\overline{\Phi}_2 k_2$), and therefore the global effect of the contact forces is a source of common turbulence as long as $\tilde{v}_1^\alpha \neq \tilde{v}_2^\alpha$. The reversible exchange remaining between k_1 and k_2 should be modeled as a function of mean velocity of phases, and of other quantities.

If we reconsider $\overline{F_{si}^{\times\alpha} v_i^\alpha} = \overline{F}_{si}^{\times\alpha}\overline{v}_{sf}^\alpha$, we see that the modeling required is, in fact, that of the "mean friction velocity" on the interface $\overline{v}_{sf}^\alpha$. We will see in Chapter 9, section 9.2, later, that a linear equation with mean velocities for each phase is suggested by the linear thermodynamics of irreversible processes. We have also seen a relation of this type appears by studying in Chapter 5, section 5.3.3, the friction between phases at the slightly curved interfaces. If we assume $\overline{v}_{sf}^\alpha = a\tilde{v}_1^\alpha + (1-a)\tilde{v}_2^\alpha$, a being a non-fluctuating value that is still to be specified, we obtain:

$$\overline{F_{s1}^{\times\alpha} v'^{\alpha}_1} = \overline{F}_{s1}^{\times\alpha}(1-a)(\tilde{v}_2^\alpha - \tilde{v}_1^\alpha),\ \overline{F_{s2}^{\times\alpha} v'^{\alpha}_2} = a\overline{F}_{s2}^{\times\alpha}(\tilde{v}_1^\alpha - \tilde{v}_2^\alpha) = a\overline{F}_{s1}^{\times\alpha}(\tilde{v}_2^\alpha - \tilde{v}_1^\alpha)$$

If a lies between 0 and 1, the two terms are of the same sign, positive, and, therefore, represent a source of fluctuating kinetic energy. If a is negative, we have a source for phase 1 and a sink for phase 2, and it is the inverse if $a > 1$.

The choice of an approximation where a is constant has been adopted in models of dense granular media, and even in Baer and Nunziato (of whom we spoke in Chapter 4, section 4.3.2) a is taken as 1 that is to say the mean velocity of the interface is supposed to be the same as that of the granular phase. But should a have a constant value or not, function of the ratio of kinetic energies of the fluctuations, of the ratio of the characteristic times, of the phases' density ratio? That probably depends on the types of flows.

In a two-phase diluted medium with dispersed particles, the dispersion by the turbulence of the fluid of small solid particles has been studied for a long time. The first study of this type has probably been done by C.M. Tchen [TCH 47]. These works have been included more recently within the general framework of the turbulent two-phase flows and have allowed phenomena and their modeling to be studied in more detail, particularly the effect of interest for us here. We can find a synthesis of these works in [SIM 00].

By directly using the instantaneous Stokes law, we can write the effects of the particles in the turbulence of the fluid as:

$$\begin{aligned}\overline{F_{sf}^{x\alpha} v'^{\alpha}_f} &= \frac{9}{2}\rho_f \bar{\Phi}_p \frac{\nu_f}{r_p^2}(\overline{v_p^{\alpha} v'^{\alpha}_f} - \overline{v_f^{\alpha} v'^{\alpha}_f} + v_{Dp}^{\alpha} V_r^{\alpha}) \\ &= \frac{\rho_p \bar{\Phi}_p}{\tau_{Sp}}(\overline{v'^{\alpha}_p v'^{\alpha}_f} - \overline{v'^{\alpha}_f v'^{\alpha}_f} + v_{Dp}^{\alpha} V_r^{\alpha})\end{aligned} \qquad [6.20a]$$

As necessary, the Stokes time of the particles appears, and it has to be noted that here the velocity of the fluid concerns the velocity at the position of the particle but without it, and not a particular velocity in the field of the real velocity around this particle. So, O. Simonin speaks of the velocity of the fluid "seen by the particle" and notes the average implied in a different way from the simple statistical average. The drift velocity also appears, multiplied by the relative mean velocity of the fluid particles, which is itself $V_r^{\alpha} = \tilde{v}_p^{\alpha} - \tilde{v}_f^{\alpha} - v_{Dp}^{\alpha}$. The second term is simply an additional dissipation of k_f due to friction with the fluid, the first term contains the correlation of particle velocity fluctuations with fluid velocity fluctuations, $q_{fp} = \overline{v'^{\alpha}_f v'^{\alpha}_p}$.

This correlation takes into account the way in which the fluid transmits its turbulence energy to the particles, and vice versa, and should be studied.

For the effect on k_p, we find:

$$\overline{F_{sp}^{x\alpha} v'^{\alpha}_p} = -\frac{9}{2}\frac{\rho_p \bar{\Phi}_p}{\tau_{Sp}}(2k_p - q_{fp}) \qquad [6.20b]$$

with still a term of loss of agitation by friction with the carrying phase, but a potential source coming from the turbulence of that phase.

6.2.2.3. *The correlation of the fluctuations of the velocities of phases*

By studying the joint probability density function of the velocities of the fluid (seen by the particles) and the particles, $f_{fp}(u_p, u_f; x, t)$, where u_p and u_f are random variables associated with the velocity vectors of the particles and the fluid, x being a vector here, O. Simonin and his colleagues been able to learn more about this correlation. It is clear that this function cannot be calculated without approximations that we will not describe here, and we will refer to this reference and the work quoted for more details.

Finally, a balance equation for $q_{fp} = \overline{v'^{\alpha}_f v'^{\alpha}_p}$ has been proposed with a structure similar to that of k_p:

$$\frac{\partial}{\partial t}(\rho_p \bar{\Phi}_p q_{fp}) + \frac{\partial}{\partial x_\alpha}(\rho_p \bar{\Phi}_p q_{fp} \tilde{v}^\alpha_p) = \frac{\partial}{\partial x_\alpha}(\Delta_{fp}) + P_{fp} + F_{fp} - \rho_p \bar{\Phi}_p \frac{q_{fp}}{\tau^t_{fp}} \qquad [6.21]$$

The term P_{fp} is a production term, due to the gradients of mean velocities of the phases; the last term is a dissipation term, proportional to q_{fp} itself with a characteristic time τ^t_{fp} to find; and F_{fp} is a term due to the forces of contact between phases. The theoretical study has permitted giving a model for all these terms.

The term due to the forces of contact is modeled by:

$$F_{fp} = -\frac{\rho_p \bar{\Phi}_p}{\tau_{Sp}} \left(q_{fp} \left(1 + \frac{\rho_p \bar{\Phi}_p}{\rho_f (1 - \bar{\Phi}_p)} \right) - 2k_f - 2\frac{\rho_p \bar{\Phi}_p}{\rho_f (1 - \bar{\Phi}_p)} k_p \right) \qquad [6.22a]$$

When [6.21] is approximated as the balance between this term and the dissipation term, situation of local equilibrium, we can find q_{fp} by a simple algebraic relation:

$$\frac{q_{fp}}{\tau^t_{fp}} = \left(\left(1 + \frac{\bar{\Phi}_p \rho_p}{\bar{\Phi}_f \rho_f} \right) q_{fp} - 2\frac{\bar{\Phi}_p \rho_p}{\bar{\Phi}_f \rho_f} k_p - 2k_f \right) / \tau_{Sp},$$

from where

$$q_{fp} \left(1 + \frac{\bar{\Phi}_p \rho_p}{\bar{\Phi}_f \rho_f} - \frac{\tau_{Sp}}{\tau^t_{fp}} \right) = 2 \left(\frac{\bar{\Phi}_p \rho_p}{\bar{\Phi}_f \rho_f} k_p + k_f \right),$$

and finally:

$$q_{fp} = 2\frac{\bar{\Phi}_p \rho_p k_p + \bar{\Phi}_f \rho_f k_f}{\bar{\Phi}_p \rho_p + \bar{\Phi}_f \rho_f - \bar{\Phi}_f \rho_f \tau_{Sp} / \tau^t_{fp}}$$

We see that the local equilibrium value of this fluctuation's correlation depends on the ratio of time τ^t_{fp} and the Stokes time τ_{Sp}. For small particles, τ_{Sp} / τ_{fp} tend toward zero, and for a diluted medium we, therefore, obtain $q_{fp} = 2k_f$.

The characteristic time τ^t_{fp} should be represented as a function of integral time of the turbulence of the fluid and of the Stokes time of the particles so as to find the effects studied since C.M. Tchen for the dispersion of particles in turbulence. The fact that the particles carried by turbulent fluctuations cannot remain in the same fluid "eddy" by the effect of their inertia, due to their larger density (this is called "crossing trajectory effect"), plays an important role here. The experimental results to this subject allow the empirical formula of Czanady to be obtained:

$$(\tau^t_{fp})^{-1} = (\tau^t_f)^{-1}(1 + C_\beta \frac{V_r^2}{2k_f / 3})^{1/2} \qquad [6.22b]$$

The "drift velocity" mentioned above appears also directly in the calculation of the mean force of contact between the two phases. We have discussed for this drift velocity in Chapter 5 an expression in function of the gradient of the volume fraction of the particulate phase:

$$v^\beta_{Dp} = -\frac{D_{tpf}}{\bar{\Phi}_p \bar{\Phi}_f} \frac{\partial \bar{\Phi}_p}{\partial x_\beta}.$$

The preceding theoretical developments (see [SIM 00]) also give a balance equation for this drift velocity, with the same structure as [6.22b]. In the conditions where the terms of production and dissipation of this equation balance, we see such an algebraic form for the drift velocity appears, and τ^t_{fp} and k_{fp} introduced here are precisely the quantities that play in the diffusion coefficient:

$$D^t_{tpf} = \frac{1}{3} k_{fp} \tau^t_{fp}.$$

For this formula, it is necessary to note that the empirical constant C_β has a different value for the diffusion in the directions parallel or perpendicular to the mean relative velocity.

Within the framework of the global representation, we do not have any possibility other than relating the dispersion coefficient D^t_{tpf} to the global turbulence, with a proportionality coefficient depending on the size of the particles. Here, with the multifluid representation, we can go further because τ^t_{fp} and q_{fp} permit the coefficient to be calculated in a more physical way.

The works of O. Simonin *et al.* also tend to propose that the coefficient of "effective viscosity" μ_{ep}, which permits calculation of the components of the Reynolds tensor relative to the particulate phase, is also related to τ^t_{fp} and q_{fp} in the following manner:

$$\mu_{ep} = \rho_p(\frac{1}{3}q_{fp}\tau^t_{fp} + \frac{1}{3}k_p\tau_{Sp})/(1+\tau_{Sp}\frac{2/5}{\tau^c_p}) \qquad [6.23]$$

The time τ^c_p is the characteristic time of the collisions between particles (which we will discuss in the following section), depending on their diameter and of their volume fraction, which also plays a role in the dissipation term. This means that the classical formulation, recalled in section 6.2.2.1, indeed comprises a coefficient $C_{\mu p}$, which is not constant, but depends on the numbers τ_{Sp}/τ^t_{fp} and τ_{Sp}/τ^c_p. The coefficient $C_{\mu f}$ for the fluid phase remains standard. It is interesting to also notice that it was found that the characteristic time of collision of the particles is modified by the presence of the turbulent fluid: the fluid introduces a correlation of velocities of close particles, which decreases the characteristic time compared to that which is usually calculated taking into account the mean separation of particles and their agitation energy. For details, we refer the reader to [LAV 95].

6.2.3. *The modeling of time or spatial scales*

As we discussed for global modeling, the problem of calculating the scales is probably different if all the phases are fluid, or if one phase consists of dispersed solid particles of known and constant diameter. Moreover, in the second situation of solid dispersed particles, we dispose of the theory of "granular gas".

– *We first envisage the case of fluid phases* and the modeling of the dissipation rate of fluctuating kinetic energy of each phase, which is the most general method. As the balance equation of the fluctuating kinetic energy of a phase is similar to the one of the turbulence of a single-phase fluid, we will use an equation of dissipation rate for each phase in the usual manner, so with the form that has been discussed in section 6.1.4. This seems "natural" if all the phases are fluids because it is simply about generalizing what is done classically for a continuous fluid phase. But we will not be able to say that the equation of the dissipation rate will have firm theoretical bases. In any case, these bases are still less demonstrated than for a continuous fluid medium. The questions that are nevertheless posed for this modeling are first evidently the same as for global modeling: first, how do we introduce the effect of other phases in this equation? Second, should the size of the phase particles appear in it?

For the first question, since there is a term due to the drag of the other phases in the balance of kinetic energy of the fluctuations, it is necessary to envisage that there will also be an additional term, of the form $C_{ctc}\frac{\varepsilon_i}{k_i}(\overline{F_{is}^{\times\alpha} v_i^{\alpha}} - \overline{F_{is}^{\times\alpha}}\tilde{v}_i^{\alpha})$, in the equation of $\varepsilon_i = C_{Di}k_i^{3/2}/l_{ti}$. This gives a general equation of the type:

$$\frac{\partial}{\partial t}(\bar{\rho}_i^*\bar{\Phi}_i\varepsilon_i)+\frac{\partial}{\partial x_\alpha}(\bar{\rho}_i^*\bar{\Phi}_i\varepsilon_i\tilde{v}_i^\alpha)=\frac{\partial}{\partial x_\alpha}(\frac{\mu_{ti}}{S_{c\varepsilon}}\frac{\partial\varepsilon_i}{\partial x_\alpha})-C_1\frac{\varepsilon_i}{k_i}(\overline{\rho_i\Phi_i v'^{\alpha}_i v'^{\beta}_i}-\frac{2}{3}\bar{\rho}_i^*\bar{\Phi}_i k_i\delta_{\alpha\beta})\frac{\partial\tilde{v}_i^\beta}{\partial x_\alpha}$$
$$-C'_1\frac{2\varepsilon_i}{3}\bar{\rho}_i^*\bar{\Phi}_i\frac{\partial\tilde{v}_i^\alpha}{\partial x_\alpha}-C_4\frac{\varepsilon_i}{k_i}\overline{\Phi_i v'^{\alpha}_i}\frac{\partial\bar{p}_i^s}{\partial x_\alpha}+C_{ctc}\frac{\varepsilon_i}{k_i}(\overline{F_{is}^{\times\alpha}v_i^\alpha}-\overline{F_{is}^{\times\alpha}}\tilde{v}_i^\alpha)-C_2\bar{\rho}_i^*\bar{\Phi}_i\frac{\varepsilon_i^2}{k_i} \quad [6.24]$$

The usual discussions on the values of the coefficients C_1–C_4, and their dependence as a function of some important parameters, are probably still valid. If the phase in question has a constant density, C'_1, C_3 and C_4 do not play a part, the term they are coupled with is 0. The coefficient C_{ctc} should be studied especially and has been indeed (again, see [SIM 00]).

For the second question, we can still note that the term due to contact forces already displays the particle sizes, and it is proportional to the mean interface area by unit volume $\bar{\sigma}_i^s$. Moreover, as this size is not fixed but is also a function of the fluctuating flow, which undergoes phenomena already taken into account in the equation and in that of $\bar{\sigma}_i^s$, it does not seem necessary to make it appear explicitly in other terms.

– *For the case of a phase of solid particles,* the situation is different. In fact, the study of a medium composed uniquely of solid spherical particles moving disorderedly in the vacuum, without gravity, known as "granular gas", was a traditional theoretical exercise for several decades, with the aim of generalizing the kinetic theory of gases of Chapman and Cowling ([CHA 60], cited in Chapter 1), for large and numerous particles and with non-elastic collisions. Such a medium often serves as a theoretical reference for granular media. These studies led to a representation of this granular gas as a continuous medium on average, with

Eulerian balance equations for mass, momentum and kinetic energy of agitation. We will refer, for example, to [LUN 84], and to [AND 11, Chapter 5].

In the balance equation of the momentum of the granular gas, we find a tensor analogous to $-\overline{\rho v'^{\alpha} v'^{\beta}}$, and theory shows (with some hypothesis on the probability function of meeting of a pair of particles) that it can be expressed by a formula analogous to that of Boussinesq [6.2]; so it still justifies the definition of a coefficient of effective viscosity μ_e. Theory finds $\mu_e = \rho_p \lambda(\overline{\Phi}_p) k^{1/2} d = \overline{\rho} \lambda(\overline{\Phi}_p) k^{1/2} d / \overline{\Phi}_p$, where d is the diameter of the particles. The balance of the kinetic energy of agitation (related to the "granular temperature") gives an equation completely similar to that found for k_i, equation [6.19b], except that there are neither pressure terms nor terms of contact with another phase. In this equation, the dissipation term is given by $\overline{\rho}\varepsilon = \rho_p \frac{12}{\sqrt{\pi}}(1-e^2)\overline{\Phi}_p^2 g_0(\overline{\Phi}_p) k^{3/2} / d$ (where e is the coefficient of restitution of the collisions, slightly less than 1), analogous formula to $\overline{\rho}\varepsilon = \overline{\rho} C_D \frac{k^{3/2}}{l_t}$.

Therefore, we find formulas very close to those discussed here, but with an essential difference: the length scale characteristic in this approach is simply related to the diameter of the grains. In fact, we still see two length scales, one for the effective viscosity coefficient: $l_P = \lambda(\overline{\Phi}_p) d$, and one for the dissipation: $l_D = d / (\overline{\Phi}_p^2 g_0(\overline{\Phi}_p))$. The two functions λ and g_0 of the volume fraction of the grains increase quite strongly with it, tending toward infinity if $\overline{\Phi}_p$ tends toward its maximal value $\overline{\Phi}_{Max}$ ($\simeq 0.64$). The length scale l_P becomes larger than the diameter of the particles when the volume fraction increases. For the dissipation, as it is natural, it is the distance between the particles that plays a role, l_D becomes smaller and smaller if the volume fraction increases. It is necessary to note, however, that the approach of the maximal volume fraction makes the probability of the lasting contacts between the particles that are not considered increase, and the theory is surely insufficient for volume fractions that are probably weaker.

The equation for k_i found here, [6.19b], is more general than that of the granular temperature but less precise, it should be completed with hypotheses, in particular concerning the implied length scales. If we complete it by relating the scales to the diameter of the particles as is done in the theory of granular gases, this theory is recovered and extended including the effect of the other phases.

We do not see in the theory of the granular gas the intervention of the geometry of the flow on this scale because this theory considers an infinite domain, and the size of the grains is, therefore, the only physical length characteristic of the problem. What interests us here, more greatly for a turbulent flow, concerns a field of mean flow that can contain instabilities in a nonlinear regime of saturation, that explain the presence of an entire range of length scales. We know by experience, at least in the single-phase turbulent flows, that the geometry of the flow plays on these scales, so that this geometry should be taken into account in the models. The granular gas equations can make instabilities appear in certain flow situations, [LUN 84] gives an example of this, but the nonlinear situations have not been studied. The objective of the modeling that we seek is to take into account the mean effect of the fluctuations that these instabilities involve; in a way that can represent these diverse situations with sufficient realism by their geometries and their boundary conditions, and it is necessary for us to go beyond the theory of granular gas in order to attain this aim.

The treatment of this problem by the usual approach of turbulent flows consists of calculating the integral length scale of fluctuations either in an algebraic way, or via a balance equation of a quantity, which is related to it. It is not useful to seek a balance equation of the kinetic energy of the instabilities because [6.19b] can contain all fluctuations. To find the length scales, we have used above, for fluid phases, the dissipation rate of the kinetic energy of the fluctuations. By continuing in the same way, we are led to consider a model equation for ε_p completely similar to [6.24], but without the pressure term. The problem now is to introduce the property of revealing the link with the diameters of the particles that compose the phase. The term due to contact forces between phases naturally contains this diameter. Here, we can write the contact force with the Stokes formulation. For example, the form [6.20] $\overline{F_{sp}^{\times\alpha} v'^{\alpha}_{p}} = -\dfrac{\rho_p \bar{\Phi}_p}{\tau_{Sp}}(\overline{v_p^{\alpha} v'^{\alpha}_{p}} - \overline{v_f^{\alpha} v'^{\alpha}_{p}})$, where $\dfrac{1}{\tau_{Sp}} = \dfrac{9}{2}\dfrac{\rho_f}{\rho_p}\dfrac{\nu_f}{r_p^2}$, and we see the radius of particles appear.

Indeed, in the case where the phase of the particles is essentially shaken by the turbulence of the fluid, the exchange term between phases, which is then a production term of fluctuations, is the most important of all the production terms for ε_p. If this term is large, there is the possibility of a local equilibrium between this term and that of dissipation, which would give:

$$-\frac{9\rho_f \nu_f \bar{\Phi}_p}{2r_p^{\,2}}(\overline{v_p^{\alpha} v'^{\alpha}_{p}} - \overline{v_f^{\alpha} v'^{\alpha}_{p}}) = C_2 \rho_p \bar{\Phi}_p \frac{\varepsilon_p}{k_p}\varepsilon_p = C_2 \rho_p \bar{\Phi}_p \frac{k_p^2}{l_{tp}^{\,2}} \qquad [6.25]$$

We would deduce from it that the integral scale of turbulence of the particulate medium is directly related to the size of these particles. But an analogous reasoning could be done for the equation of kinetic energy itself, since the implied terms appear there in the same way, and the result would be the same but with the number 1 instead of C_2. In order that the two relations can be compatible, we arrive at the conclusion that C_2 should be a function of l_{tp} / r_p.

It still remains to find the right formulation of C_2, function on the one hand of l_{tp} / r_p, probably with a weighting of the volume fraction of the particles, and also eventually of $\rho_p / \bar{\rho}$ or ρ_p / ρ_f.

For the phase of particles, we have just considered a "shaking equilibrium", in the case where the principal source of it is the agitation of the fluid. Such an equilibrium cannot be invoked simultaneously for the fluid, which should see its velocity fluctuations generated by another phenomenon, its mean velocity gradient the most often, and its scale, therefore, will be very different. Another situation is that where there are velocity fluctuations in the fluid, which could be quasi-exclusively generated by these particles. In this case, the balance previously considered would occur on the fluid, and then, this would be the integral scale of the fluctuations of velocity of the fluid, which would be directly related to the size of the particles, since the same term due to the contact forces can be found also in the balance equation of the dissipation rate of the kinetic energy of the fluid fluctuations.

The k-ε models for a multifluid representation have often been used. In general, for the fluid-carrying phase, the additional production terms due to the effect of the other phase have been considered with a constant value of C_2 (see, for example, [ELG 94, SIM 00]), this for very diluted media. For the case of modeling "fluidized beds", the use for the solid phase of a direct relation between the dissipation length scale and the size of the grains is most often chosen. It could be justified because, as we have said previously, there is a sort of equilibrium, in the equation of the dissipation rate, between the production by friction with turbulent fluid and the dissipation (by collisions between particles). This hypothesis remains to be assessed by further studies of turbulence flows highly loaded with solid particles.

6.2.4. *Modeling of the Reynolds tensor for every phase*

We can also use a model where all the components of the generalized Reynolds tensors are calculated by balance equations. This makes it possible to be freed from the formulation of Boussinesq (see section 6.2.1) and to be able to take into account in a much more physical way an anisotropy of velocity fluctuations. This anisotropy was already noted, especially for the particulate phase (see [SIM 00]). The adequate length scale for each phase should also be given by a separate equation, as was done previously.

If we follow the same approach as for the kinetic energy, but without contracting the product, we can find one equation for the tensor $\overline{\rho}_i^* \overline{\Phi}_i R_i^{\beta\gamma} = \overline{\rho_i \Phi_i v'^{\beta}_i v'^{\gamma}_i}$. Without detailing the contact force between the phases, as [6.19a], we obtain:

$$\begin{aligned}
&\frac{\partial}{\partial t}(\overline{\rho}_i^* \overline{\Phi}_i R_i^{\beta\gamma}) + \frac{\partial}{\partial x_\alpha}(\overline{\rho}_i^* \overline{\Phi}_i \tilde{v}_i^{\alpha} R_i^{\beta\gamma}) = \frac{\partial}{\partial x_\alpha}(-\overline{\rho_i \Phi_i v'^{\beta}_i v'^{\gamma}_i v'^{\alpha}_i} - \overline{p_i \Phi_i v'^{\beta}_i}\delta^{\alpha\gamma} - \overline{p_i \Phi_i v'^{\gamma}_i}\delta^{\alpha\beta} + \overline{\Phi_i v'^{\beta}_i \tau_i^{\alpha\gamma}} + \overline{\Phi_i v'^{\gamma}_i \tau_i^{\alpha\beta}}) \\
&-\overline{\rho}_i^* (\overline{\Phi_i v'^{\gamma}_i v'^{\delta}_i}\frac{\partial \tilde{v}_i^{\beta}}{\partial x_\delta} + \overline{\Phi_i v'^{\beta}_i v'^{\delta}_i}\frac{\partial \tilde{v}_i^{\gamma}}{\partial x_\delta}) + (\overline{\Phi_i p_i \frac{\partial v'^{\beta}_i}{\partial x_\gamma}} + \overline{\Phi_i p_i \frac{\partial v'^{\gamma}_i}{\partial x_\beta}}) \\
&-(\overline{\Phi_i \tau_i^{\alpha\gamma}\frac{\partial v'^{\beta}_i}{\partial x_\alpha}} + \overline{\Phi_i \tau_i^{\alpha\beta}\frac{\partial v'^{\gamma}_i}{\partial x_\alpha}}) + \overline{(m_i v_i^{\beta} v_i^{\gamma} - m_i \tilde{v}_i^{\beta}\tilde{v}_i^{\gamma})}^s \overline{\sigma}_i^s + (\overline{\sigma_i^{\alpha\gamma} n_i^{\alpha} v'^{\beta}_i}^s + \overline{\sigma_i^{\alpha\beta} n_i^{\alpha} v'^{\gamma}_i}^s)\overline{\sigma}_i^s
\end{aligned} \quad [6.26]$$

Using $\overline{p}_i^s$, as in [6.19b], we obtain:

$$\begin{aligned}
&\frac{\partial}{\partial t}(\overline{\rho}_i^* \overline{\Phi}_i R_i^{\beta\gamma}) + \frac{\partial}{\partial x_\alpha}(\overline{\rho}_i^* \overline{\Phi}_i \tilde{v}_i^{\alpha} R_i^{\beta\gamma}) = \frac{\partial}{\partial x_\alpha}(-\overline{\rho_i \Phi_i v'^{\beta}_i v'^{\gamma}_i v'^{\alpha}_i} - \overline{p_i \Phi_i v'^{\beta}_i}\delta^{\alpha\gamma} - \overline{p_i \Phi_i v'^{\gamma}_i}\delta^{\alpha\beta} + \overline{\Phi_i v'^{\beta}_i \tau_i^{\alpha\gamma}} + \overline{\Phi_i v'^{\gamma}_i \tau_i^{\alpha\beta}}) \\
&-\overline{\rho}_i^* \overline{\Phi}_i (\overline{v'^{\gamma}_i v'^{\delta}_i}\frac{\partial \tilde{v}_i^{\beta}}{\partial x_\delta} + \overline{v'^{\beta}_i v'^{\delta}_i}\frac{\partial \tilde{v}_i^{\gamma}}{\partial x_\delta}) \\
&+(\overline{\Phi_i p''_i \frac{\partial v'^{\beta}_i}{\partial x_\gamma}} + \overline{\Phi_i p''_i \frac{\partial v'^{\gamma}_i}{\partial x_\beta}}) - \overline{\Phi_i v'^{\gamma}_i}\frac{\partial \overline{p}_i^s}{\partial x_\beta} - \overline{\Phi_i v'^{\beta}_i}\frac{\partial \overline{p}_i^s}{\partial x_\gamma} - (\overline{\Phi_i \tau_i^{\alpha\gamma}\frac{\partial v'^{\beta}_i}{\partial x_\alpha}} + \overline{\Phi_i \tau_i^{\alpha\beta}\frac{\partial v'^{\gamma}_i}{\partial x_\alpha}}) \\
&+\overline{(m_i v_i^{\beta} v_i^{\gamma} - m_i \tilde{v}_i^{\beta}\tilde{v}_i^{\gamma})}^s \overline{\sigma}_i^s + (\overline{\tau_i^{\times\alpha\gamma} n_i^{\alpha} v'^{\beta}_i}^s + \overline{\tau_i^{\times\alpha\beta} n_i^{\alpha} v'^{\gamma}_i}^s)\overline{\sigma}_i^s
\end{aligned} \quad [6.27]$$

This equation is similar to the equation of the Reynolds tensor for turbulence in a fluid, with additional terms due to the exchanges at the interfaces.

The problem of mean pressures is still to be discussed, and the discussions have already remained applicable: more often, we have supposed $\overline{p}_i^s = \overline{p}_i^*$. The modeling of usual terms can still be done in a standard way that is the case when every phase is at constant density. We will refer to [SIM 00].

The dissipation term, $(\overline{\Phi_i \tau_i^{\alpha\gamma}\frac{\partial v'^{\beta}_i}{\partial x_\alpha}} + \overline{\Phi_i \tau_i^{\alpha\beta}\frac{\partial v'^{\gamma}_i}{\partial x_\alpha}}) = \overline{\rho}_i^* \overline{\Phi}_i \varepsilon_i^{\beta\gamma}$, is usually assumed to be isotropic; therefore, $\varepsilon_i^{\beta\gamma} = \varepsilon_i \delta^{\beta\gamma}$, and the dissipation rate ε_i is given by its balance equation or by specifying a characteristic length, as we have seen. The fluctuating pressure terms $\bar{\Phi}_i \Pi_i^{\beta\gamma} = (\overline{\Phi_i p''_i \frac{\partial v'^{\beta}_i}{\partial x_\gamma}} + \overline{\Phi_i p''_i \frac{\partial v'^{\gamma}_i}{\partial x_\beta}})$, if the phase is an incompressible fluid, are supposed to involve the return to isotropy of the fluctuations of velocity, and so of $R_i^{\beta\gamma}$. We usually write that $\Pi_i^{\beta\gamma} = \Pi_{i1}^{\beta\gamma} + \Pi_{i2}^{\beta\gamma}$, with a "slow" part, which follows the characteristic time k_i / ε_i, and a "fast" part, which follows a characteristic time related to the gradient of the mean velocity. If the phase in question is not incompressible, the fact that these pressure terms play as a redistribution effect is in general preserved (except for flows with quite an elevated Mach number, as we have seen previously for global modeling), but we can consider that the fast return to isotropy can also be influenced by a third characteristic time, which is related to the mean pressure, because the terms that contain it in [6.22] are no longer zero. We will consult [DEM 07] on this subject.

To extend this traditional model, recent works [BEI 07, MAS 02] have envisaged in the same way the rapid redistribution term related to a characteristic exchange of momentum with other phases. This seems to slightly improve the agreement with experimental results.

When we consider a phase constituted of dispersed solid particles, the form of the balance equation of $R_i^{\beta\gamma}$ does not change, but it is necessary to re-examine its modeling. The dissipation of kinetic energy of agitation of the particles is done here during non-elastic collisions between particles, and the (elastic) fluctuations of pressure are still here due to collisions between particles. This does not obviously change the fact that this creates an isotropization of velocity fluctuations, nor the intervention of characteristic times that we have discussed; a more detailed study is necessary in order to clarify this, and to also give numerical constants that are necessary to the model. For diluted media, the works of O. Simonin *et al.* have also considered this type of model, with some success (see [SIM 00]). We will discuss that for granular media as well in the last chapter.

With regard to terms of exchange of this equation, that of mass exchange is still very often negligible. The term due to the friction is a simple generalization of that which appears in the equation of kinetic energy of turbulence of phase *i*, and the discussions that have been had on this subject are qualitatively transposable here.

Even by using the balance equation of $R_i^{\beta\gamma}$, we should, anyway, use an equation to find a characteristic length of fluctuations, and previous discussions on this subject remain valuable.

Chapter 7

Modeling the Mean Gas–Liquid Interface Area per Unit Volume

7.1. Introduction

When the multiphase medium contains particles of varying shapes and sizes – which is most often the case – or when a gas–liquid medium contains either bubbles or drops, and also bubbles inside drops, drops in bubbles and a very spread out interface that is in contact with the walls between which the medium flows, it is no longer possible to define precise categories of particles nor the number of particles in a category. The notion of a "number of particles" is not able to describe the distribution of phases in the medium. Nevertheless, the mean interface area per unit volume $\overline{\sigma}^s_{ii'}$ remains perfectly defined, and as previously mentioned, it is this quantity that plays a crucial role during the exchange of mass and heat between phases. For the exchange of momentum, the particular shape of each inclusion is theoretically important, but if this shape varies only relatively slightly, again the interface area matters more. Whether $\overline{\sigma}^s_{ii'}$ can be calculated using a model equation that can be applied to various different situations of multiphase flow is, therefore, a potentially interesting question. This importance has been known for over 20 years (see [MOR 07]), but practical application has only recently been attempted for the case of a liquid jet that was atomized into fine droplets [VAL 01]. Extension to the case where the drops would be vaporized has also been suggested; generalization of the concept to the case of bubbles seems possible, as well as for all suspensions containing immiscible liquids. This chapter aims to describe the work done on this subject. Flows where aerosol particles of various shapes are agglomerated can, in principle, also be studied this way; however, the phenomena of agglomeration being entirely different, all the work would need to be redone.

7.2. Initial equation for the mean interface area per unit volume

It has already been mentioned in Chapter 5, section 5.2, that a balance equation for $\bar{\sigma}_i^s$ can be established using the instantaneous equation [2.15a], derived from the equation giving the corresponding phase index. Thus, the unclosed form of this equation can be written as:

$$\frac{\partial \bar{\sigma}_i^s}{\partial t} + \frac{\partial}{\partial x^\alpha}(\bar{\sigma}_i^s \bar{v}_s^\alpha) = +\overline{\sigma_i^s (\delta^{\alpha\beta} - n_i^\alpha n_i^\beta)\frac{\partial v_s^\alpha}{\partial x^\beta}} \qquad [7.1a]$$

It could also be written as:

$$\frac{\partial \bar{\sigma}_i^s}{\partial t} + \frac{\partial}{\partial x^\alpha}(\bar{\sigma}_i^s \tilde{v}_i^\alpha) = \frac{\partial}{\partial x^\alpha}(-\overline{\sigma_i^s v'^\alpha_{is}}) + \overline{\sigma_i^s (\delta^{\alpha\beta} - n_i^\alpha n_i^\beta)\frac{\partial v_s^\alpha}{\partial x^\beta}} \qquad [7.1b]$$

if we define $v'^\alpha_{si} = v_s^\alpha - \tilde{v}_i^\alpha$, or even:

$$\frac{\partial \bar{\sigma}_i^s}{\partial t} + \frac{\partial}{\partial x^\alpha}(\bar{\sigma}_i^s \tilde{v}^\alpha) = \frac{\partial}{\partial x^\alpha}(-\overline{\sigma_i^s v'^\alpha_{s}}) + \overline{\sigma_i^s (\delta^{\alpha\beta} - n_i^\alpha n_i^\beta)\frac{\partial v_s^\alpha}{\partial x^\beta}} \qquad [7.1c]$$

with $v'^\alpha_s = v_s^\alpha - \tilde{v}^\alpha$.

Indeed, while the velocity of the interface is always equal to the velocity of the two touching phases, the statistical average at a given point of the velocity of the interface is usually not the mean velocity of phase i or of the other phase, since at this point, there is flow velocity also when the interface is not present. As discussed in the beginning of Chapter 5, it is only when phase p consists of identical solid particles, all of the same radius, that the mean velocity of the interface is equal to the mean velocity of the particles, $\bar{v}_s^\alpha = \tilde{v}_p^\alpha$. This equality can still be considered applicable in the case of liquid droplets with a very small volume fraction and with very small volumes, in a carrier medium, as long as the shape and size of the droplets vary little during their movement, avoiding fracture and coalescence phenomena.

In the literature, some simple models of two-phase flow state that the mean velocity of the interface is some weighted arithmetical average of the mean velocities of the two interacting phases. In Chapter 5, section 5.3.3, a certain justification of this statement was observed by studying the exchange of momentum at a slightly curved liquid–gas interface. Recall what we have said at the limit of a highly dilute medium. This relation can be used if it is possible to calculate the mean velocity of each phase. Another modeling uses the definition of a "dispersion by velocity fluctuations", as was briefly discussed in Chapter 5. Either the fluctuations with respect to the mean velocity of phase i are considered, as in equation [7.1b], which allows us to write: $-\overline{\sigma_i^s v'^{\alpha}_{si}} = D_{si}\dfrac{\partial \bar{\sigma}_i^s}{\partial x_\alpha}$, where D_{si} is the diffusivity in phase i, or the fluctuations with respect to the mean velocity of the global medium are considered, as in equation [7.1c], which then allows us to write: $-\overline{\sigma_i^s v'^{\alpha}_{s}} = D'_{si}\dfrac{\partial \bar{\sigma}_i^s}{\partial x_\alpha}$, where D'_{si} is the global diffusivity. These terms are clearly dispersion terms, due to the fact that the definition of the mean medium (either of the phase or the global medium) loses information about the exact position of the interface. Chapter 6 discussed methods to calculate similar diffusivities for quantities other than the mean interface area, as a function of the kinetic energy of the fluctuations, as well as their length scale, either in phase i or in the global medium.

The last term of equation [7.1a] or [7.1b] represents, in a very condensed manner, both the mean production and the mean destruction of the interface caused by its fluctuating movements. When studying situations where phase i is dispersed in the form of a given number of inclusions of relatively constant shape (meaning that there is no fracturing or coalescence of these inclusions), then this term is zero and the balance equation of the mean interface area is a simple convection-diffusion equation. Building a mathematical model for this dispersion is the only problem, yet relatively simple. These conditions are encountered in many situations, including that of fluidized beds.

There are also a number of situations where phenomena of coalescence and fracturing of interfaces are important. In this case, the modeling of this latter term has to be studied. To start with this modeling, the effects of the mean values and correlations of fluctuations can be made apparent in [7.1c]:

$$\frac{\partial \bar{\sigma}_i^s}{\partial t} + \frac{\partial}{\partial x^\alpha}(\bar{\sigma}_i^s \bar{v}_s^\alpha) = +\bar{\sigma}_i^s(\delta^{\alpha\beta} - \overline{n_i^\alpha n_i^\beta}^s)\frac{\partial \tilde{v}^\alpha}{\partial x^\beta} + \overline{\sigma_i^s(\delta^{\alpha\beta} - n_i^\alpha n_i^\beta)\frac{\partial v'^{\alpha}_s}{\partial x^\beta}} \qquad [7.2]$$

In order to model the first term on the right-hand side of the above equation, we have to study the mean anisotropy tensor of the interface $\overline{q_i^{\alpha\beta}} = \overline{\sigma_i^s (n_i^\alpha n_i^\beta - \delta^{\alpha\beta})} = \overline{\sigma}_i^s (\overline{n_i^\alpha n_i^\beta}^s - \delta^{\alpha\beta})$, which is dependent on the velocity field and the surface tension, as both of these influence the normal vectors. The way to link this tensor to the strain tensor of the velocity field, in a case with no turbulence, where $\overline{q_i^{\alpha\beta}} = \overline{\sigma}_i^s (\overline{n_i^\alpha}^s \, \overline{n_i^\beta}^s - \delta^{\alpha\beta})$, is discussed by Morel [MOR 07] and is intricately dependent on the structure of the flow.

It can also be rewritten as:

$$\begin{aligned} &\frac{\partial \overline{\sigma}_i^s}{\partial t} + \frac{\partial}{\partial x^\alpha}(\overline{\sigma}_i^s \overline{v}_s^\alpha) = +\overline{\sigma}_i^s (\delta^{\alpha\beta} - \overline{n_i^\alpha}^s \, \overline{n_i^\beta}^s) \frac{\partial \tilde{v}^\alpha}{\partial x^\beta} - \overline{\sigma}_i^s \overline{n'^\alpha_i n'^\beta_i}^s \frac{\partial \tilde{v}^\alpha}{\partial x^\beta} \\ &\overline{+\sigma_i^s (\delta^{\alpha\beta} - n_i^\alpha n_i^\beta) \frac{\partial v'^\alpha_s}{\partial x^\beta}} \end{aligned} \qquad [7.2a]$$

The first term is the non-turbulent contribution, where the only goal of statistical averages is to take account of the random nature of the entrance or initial conditions of the multiphase flow. Following the definition of the interface area per unit volume: $-n_i^\alpha \sigma_i^s = \dfrac{\partial \Phi_i}{\partial x_\alpha}$, the mean of the normal directly results from $\overline{n_i^\alpha}^s \, \overline{\sigma}_i^s = -\dfrac{\partial \overline{\Phi}_i}{\partial x_\alpha}$. This means that it can be calculated at any point from the equation of the volume fraction of i and that of $\overline{\sigma}_i^s$ itself. The two other terms in [7.2a] take account of the possible turbulence in one of the two media or in both media.

In any case, it must be remembered that this equation must be completed to be able to take into consideration the destruction of the interface due to disappearance of a too thin sheet or ligament of a phase into another, and this will be an important element of the model.

Note that this equation is very general, applicable to gas, liquid and even solid phases. In fact, the differences between these different cases become apparent in the momentum balance equation, which allows the calculation of the velocity field of the phases. Some interesting studies regarding these quantities are also presented in [OES 06].

7.3. Model of the mean interface area during the "atomization" of a liquid jet

This model was proposed by Vallet *et al.* [VAL 01] for the atomization of a liquid jet in a gas medium, either at rest or at motion. There are, therefore, only two phases and no noteworthy vaporization phenomena; the velocity of the flow is large, which results in the presence of turbulence, at least in the area close to the injection duct that produces the jet of liquid in the gas medium.

In these conditions, the dispersion of the liquid in the medium has been represented as a phenomenon of turbulent mixing, with a turbulent liquid dispersion coefficient that can be calculated using the global turbulence kinetic energy of the medium. This approach is fully in accordance with the discussions in the previous section and Chapter 5.

The mathematical model, however, is established slightly differently, by using the mean interface area per unit mass of the medium, and not per unit volume, then:

$$\overline{\sigma_i^s v'^{\alpha}_s} = -\bar{\rho} D_{ts} \frac{\partial}{\partial x^{\alpha}} (\bar{\sigma}_i^s / \bar{\rho})$$

An interface's effective diffusion coefficient must be calculated – just like the turbulent diffusion coefficient of liquid – by taking into account the modeled characteristics of the fluctuations. Its dependency on the size of the liquid parcels or their volume fraction has not been considered. It has been suggested that the size was small enough to no longer play a role, because of the high Weber numbers. Considering the volume fraction of the liquid, the appearance of the mean density as the first factor, which takes into account this volume fraction, was assumed to be sufficient.

Modeling of the effects of interface creation and destruction is done here without considering the non-turbulent term. It is self-evident that this model is meant to take into account all the effects of the surface tension, even those that are not included in the initial equation, as mentioned in Chapter 2.

The effects of interface creation are considered as being generated by the stretching caused by the flow itself; this is always proportional to the interface area already existing. The stretching is partly due to the mean flow and partly due to the turbulent fluctuations. The effect of the stretching due to the mean velocity field can be found in the second term to be modeled in equation [7.2a]: $-\bar{\sigma}_i^s \overline{n'^{\alpha}_i n'^{\beta}_i} \frac{\partial \tilde{v}^{\alpha}}{\partial x^{\beta}}$ where the orientation tensor of the interface $\overline{n'^{\alpha}_i n'^{\beta}_i} \approx \overline{n^{\alpha}_i n^{\beta}_i}$ appears. Since the orientations of the normal vectors of the interface are directly linked to the flow turbulence, it can be assumed that $\overline{n'^{\alpha}_i n'^{\beta}_i} \approx \overline{n^{\alpha}_i n^{\beta}_i} = -A \frac{\overline{\rho v'^{\alpha} v'^{\beta}}}{\bar{\rho} k}$, in such a way that the creation term can be modeled by $A \bar{\sigma}_i^s \frac{\overline{\rho v'^{\alpha} v'^{\beta}}}{\bar{\rho} k} \frac{\partial \tilde{v}^{\alpha}}{\partial x^{\beta}}$, with A being a constant to be fixed through experimentation.

Stretching due to turbulence, which is included in the third term, is represented by utilizing the reciprocal of a time constant instead of the mean velocity gradient. It is modeled as $\frac{\overline{\sigma}_i^s}{\tau_P}$, and the time constant must be chosen in relation to the physical phenomena that may be responsible for the appearance of the interface.

Effects of destruction of an interface, whether they are due to the surface tension that avoids too strong curvature to the interface or due to annihilation by the interface interaction during coalescence of droplets or disappeareance of too thin ligaments (not covered by equation [7.2a] as mentioned previously), must at least be considered in a last term. With the goal of ultimately achieving a stable equilibrium between the two effects of creation and destruction, the term of interface destruction has been assumed to be proportional to the square of the mean interface area per unit volume itself, whereas the creation term was linear. In these conditions, this term can be written as $-\frac{\overline{\sigma}_i^s}{\sigma_{eq}}\frac{\overline{\sigma}_i^s}{\tau_P}$, with the value σ_{eq} being nothing but the value of the mean interface area per unit volume that would result from equilibrium, which must be more precisely defined by taking into account the physical phenomena directly involved.

The model can, therefore, be written in the general form:

$$\frac{\partial\overline{\sigma}_i^s}{\partial t}+\frac{\partial}{\partial x^\alpha}(\overline{\sigma}_i^s\tilde{v}^\alpha)=\frac{\partial}{\partial x^\alpha}(\overline{\rho}D_{ts}\frac{\partial}{\partial x^\alpha}(\overline{\sigma}_i^s/\overline{\rho}))+A\overline{\sigma}_i^s\frac{\overline{\rho v'^\alpha v'^\beta}}{\overline{\rho}k}\frac{\partial\tilde{v}^\alpha}{\partial x^\beta}+\frac{\overline{\sigma}_i^s}{\tau_P}(1-\frac{\overline{\sigma}_i^s}{\sigma_{eq}}) \quad [7.3]$$

The quantities τ_P and σ_{eq} still need to be defined.

The second of these quantities represents the mean interface area, per unit volume of the medium, toward which the medium equilibrium will tend, as a result of all the physical phenomena involved: collisions between bubbles and/or between liquid pockets, stretching and friction between liquid pockets and the surrounding gas or between gas bubbles and the surrounding liquid. This interface area can be associated with the radius of a droplet or bubble. In a gas–liquid medium with low volume fractions of liquid (which is what would be the case ultimately during the atomization of a liquid jet), with a given mass of liquid per unit volume, it would give: $\sigma_{eq}=\frac{3\overline{\Phi}_l}{r_{eq}}=\frac{3\overline{\rho}\tilde{\Phi}_l}{\rho_l r_{eq}}$. The final equilibrium between the phenomena of surface tension and the turbulent shaking of the gas and the liquid, without going into too much detail, will occur when a given value is reached by a global turbulent Weber number, defined using a kinetic energy scale and a density scale for shaking:

$\frac{\rho^* k^* r_{eq}}{\eta_c} = W_{eq}$. The quantity η is the surface tension constant of the interface in question. An estimation of the characteristic kinetic energy k^* can be more or less precise, but is always proportional to the kinetic energy of the fluctuations in the case of a global two-phase model. It is also worth noting that the stretching of droplets by the surrounding turbulent gas is only affected by the fraction of the kinetic energy belonging to a scale that is smaller than the droplets themselves; the bigger scales are only used to transport the droplets. The density ρ^* that is to be taken into account then is that of the liquid only if $\rho_l \gg \rho_g$. However, collisions between droplets are only affected by scales ranging between the size of the droplets and the mean separation between the droplets (which is $(\frac{12\pi}{27})^{1/3} \frac{r_{eq}}{\bar{\Phi}_l^{1/3}}$), and the characteristic density is that of the liquid.

The time constant τ_P is a mean time for the different phenomena involved, always linked to the time constant of the global medium's turbulence τ_t, but with a proportionality coefficient that can also be dependent on the fact that different length scales must also be taken into account. Slightly different estimations of these time and radius of equilibrium have been used, on the basis of the previous discussion, in various works (see [VAL 01, BLO 02, BEH 07, KAD 08]).

These works, through comparison with experimental data, have resulted in convenient values for the constants of the model. The general trends that have been observed experimentally are well reproduced by the model. Nevertheless, the measurements of the sizes and volume fraction of the droplets and liquid parcels, and even the measurements of velocity and velocity fluctuations, in dense two-phase media are not easily performed. There are very few well-documented experimental data sets for this matter. Good knowledge of the characteristics of gas and liquid flows inside an atomization device is important also, and this knowledge is difficult to achieve by experiments. However, the model does have the advantage of also being able to calculate the flows inside the atomization device itself. An example can be found in the work of [DE 09]. Nevertheless, this is often associated with important numerical calculation difficulties. Work using this model must continue to establish the range of its validity.

It would be interesting to adapt this model of the interface area to a multifluid representation of the turbulent two-phase medium. The main problem is finding out how the fluctuating stresses imposed on the interface can be calculated from the information on the turbulence of the two phases.

7.4. Effects of vaporization on the interface area

In the presence of vaporization, we should expect the balance equation of the interface area to include an additional sink term, resulting in a decrease in surface area, $\dot{W}^{\Sigma}_{vap}$. Calculating this term is very simple in the case where the liquid phase is present in the form of identical, spherical droplets.

If the liquid phase is present in the form of droplets, it has already been studied in Chapter 5, section 5.6.2, how a droplet is vaporized. The vaporized mass flow rate is still a function of the radius of the droplet and the characteristics of the liquid and surrounding gas, similarly to the formula corresponding to the simple case of an isolated droplet at rest:

$$\frac{d}{dt}m_p = 4\pi r^2 \rho_p \frac{dr}{dt} = -4\pi r^2 \dot{m} = -4\pi r \rho_g D_V Ln\left(\frac{1-Y_{V\infty}}{1-Y_{Vsat}(T_s)}\right)$$

From this, a rate of disappearance of the interface area can also be deduced:

$$\frac{d}{dt}(4\pi r^2) = 8\pi r \frac{dr}{dt} = -8\pi \frac{\rho_g}{\rho_p} D_V Ln\left(\frac{1-Y_{V\infty}}{1-Y_{Vsat}(T_s)}\right)$$

In an homogeneous medium containing C^N droplets per unit volume, and assuming that all the droplets are identical, the disappearance rate of interface area, per unit volume, can be calculated as:

$$\dot{W}^{\Sigma}_{vap} = -8\pi C^N \frac{\rho_g}{\rho_p} D_V Ln\left(\frac{1-Y_{V\infty}}{1-Y_{Vsat}(T_s)}\right) = -\frac{9}{2}\frac{(\bar{\sigma}^s_p)^3}{(\bar{\Phi}_p)^2}\frac{\rho_g}{\rho_p} D_V Ln\left(\frac{1-Y_{V\infty}}{1-Y_{Vsat}(T_s)}\right)$$

If the liquid phase is in the form of plane layers, as in Chapter 5, section 5.6.2, the disappearance of the interface area is not continuous, but occurs suddenly when the sheet of liquid is entirely vaporized, and the vaporization time of a liquid sheet of thickness 2H can be calculated. As the medium consists of sheets of different thicknesses, the rate of disappearance of mean interface area can be approximated as the ratio between the mean thickness of sheets and its consumption time. This calculation gives the following:

$$\dot{W}^{\Sigma}_{vap} = -2\frac{(\bar{\sigma}^s_p)^3}{\bar{\Phi}_g \bar{\Phi}_p}\frac{\rho_g}{\rho_p} D_V Ln\left(\frac{1-Y_V(H)}{1-Y_{Vsat}(T_s)}\right)$$

These two extreme geometrical situations give very similar formulas. This rate of interface disappearance could, therefore, be represented in any case as:

$$\dot{W}_{vap}^{\Sigma} = -C_{\Sigma}(\overline{\Phi}_p)\frac{(\overline{\sigma}_p^s)^3}{\overline{\Phi}_g\overline{\Phi}_p}\frac{\rho_g}{\rho_p}D_V Ln\left(\frac{1-Y_V(H)}{1-Y_{Vsat}(T_s)}\right)$$

The law $C_{\Sigma}(\overline{\Phi}_p)$ can be empirically adjusted, so as to recover the two previous formulas in the corresponding situations.

Chapter 8

"Large Eddy Simulation" Style Models

8.1. Introduction

It has already been pointed out previously that the particular advantage of "large eddy simulation" (LES) models is that they are able to allow numerical simulations of the time evolution of large geometrical flow structures. These simulations have proven their capabilities for studying the morphology of these structures in the case of classical turbulent flow (i.e. of a liquid or a gas of constant density). A significant practical result was the recovering of the discrete frequencies of low-frequency fluctuations in these flows. Several calculations of this type have already been made for two-phase flows, without precisely justifying the equations or the models used, but with results that seem very interesting indeed. Without going into an in-depth examination of these works, the basis of these simulations will be examined here, along with a discussion of the adequate precautions that must be taken when using the models and techniques from classical turbulent flows in the context of multiphase flows.

It was already mentioned in Chapter 3 that the definition of "filtered" averages is based on a filter kernel that has a characteristic spatial scale (and sometimes a temporal scale); it must allow the definition of mean values that are able to display the "large structures", while smoothing out the fluctuations that are smaller than its spatial scale. A practical example of a filter can quite simply be that which gives the volumetric average; in this case, there is only a characteristic spatial scale $\Delta = (\mathrm{Vol})^{1/3}$ and the temporal scale is equal to zero. With this kind of filter, it would be ideal to be able to say that the fluctuations with (spatial) wavelengths smaller than Δ are smoothed out, and those wavelengths larger than Δ are preserved. Unfortunately, this is not exactly the case, as the dampening of the fluctuations is not equal to zero if the wavelengths are superior to Δ and not very high if the wavelengths are smaller than Δ (fluctuations of wave number $2\pi/\Delta$ are only dampened by roughly 90% and the transfer function has negative parts). To be sure that only the fluctuations of wavelengths inferior to a certain value are filtered, a kernel that employs the Fourier transform must be used – a low-pass filter; the kernel in the physical space has oscillating wings (see [SAG 06]).

It was also seen in Chapter 3 that all Reynolds relationships cannot be satisfied using these filtered averages. Nevertheless, it will be established that the use of filtered averages leads, as with statistical averages, to the presence of additional fluxes in the filtered balance equations, although these additional fluxes are slightly different. This will be demonstrated in next section 8.2. The problem of modeling these additional fluxes arises, in a similar way to the previous case, but with one fundamental difference: here, the models will explicitly have to depend on the characteristic scale of the filter, and the precise choice of the filter will later have to be made to meet the objectives of the study and the geometrical characteristics of the studied flow. Reasoning based on the properties of the turbulent fluctuations as a function of their wavelength will lead to the proposal of this type of model, as a kind of extension of the classical models. It must be noted that the use of such a filter has also been suggested in contexts other than the LES [JAC 00], but the models have always been used as classic models, as in no case they have been dependent on the size of the filter, which was always presumed to be very small in comparison to the various scales of the flow. The basis of the already used LES models will be presented in section 8.3. On this subject, it must be noted that these filters do not take into account the random fluctuations in the boundary conditions nor in the initial conditions between several experiments, but can take into account random fluctuations of these conditions during one experiment. To a certain extent, this can simplify the modeling process, but can complicate the practical usage of the model, especially because the effects of the initial conditions in particular are very often important.

For the use of these models in non-homogenous turbulent media, particularly close to the walls, it quickly becomes apparent that it is necessary to use (spatial) filter sizes that vary in function of the position within the medium. It was observed that this could additionally affect the validity of Reynolds relationships and better-adapted filters to this effect have been suggested (see [SAG 06]). In any case, this complicates the modeling of the additional fluxes, as it must take into account the more varied shapes of the spectrum of the velocity fluctuations. Another way of adapting this method to a non-homogenous medium is therefore to combine an approach that uses the filtered averages with an approach using statistical averages. Among these approaches, the partially integrated turbulence modeling (PITM) method, proposed by Schiestel and Chaouat [CHA 05, CHA 09a], depends on the safest theoretical bases, and is discussed in section 8.5. Before this, the problems raised by the LES model of the mean interface area by unit of volume will be discussed in section 8.4.

8.2. Filtered equations and the nature of the models to be provided

Whatever the filter used, the filtered average will always here be marked by an overbar, even if the end result is dependent on the selected filter, and, in particular, on the characteristic scales of this filter. As long as the characteristic scales of the filter are constant in the studied flow, the first Reynolds relationships are valid, as seen in Chapter 3. These are:

$$\overline{(\lambda g)} = \lambda \overline{g}\,,\ \overline{g+h} = \overline{g} + \overline{h}\,,\ \overline{\left(\frac{\partial g}{\partial x_\alpha}\right)} = \left(\frac{\partial \overline{g}}{\partial x_\alpha}\right), \overline{\left(\frac{\partial g}{\partial t}\right)} = \left(\frac{\partial \overline{g}}{\partial t}\right)$$

where g and h are functions of space and time, and λ is a scalar constant. However, it remains that $\overline{\overline{g}} \neq \overline{g}$ (except, as will be seen later on, for certain special filters). If the fluctuations of g are again defined by $g' = g - \overline{g}$, which is therefore a function of space and time, it is not possible to say that $\overline{g'} = 0$, nor $\overline{(gh)} = \overline{g}\overline{h} + \overline{g'h'}$. It is also possible to define averages weighted by the density, which can be written as $\tilde{g} = \overline{\rho g} / \overline{\rho}$, as well as fluctuations relative to this weighted average.

The previous definition of a multiphase medium included the index of phase to extend the definitions of phase characteristics onto the remaining domain, which is not concerned with this phase. The filtered mean of this indicative function of phase no longer means the probability of the presence of the phase; it only means the volume fraction of the phase if the filter is the filter of volumetric average (and thus the average weighted by density is the mass fraction), but this filtered average is, nevertheless, always defined precisely.

Therefore, it is not a problem to write out the average of the balance instantaneous equations of the mass, momentum and energy, for each phase, in a similar fashion. In the simplest case, it is not even useful to explicitly bring into the equation the fluctuations, which results in the following:

$$\frac{\partial}{\partial t}(\overline{\rho}_i^* \overline{\Phi}_i) + \frac{\partial}{\partial x_\alpha}(\overline{\rho}_i^* \tilde{v}_i^\alpha \overline{\Phi}_i) = -\overline{\rho_i v_{ri}^\alpha n_i^\alpha \sigma_i^s} \qquad [8.1]$$

$$\frac{\partial}{\partial t}\left(\overline{\rho}_i^*\overline{\Phi}_i\tilde{v}_i^\beta\right)+\frac{\partial}{\partial x_\alpha}(\overline{\rho}_i^*\overline{\Phi}_i\tilde{v}_i^\alpha\tilde{v}_i^\beta)=\frac{\partial}{\partial x_\alpha}(\overline{\Phi}_i\overline{\sigma}_i^{*\alpha\beta}-\overline{\rho_i\Phi_i v_i^\alpha v_i^\beta}+\overline{\rho}_i^*\overline{\Phi}_i\tilde{v}_i^\alpha\tilde{v}_i^\beta)+\overline{\rho}_i^*\overline{\Phi}_i g_\beta$$
$$+\overline{\rho_{is}\dot{w}_i^{\sec h}v_i^\beta}+\overline{p}_i^s\frac{\partial\overline{\Phi}_i}{\partial x_\beta}+\overline{F}_{si}^{'\beta} \qquad [8.2]$$

$$\frac{\partial}{\partial t}\left(\overline{\rho}_i^*\overline{\Phi}_i\tilde{h}_{ti}\right)+\frac{\partial}{\partial x_\alpha}(\overline{\rho}_i^*\overline{\Phi}_i\tilde{v}_i^\alpha\tilde{h}_{ti})=\frac{\partial}{\partial x_\alpha}(\overline{\Phi}_i(-\overline{j_{Qi}^\alpha}^*+\overline{\tau_i^{\alpha\beta}v_i^\beta}^*)$$
$$-\overline{\rho_i\Phi_i v_i^\alpha h_{ti}}+\overline{\rho}_i^*\overline{\Phi}_i\tilde{v}_i^\alpha\tilde{h}_{ti})+\overline{\rho}_i^*\overline{\Phi}_i\tilde{v}_i^\beta g_\beta+(-\overline{\rho_{is}\dot{w}_i^{\sec h}h_{ti}}+\overline{\tau_i^{\alpha\beta}n_i^\alpha v_i^\beta\sigma_i^s}+\overline{q_{is}\sigma_i^s})+\overline{\Phi}_i\frac{\partial\overline{p}_i^*}{\partial t} \qquad [8.3]$$

by neglecting the effect of the surface tension forces for [8.3].

The equations can again be used to represent the balance of the filtered averages that are being studied, just like for the statistical averages, as long as it is possible to know the terms of the additional fluxes that here are simply:

$$-\overline{\rho_i\Phi_i v_i^\alpha v_i^\beta}+\overline{\rho}_i^*\overline{\Phi}_i\tilde{v}_i^\alpha\tilde{v}_i^\beta=\overline{\rho}_i^*\overline{\Phi}_i T_{LESi}^{\alpha\beta}$$
$$-\overline{\rho_i\Phi_i v_i^\alpha h_{ti}}+\overline{\rho}_i^*\overline{\Phi}_k\tilde{v}_i^\alpha\tilde{h}_{tk}+\overline{\Phi}_i(\overline{\tau_i^{\alpha\beta}v_i^\beta}^*-\overline{\tau_i^{\alpha\beta}}^*\tilde{v}_i^\beta)=\overline{\rho}_i^*\overline{\Phi}_i F_{LESi}^{H\alpha}$$

It is possible to make explicitly apparent the fluctuations again, which gives:

$$\overline{\rho}_i^*\overline{\Phi}_i T_{LESi}^{\alpha\beta}=-\overline{\rho_i\Phi_i\tilde{v}_i^\alpha\tilde{v}_i^\beta}+\overline{\rho}_i^*\overline{\Phi}_i\tilde{v}_i^\alpha\tilde{v}_i^\beta-\overline{\rho_i\Phi_i v'^\alpha_i\tilde{v}_i^\beta}-\overline{\rho_i\Phi_i\tilde{v}_i^\alpha v'^\beta_i}-\overline{\rho_i\Phi_i v'^\alpha_i v'^\beta_i}$$

and

$$\overline{\rho}_i^*\overline{\Phi}_i F_{LESi}^{H\alpha}=-\overline{\rho_i\Phi_i\tilde{v}_i^\alpha\tilde{h}_{ti}}+\overline{\rho}_i^*\overline{\Phi}_i\tilde{v}_i^\alpha\tilde{h}_{ti}-\overline{\rho_i\Phi_i\tilde{v}_i^\alpha h'_{ti}}-\overline{\rho_i\Phi_i v'^\alpha_i h_{ti}}+\overline{\Phi}_i(\overline{\tau_i^{\alpha\beta}\tilde{v}_i^\beta}^*-\overline{\tau_i^{\alpha\beta}}^*\tilde{v}_i^\beta+\overline{\tau_i^{\alpha\beta}v'^\beta_i}^*)$$

The last term here is familiar, it is due to the correlations between the fluctuations that are eliminated by the filter, just like with the statistical averages; several other terms also reappear that here are not equal to zero. This decomposition is called "Leonard decomposition".

If the last Reynolds relationship was fulfilled, all these first terms would be equal to zero, and the fluxes would be reduced to their classic expression, as in [4.5] and [4.7].

If the densities of the phases are constant (and, therefore, $\overline{\rho}_i^*=\rho_i$), they can be extracted as done in the following. The first two terms only use filtered quantities and therefore can be calculated explicitly, even if the first one does not simply require algebraic operations and depends on the exact shape of the filter. However, the three other terms do contain fluctuations on a smaller scale than that of the filter, which are not known, and must be modeled. They can be distinguished by writing:

$$\rho_i\overline{\Phi}_i T_{LESi}^{\alpha\beta}=-\rho_i\overline{\overline{\Phi}_i\tilde{v}_i^\alpha\tilde{v}_i^\beta}+\rho_i\overline{\Phi}_i\tilde{v}_i^\alpha\tilde{v}_i^\beta+\rho_i\overline{\Phi}_i T_{SGSi}^{\alpha\beta}$$

and

$$\rho_i \bar{\Phi}_i F_{LESi}^{H\alpha} = -\rho_i \overline{\bar{\Phi}_i \tilde{v}_i^\alpha \tilde{h}_{ti}} + \rho_i \bar{\Phi}_i \tilde{v}_i^\alpha \tilde{h}_{ti} + \rho_i \bar{\Phi}_i F_{SGSi}^{H\alpha}$$

In general, densities also can vary and fluctuate, and only their filtered average can be known. Therefore, to observe the effects of their fluctuations and of their mean values, the latter can be made apparent in the equations. We then obtain the following:

$$\bar{\rho}_i^* \bar{\Phi}_i T_{LESi}^{\alpha\beta} = -\overline{\bar{\rho}_i^* \bar{\Phi}_i \tilde{v}_i^\alpha \tilde{v}_i^\beta} + \bar{\rho}_i^* \bar{\Phi}_i \tilde{v}_i^\alpha \tilde{v}_i^\beta - \overline{(\rho_i \Phi_i)' \tilde{v}_i^\alpha \tilde{v}_i^\beta} - \overline{\rho_i \Phi_i v'^\alpha_i \tilde{v}_i^\beta} - \overline{\rho_i \Phi_i \tilde{v}_i^\alpha v'^\beta_i} - \overline{\rho_i \Phi_i v'^\alpha_i v'^\beta_i}$$

and

$$\bar{\rho}_i^* \bar{\Phi}_i F_{LESi}^{H\alpha} = -\overline{\bar{\rho}_i^* \bar{\Phi}_i \tilde{v}_i^\alpha \tilde{h}_{ti}} + \bar{\rho}_i^* \bar{\Phi}_i \tilde{v}_i^\alpha \tilde{h}_{ti} - \overline{(\rho_i \Phi_i)' \tilde{v}_i^\alpha \tilde{h}_{ti}} - \overline{\rho_i \Phi_i \tilde{v}_i^\alpha h'_{ti}} - \overline{\rho_i \Phi_i v'^\alpha_i h_{ti}} + \ldots$$

Under these conditions, the definition of the terms to be modeled is slightly different, and it is clear that they contain the effect of the density fluctuations. The general definition is:

$$\bar{\rho}_i^* \bar{\Phi}_i T_{LESi}^{\alpha\beta} = -\overline{\bar{\rho}_i^* \bar{\Phi}_i \tilde{v}_i^\alpha \tilde{v}_i^\beta} + \bar{\rho}_i^* \bar{\Phi}_i \tilde{v}_i^\alpha \tilde{v}_i^\beta + \bar{\rho}_i^* \bar{\Phi}_i T_{SGSi}^{\alpha\beta} \qquad [8.4]$$

$$\bar{\rho}_i^* \bar{\Phi}_i F_{LESi}^{H\alpha} = -\overline{\bar{\rho}_i^* \bar{\Phi}_i \tilde{v}_i^\alpha \tilde{h}_{ti}} + \bar{\rho}_i^* \bar{\Phi}_i \tilde{v}_i^\alpha \tilde{h}_{ti} + \bar{\rho}_i^* \bar{\Phi}_i F_{SGSi}^{H\alpha} \qquad [8.5]$$

In both cases, the terms $T_{SGSi}^{\alpha\beta}$ and $F_{SGSi}^{H\alpha}$ only represent the contributions of the fluctuations to be modeled of a smaller scale than that of the filter (noted as subgrid scale). These are different from the terms $F_{LESi}^{\alpha\beta}$ and $F_{LESi}^{H\alpha}$; this difference is only due to the fact that the filter does not satisfy $\bar{\bar{g}} = \bar{g}$.

Filtered equations for an LES model of two-phase gas-droplet flow have been studied recently in [SIR 05]. Like here, the study highlighted the terms of the additional fluxes and the physical interpretation of the mean terms of interphase exchanges. Here, the conclusions are the same, but the notation used is not, so extra care must be taken while reading this chapter. As noted by Sirignano [SIR 05], the additional terms are sometimes forgotten in simulations of two-phase flow, and the exact definition of the terms of interphase exchanges is not often known, therefore the models used for this purpose are often lacking in precision.

If the focus is once again on the global modeling of the multiphase medium, i.e. without taking into account the mean velocities or the mean energies of each phase necessary in the "multifluid" model, the equations are the following:

$$\frac{\partial}{\partial t}\left(\bar{\rho}\tilde{\Phi}_i\right)+\frac{\partial}{\partial x_\alpha}(\bar{\rho}\tilde{v}^\alpha\tilde{\Phi}_i)=\frac{\partial}{\partial x_\alpha}(-\overline{\rho_i v_i^\alpha \Phi_i}+\bar{\rho}\tilde{v}^\alpha\tilde{\Phi}_i)-\overline{\rho_i v_{ri}^\alpha n_i^\alpha \sigma_i^s} \quad [8.6]$$

$$\frac{\partial}{\partial t}\left(\bar{\rho}\tilde{v}^\beta\right)+\frac{\partial}{\partial x_\alpha}(\bar{\rho}\tilde{v}^\alpha\tilde{v}^\beta)=\frac{\partial}{\partial x_\alpha}(\bar{\sigma}^{\alpha\beta}-\overline{\rho v^\alpha v^\beta}+\bar{\rho}\tilde{v}^\alpha\tilde{v}^\beta)+\bar{\rho}g_\beta+\overline{F_{cap}^\beta\sigma^s} \quad [8.7]$$

and by neglecting the influence of surface forces:

$$\frac{\partial}{\partial t}(\bar{\rho}\tilde{e}_t)+\frac{\partial}{\partial x_\alpha}(\bar{\rho}\tilde{v}^\alpha\tilde{h}_t)=\frac{\partial}{\partial x_\alpha}(-\bar{j}_\alpha^Q+\overline{\tau^{\alpha\beta}v^\beta}-\overline{\rho v^\alpha h_t}+\bar{\rho}\tilde{v}^\alpha\tilde{h}_t)+\bar{\rho}g_\beta\tilde{v}^\beta \quad [8.8]$$

The situation is similar and there are three new additional fluxes:

$$-\overline{\rho_i v_i^\alpha \Phi_i}+\bar{\rho}\tilde{v}^\alpha\tilde{\Phi}_i=\bar{\rho}F_{LES}^{i\alpha},\; -\overline{\rho v^\alpha v^\beta}+\bar{\rho}\tilde{v}^\alpha\tilde{v}^\beta=\bar{\rho}T_{LES}^{\alpha\beta},$$

and also

$$-\overline{\rho v^\alpha h_t}+\bar{\rho}\tilde{v}^\alpha\tilde{h}_t+\overline{\tau^{\alpha\beta}v^\beta}-\overline{\tau^{\alpha\beta}}\tilde{v}^\beta=\bar{\rho}F_{LES}^{H\alpha}$$

These three fluxes each contain a term that can be calculated as a function of the filter, and another term (subscript as “SGS” above, an acronym Sub-Grid Scale) that is dependent on the fluctuations eliminated by the filter, to be modeled, as previously for [8.4] and [8.5]. With this approach, it is clear that the density of the global multiphase medium is most often variable, and the contributions of the fluctuations of this global density (which include the fluctuations of the phase indices) are most likely significant.

Whatever the approach, the newly defined fluxes have a very similar role to the fluxes obtained with statistical averages. The role of $T_{LESi}^{\alpha\beta}$ and $F_{LESi}^{H\alpha}$ in the balance equations of the filtered values is exactly the same as the role of the classic additional fluxes in the balance equations for the statistical averages. The fluxes $T_{SGSi}^{\alpha\beta}$ and $F_{SGSi}^{H\alpha}$, just like the classic additional fluxes, are influenced by the fluctuations, the only change is the definition of the fluctuations: here, they are fluctuations relative to the filtered average, and previously they were fluctuations relative to the statistical average, which contained the first ones, but also others. Logically, the modeling must therefore be very similar, but what must imperatively be taken into account is the size of the filter, which makes distinctions between the different fluctuations that must be taken into account.

Finally, in any case, it is important to note that the mean terms of exchanges between phases must also be modeled. These are:

$$\overline{F'^{\beta}_{is}} = -\overline{\left(p"_i \delta^{\alpha\beta} - \tau_i^{\alpha\beta}\right)\sigma_i^s n_i^\alpha} = -\overline{\left(p"_i \delta^{\alpha\beta} - \tau_i^{\alpha\beta}\right)n_i^\alpha}^s \overline{\sigma}_i^s$$

$$\overline{q_{is}\sigma_i^s} = -\overline{j_i^{Q\alpha} n_i^\alpha}^s \overline{\sigma_i^s}, \ \overline{f_{i",s}^\beta v_i^\beta \sigma_i^s} = \overline{\sigma_i^{\alpha\beta} v_i^\beta n_i^\alpha}^s \overline{\sigma}_i^s$$

and the terms linked to the exchange of mass between phases:

$$\overline{\rho_i \dot{w}_i^{\sec h}} = -\overline{\rho_i v_{ri}^\alpha n_i^\alpha}^s \overline{\sigma_i^s}, \qquad \overline{\rho_i \dot{w}_i^{\sec h} v_i^\beta} = -\overline{\rho_i v_{ri}^\alpha n_i^\alpha}^s \overline{\sigma_i^s}\overline{v}_i^{\beta s}$$

$$\overline{\rho_i \dot{w}_i^{\sec h} h_{ti}} = -\overline{\rho_i v_{ri}^\alpha n_i^\alpha}^s \overline{\sigma_i^s}\overline{h}_{ti}^s$$

These terms are expressed in exactly the same way as the statistical averages, but the physical significance is not the same. The value $\overline{\sigma_i^s}$ is the filtered interface area per unit of volume, and if the filter is the volume filter, it exactly matches the interface area in the volume of the filter (for a multiphase medium, this is the interface between phase *i* and any other phase). The quantities filtered per unit of interface, such as $\overline{h_{ti}}^s$, if the filter is the volume filter, are the considered values (here the total enthalpy of phase *i*) integrated on the area of the interface that is within the volume of the filter (for a multiphase medium, this quantity must take into account the fact that several phases can be in contact with phase *i*, which can lead to complications). They require yet another model, as they are dependent on the fluctuations that are smoothed by the filter. As for the fluxes, this can be done with a simple transposition of the reasoning that was explained for the use of statistical averages in Chapter 5. The filtered interface area per unit of volume is again a crucial quantity, and will be discussed in section 8.4.

8.3. Classic LES modeling for SGS additional fluxes

The goal here is to quickly demonstrate how the classic approach of LES modeling can be transposed to multiphase flows. To do this, first of all the principles used in this model will be explained in the case of classic turbulent flow, i.e. single phase and with a constant mass per unit of volume. The following discussion will revolve around issues encountered when applying these principles to multiphase flows, both for the global modeling as well as for the complete "multifluid" modeling.

8.3.1. *Reminder of LES in single-phase, constant density turbulent flows*

The simplest model, known as the "Smagorinsky model", uses a Boussinesq formula for the SGS fluxes, with an effective diffusion coefficient, dependent only on a length scale and a kinetic energy characteristic of the fluctuations, as well as a formula for the coefficient similar to the well-known "mixing length formula" (proposed by Prandtl in 1926 for the turbulent boundary layers). The fact that a Boussinesq formula is used means that the additional SGS fluxes are presumed to be of diffusion type, without any possibility of "counter-gradient diffusion" or phenomena of propagation acting on these terms. This is actually in concordance with the classic representation of the effects of small-scale turbulence (as has been explicitly described very early, for example, by Heisenberg). This property is not always observed for turbulent fluxes defined in the statistical approach, and this is what warrants the use of Reynolds stress models (RSM), but it is expected that this is valid only when the fluctuations are of a small enough scale in the spectrum.

Moreover, this model is based on the first hypothesis that the filtered fluctuations, called "subgrid", whose spatial scales are all smaller than the size of the filter, have a given repartition of kinetic energies, which can be plotted by only knowing two quantities: the global kinetic energy of the fluctuations and the size of the filter Δ, which is therefore a characteristic spatial scale for these fluctuations. This means that the "spectrum" of these fluctuations (spatial spectrum, tridimensional or integrated on spheres around the considered point) is self-similar in different points and for different conditions of the medium. This similarity is often observed, and explained theoretically, in classic turbulent media, for the central region of the spectrum, i.e. by excluding the larger spatial scales and by excluding the viscous dissipation region, on the side of the very small scales. To take into account this dissipation zone, it is necessary to add as a parameter the "turbulence Reynolds number ", the value of which effectively changes the location of the dissipation region. However, for a high enough Reynolds number, the total kinetic energy contained within the part of the spectrum on the side of the very small scales is completely negligible, and the similarity can be observed in nearly all points. This hypothesis, justified by this classic reasoning, allows Δ to be used as the only characteristic scale of the fluctuations involved.

As a simplification, the Smagorinski model adds the second hypothesis that states that the kinetic energy of the fluctuations, in a given point, is proportional to the square of a norm of the gradient of the filtered velocity at that point, which is the same hypothesis as in Prandtl's "mixing length" theory. The reasoning behind this theory requires to assume that the velocity gradients are the only turbulence-producing process, or at least the principal process, and that there is very fast equilibrium between the processes of production and dissipation; this is actually meant to be the characteristic of the inertial region of the spectrum, which is found in, or can be identified by, the central region.

All of this allows the writing of the algebraic formulas for the fluxes:

$$\bar{\rho}T_{SGS}^{\alpha\beta} + \frac{2}{3}\bar{\rho}k_{SGS}\delta_{\alpha\beta} = \mu_e(\frac{\partial\tilde{v}_\alpha}{\partial x_\beta} + \frac{\partial\tilde{v}_\beta}{\partial x_\alpha} - \frac{2}{3}\frac{\partial\tilde{v}_\gamma}{\partial x_\gamma}\delta_{\alpha\beta}) \qquad [8.9]$$

$$\bar{\rho}F_{SGS}^{H\alpha} = \frac{\mu_e}{\Pr_e}\frac{\partial\tilde{h}_t}{\partial x_\alpha} \qquad [8.10]$$

and

$$\bar{\rho}F_{SGS}^{i\alpha} = \frac{\mu_e}{Sc_e}\frac{\partial\tilde{\Phi}_i}{\partial x_\alpha} \qquad [8.11]$$

with

$$\mu_e = \bar{\rho}C_S\Delta k_{SGS}^{1/2},$$

$$\text{with} \quad k_{SGS} = C_k\Delta^2(\frac{\partial\tilde{v}_\alpha}{\partial x_\beta} + \frac{\partial\tilde{v}_\beta}{\partial x_\alpha} - \frac{2}{3}\frac{\partial\tilde{v}_\gamma}{\partial x_\gamma}\delta_{\alpha\beta})(\frac{\partial\tilde{v}_\alpha}{\partial x_\beta} + \frac{\partial\tilde{v}_\beta}{\partial x_\alpha} - \frac{2}{3}\frac{\partial\tilde{v}_\gamma}{\partial x_\gamma}\delta_{\alpha\beta}) \qquad [8.12]$$

where C_S and C_k are constant.

This model was proposed for turbulent flows with constant density, which justifies why the second invariant of the tensor gradient of filtered velocity was chosen as the only important norm.

The values of C_S and C_k (actually of $C_S C_k^{1/2}$) in reality were observed to be variable in a flow, even in the classic case, and it has been possible to take this into account through improvements to the model.

These improvements, and the other proposed models, will not be detailed any further here. If needed, a book on the LES can be read, for example the book by Sagaut [SAG 06]. The next section is focused on the application of this approach in the case of multiphase flows. The balance equations of the filtered values in a multiphase medium obviously pave the way for modeling the terms of additional fluxes following the same principles.

8.3.2. *Toward an extension for multiphase flows*

To start with, if a complete multifluid approach is considered, it can be thought after the discussions of Chapter 6 that the turbulent medium in each phase probably has properties very similar to those of turbulent single-phase medium; the difference being, if a phase of solid particles is present, the role of the mean size of the particles and the average distance between them, represented by the variable $\lambda(\overline{\Phi}_p)r_p$ (r_p being the radius of the particles). If the medium contains droplets or bubbles, the size of these structures is no longer imposed; it is modified by the velocity fluctuations within the two phases, and depends on the capillarity and the viscosity, which in any case change the shape of the spectrum of velocity fluctuations. In any situation, the velocity fluctuations must display a whole spectrum of scales, the largest being linked yet again to the geometrical characteristics of the flow and the smallest being again fed into by a "cascade" of vortices. It is important that the influence of possible solid particles does not overly disturb the justification of the hypotheses used in the classic subgrid models, so that these may be transposed.

The first issue raised is that of the similarity of the spectrum, both in its central part and at small scales. There is very little experimental data for this matter, and they are hard to obtain. Hopefully, soon the results of direct numerical simulations will become available. It is possible to imagine that formulas of the Smagorinski type can prove flaws, and that the influence of the size of the particles and the distance between them (and therefore of their volume fraction) must be explicitly taken into account. It would then be necessary to modify [8.12] and envisage making coefficient C_S a function of $\Delta/\lambda(\overline{\Phi}_p)r_p$, in a way that would have to be studied by direct numerical simulations. Moreover, importantly, the presence of the term due to friction with the other phase in the equation of k_{SGS} does cast some doubt over the fact implied by [8.12], namely that this kinetic energy is simply linked to the gradient of the filtered velocity. In the cases where this term is not minor, it is necessary to calculate k_{SGS} directly using its balance equation; this equation will therefore contain the influence of $\Delta/\lambda(\overline{\Phi}_p)r_p$ in modeling its dissipation term. This possibility has already been mentioned for the model using statistical averages. This has never been considered up until now, but the vast majority of studies have been conducted in highly diluted media.

If a global approach is now considered, based on equations [8.6]–[8.8], the questions raised are the same as before, with the added fact that the global medium is now more likely to have a varying density. The mean global medium not only undergoes a modification of its rate of dissipation of kinetic energy of the fluctuations (or rather of the transfer rate toward the smaller scales), but also receives a new source of fluctuations linked to the pressure fluctuations acting on a medium with a fluctuating density. For this reason, it is advisable again to avoid formulas of the Smagorinski type and to look toward using a balance equation for k_{SGS}. The discussions in Chapter 6 provide an interesting base for this.

Nevertheless, at first glance, there is no obvious contradiction in the LES model for a multiphase medium. It could even be that the success of models based on statistical averages, already frequently used, might be predictive of the success of LES models. In any case, this domain remains very promising and there should be an increase in the research surrounding it. This will allow numerical simulations of flows that are closer to practical or industrial applications than the ones that the direct numerical simulations direct numerical simulations (DNS, i.e. without any turbulent model) can study now. The computer codes using such an LES model are simple extensions of the existing DNS codes, used today with clear success, see, for instance, [TRY 11] and [MÉN 07].

8.4. Subgrid modeling of the interface area per unit volume

The next step is to look at how it is possible to calculate the filtered interface area per unit volume $\overline{\sigma_i^s}$. As mentioned previously, in the case where the filter is simply the volume filter, this is simply the interface area within the considered filter volume, at a given instant. It is clear that this area must contain all the folds and wrinkles that are present in this volume, from the largest curvature radius to the smallest curvature radius that can be found at any point of the interface. If the filtering volume is very small, it is clear that the interface goes through it by only displaying one single curvature radius, the smallest of all, but if the volume is larger, other larger curvature radii (but smaller than the size of the volume) will progressively become visible.

If a numerical simulation is performed directly with the instantaneous equations from Chapter 2 (then a DNS), all these folds of different curvature radii must be described, therefore the numerical grid (for example with a finite volume method) must be such that the interface displays within a single mesh a single curvature radius, and this interface can therefore be correctly "rebuilt". In these conditions, the calculation can directly give the interface area present in the numerical grid, without requiring an additional equation.

The same property holds by using the filtered equations with a volume filter whose scale is precisely the size of the numerical grid used previously; because of an interface reconstruction algorithm, these equations do not need a particular equation for $\overline{\sigma_i^s}$. Indeed, in these conditions, there is no fold in the interface with a curvature radius smaller than the size of the filter. Moreover, the SGS terms, with such a small filter, are negligible as the subgrid fluctuations are therefore themselves completely negligible. However, if the equations are filtered using a filter with a larger characteristic scale, there will be interface folds in the volume of the filter, at the same time as the subgrid fluctuations will give rise to non-negligible SGS terms.

This happens independently of the size of the numerical grid. And to obtain a correct numerical convergence, it is enough to choose a mesh size similar to the size of the filter – so, much larger than that required for a direct simulation of the non-filtered equations. In these conditions, the reconstruction of the interface that can take place, based on the filtered variables, will not be able to give all the existing folds under the scale of the filter, only the largest curvature radius, similar to the scale of the filter, will be found. Therefore, it is necessary to use an additional equation to obtain the actual values of $\overline{\sigma_i^s}$.

It must be written that:

$$\overline{\sigma_i^s} = \overline{\sigma_i^s}^{\Delta} + \overline{\sigma_i^s}^{SGS} \qquad [8.13]$$

The first term on the right side is the value obtained through the reconstruction of the interface using the calculations of the filtered variables (with a filter size Δ), and the second term is the area which is caused by folds of a size smaller than Δ. The first term can be obtained through classic numerical processing, using an interface reconstruction process; a method is needed to calculate the second term, as a function of the size of the filter Δ, so that if Δ became smaller or equal to the smallest curve radius of the interface, $\overline{\sigma_i^s}^{SGS}$ would be equal to zero. Is it possible to generalize the Vallet *et al.* model [VAL 01]?

In an infinitely large homogeneous medium where the ergodic hypothesis is valid, it is possible to identify a statistical average and a volume average, and, in these conditions, it is possible to directly apply the Vallet *et al.* equation to $\overline{\sigma_i^s}^{\infty}$, where the infinite exponent shows that the filter here has a very large volume, and therefore contains all the possible folds of the interface, in all the possible experimental conditions. It is important to note that $\overline{\sigma_i^s}$ is different from $\overline{\sigma_i^s}^{\infty}$, as it does not contain all the possible folds of curvature radii superior to Δ. The classic Vallet *et al.* equation can, therefore, be written for $\overline{\sigma_i^s}^{\infty}$ in the following way:

$$\frac{\partial \overline{\sigma_i^{s\infty}}}{\partial t} + \frac{\partial}{\partial x^\alpha}(\overline{\sigma_i^{s\infty}}\tilde{v}^\alpha) = \frac{\partial}{\partial x^\alpha}(\bar{\rho} D_{t\infty s} \frac{\partial}{\partial x^\alpha}(\overline{\sigma_i^{s\infty}} / \bar{\rho})) + A\overline{\sigma_i^{s\infty}} \frac{\overline{\rho v'^\alpha v'^\beta}}{\bar{\rho} k} \frac{\partial \tilde{v}^\alpha}{\partial x^\beta} + \frac{\overline{\sigma_i^{s\infty}}}{\tau_P}\left(1 - \frac{\overline{\sigma_i^{s\infty}}}{\sigma_{eq}}\right)$$

The values of τ_P and σ_{eq} still need to be defined: $\sigma_{eq} = \frac{3\bar{\Phi}_l}{r_{eq}} = \frac{3\bar{\rho}\tilde{\Phi}_l}{\rho_l \eta_c W_{eq}} \rho^* k^*$, where the quantities marked by an asterisk represent a density and the kinetic energy of the turbulent fluctuations acting on the fracture or the coalescence of droplets, and $\tau_P \propto \tau_t$. The turbulent quantities (for the global flow of both phases) k and τ_t, as well as D_{ts}, which is a function of these quantities, concern all the wavelengths of the turbulence, as in this case $\overline{\sigma_i^s}^{\infty}$ is considered.

But the reasoning has to be transposed by only considering the effect of the fluctuations smaller than the size of the filter, in any medium. Since the Vallet *et al.* equation neglects non-turbulent stretch, the transposition therefore may deal directly with $\overline{\sigma_i^s}^{SGS}$, whereas $\overline{\sigma_i^s}^{\Delta}$ does take into account this stretch at a larger scale than the fluctuations considered. This gives:

$$\frac{\partial \overline{\sigma_i^s}^{SGS}}{\partial t} + \frac{\partial}{\partial x^\alpha}(\overline{\sigma_i^s}^{SGS} \tilde{v}^\alpha) = \frac{\partial}{\partial x^\alpha}(\bar{\rho} D_{ts} \frac{\partial}{\partial x^\alpha}(\overline{\sigma_i^s}^{SGS} / \bar{\rho})) + A \overline{\sigma_i^s}^{SGS} \frac{\overline{\rho v'^\alpha v'^\beta}^{SGS}}{\bar{\rho} k^{SGS}} \frac{\partial \tilde{v}^\alpha}{\partial x^\beta} + \frac{\overline{\sigma_i^s}^{SGS}}{\tau_P^{SGS}}(1 - \frac{\overline{\sigma_i^s}^{SGS}}{\sigma_{eq}}) \quad [8.14]$$

The filtered velocity field, dispersed by the fluctuations of a smaller scale, transports this quantity; thus, D_{ts} now depends on k^{SGS} and τ_t^{SGS}, and $\overline{\sigma_i^s}^{SGS}$ is increased by the gradient of filtered velocity, or by the turbulence under the "characteristic scale" of the filter, and decreased until reaching the value of the equilibrium.

The characteristic scale of the filter comes into play here by the fact that $\tau_P^{SGS} \propto \tau_t^{SGS} = C_t \frac{\Delta}{(k^{SGS})^{1/2}}$, and therefore τ_t^{SGS} increases as the subgrid kinetic energy tends toward zero.

What this equation gives when the subgrid turbulence becomes equal to zero must be examined: in principle, $\overline{\sigma_i^s}^{SGS}$ should be found to be equal to zero. This vanishing turbulence limit does not necessarily go hand in hand with the fact that the gradient of the filtered velocity is equal to zero, so k^{SGS} should not be considered proportional to the square of a norm of this gradient, as is the case in the Smagorinski model. Nevertheless, a balance equation for k^{SGS} can be considered, as proposed in Chapter 6, without major difficulty. The original Vallet *et al.* model can be applied to situations where the turbulence Reynolds number is very large, and must therefore be extended in the direction of the small and very small Reynolds numbers. In this limit, the first term of production shows a linear amplification since $\frac{\overline{\rho v'^\alpha v'^\beta}^{SGS}}{\bar{\rho} k^{SGS}} \frac{\partial \tilde{v}^\alpha}{\partial x^\beta}$ has to tend toward a finite constant, if A is also a constant. The second term of production and the term of destruction are negligible in comparison to the first term, as τ_t^{SGS} tends toward infinity. This would not lead to $\overline{\sigma_i^s}^{SGS}$ equal to zero. However, if A is assumed to be a function of the Reynolds number, which becomes proportional to the Reynolds number when this number is small, an equilibrium between the three terms can be obtained, resulting in: $\overline{\sigma_i^s}^{SGS} = (A_0 C_t \frac{\Delta^2}{\nu} \frac{\overline{\rho v'^{\alpha} v'^{\beta}}^{SGS}}{\bar{\rho} k^{SGS}} \frac{\partial \tilde{v}^{\alpha}}{\partial x^{\beta}} + 1)\sigma_{eq}$, which finally results in $\overline{\sigma_i^s}^{SGS}$ equal to zero if σ_{eq} is also equal to zero. In the Vallet *et al.* model, it is supposed that $\sigma_{eq} = \frac{3\bar{\Phi}_l}{r_{eq}} = \frac{3\bar{\rho}\tilde{\Phi}_l}{\rho_l \eta_c W_{eq}} \rho^* k^*$, and therefore it is sufficient for k^* to be equal to zero, which would obviously be the case if k^{SGS} was.

We must be reminded that in this model, it is the turbulence of the global medium that is taken into account, as defined in Chapter 6. Therefore, it is to be reconsidered using the turbulence in both phases if a complete multifluid model is to be used.

A direct transposition of the Vallet *et al.* model for $\overline{\sigma_i^s}^{SGS}$, therefore, seems possible.

8.5. Partially Integrated Turbulence Modeling

It was seen previously that the Smagorinski model – which directly links the length scale of the model for the SGS fluxes to the one of the filter – lacked generality and also that the presence of walls required a variation in the length scale of the filter in the neighboring regions. In these conditions, numerous studies now attempt to combine the LES approach (far from the walls) and the Reynolds averaged Navier-Stokes approach using statistical averages (near the walls). One way of doing this, without introducing two separate geometrical regions for the two separate approaches, would involve creating an LES model that does not directly link the length scale of the model to that of the filter. This is the proposition by Schiestel and Chaouat [CHA 05] who suggest the use of an equation for the dissipation rate of SGS kinetic energy, jointly with an equation for the SGS kinetic energy itself, or even better, an equation for the tensor $T_{SGS}^{\alpha\beta}$ itself. In this framework, it is possible to maintain a constant filter length scale throughout the whole calculation domain, with the necessary length scale of the model changing within the flow through the use of the new equation, all the while taking into account the length scale of the filter (which is already accounted for in the model of its different terms). Of course, if the scale of the filter becomes larger that the distance to the wall, that means the filter kernel is not in the physical space and is defined conveniently.

Obtaining an equation for the dissipation rate has been a source of difficulty in classic turbulence models very early; one of the advantages of the LES models was that it was no longer required. The question must, therefore, be studied again in detail, which will be useful even for classic turbulence models. The Partially Integrated Turbulence Modeling approach allows such a study. It is based on the studies on two-point correlations within inhomogeneous turbulence done since many years in Lyon (see [LAP 95]). Through integration onto a sphere, it is possible to deduce information on the repartition of energy as a function of a single spatial scale. This enables the obtaining of an equation for dissipation rate ε_{SGS}, which becomes identical to ε if the spectrum is well developed, but it also entails the consideration of an equation for k_{SGS}, or, even better, for $T^{\alpha\beta}_{SGS}$ [CHA 05].

Moreover, they propose a particular filter shape that links the filtering with the statistical average. The filter proposed here is a filter defined in the spectral space, i.e. in the space of the wave numbers κ_β associated with the spatial coordinates ξ_β. The instantaneous velocity field v^α can be decomposed in a classical way into $v^\alpha = \langle v^\alpha \rangle + v'^\alpha$, by making apparent the statistical average and the fluctuation. The spectral decomposition defines contributions for the fluctuations in wavelength bands: $v'^\alpha = \sum_{m=1,..M} v'^\alpha_m$ where $v'^\alpha_m(\xi) = \int_{\kappa_{m-1}<|\kappa|<\kappa_m} \hat{v}'^\alpha(\kappa)\exp(j\kappa\xi)d\kappa$, with the function $\hat{v}'^\alpha(\kappa)$ being the Fourier transform of $v'^\alpha(\xi)$. The simplest version of this decomposition defines a single "cut-off wave number", κ_C. This results in:

$$v^\alpha = \langle v^\alpha \rangle + v^\alpha_< + v^\alpha_>$$

where $v^\alpha_< = \int\limits_{|\kappa|<\kappa_C} \hat{v}'^\alpha(\kappa)\exp(j\kappa\xi)d\kappa$ and $v^\alpha_> = \int\limits_{|\kappa|\geq\kappa_C} \hat{v}'^\alpha(\kappa)\exp(j\kappa\xi)d\kappa$

Now, if the filtering is defined as:

$\bar{v}^\alpha = \langle v^\alpha \rangle + \int\limits_{|\kappa|\leq\kappa_C} \hat{v}'^\alpha(\kappa)\exp(j\kappa\xi)d\kappa$, this results in $v^\alpha = \bar{v}^\alpha + v^\alpha_>$

The properties of this particular filter, defined differently from the classic spatiotemporal filters of Chapter 3, make it easier to find the balance equations for $\bar{v}^\alpha, v^\alpha_>, \langle v^\alpha_< v^\alpha_< \rangle, \langle v^\alpha_< v^\beta_< \rangle$. In a turbulent, non-homogeneous and anisotropic medium approximated as locally homogeneous and anisotropic, it is possible to show that the fluctuations above and under the cut-off are not correlated: $\langle v_<^\alpha v_>^\alpha \rangle = 0$ (see [CHA 07, CHA 09a]).

With this particular filter, from the classic equation of the two-point velocity correlations tensor, the two points being separated by the vector ξ:

$$R_2^{\alpha\beta}(x,\xi) = \langle v'^\alpha (x+\xi/2) v'^\beta (x-\xi/2) \rangle$$

the equations for $\langle v_<^\alpha v_<^\beta \rangle = k_{SGS}$ and ε can be found. These equations are modeled in the Fourier space associated with x using an approximated form of the tridimensional spectrum $E(|\kappa|)$, valid for all the values of the wave number.

This new approach has only been applied currently to single-phase turbulent flows with a constant density. It can be questioned whether this approach can be adapted without too much extra difficulty to situations where the density varies, but this is yet to be established. In any case, it is clear that there is no theory similar to the one from Lyon for a global two-phase turbulent medium. Its application in that case would, therefore, be more heuristic and would require an equation for ε_{SGS} that would have to be derived phenomenologically, following the discussion of Chapter 6. With the multifluid approach, this appears to be simpler for the cases where the two fluids involved have a constant density, but the issue of how to include the effects of the other phases needs to be studied. In both cases, detailed studies by direct numerical simulations are required to get a clearer picture.

Chapter 9

Contribution of Thermodynamics of Irreversible Processes

Terms of additional "turbulent" dispersion, as well as, in the majority of cases, terms of exchange between phases, take into account irreversible phenomena. Their modeling must not contradict thermodynamics and its famous second law. Nevertheless, making sure this does not happen, or at least ensuring the conditions such that this does not happen, is not easy. It is not possible to apply what is classically termed "thermodynamics of irreversible processes" (TIP), i.e. dealing with media and flows that can be represented without defining mean variables. The instantaneous equations that are the basis of the models here have already been obtained in accordance with this classic TIP: these are not equations we will attempt to validate. Those that *are* to be validated are the result of a model for the mean medium, namely those that represent such a medium with averaged values and then less detailed variables, and therefore need closure approximations to compensate for the loss of details resulting from averaging.

The situation is similar to that of systemic modeling, where the variables considered are also global characteristics that do not permit the representation of all the details in these systems; it is "global" thermodynamics, adapted to these systems, that is applied (see [BOR 08a] and [BOR 08b]). Here, for media with random local structures (the media that are relevant here), as well as for single-phase turbulent media, this global thermodynamics must be generalized so that it may be adapted to the "macroscopic" variables on which the models are based, which are the well-defined statistical averages here. Moreover, to be able to deal with the most recent models, the generalized "extended irreversible thermodynamics" must be used (see [JOU 01]).

This generalized thermodynamic point of view refers to the definition of macroscopic (per unit of mass) entropy, not directly related to any microscopic statistical representation of the phenomena involved, which is a function of the variables (per unit of mass) chosen to completely represent the state of the medium, themselves obtained from their balance equations. Thus, entropy depends on the selection of this set of variables, on their number and on their nature. Two different models of the same medium therefore have different entropies if they are based on different variables. This entropy function is completely determined when a certain number of state equations can be known, prescribing the primary partial derivatives of the entropy with respect to the different variables, as a function of these different variables.

Once these equations of state have been chosen, taking account of the physical nature of the problem and with certain approximations that are to be validated through the experimentation, it is therefore possible to extract from the basic balance equations of the variables a balance equation of entropy, and to verify that the entropy production term is always positive or null. This is the verification of the thermodynamic coherence of the model. This is exactly the same procedure as for the case of a simple continuous fluid medium, where the local variables are the volume, momentum and internal energy (all per unit of mass), or for a mixture of fluids where mass fraction in different constituents is added, but here it must be applied to other groups of variables so as to find new expressions for entropy and for equations of state. Justification of the equations of state can initially be made not only through internal structure of the medium, but also by taking into account the approximations and hypotheses that result in the considered model. Additional approximations may eventually be necessary, but in any case all these must be validated through experimentation, either directly or indirectly.

In this chapter, an attempt will be made to apply this generalized thermodynamic approach to see what it can add in comparison to the models in Chapters 5 and 6.

9.1. Global two-phase medium models

When the multiphase medium is represented globally, the mean balance equations of the total mass, momentum, total energy of the medium and mass fractions of the phases are no different from those in a single-phase turbulent medium, apart from terms of surface tension, and of course they lack terms of exchange between phases. Superficial effects are neglected again, and these are the equations [4.1], [4.2], [4.3b] and [4.4b].

It is important to note that in this case, it is the equations of state, allowing the determination of the pressure and temperature as a function of the quantities

$\bar{\rho}, \tilde{e}, \widetilde{\Phi}_i, i = 1...N-1$, which contain the specificities of the multiphase mixture; these are therefore crucial and must be studied carefully. Consequently, thermodynamics will offer indications on how to model terms of additional turbulent dispersion fluxes.

Starting with [4.3b], and taking [4.2] into account, as well as the balance equation of the kinetic energy of the fluctuations [6.5a], it is possible to obtain the balance equation of the mean internal energy (already done in Chapter 4, section 4.2). This results in:

$$\frac{\partial}{\partial t}(\bar{\rho}\tilde{e})+\frac{\partial}{\partial x_\alpha}(\bar{\rho}\tilde{v}^\alpha\tilde{e})=\frac{\partial}{\partial x_\alpha}(-\bar{j}_\alpha^Q-\overline{\rho v'^\alpha e})-\overline{p\frac{\partial v^\beta}{\partial x_\beta}}+\bar{\tau}^{\alpha\beta}\frac{\partial \tilde{v}^\beta}{\partial x_\alpha}+\overline{\tau^{\alpha\beta}\frac{\partial v'^\beta}{\partial x_\alpha}}$$

As before, total mass is written as:

$$\frac{\partial}{\partial t}(\bar{\rho})+\frac{\partial}{\partial x_\alpha}(\bar{\rho}\tilde{v}^\alpha)=0$$

The balance equation of the mass of phase i is written as:

$$\frac{\partial}{\partial t}(\bar{\rho}\tilde{\Phi}_i)+\frac{\partial}{\partial x_\alpha}(\bar{\rho}\tilde{v}^\alpha\tilde{\Phi}_i)=\frac{\partial}{\partial x_\alpha}(-\overline{\rho v'^\alpha \Phi_i})-\overline{\rho_i v_{ri}^\alpha n_i^\alpha \sigma_i^s}$$

It can be observed that the use of the mass fraction $\tilde{\Phi}_i$ is more practical here than the use of the volume fraction (or presence probability) $\bar{\Phi}_i$.

With the convective derivative $\frac{D}{Dt}(.)=\frac{\partial}{\partial t}(.)+\tilde{v}^\alpha\frac{\partial}{\partial x_\alpha}(.)$, it can be written:

$$\frac{D}{Dt}(\bar{\rho})=-\bar{\rho}\frac{\partial \tilde{v}^\alpha}{\partial x_\alpha} \tag{9.1}$$

$$\bar{\rho}\frac{D\tilde{e}}{Dt}=\frac{\partial}{\partial x_\alpha}(-\bar{j}_\alpha^Q-\overline{\rho v'^\alpha e})-\overline{p\frac{\partial v^\beta}{\partial x_\beta}}+\bar{\tau}^{\alpha\beta}\frac{\partial \tilde{v}^\beta}{\partial x_\alpha}+\overline{\tau^{\alpha\beta}\frac{\partial v'^\beta}{\partial x_\alpha}} \tag{9.2}$$

$$\bar{\rho}\frac{D}{Dt}(\tilde{\Phi}_i)=\frac{\partial}{\partial x_\alpha}(-\overline{\rho v'^\alpha \Phi_i})-\overline{\rho_i v_{ri}^\alpha n_i^\alpha \sigma_i^s} \tag{9.3}$$

These equations are the balance equations available for the global mean representation of a multiphase medium after modeling of the terms of the dispersion fluxes involved. Here, the irreversible thermodynamics of a two-phase medium (containing the phase of mass fraction $\tilde{\Phi}$ and the other phase) will be studied by first assuming that the dispersion fluxes follow the model of Prandtl's mixing length. As this model requires neglecting the effects of the density fluctuations, and in particular the coupling with a mean pressure gradient, next the classic "k–ε" model will be studied, which will be better suited to the general case.

9.1.1. *Entropy of a mean two-phase medium using the Prandtl model*

In this case, the medium is supposed to be completely represented by the mean variables $\bar{\rho},\tilde{e},\tilde{\Phi},\tilde{v}^{\alpha}$, which are given by their own classical balance equations, the main variables of which are given above. Since the model being used is that of Prandtl's mixing length, this is equivalent to assuming that the balance equation of the kinetic energy of the turbulence, given in [6.5a], can be simplified by simply writing that the production term is everywhere and always equal to the dissipation term, which is:

$$-\overline{\rho v'^{\alpha} v'^{\beta}}\frac{\partial \tilde{v}^{\beta}}{\partial x_{\alpha}}+\overline{p'\frac{\partial v'^{\alpha}}{\partial x_{\alpha}}}+\bar{p}\frac{\partial \overline{v'^{\alpha}}}{\partial x_{\alpha}}=\overline{\tau^{\alpha\beta}\frac{\partial v'^{\beta}}{\partial x_{\alpha}}} \qquad [9.4]$$

Furthermore, the influence of the pressure is assumed negligible in the production, the left-hand side of the equation, as in the situation where the density is constant. This approximation was discussed in Chapter 6. The dissipation rate (the right-hand side of the equation) must then be calculated only as a function of k and l_t, a characteristic length scale of turbulence (the "mixing length"), or $\bar{\rho}\varepsilon = C_d k^{3/2}/l_t$, where C_d is a dimensionless constant. This can also be written as:

$$\bar{\rho}C_d\frac{k^{3/2}}{l_t}=-\overline{\rho v'^{\alpha} v'^{\beta}}\frac{\partial \tilde{v}^{\beta}}{\partial x_{\alpha}} \qquad [9.5]$$

Obviously, the model is completed by giving the turbulence length scale, depending on the geometry of the flow, and the expression of the Reynolds tensor as a function of the variables already defined; what does the TIP say about this?

As the four variables $\bar{\rho},\tilde{e},\tilde{\Phi},\tilde{v}^{\alpha}$ are known, the mass entropy must be found as only $\varsigma=\varsigma(\tilde{e},1/\bar{\rho},\tilde{\Phi})$, as velocity is not an objective variable. The Gibbs equations and the Gibbs–Euler equation must be written as:

$$\tilde{e} = \theta\varsigma - \pi\frac{1}{\overline{\rho}} - \theta y\tilde{\Phi} + \mu, \quad d\tilde{e} = \theta d\varsigma - \pi d(\frac{1}{\overline{\rho}}) - \theta y d\tilde{\Phi}$$

The three functions θ, $-\pi$ and $-\theta y$ are the partial derivatives of $\tilde{e}$ respective to ς, $1/\overline{\rho}$ and $\tilde{\Phi}$, or otherwise $1/\theta$, π/θ and y are the partial derivatives of ς respective to $\tilde{e}$, $1/\overline{\rho}$ and $\tilde{\Phi}$, and the relations $\theta = \theta(\tilde{e}, 1/\overline{\rho}, \tilde{\Phi})$, $\pi = \pi(\tilde{e}, 1/\overline{\rho}, \tilde{\Phi})$ and $y = y(\tilde{e}, 1/\overline{\rho}, \tilde{\Phi})$ are the equations of state to be found. Next, $\varsigma = \varsigma(\tilde{e}, 1/\overline{\rho}, \tilde{\Phi})$ can be deduced (with an additive unknown constant). Obviously, these equations of state comply with Maxwell's equations, which translate the fact that the secondary cross partial derivatives of the entropy do no depend on the order of the derivations:

$$\frac{\partial^2\varsigma}{\partial\tilde{e}\partial(1/\overline{\rho})} = \frac{\partial(1/\theta)}{\partial(1/\overline{\rho})} = \frac{\partial^2\varsigma}{\partial(1/\overline{\rho})\partial\tilde{e}} = \frac{\partial(\pi/\theta)}{\partial\tilde{e}}, \quad \frac{\partial^2\varsigma}{\partial\tilde{e}\partial(\tilde{\Phi})} = \frac{\partial(1/\theta)}{\partial(\tilde{\Phi})} = \frac{\partial^2\varsigma}{\partial(\tilde{\Phi})\partial\tilde{e}} = \frac{\partial(y)}{\partial\tilde{e}}$$

and

$$\frac{\partial^2\varsigma}{\partial\tilde{\Phi}\partial(1/\overline{\rho})} = \frac{\partial(y)}{\partial(1/\overline{\rho})} = \frac{\partial^2\varsigma}{\partial(1/\overline{\rho})\partial\tilde{\Phi}} = \frac{\partial(\pi/\theta)}{\partial\tilde{\Phi}}$$

How can these equations of state be found? To start with, it must be noted that the phases present in the medium themselves possess their equations of state. Taking, for example, the case of an "ideal" gas, containing droplets of liquid, or vice versa, $\tilde{\Phi}$ is the volume fraction of the liquid. For the gas: $e_g = C_g T_g + q^0, p_g = \rho_g R T_g$, with three constants: C_g, q^0, R. From this, the following can be deduced: $\tilde{e}_g = C_g \tilde{T}_g + q^0, \overline{p}_g^* = \overline{\rho}_g^* R \tilde{T}_g$, simply by taking the average (weighted by the density for the first one, and non-weighted for the second one). For the liquid, it can be assumed that $e_l = C_l T_l + q_l^0, \rho_l = cste = \rho_L$, therefore $\tilde{e}_l = C_l \tilde{T}_l + q_l^0$. By using the simplifying assumptions $\tilde{T}_l = \tilde{T}_g = \tilde{T}, \overline{p}_g^* = \overline{p}_l^* = \overline{p}$, it follows: $\overline{p} = \dfrac{(1-\tilde{\Phi})}{(1/\overline{\rho} - \tilde{\Phi}/\rho_L)} R\tilde{T}$ (see Chapter 4), $\tilde{e} = (C_l\tilde{\Phi} + C_g(1-\tilde{\Phi}))\tilde{T} + \tilde{\Phi}(q_l^0 - q^0) + q^0$.

This results in the partial entropies: $\varsigma_g - \varsigma_{gr} = C_g Ln(\dfrac{\tilde{T}}{\tilde{T}_r}) - RLn\left(\dfrac{\overline{\rho}_g^*}{\overline{\rho}_{gr}^*}\right)$ and $\varsigma_l - \varsigma_{lr} = C_l Ln\left(\dfrac{\tilde{T}}{\tilde{T}_r}\right)$. In the mixture, if all contribution of the interface is neglected, we obtain $\varsigma=\varsigma_l\tilde{\Phi}+\varsigma_g(1-\tilde{\Phi})$. Knowing that: $\bar{\rho}_g^*=\dfrac{\bar{p}}{R\tilde{T}}=\dfrac{(1-\tilde{\Phi}\ )}{(1/\bar{\rho}-\tilde{\Phi}/\rho_L)}$, the entropy is simply:

$$\varsigma=(\tilde{\Phi}C_l+(1-\tilde{\Phi})C_g)Ln(\frac{\tilde{T}}{\tilde{T}_r})-R(1-\tilde{\Phi})Ln(\frac{(1-\tilde{\Phi}\ )}{(1/\bar{\rho}-\tilde{\Phi}/\rho_L)\rho_r})+\tilde{\Phi}(\varsigma_{lr}-\varsigma_{gr})+\varsigma_{gr}$$

The partial derivative of $\varsigma=\varsigma(\tilde{e},1/\bar{\rho},\tilde{\Phi})$ respective to $\tilde{\Phi}$ can be calculated, resulting in:

$$y=(C_l-C_g)Ln(\frac{\tilde{T}}{\tilde{T}_r})-\frac{(q_l^0-q^0)}{\tilde{T}}+RLn(\frac{(1-\tilde{\Phi}\ )}{(1/\bar{\rho}-\tilde{\Phi}/\rho_L)\rho_r})-R\frac{1-\tilde{\Phi}}{(1/\bar{\rho}-\tilde{\Phi}/\rho_L)\rho_L}+cst \qquad [9.6a]$$

By identifying $\theta=\tilde{T}$ and $\pi=\bar{p}$, it is possible to verify that these equations satisfy Maxwell's laws (because they were built from a mixture for which the constituents had equations of state that satisfy these relations). Therefore, these laws constitute suitable equations of state and the thermodynamic description is complete (provided that a few constants of no importance are given).

The situations could have been simplified by assuming that the volume fraction of the liquid was small (so $\dfrac{\bar{\rho}}{\rho_L}\tilde{\Phi}=\bar{\Phi}$ much smaller than 1), and so $\bar{p}=\bar{\rho}(1-\tilde{\Phi})R\tilde{T}$, which would have resulted in:

$$y=(C_l-C_g)Ln(\frac{\tilde{T}}{\tilde{T}_r})-\frac{(q_l^0-q^0)}{\tilde{T}}+RLn(\frac{\bar{\rho}}{\rho_r}(1-\tilde{\Phi}\ ))+cst \qquad [9.6b]$$

Once the equations of state have been obtained, the balance equations of the variables can be used to deduce the equation for the entropy. We start with the Gibbs–Euler equation, expressed in terms of convective derivatives:

$$\frac{D}{Dt}\tilde{e}=\tilde{T}\frac{D}{Dt}\varsigma-\bar{p}\frac{D}{Dt}(\frac{1}{\bar{\rho}})-y\tilde{T}\frac{D\tilde{\Phi}}{Dt} \qquad [9.7]$$

With the averaged balance equations for $\tilde{e},\bar{\rho},\tilde{\Phi}$ [9.1]–[9.3], and by using [9.4], the following can be deduced:

$$\bar{\rho}\frac{D\varsigma}{Dt}=\frac{\partial}{\partial x_\alpha}\left(-\frac{\bar{j}^Q_\alpha+\overline{\rho v'^\alpha e}}{\tilde{T}}-y\overline{\rho v'^\alpha\Phi}\right)$$

$$+\frac{1}{\tilde{T}}\left(\overline{p'\frac{\partial v'^\beta}{\partial x_\beta}}+\bar{p}\overline{\frac{\partial v'^\beta}{\partial x_\beta}}\right)+\left(\frac{\bar{\tau}^{\alpha\beta}-\overline{\rho v'^\alpha v'^\beta}}{\tilde{T}}\right)\frac{\partial\tilde{v}^\beta}{\partial x_\alpha}$$

$$+(\bar{j}^Q_\alpha+\overline{\rho v'^\alpha e})\frac{\partial 1/\tilde{T}}{\partial x_\alpha}+\overline{\rho v'^\alpha\Phi}\frac{\partial y}{\partial x_\alpha}-\overline{\rho_\phi v^\alpha_{r\phi}n^\alpha_\phi\sigma^s y} \qquad [9.8]$$

The pressure terms do not imply any irreversible phenomena and should be able to disappear from the production of entropy. Obviously, this can be the case if $\overline{v'^\beta}=0$ and $\frac{\partial v'^\beta}{\partial x_\beta}=0$, i.e. if the global density is constant. This is only the case if the densities of the phases are practically the same (and constant), which is obviously a very unique situation. They have already been assumed negligible in the equation of the kinetic energy of the fluctuations, and the validity of this approximation has already been discussed, we continue to disregard them.

The usual terms of entropy production are the last four terms: the last term is due to mass exchange between phases, the second last term is due to the turbulent dispersion of the liquid phase, the second term is relative to the heat transfers in the medium due to the gradients of the mean temperature and the first term is due to the total friction in the medium. The diagonal contributions of the tensors disappear if the divergence of the mean velocity is equal to zero, which is still the case if the total density is more or less constant.

The term of mass transfer between phases – if it is not physically null – can nevertheless be reversible, i.e. it does not lead to production of entropy, if at any time $y(\tilde{T},\bar{\rho},\tilde{\Phi})=0$. Looking further into this situation (this is the case for studying cavitation), it is necessary to be sure that the chosen laws of state do lead to a realistic approximation on this matter. Laws that were chosen here (given higher up) were not made for this goal as the possible equilibrium of vaporization–condensation is not part of the applicable domain of the model.

By classically applying the linear approach of TIP, a form of laws for the additional fluxes of heat dispersion can be deduced, as well as for dispersion of the liquid-phase mass and the momentum.

To determine friction (i.e. dispersion of momentum), it is sufficient to define a positive "total effective viscosity coefficient", μ_{et}, that permits the expression of the Reynolds tensor in the following manner:

$$\overline{\tau}^{\alpha\beta} - \overline{\rho v'^{\alpha} v'^{\beta}} + \frac{2}{3}\overline{\rho} k \delta^{\alpha\beta} = \mu_{et}\left(\frac{\partial \tilde{v}_{\alpha}}{\partial x_{\beta}} + \frac{\partial \tilde{v}_{\beta}}{\partial x_{\alpha}} - \frac{2}{3}\frac{\partial \tilde{v}_{\gamma}}{\partial x_{\gamma}}\delta_{\alpha\beta}\right)$$

By neglecting the molecular viscosity $\overline{\tau}^{\alpha\beta}$, or by assuming that this term is represented in the same way as for a mean Newtonian fluid, the following law is obtained:

$$-\overline{\rho v'^{\alpha} v'^{\beta}} + \frac{2}{3}\overline{\rho} k \delta^{\alpha\beta} = \mu_{e}\left(\frac{\partial \tilde{v}_{\alpha}}{\partial x_{\beta}} + \frac{\partial \tilde{v}_{\beta}}{\partial x_{\alpha}} - \frac{2}{3}\frac{\partial \tilde{v}_{\gamma}}{\partial x_{\gamma}}\delta_{\alpha\beta}\right) \quad [9.9]$$

This formula is simply the Boussinesq formula, proposed at the end of the 19th Century and widely used since then. An effective volume viscosity coefficient could also be defined for the diagonal part.

The mass and heat fluxes must be treated together, leading to:

$$\overline{j}_{\alpha}^{Q} + \overline{\rho v'_{\alpha} e'} = D_{T}\frac{\partial 1/\tilde{T}}{\partial x_{\alpha}} + D_{c}\frac{\partial y}{\partial x_{\alpha}}$$

$$\overline{\rho v'_{\alpha}\Phi} = D_{c}\frac{\partial 1/\tilde{T}}{\partial x_{\alpha}} + D_{\Phi}\frac{\partial y}{\partial x_{\alpha}}$$

The matrix of the coefficients D is symmetrical and positive definite.

Without taking into account the coupling, $D_{c} = 0$, and if the molecular conduction is again neglected, the classic shape of the law for heat flux is obtained, with the effective Prandtl number:

$$-\overline{\rho v'^{\alpha} e'} = \frac{\mu_{e}}{\Pr_{e}}\frac{\partial \tilde{T}}{\partial x_{\alpha}} \quad [9.10]$$

This law is often given with the derivative of the enthalpy instead of that of the temperature, as is the case in [6.2], where even the total enthalpy is taken into account. Thermodynamics is in favor of the temperature alone, which is the same thing if the enthalpy is a function of only temperature, but for a mixture it is also a function of the mass fraction.

By also neglecting the coupling, the dispersion flux of the mass of the phase is given as: $\overline{\rho v'_\alpha \Phi} = D_\Phi \dfrac{\partial y}{\partial x_\alpha}$. As $y = y(\tilde{T}, \bar{\rho}, \tilde{\Phi})$, it entails that:

$$\frac{\partial y}{\partial x_\alpha} = \frac{R}{\tilde{T}}\left(\frac{C_g - C_l}{R} + \frac{q_l^0 - q^0}{R\tilde{T}}\right)\frac{\partial \tilde{T}}{\partial x_\alpha} - \frac{R(1/\bar{\rho} - 1/\rho_L)}{(1/\bar{\rho} - \tilde{\Phi}/\rho_L)^2}\frac{\partial 1/\bar{\rho}}{\partial x_\alpha}$$

$$-R\frac{(1/\bar{\rho} - 1/\rho_L)^2}{(1-\tilde{\Phi})(1/\bar{\rho} - \tilde{\Phi}/\rho_L)^2}\frac{\partial \tilde{\Phi}}{\partial x_\alpha}$$

The coefficient of the gradient of the mass fraction is always negative, as can be observed. As the mean density was assumed to be practically constant, the mean density gradient term disappears. The coefficient of the temperature gradient is only due to thermodynamic characteristic differences; its order of magnitude is still R, whereas the order of magnitude of the coefficient of the gradient of the mass fraction can be significantly larger, following the values of the mass fraction. It can be written as:

$$-\overline{\rho v'^\alpha \Phi} = D_{\Phi T}\frac{\partial \tilde{T}}{\partial x_\alpha} + \frac{\mu_e}{Sc_e}\frac{\partial \tilde{\Phi}}{\partial x_\alpha} \qquad [9.11]$$

This equation contains the classic formula where only the derivative of $\tilde{\Phi}$ appears; it would be valid for cases where the temperature gradients are low. But thermodynamics shows that there is also a need to take into account the effects of the temperature gradient, if it exists. It is clear that this directly depends on the equations of state chosen, and in particular on the form of $y(\tilde{T}, \bar{\rho}, \tilde{\Phi})$. It is necessary to determine the physical justification of the influence of the temperature gradients, which can present itself also in the coupling between fluxes. For an aerosol medium, the phenomenon of thermophoresis is known; here, it would be a sort of turbulent thermophoresis, which has never been studied closely before.

Since the matrix of the D is not necessarily constant, the coefficients $D_{\Phi T}, \mu_e, \Pr_e, Sc_e$ can be functions of three variables $\bar{p}, \tilde{T}, \tilde{\Phi}$. The first two, moreover, also depend necessarily on turbulent conditions, which will be discussed later.

The terms due to viscosity and molecular conductivity processes, $\overline{\tau}^{\alpha\beta}, \overline{j}_\alpha^Q$, when they are not negligible, are difficult to calculate as a function of the composition of the two-phase medium. It can nevertheless be seen that a global approximation, which uses a global viscosity coefficient and a global conductivity constant, whose relation to $\tilde{\Phi}$ is still to be determined, is in agreement with the second principle.

The eddy viscosity coefficient must now be given as a function of the variables used. Simply through considering its dimensions, it can be seen that $\mu_e = \bar{\rho} C k^{1/2} l_t$, with [9.4] and [9.5], so that:

$$k = \frac{C}{C_d} l_t^2 \left(\frac{\partial \tilde{v}^\alpha}{\partial x_\beta} + \frac{\partial \tilde{v}^\beta}{\partial x_\alpha}\right) \frac{\partial \tilde{v}^\beta}{\partial x_\alpha} = \frac{C}{2C_d} l_t^2 \left(\frac{\partial \tilde{v}^\alpha}{\partial x_\beta} + \frac{\partial \tilde{v}^\beta}{\partial x_\alpha}\right)\left(\frac{\partial \tilde{v}^\alpha}{\partial x_\beta} + \frac{\partial \tilde{v}^\beta}{\partial x_\alpha}\right),$$

and finally Prandtl's classic "mixing length" formula is obtained:

$$\mu_e = \bar{\rho} C_{ML} l_t^2 \left(\left(\frac{\partial \tilde{v}^\alpha}{\partial x_\beta} + \frac{\partial \tilde{v}^\beta}{\partial x_\alpha}\right)\left(\frac{\partial \tilde{v}^\alpha}{\partial x_\beta} + \frac{\partial \tilde{v}^\beta}{\partial x_\alpha}\right)\right)^{1/2} \qquad [9.12]$$

It can be seen in this formula that, for this situation of two-phase mixing, if it is assumed that the model "constant" C_{ML} is really a constant, the composition of the medium and its thermodynamic state are taken into account only through the mean density, like a gas, all things considered. This has not been fully proven above, as it was seen that phenomenological coefficients of the matrix D can be functions of the state variables. In this case, the dependency of C_{ML} on $\tilde{\Phi}$ is yet to be established experimentally.

To conclude, it can be observed that application of TIP, in the framework of appropriate variables, completely justifies the classic Boussinesq and Prandtl models, who had not considered this sort of argument. However, the influence of pressure and its fluctuations on kinetic energy of fluctuations had been neglected, and this is strictly true only for flows with a constant density. Moreover, it must be noted that relations linking heat and mass of the phase fluxes to corresponding gradients are not completely classic.

9.1.2. *Entropy for the k–ε model, in a medium with a variable density*

The goal here is to use the well-known "k–ε" model, which contains an equation for k, based on [6.5a], which is rewritten here:

$$\frac{\partial}{\partial t}(\bar{\rho}k)+\frac{\partial}{\partial x_\alpha}(\bar{\rho}k\tilde{v}^\alpha)=\frac{\partial}{\partial x_\alpha}(\overline{-\rho v'^\beta v'^\beta v'^\alpha/2}-\overline{pv'^\alpha}+\overline{v'^\beta\tau^{\alpha\beta}})$$

$$-\overline{\rho v'^\alpha v'^\beta}\frac{\partial\tilde{v}^\beta}{\partial x_\alpha}+\overline{p\frac{\partial v'^\alpha}{\partial x_\alpha}}-\overline{\tau^{\alpha\beta}\frac{\partial v'^\beta}{\partial x_\alpha}}$$

and an equation for $\varepsilon^*=\varepsilon-2\nu\frac{\partial k^{1/2}}{\partial x_\alpha}\frac{\partial k^{1/2}}{\partial x_\alpha}$, less strictly based on physics. The quantity ε^* is only equal to the dissipation rate of turbulence, which is $\varepsilon=\frac{1}{\bar{\rho}}\overline{\tau^{\alpha\beta}\frac{\partial v'^\beta}{\partial x_\alpha}}$, when turbulence is well established, the additional term is not negligible near solid walls and ε^* is null on solid walls, while the dissipation rate itself is not. A general form is used for the balance equation of ε^*, which is found in all the versions of the model, with a diffusion term and production and destruction terms:

$$\frac{\partial}{\partial t}(\bar{\rho}\varepsilon^*)+\frac{\partial}{\partial x_\alpha}(\bar{\rho}\tilde{v}^\alpha\varepsilon^*)=\frac{\partial}{\partial x_\alpha}(-J_\varepsilon^\alpha)-\bar{\rho}C_3\frac{2\varepsilon^*}{3}\frac{\partial\tilde{v}^\alpha}{\partial x_\alpha}-\bar{\rho}C_1\frac{\varepsilon^*}{\tilde{k}}R_d^{\alpha\beta}\frac{\partial\tilde{v}^\beta}{\partial x_\alpha}$$

$$+C_4\frac{\varepsilon^*}{\tilde{k}}\overline{p\frac{\partial v'^\alpha}{\partial x_\alpha}}-\bar{\rho}C_2\frac{\varepsilon\varepsilon^*}{\tilde{k}} \qquad [9.13]$$

The C values are constants, without dimension, or in some cases algebraic functions of the variables (the first two being positive in any case), depending on the exact version of the chosen model. There is $C_1=1.44, C_2=1.92, C_3=1.44, C_4=0$ in the classical version.

In this model, the entropy function must be built with the variables $\bar{\rho},\tilde{e},\tilde{\Phi},k,\varepsilon^*$, and $\varsigma=\varsigma(\tilde{e},1/\bar{\rho},\tilde{\Phi},k,\varepsilon^*)$ must be found, with:

$$\tilde{e}=\theta\varsigma-\pi\frac{1}{\bar{\rho}}-\theta y\tilde{\Phi}-y_k\theta k-y_\varepsilon\theta\varepsilon^*+\mu$$

$$d\tilde{e}=\theta d\varsigma-\pi d(\frac{1}{\bar{\rho}})-\theta y_* d\tilde{\Phi}-y_k\theta dk-y_\varepsilon\theta d\varepsilon^*$$

The functions $\theta=\theta(\tilde{e},1/\bar{\rho},\tilde{\Phi},k,\varepsilon^*)$, $\pi=\pi(\tilde{e},1/\bar{\rho},\tilde{\Phi},k,\varepsilon^*)$, $y_*=y_*(\tilde{e},1/\bar{\rho},\tilde{\Phi},k,\varepsilon^*)$, $y_k=y_k(\tilde{e},1/\bar{\rho},\tilde{\Phi},k,\varepsilon^*)$, $y_\varepsilon=y_\varepsilon(\tilde{e},1/\bar{\rho},\tilde{\Phi},k,\varepsilon^*)$ are the new equations of state to be found, satisfying the 10 corresponding Maxwell relations. Method for finding these will be discussed later.

By using [9.1]–[9.3], [6.5a] and [9.13], it is possible to find as before the balance equation of entropy, which can be written as:

$$\bar{\rho}\frac{D\varsigma}{Dt}=\frac{\partial}{\partial x_\alpha}\left(-\frac{\bar{j}^Q_\alpha+\overline{\rho v'^\alpha e}}{\theta}-y_k J^\alpha_k-y_\varepsilon J^\alpha_\varepsilon-\overline{\rho v'^\alpha\Phi y_*}\right)+(\bar{j}^Q_\alpha+\overline{\rho v'^\alpha e})\frac{\partial 1/\theta}{\partial x_\alpha}$$

$$+\overline{\rho v'^\alpha\Phi}\frac{\partial y_*}{\partial x_\alpha}+J^\alpha_k\frac{\partial y_k}{\partial x_\alpha}+J^\alpha_\varepsilon\frac{\partial y_\varepsilon}{\partial x_\alpha}+\frac{1}{\theta}\left(\pi-\bar{p}-\frac{2}{3}\bar{\rho}\tilde{k}\theta\left(y_k+\frac{C_3\varepsilon^* y_\varepsilon}{\tilde{k}}\right)\right)\frac{\partial\tilde{v}^\gamma}{\partial x_\gamma}$$

$$-\left(\frac{1}{\theta}-y_k-C_4\varepsilon^* y_\varepsilon/\tilde{k}\right)p\overline{\frac{\partial v'^\beta}{\partial x_\beta}}-(y_k+C_1\varepsilon^* y_\varepsilon/\tilde{k})\bar{\rho}R^{\alpha\beta}_d\frac{\partial\tilde{v}^\beta}{\partial x_\alpha}+\frac{1}{\theta}\bar{\tau}^{\alpha\beta}_d\frac{\partial\tilde{v}^\beta}{\partial x_\alpha}$$

$$+\frac{1}{3\theta}\bar{\tau}^{\gamma\gamma}\frac{\partial\tilde{v}^\alpha}{\partial x_\alpha}+\left(\frac{1}{\theta}-y_k-C_2 y_\varepsilon\varepsilon^*/\tilde{k}\right)\bar{\rho}\varepsilon \qquad [9.14]$$

The deviators of $\tau^{\alpha\beta}$ and $R^{\alpha\beta}$ have been marked as $\tau^{\alpha\beta}_d$ and $R^{\alpha\beta}_d$, respectively, and $-J^\alpha_k$ is the diffusion term of [6.5a]. The first term on the right-hand side is entropy diffusion, and the other terms on the right side are the entropy production terms.

Before discussing the implications of the second principle, it is important to discuss the physical significance of the various terms, and to give precisions about the state equations. The first thing to note is that pressure must not appear as an irreversible phenomenon, and therefore the other terms where it appears must cancel out each other. This implies that:

$$\pi=\bar{p}+\frac{2}{3}\bar{\rho}k\theta y_k+\frac{2}{3}\bar{\rho}C_3 y_\varepsilon\varepsilon^*\theta$$

and

$$\frac{1}{\theta}-y_k-C_4\frac{\varepsilon^*}{\tilde{k}}y_\varepsilon=0$$

These equations now contain constraints that can be taken into account in the new state equations, something that was not possible in the Prandtl model.

Moreover, in the last term in [9.14], which is the scalar part of the production of entropy, there is a contribution not only from the energy equation but also from the equation of k and ε^*, which is $-\left(y_k + C_2 \frac{\varepsilon^*}{\tilde{k}} y_\varepsilon\right)$. This last contribution is likely to render the whole term negative, which must always be positive. This can be avoided by choosing $y_k + C_2 \frac{\varepsilon^*}{\tilde{k}} y_\varepsilon = 0$. This results in a sort of mutual equilibrium between k and ε^*, which ensures that there is no additional irreversibility brought by this new model.

These three relations constitute a constraint for state equations, or for the values C. The relations existing between $\overline{p}, \overline{\rho}, \tilde{T}, \tilde{\Phi}, \tilde{e}$ must also be taken into account, due to the state equations of the gas and the liquid that make up the two-phase medium chosen as an example. Here, they are the same as before.

It is not difficult to verify that state equations can be found that satisfy the 10 Maxwell relations, simply by identifying $\theta = \tilde{T}$ and $y = y_*$, and by choosing $y_k = \frac{R_k}{\tilde{k}}$, $y_\varepsilon = -\frac{R_k}{C_2 \varepsilon^*}$, where $R_k = cst.$ Indeed, this entails that $\pi = \overline{\rho} R \tilde{T} + \frac{2}{3} \overline{\rho} \tilde{k} y_k \tilde{T} + \frac{2}{3} \overline{\rho} C_3 y_\varepsilon \varepsilon^* \tilde{T} = \overline{\rho} R^* \tilde{T}$ where $R^* = R + \frac{2}{3} R_k \left(1 - \frac{C_3}{C_2}\right)$, and the additional Maxwell equations are satisfied simply on the condition that the ratio $\frac{C_3}{C_2}$ be constant:

$$\frac{\partial(\pi/\tilde{T})}{\partial \varepsilon^*} = \frac{\partial y_\varepsilon}{\partial(1/\overline{\rho})}, \quad \frac{\partial(\pi/\tilde{T})}{\partial k} = \frac{\partial y_k}{\partial(1/\overline{\rho})}, \quad \frac{\partial y_k}{\partial \varepsilon^*} = \frac{\partial y_\varepsilon}{\partial k} = 0, \quad \frac{\partial y_k}{\partial \tilde{e}} = \frac{\partial(1/\tilde{T})}{\partial k} = 0,$$

$$\frac{\partial y_\varepsilon}{\partial \tilde{e}} = \frac{\partial(1/\tilde{T})}{\partial \varepsilon^*} = 0, \quad \frac{\partial y_k}{\partial \tilde{\Phi}} = \frac{\partial y}{\partial k} = 0, \quad \frac{\partial y_\varepsilon}{\partial \tilde{\Phi}} = \frac{\partial y}{\partial \varepsilon^*} = 0$$

$R_k = cst$ can be taken as positive without losing any generality, and as the temperature must always be positive, R^* must also be positive, which implies $1 > \frac{2R_k}{3R}(1 - \frac{C_3}{C_2})$.

The implications of the second law of thermodynamics can now be addressed. It requires that the entropy production term (the big right-hand term of [9.15] below) always be positive or equal to zero on equilibrium. Curie's theorem even requires that terms that make it up and that are tensorial products of different orders be separately positive.

$$\bar{\rho}\frac{D\varsigma}{Dt}-\frac{\partial}{\partial x_\alpha}\left(\frac{\bar{j}_\alpha^Q+\overline{\rho v'^\alpha e}}{\tilde{T}}+y_k J_k^\alpha+y_\varepsilon J_\varepsilon^\alpha-\overline{\rho v'^\alpha \Phi y}\right)=(\bar{j}_\alpha^Q+\overline{\rho v'^\alpha e})\frac{\partial 1/\tilde{T}}{\partial x_\alpha}$$

$$+J_k^\alpha\frac{\partial y_k}{\partial x_\alpha}+J_\varepsilon^\alpha\frac{\partial y_\varepsilon}{\partial x_\alpha}+\overline{\rho v'^\alpha \Phi}\frac{\partial y}{\partial x_\alpha}-\left(\frac{1}{\tilde{T}}-y_k(1-C_4/C_2)\right)\overline{p\frac{\partial v'^\beta}{\partial x_\beta}}$$

$$-y_k(1-C_1/C_2)\bar{\rho}R_d^{\alpha\beta}\frac{\partial\tilde{v}^\beta}{\partial x_\alpha}+\frac{1}{\tilde{T}}\bar{\tau}_d^{\alpha\beta}\frac{\partial\tilde{v}^\beta}{\partial x_\alpha}+\frac{1}{3\tilde{T}}\bar{\tau}^{\gamma\gamma}\frac{\partial\tilde{v}^\alpha}{\partial x_\alpha}+\frac{1}{\tilde{T}}\bar{\rho}\varepsilon \qquad [9.15]$$

We remind ourselves that $\bar{\rho}\varepsilon=\overline{\tau^{\alpha\beta}\frac{\partial v'^\beta}{\partial x_\alpha}}$, so that the last three terms are grouped together to give $\frac{1}{\tilde{T}}\overline{\tau^{\alpha\beta}\frac{\partial v^\beta}{\partial x_\alpha}}$. This group must obviously always be positive, even if the two first terms are not always, as a function of the density fluctuations.

For a single-phase fluid, $\tau^{\gamma\gamma}=\mu_v\frac{\partial v^\alpha}{\partial x_\alpha}$, $\tau_d^{\alpha\beta}=\mu\left(\frac{\partial v^\beta}{\partial x_\alpha}+\frac{\partial v^\beta}{\partial x_\alpha}-\frac{\delta^{\alpha\beta}}{3}\frac{\partial v^\gamma}{\partial x_\gamma}\right)$; μ and μ_v are positive, in such a way that $\tau^{\alpha\beta}\frac{\partial v^\beta}{\partial x_\alpha}$ is always positive or equal to zero, in accordance with the previous statements. For a two-phase medium, as mentioned previously, the same formula is still adequate, with global (pseudo) viscosity coefficients μ and μ_v as positive, probably as a function of at least $\tilde{\Phi}$.

The term $-y_k(1-C_1/C_2)\bar{\rho}R_d^{\alpha\beta}\frac{\partial\tilde{v}^\beta}{\partial x_\alpha}$, which describes the influence of turbulent friction, must be positive or equal to zero in the conditions in which it dominates the final three terms, i.e. when fluctuations are strong. At the same time, $-\bar{\rho}R_d^{\alpha\beta}\frac{\partial\tilde{v}^\beta}{\partial x_\alpha}$ is the production term for k and essentially positive, but not necessarily positive everywhere. With C_2, C_1 constant, it is then clear that $1-C_1/C_2$ is necessarily positive, and therefore that linear irreversible thermodynamics still justifies the Boussinesq equation, in the general form for flows with variable density: $-\bar{\rho}R_d^{\alpha\beta}=\mu_e\left(\frac{\partial\tilde{v}^\beta}{\partial x_\alpha}+\frac{\partial\tilde{v}^\beta}{\partial x_\alpha}-\frac{1}{3}\frac{\partial\tilde{v}^\gamma}{\partial x_\gamma}\delta^{\alpha\beta}\right)$, with μ_e as positive.

Here, again it must be written $\mu_e = \bar{\rho} C_l \tilde{k}^{1/2} l_t = \bar{\rho} C_\mu \tilde{k}^2 / \varepsilon^*$, C_μ being positive, constant or a function of the Reynolds number of the turbulence, defined here as $\mathrm{Re}_t = \tilde{k}^2 / \varepsilon^* \mu$, if this number is large enough. This is exactly what is used in all the versions of the k–ε model. For the versions that use small Reynolds numbers, $1 - C_1 / C_2$ can become negative, and in this case the term must be smaller than the last three terms corresponding to $\frac{1}{\tilde{T}} \overline{\tau^{\alpha\beta} \frac{\partial v^\beta}{\partial x_\alpha}} \approx \frac{1}{\tilde{T}} \overline{\tau^{\alpha\beta}} \frac{\partial \tilde{v}^\beta}{\partial x_\alpha}$.

The final point concerns the terms:

$$(\bar{j}_\alpha^Q + \overline{\rho v'^\alpha e}) \frac{\partial 1/\tilde{T}}{\partial x_\alpha} + J_k^\alpha \frac{\partial y_k}{\partial x_\alpha} + J_\varepsilon^\alpha \frac{\partial y_\varepsilon}{\partial x_\alpha} + \overline{\rho v'^\alpha \Phi} \frac{\partial y}{\partial x_\alpha}$$

Once again, the expression of $-J_k^\alpha$ and $-J_\varepsilon^\alpha$, as well as $-\left(\bar{j}_\alpha^Q + \overline{\rho v'^\alpha e}\right)$ and $\overline{\rho v'^\alpha \Phi}$, can be given proportionally to $\frac{\partial k}{\partial x_\alpha}$, $\frac{\partial \varepsilon^*}{\partial x_\alpha}$, $\frac{\partial \tilde{T}}{\partial x_\alpha}$ and $\frac{\partial y}{\partial x_\alpha}$, with possible couplings that have not often been taken into account. The same situation as before concerning the dependency of $\frac{\partial y}{\partial x_\alpha}$ on $\frac{\partial \tilde{T}}{\partial x_\alpha}$ arises, and even here on $\frac{\partial \bar{\rho}}{\partial x_\alpha}$, or on $\frac{\partial \bar{p}}{\partial x_\alpha}$. The diffusion coefficients of the kinetic energy and the dissipation rate of turbulence are all proportional to μ_e. Just as before, the dispersion flux of the phase mass is not only proportional to $\frac{\partial \tilde{\Phi}}{\partial x_\alpha}$, but also to $\frac{\partial \bar{\rho}}{\partial x_\alpha}$ and $\frac{\partial \tilde{T}}{\partial x_\alpha}$, or to $\frac{\partial \bar{p}}{\partial x_\alpha}$ and $\frac{\partial \tilde{T}}{\partial x_\alpha}$.

Once again, irreversible thermodynamics justifies the classical aspects of the k–ε models. $C_2 > C_1$ is well respected by the models. This thermodynamic approach also shows that the situations where the density of the medium is not constant can be treated as well, providing the reason for choosing $C_3 / C_2 = cst$, even if the C values vary. With regard to C_4, the following equation has been found: $1 - R_k \frac{\theta}{\tilde{k}}\left(1 - \frac{C_4}{C_2}\right) = 0$. This implies a condition for $C_4, C_4 = C_2 - \frac{\tilde{k} C_2}{\tilde{T} R_k}$, which also depends on R_k. The choice of this constant is still to be made, so that $1 > \frac{2R_k}{3R}(1 - \frac{C_3}{C_2})$, but in any case, for values of R_k higher than a certain limit, this condition is saturated and will result in only $C_4 = C_2$. However, the model of $\overline{p\frac{\partial v'_\alpha}{\partial x_\alpha}} = \overline{p}\frac{\partial \overline{v'_\alpha}}{\partial x_\alpha} + \overline{p'\frac{\partial v'_\alpha}{\partial x_\alpha}}$ is still to be built.

9.2. Contribution of thermodynamics to multifluid models

In this section, we discuss a model similar to Prandtl's model for each of the two fluids for a *two-phase medium with no mass exchange* between phases. The variables to be considered, represented by their own balance equations, are therefore $\overline{\rho}_i^*, \tilde{v}_i^\alpha, \tilde{e}_i, \overline{\Phi}_i, i = 1,2$. The approach of section 9.1 is used below.

The kinetic energies of the fluctuations, $k_i = +\overline{\rho_i \Phi_i v'^\alpha_i v'^\alpha_i} / 2\overline{\rho}_i, i = 1,2$, follow a balance equation that is studied in Chapter 6. Using [6.19b], assuming no mass exchange between phases, it can be written:

$$\frac{\partial}{\partial t}\left(\overline{\rho}_i^*\overline{\Phi}_i k_i\right) + \frac{\partial}{\partial x_\alpha}\left(\overline{\rho}_i^*\overline{\Phi}_i k_i \tilde{v}_i^\alpha\right) = \frac{\partial}{\partial x_\alpha}\left(-\overline{\rho_i \Phi_i v'^\beta_i v'^\beta_i v'^\alpha / 2} + \overline{\Phi_i v'^\beta_i \sigma_i^{\alpha\beta}}\right)$$

$$-\overline{\rho_i \Phi_i v'^\alpha_i v'^\beta_i}\frac{\partial \tilde{v}_i^\beta}{\partial x_\alpha} + \overline{p_i \Phi_i \frac{\partial v'^\alpha_i}{\partial x_\alpha}} - \overline{\rho}_i^*\overline{\Phi}_i \varepsilon_i + \overline{\sigma_i^{\alpha\beta} n_i^\beta v'^\alpha_i \sigma_s}$$

As seen before, it will not be considered that each k_i follows such an equation, but instead can be reduced to:

$$\overline{\rho}_i^*\overline{\Phi}_i \varepsilon_i = -\overline{\rho_i \Phi_i v'^\alpha_i v'^\beta_i}\frac{\partial \tilde{v}_i^\beta}{\partial x_\alpha} + \overline{p_i \Phi_i \frac{\partial v'^\alpha_i}{\partial x_\alpha}} + \overline{\sigma_i^{\alpha\beta} n_i^\beta v'^\alpha_i \sigma_s} \qquad [9.16]$$

The last terms of these equations are due to the exchange of energy with the other phase, caused by contact forces.

An entropy must now be built with the variables $\overline{\rho}_i^*, \tilde{v}_i^\alpha, \tilde{e}_i, \overline{\Phi}_i, i = 1,2$, so that $\overline{\rho}\varsigma = \sum_{i=1,2} \overline{\rho}_i^* \overline{\Phi}_i \varsigma_i$, because entropy is an extensive variable. The entropy of each phase is function of the variables relative to this phase (as opposed to the case of a single-phase mixture), therefore: $\varsigma_i = \varsigma_i(\overline{\rho}_i^*, \tilde{e}_i)$, and these entropies satisfy the usual Gibbs–Euler equations, therefore: $\tilde{e}_i = \theta_i \varsigma_i - \pi_i^* \frac{1}{\overline{\rho}_i^*} + \mu_i$, or $\varsigma_i = \frac{1}{\theta_i}\tilde{e}_i + \frac{\pi_i^*}{\theta_i}\frac{1}{\overline{\rho}_i^*} - \frac{\mu_i}{\theta_i}$, and $d\tilde{e}_i = \theta_i d\varsigma_i - \pi_i^* d(\frac{1}{\overline{\rho}_i^*})$ or $d\varsigma_i = \frac{1}{\theta_i} d\tilde{e}_i + \frac{\pi_i^*}{\theta_i} d(\frac{1}{\overline{\rho}_i^*})$. The Maxwell equations follow, for $i = 1,2$: $\frac{\partial(1/\theta_i)}{\partial(1/\overline{\rho}_i^*)})_{\tilde{e}_i} = \frac{\partial(\pi_i^*/\theta_i)}{\partial(\tilde{e}_i)})_{\overline{\rho}_i^*}$.

If simple equations of state are taken for the fluids of each phase, it is enough to average them to obtain $\tilde{e}_i = C_{vi}\tilde{T}_i + e_i^0, \overline{p}_i^* = \overline{\rho}_i^* R_i \tilde{T}_i$ for a gas phase, and $\tilde{e}_s = C_s \tilde{T}_s + e_s^0, \overline{p}_s^* = K_s(\overline{\rho}_s^* - \rho_0)$ for a solid or liquid phase (by neglecting the influence of the compressibility in the internal energy). These equations can be considered as the state equations sought simply by identifying $\theta_i = \tilde{T}_i, \pi_i^* = \overline{p}_i^*$, since the Maxwell equations mentioned are satisfied, and no variable other than those listed above appear in them. To calculate the entropy of the two-phase medium system, first it can be shown that the convective derivative of this entropy satisfies the following:

$$\overline{\rho}\frac{D\varsigma}{Dt} = \sum_i \overline{\rho}_i^* \overline{\Phi}_i \frac{D\varsigma_i}{Dt_i} - \frac{\partial}{\partial x_\alpha}(\sum_i \overline{\rho}_i^* \overline{\Phi}_i \varsigma_i (\tilde{v}_i^\alpha - \tilde{v}^\alpha)) \qquad [9.17]$$

In this equation, the convective derivative of the entropy of the medium follows the mean velocity of the medium while the convective derivatives of the entropies of the phases follow the velocities of the phases, as they should. For each phase, the Euler–Gibbs equation is written as:

$$\frac{D}{Dt_i}\tilde{e}_i = \tilde{T}_i \frac{D}{Dt_i}\varsigma_i - \overline{p}_i^* \frac{D}{Dt_i}(\frac{1}{\overline{\rho}_i^*}) \qquad [9.18]$$

It is easy to verify that, with no exchange of mass between phases, the total mass of phase [4.4a] allows the following:

$$\overline{\rho}_i^* \overline{\Phi}_i \frac{D}{Dt_i}(\frac{1}{\overline{\rho}_i^*}) = -\frac{\overline{\Phi}_i}{\overline{\rho}_i^*}\frac{D\overline{\rho}_i^*}{Dt_i} = \overline{\Phi}_i \frac{\partial \tilde{v}_i^\alpha}{\partial x_\alpha} + \frac{D\overline{\Phi}_i}{Dt_i}$$

This with the total phase volume [4.9a] and $v_s^\alpha = v_i^\alpha$ results in the following:

$$\overline{\rho}_i^* \overline{\Phi}_i \frac{D}{Dt_i}(\frac{1}{\overline{\rho}_i^*}) = \overline{\Phi}_i \frac{\partial \tilde{v}_i^\alpha}{\partial x_\alpha} - \overline{v_i'^\alpha \frac{\partial \Phi_i}{\partial x_\alpha}} \qquad [9.19]$$

At this point, it is necessary to obtain the balance equation of the internal energy of the phases. From the total energy equation [4.7a], we substract the balance equation of the kinetic energy of the mean motion, already written as [6.18b], as well as the equation of the kinetic energy of the fluctuations [6.19b]. This results in:

$$\bar{\rho}_i^* \bar{\Phi}_i \frac{D\tilde{e}_i}{Dt_i} = \frac{\partial}{\partial x_\alpha}(-\bar{\Phi}_i \overline{j_i^{Q\alpha}}^* - \overline{\rho_i \Phi_i e_i v'^{\alpha}_i}) - \bar{\Phi}_i \bar{p}_i^* \frac{\partial \tilde{v}_i^\alpha}{\partial x_\alpha} + \bar{\Phi}_i \overline{\tau_i^{\alpha\beta}}^* \frac{\partial \tilde{v}_i^\beta}{\partial x_\alpha}$$

$$-\overline{p_i \Phi_i \frac{\partial v'^{\beta}_i}{\partial x_\beta}} - \overline{j_i^{Q\alpha} n_i^\alpha \sigma_s} + \bar{\rho}_i^* \bar{\Phi}_i \varepsilon_i$$

With [9.16], this results in:

$$\bar{\rho}_i^* \bar{\Phi}_i \frac{D\tilde{e}_i}{Dt_i} = \frac{\partial}{\partial x_\alpha}(-\bar{\Phi}_i \overline{j_i^{Q\alpha}}^* - \overline{\rho_i \Phi_i e_i v'^{\alpha}_i}) - \bar{\Phi}_i \bar{p}_i^* \frac{\partial \tilde{v}_i^\alpha}{\partial x_\alpha} + \bar{\Phi}_i (\overline{\tau_i^{\alpha\beta}}^* - \bar{\rho}_i^* R_i^{\alpha\beta}) \frac{\partial \tilde{v}_i^\beta}{\partial x_\alpha}$$

$$-\overline{j_i^{Q\alpha} n_i^\alpha \sigma_s} + \overline{\sigma_i^{\alpha\beta} n_i^\beta v'^{\alpha}_i \sigma_s}$$

where $R_{id}^{\alpha\beta}$ is the Reynolds tensor deviator of phase i. The two last terms take into account the exchange of heat by conduction $\bar{\varphi}_i^Q = -\overline{j_i^{Q\alpha} n_i^\alpha \sigma_s}$ coming from the other phase, and the exchange of kinetic energy by contact force $\overline{\sigma_i^{\alpha\beta} n_i^\beta v'^{\alpha}_i \sigma_s}$, respectively. In the total momentum calculation, we have seen that the contact force can be split as $\overline{\sigma_i^{\alpha\beta} n_i^\beta \sigma_s} = +\bar{p} \frac{\partial \bar{\Phi}_i}{\partial x_\alpha} + \bar{F}_{si}^{\times\beta}$. The first part is simply the influence of the volume variation of the phase (the generalized Archimede force), while $\bar{F}_{si}^{\times\beta} = \overline{\tau_i^{\times\alpha\beta} n_i^\alpha \sigma_s}$ is the real force exerted by the other phase, and the equilibrium condition of the forces (without any effects from surface forces) gives $\sum_{i=1,2} \bar{F}_{si}^{\times\beta} = 0$.

This leads to $\overline{\sigma_i^{\alpha\beta} n_i^\beta v'^{\alpha}_i \sigma_s} = +\bar{p} \overline{v'^{\alpha}_i \frac{\partial \Phi_i}{\partial x_\alpha}} + \overline{F_{si}^{\times\alpha} v'^{\alpha}_i}$, which finally results in:

$$\bar{\rho}_i^* \bar{\Phi}_i \frac{D\tilde{e}_i}{Dt_i} = \frac{\partial}{\partial x_\alpha}(-\bar{\Phi}_i \overline{j_i^{Q\alpha}}^* - \overline{\rho_i \Phi_i e_i v'^{\alpha}_i}) - \bar{\Phi}_i \bar{p}_i^* \frac{\partial \tilde{v}_i^\alpha}{\partial x_\alpha} + \bar{\Phi}_i (\overline{\tau_i^{\alpha\beta}}^* - \bar{\rho}_i^* R_i^{\alpha\beta}) \frac{\partial \tilde{v}_i^\beta}{\partial x_\alpha}$$

$$-\overline{j_i^{Q\alpha} n_i^\alpha \sigma_s} + \bar{p} \overline{v'^{\alpha}_i \frac{\partial \Phi_i}{\partial x_\alpha}} + \overline{F_{si}^{\times\alpha} v'^{\alpha}_i} \qquad [9.20]$$

With [9.17], [9.19] and [9.20], the total entropy for the two-phase medium can be obtained:

$$\bar{\rho}\frac{D\varsigma}{Dt}=\frac{\partial}{\partial x_\alpha}\Big(-\sum_i\Big(\bar{\rho}_i^*\bar{\Phi}_i\varsigma_i(\tilde{v}_i^\alpha-\tilde{v}^\alpha)+\frac{\bar{\Phi}_i}{\tilde{T}_i}\overline{j_i^{Q\alpha}}^*+\frac{\overline{\rho_i\Phi_i e_i v'^\alpha_i}}{\tilde{T}_i}\Big)\Big)+\sum_i-\Big(\bar{\Phi}_i\overline{j_i^{Q\alpha}}^*+\overline{\rho_i\Phi_i e_i v'^\alpha_i}\Big)\frac{\partial}{\partial x_\alpha}\Big(\frac{1}{\tilde{T}_i}\Big)$$
$$+\sum_i\frac{\bar{\Phi}_i}{\tilde{T}_i}\Big(\overline{\tau_i^{\alpha\beta}}^*-\bar{\rho}_i^*R_i^{\alpha\beta}\Big)\frac{\partial\tilde{v}_i^\beta}{\partial x_\alpha}+\sum_i\frac{\overline{F_{si}^{\times\beta}v'^\beta_i}}{\tilde{T}_i}-\sum_i\frac{\bar{p}_i^*-\bar{p}}{\tilde{T}_i}\overline{v'^\alpha_i\frac{\partial\Phi_i}{\partial x_\alpha}}+\sum_i\frac{\bar{\varphi}_i^Q}{\tilde{T}_i}\qquad[9.21]$$

where the rate of production can be recognized, and must be positive or equal to zero in any case.

Terms related to the additional fluxes can be seen to appear in the rate of entropy production, as well as terms due to the exchange of heat and momentum between phases. How to maintain the positivity of the rate can then be addressed classically, as a linear approximation, by separating the terms of the different tensorial structures.

The first two terms of the production of entropy are relative to the heat dispersion fluxes and to friction in the phases. Without considering couplings with other terms, the two must always be positive or equal to zero. For the friction, when each phase is incompressible (which is initially the assumption of the Prandtl model), a formula similar to the turbulent single-phase flows is obtained, with a Boussinesq-type formula. In a similar way, the heat flux, if considered to be uncoupled from the other fluxes, is expressed in this case by classic temperature gradient law.

The problem of providing an expression for the "effective viscosity" constant, μ_{ei}, for each phase still exists. The kinetic energy of the fluctuations in each phase is produced not only by the mean velocity gradients of the phase, but also by the energy from the contact forces (mainly friction) with the other phases. It is only if this effect is negligible – for example if the mean velocities of the two phases are close, with fluctuations larger than the difference of mean velocities (the domain of application of the global model) – that a law similar to the Prandtl classic mixing length law can be justified, directly linking the turbulent viscosity coefficient to the mean velocity gradient.

The other terms are specific to the two-fluid models. The last term corresponds to the exchange of heat between phases. Like $\sum_{i=1,2}\bar{\varphi}_i^Q=0$, this term is written as: $\bar{\varphi}_1^Q\left(\frac{1}{\tilde{T}_1}-\frac{1}{\tilde{T}_2}\right)$, and linear thermodynamics classically proposes that this term be proportional to the difference between the inverses of the temperature, or simply to the difference between the temperature:

$$\overline{\varphi}_1^Q = K_Q\left(\frac{1}{\tilde{T}_1} - \frac{1}{\tilde{T}_2}\right) = K'_Q(\tilde{T}_2 - \tilde{T}_1)$$

The coefficient K_Q is not given and depends on the variables of the model, but is always positive. By referring to Chapter 5, section 5.4, where the model of the heat transfers between phases was discussed, it can be remarked that the formulas proposed are indeed of this type. The sign is correct, and the coefficient K'_Q is more explicitly given in Chapter 5.

The second last term contains $\overline{v'^{\alpha}_i \frac{\partial \Phi_i}{\partial x_\alpha}}$, so it is necessary to be able to give an approximation of this term to close the balance equation for the phase volume [4.9a]. Assuming that $\overline{p}_i^* = \overline{p}$ eliminates this term, and at the same time the need for [4.9a]. The equilibrium of the mean pressures therefore removes a source of irreversibility. Instead of this hypothesis, irreversible linear thermodynamics would propose $-\overline{v'^{\alpha}_i \frac{\partial \Phi_i}{\partial x_\alpha}} \propto (\overline{p}_i^* - \overline{p})$, which gives a closure assumption for [4.9a]. The balance equation for the phase volume then becomes: $\frac{\partial}{\partial t}(\overline{\Phi}_i) + \tilde{v}_i^\alpha \frac{\partial}{\partial x_\alpha}(\overline{\Phi}_i) = -\overline{v'^{\alpha}_i \frac{\partial \Phi_i}{\partial x_\alpha}} = K(\overline{p}_i^* - \overline{p})$, and it can be observed that this equation is practically identical to the equation proposed by Baer and Nunziato [BAE 86] (see Chapter 4, section 4.3).

Finally, the term $\sum_i \frac{\overline{F_{si}^{\times\beta} v'^{\beta}_i}}{\tilde{T}_i} = \sum_i \frac{1}{\tilde{T}_i}(\overline{F_{si}^{\times\beta} v_i^\beta} - \overline{F}_{si}^{\times\beta}\tilde{v}_i^\beta)$ is linked to the exchange of momentum between phases. In the same way as $\sum_{i=1,2} \overline{F}_{si}^{\times\beta} = 0$, there is $\sum_{i=1,2} \overline{F_{si}^{\times\beta} v_i^\beta} = 0$ since the velocities of the two phases on the contact area are equal (and equilibrium of the total energy exchanges then results in $\sum_{i=1,2} \overline{\varphi}_i^Q = 0$ as stated previously). These terms can be written by defining a mean friction velocity over the interface between phases (and only one, since the velocity of the two phases in this point is the same), $\overline{v}_{sf}^\beta$, and then $\overline{F_{si}^{\times\beta} v_i^\beta} = \overline{F}_{si}^{\times\beta}\overline{v}_{sf}^\beta$. The source term of entropy is then

$\sum_i \frac{1}{\overline{T}_i} \overline{F}_{si}^{\times\beta} (\overline{v}_{sf}^{\beta} - \tilde{v}_i^{\beta})$. Classic irreversible linear thermodynamics leads to $\overline{F}_{si}^{\times\beta} = K_i (\overline{v}_{sf}^{\beta} - \tilde{v}_i^{\beta})$, and with $\overline{F}_{s1}^{\times\beta} = -\overline{F}_{s2}^{\times\beta}$, this results in an approximation of this mean interface velocity: $\overline{v}_{sf}^{\beta} = \frac{K_1 \tilde{v}_1^{\beta} + K_2 \tilde{v}_2^{\beta}}{K_1 + K_2}$, which finally means: $\overline{F}_{s1}^{\times\beta} = \frac{K_1 K_2}{K_1 + K_2} (\tilde{v}_2^{\beta} - \tilde{v}_1^{\beta})$ and $\overline{F}_{s2}^{\times\beta} = \frac{K_1 K_2}{K_1 + K_2} (\tilde{v}_1^{\beta} - \tilde{v}_2^{\beta})$.

As previously for K_Q, the K_i values are coefficients that depend on the characteristics of the two phases in contact, and on the manner in which this contact is made. They must be positive for the source of entropy to be positive, no matter what the temperature of the phases, but they can also be negative, keeping the source of entropy positive if they are functions of the temperature of the phases. For identical phase temperatures, it is clear that negative values can be allowed without the source of entropy being negative as long as $K_1 K_2 (K_1 + K_2) > 0$.

To more precisely deal with the case where there is a difference in temperature between phases, a coupling between the various sources of irreversibility must be considered, in accordance with Curie's theorem. In particular, it then becomes clear that it is possible that the force exerted by the other phase on phase i, $\overline{F}_{si}^{\times\beta}$, is also linked to the temperature gradient in the phases. Inversely – and symmetrically according to Onsager – the heat fluxes in the phases could depend on these contact forces. This is also a sort of thermophoresis effect.

A certain number of models for this force between two phases were seen in Chapter 5, section 5.3, in different situations. The Stokes theory, for the drag of a spherical solid particle in a homogenous medium, gives a formula of the same type as the formula obtained above, proportional to the difference between the velocities, with the correct sign:

$$\overline{F}'^{\alpha}_{sp} = -\frac{1}{2} \rho_f v_f \frac{\overline{\sigma}^s_{pf}}{\overline{\Phi}_p} (\tilde{v}_p^{\alpha} - \tilde{v}_f^{\alpha}) \overline{\sigma}^s_{pf}$$

However, irreversible thermodynamics developed here does not allow the incorporation of a "drift velocity", because the gradients of the volume fractions do not appear in the production of entropy. The approach probably has to be pushed further by taking into account at least the correlation q_{fp} in the definition of entropy.

We have emphasized the interesting implications of the thermodynamic approach here, which justify practically all the approximations that are classically used. These consequences can also be generalized by considering more evolved models taking into account the fluctuations within each phase, like k–ε, for example.

Chapter 10

Experimental Methods

10.1. Introduction

As we have discussed in previous chapters, the balance equations for multiphase flows contain a number of terms that require modeling. In this respect, the situation is similar to problems encountered in the modeling of single-phase turbulent flows, for which experiments are essential. Therefore, it seemed important to devote a separate chapter to the different experimental methods available, as they not only form the basis for the characterization of multiphase flow properties, but also provide the ability of inferring modeling hypotheses that allow the balance equations to be solved, and, therefore, enable the setup of pertinent numerical modeling. However, most of the methods presented in this chapter will only deal with the study of two-phase flows, consisting of one continuous phase and one dispersed phase. The combination of several of these methods may be used to study situations where there are more than two phases. This is the reason why "two-phase media" will be predominantly discussed in this chapter.

Various variables or quantities must be measured. These include the velocities, sizes and locations of particles in the dispersed phase, the velocity of the continuous phase, the concentration of the dispersed phase (void fraction or fluid content), temperature and other variables linked to the chemical composition (concentration of certain gaseous elements or specific liquids). The complexity of flows often originates from combinations of these variables, resulting in nonlinear equations.

The presentation will start with the most classical methods, which are still the most widely used, before more modern optical methods, most of which are still at the developmental/perfection stage. It is important to remember that there is no universal, general technique (possibly besides particle image velocimetry (PIV), provided the particles are large enough) that may be used for all types of flows. The various measuring methods differ according to the type of the flow being considered (for example gas bubbles in a liquid, or droplets of liquid in a gas), and some of them are only applicable when the dispersed phase is present in very small quantities. Significant progress has been made in the past few years in optical methods, in particular, due to the development of cameras (in spatial and frequency resolutions especially) and of software tools adapted for the storage and processing of the obtained data. It is predicted that the development of optical methods will therefore expand despite the cost being significantly higher than that of intrusive methods. Furthermore, in view of the complexity of the situations being studied, and therefore of the variables involved, it is often necessary to perform simultaneous measurements that combine several methods.

The information will be presented in a fairly standard order, discussing intrusive methods before non-intrusive methods. This order also corresponds to the progression from simple to more complex methods. As a result, the presentation is broadly similar to the evolution going from the more robust methods that are used for industrial applications (notably in nuclear and oil industries) to methods that are still being developed in university laboratories or those of larger research organizations (such as the French Aeronautics and Space Center, the French Atomic Center, and their equivalents in other countries). Only the methods concerning "local measurements" will be considered as they allow the characterization of physical phenomena, and therefore help in their modeling and interpretation in the form of balance equations. "Global measurement" methods (used to determine the flow rate in ducts, for example) will not be considered even though they are of great significance for practical applications as they enable installed devices to be monitored in industry; refer to [ODD 04] or [FAL 09] for further information. The subject of aerosols (where particles are present in very weak concentrations, with diameters ranging from less than 1 µm to a about one hundred micrometers), which require the setup of specific measurement methods, will not be discussed either in the remaining book.

10.2. Intrusive methods

In comparison to standard methods, based on and developed for single-phase flows, the application of intrusive methods to two-phase flows requires a special treatment for the signal because of the need to identify contributions from the dispersed and continuous phases, which are then treated in distinct ways.

10.2.1. *Pitot tubes*

Similarly to the single-phase flows, whether they are sub- or supersonic, Pitot tubes are the simplest technique available to measure velocity within a flow. On the other hand, they are only able to provide information on the mean velocity. In the case of two-phase flows, mainly liquid–gas, gas–liquid or liquid–liquid configurations, Pitot tubes measure the mean velocity of the continuous phase. The operating principle is very simple (see Figure 10.1): a double-walled tube is placed in front of the mean flow. The tube's central duct leads to the upstream face in order to determine the mean total pressure (also called stagnation pressure), $P_t = P_s + \frac{1}{2}\rho U^2$, whereas the annular section of the tube, probing the flow by means of several lateral holes, enables the average static pressure, Ps, to be determined. The tubes connecting both these ducts are linked to a differential manometer and the velocity U can therefore be found via the measurement of $\Delta P = P_t - P_s$. Although this method is very simple to carry out in a fluid that has a constant density, this is not so for two-phase flows as even the mean density depends on whether the tube is in contact with the continuous or the dispersed phase. In this case, it is necessary to pair the Pitot tube with another device to have direct or indirect access to the value of ρ. A simple hot film (see below) or more sophisticated devices such as gamma-densitometer (see further sections) can be used in this way and the following relationship is obtained:

$$U = \sqrt{\frac{\Delta P}{0.5\bar{\alpha}\rho_c + J(1-\bar{\alpha})\rho_d}} \qquad [10.1]$$

where ρ_c and ρ_d, respectively, represent the density of the continuous and dispersed phases, $\bar{\alpha}$ is the mean presence rate of the dispersed phase and J is the momentum transfer coefficient, having a value of 0.5 for liquids transporting gas bubbles (as they closely follow the current lines that go around the front face of the tube). However, the following form of the equation is sometimes used, in particular for liquid–vapor flows where the density of the vapor is negligible in comparison to that of the liquid (ρ_l):

$$U = \frac{1}{k}\sqrt{\frac{2\Delta P}{(1-\bar{\alpha}^2)\rho_l}} \qquad [10.2]$$

The constant k is obtained through calibration and has a value close to 1.

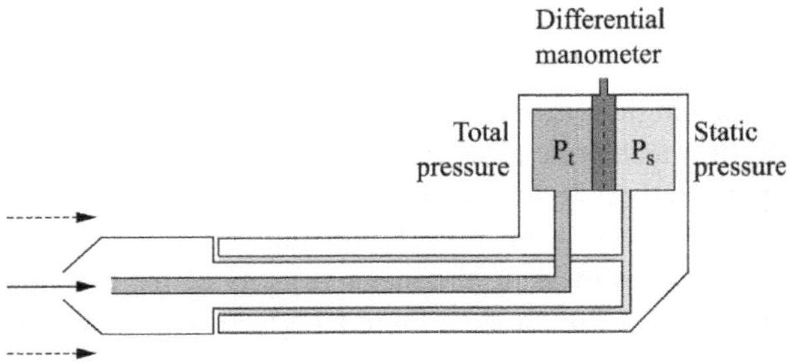

Figure 10.1. *Diagram of the principle of Pitot tubes*

The diameter of the section, where the measurements for static pressure are made, is usually a few millimeters (1–10), and its length is approximately 15 times this diameter, the holes that measure the static pressure are distributed along the circumference of the section located three diameters away from the stagnation point.

In addition, it should be noted that the simple measurement of the differential static pressure Δh acting on the wall between two sections of a vertical tube allows the void fraction $<\bar{\alpha}>$, averaged along the height L that separates the two sections, using $<\bar{\alpha}> = \Delta h/L$, to be found for a flow of liquid containing gas bubbles.

10.2.2. *Hot films*

Measurement by hot wire (or film) anemometry is commonly used for single-phase flows. They enable one or several velocity components to be determined and split into mean variables and fluctuations depending on the number of sensors used. For two-phase flows, the same method is also used even though its implementation is more difficult as the sensor response becomes very complex. This requires the signal to be analyzed in a particular way for which improvements are still being researched.

In its most basic form, the hot film technique is based on the use of sensors having a fine metallic element (wire or film) heated by means of a perfectly controlled current. The aim is to maintain a constant temperature difference between the film and the flow, despite the fact that the flow takes away some heat by forced convection and this depends primarily on the flow velocity at that precise location. The constant temperature difference is maintained by a specific electronic circuit (termed the feedback loop) that varies the current in the wire to maintain a constant resistance. Only the voltage at the wire terminals therefore needs to be measured in order to determine the flow velocity. The relationship between the voltage and the flow velocity can be found theoretically, but in practice it is usually found by calibration due to various external factors that are difficult to determine and that modify the wire's response (soiling by microdust, ageing of welds on the support and electrical connections, etc.).

For two-phase flows, often consisting of one or two liquid phases with a density close to 1,000 kg/m^3, a straight film is often used instead of a straight wire tightened between two metal pins (Figure 10.2(a)) so as to increase the mechanical resistance. In some cases, a ring of metallic deposit is placed on the tip of a quartz conical support (Figure 10.2(b)) and operation is the same.

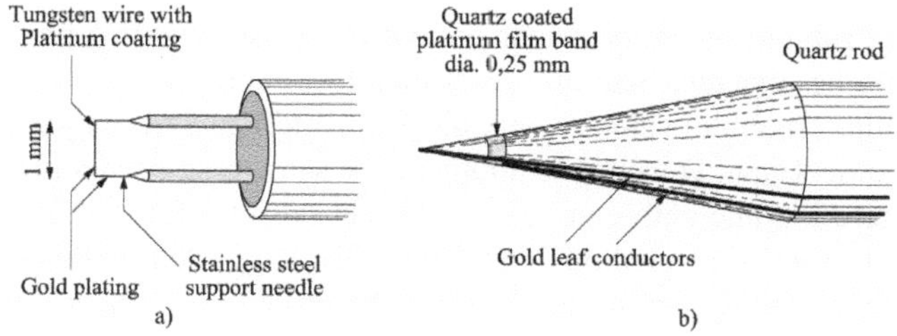

Figure 10.2. *Hot wire sensors: a) straight and b) conical*

As it has been explained earlier, the probe response is clearly more complex and difficult to interpret quantitatively when it is placed in two-phase flows. By considering flows that have one gas phase and one liquid phase, the heat conduction will be very different depending on whether the sensitive part of the probe is in the liquid or in the gas. Figure 10.3 shows a typical example of the signal obtained, while taking measurements in a liquid flow containing gas bubbles, focusing on the impact that the presence of a gas bubble has on the signal (the voltage between the probe's terminals tips) represented in the time series by a sudden drop.

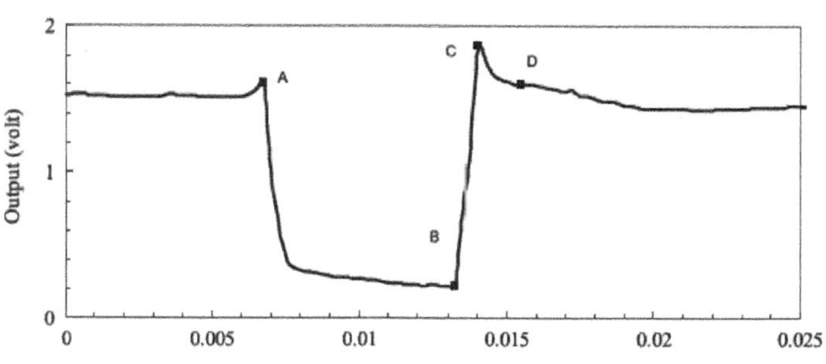

Figure 10.3. *Voltage measured at the terminals tips of a straight hot film probe when a gas bubble passes on it [HAM 10]*

Conventional (see, for example, [HAM 10]), but more complex analyses are still being developed, which include the use of fast cameras to simultaneously obtain pictures of the passage of the bubble in order to take into account its size and the direction of its movement relative to the hot film, four specific instants are considered (listed as A, B, C and D) and the signal processing relies on these four instants. Point A corresponds to the instant when the upstream side of the bubble touches the film, causing a sudden decrease in the voltage at the terminals tips. This drop extends until point B, corresponding to the instant when the downstream side of the bubble touches the film. This stage continues until point C, when the hot film is no longer in contact with the bubble, meaning that the voltage increases fairly rapidly. The section between C and D corresponds to the relaxation that follows the formation and rupture of the meniscus around the sensor and this part of the signal must therefore be rejected. Section CD is therefore more or less visible depending on the situation (bubble velocity, bubble size, etc.) and it may be difficult to distinguish points C and D from one another. The same case occurs in the vicinity of point A as there is some ambiguity as to its exact position for the same reason. The various signal processing methods that have been suggested to this date, and that some researchers are attempting to improve, therefore aim to clearly determine the positions of points A and C. To achieve this, it is possible to use threshold methods, first- or second-order derivative calculations, two staggered sensors, etc. The aim of the current discussion is of course not to present this in detail. For further reading, the reader may refer to [HOG 97] or [REN 05b].

The signal processing step is very important as the gas phase's mean presence rate can be simply obtained from the time separation between A and B by: $\overline{\alpha} = \frac{\sum \Delta t_{AB}}{T}$. The values of the fluctuating velocity in the liquid phase, and the resulting statistics, such as the mean velocity, its standard deviation or its spectral distribution, are obtained by using the signal from which sections between points A and D are removed. For the spectral distribution, the question of how to gather the signal pieces corresponding to the liquid's velocity measurements in order to evaluate its spectrum is also an important issue that is the subject of numerous studies, based either on modified Fourier calculations or time correlation calculations. There is no particular problem with the calibration of these probes as it is carried out in the liquid via the standard method for single-phase flows.

In addition to standard signal analyses by hot films, it is also possible to implement more sophisticated treatments. By correctly estimating the signal's time derivative in the decreasing section after point A, it is also possible, for instance, to obtain the velocity of inclusions (bubbles or droplets) and therefore the slip velocity between these inclusions and the continuous phase.

10.2.3. *Optical needle probes (single probes, bi-probes and quadri-probes)*

These sensors were mainly developed and perfected over the years to determine the characteristics of gas bubbles in carrier liquid flows and to measure the characteristics (velocity, size, etc.) of inclusions (most often gas bubbles, but also droplets). They are relatively simple to implement and are often paired to hot films. In the latter case, the needle probe provides information on the gas phase, and the hot film on the liquid phase.

The principle of the method is based on the refraction index difference of the medium present at the tip of the optical fiber (the active part of the fiber is also called the *optrode*). The optrode is linked to an optoelectronic module via an optical cable. The light emitted by the module's light-emitting diode is guided by the cable to the optrode's tip. The light is partially or totally reflected (depending on the medium surrounding the probe), goes back through the reverse route, is separated by a coupler and lights the photo-detector as shown in Figure 10.4. The analog electrical signal provided by the photo-detector is then amplified and converted into a two-level logical signal representing the state of the medium phase. The tip of the optic fiber is cut so as to have a low radius of curvature that enables total internal reflection in air, and the detection of small bubbles (up to 10 μm in diameter, depending on the manufacturer). The method was initially developed and perfected to analyze gas bubbles in carrier liquids but has been equally applicable, due to recent developments, to the analysis of droplet properties in gases. Optic fibers are made from quartz. They are wrapped in a protective sheath (usually Teflon) and the end is stripped along its length (over a length of approximately 1 cm) to enable measurements to be taken. Their diameter is always under 200 μm and often less than 50 μm.

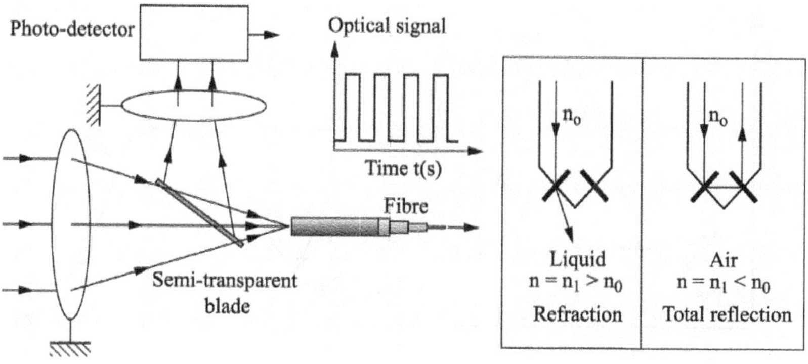

Figure 10.4. *Operating principle of a single optical probe (detection of gas bubbles in a liquid: $n_0 = 1.15$, $n_{air} = 1$, $n_{water} = 1.33$)*

The use of single probes provides the indicator function of the phase, which provides the local time-averaged void fraction. However, optical bi-probes are more commonly used as they also provide information on the velocity of the bubbles. They integrate two simple optical probes, their sensitive parts separated by a distance a. The time lag between the two signals and the bubbles' residence time allows the local velocity u and the chord length l of the bubbles to be found. Most of the time, these dynamic variables require the two optical signals to be numerically processed. In general, the study of flows containing bubbles of mean diameters between 2 and 10 mm uses bi-probes separated by a distance a of approximately 1–5 mm so as to minimize the likelihood of a bubble "pricked" by the upstream fiber to be missed by the downstream fiber.

For each probe, the delivered signal is similar in shape to that shown in Figure 10.5. Similarly to hot films, sudden changes in the voltage provided by the probe are observed and are associated with the transitions linked to the passage of bubbles on the probe (see, for example, [BAR 99]). For case type I, the increase starting at point A corresponds to the arrival of a bubble: the voltage is much higher than the low voltage V_L, corresponding to the situation when the probe is in complete contact with the liquid. The departure of the bubble is recognizable at point B, corresponding to a significant decrease in the voltage relative to the level V_G reached when the probe was completely surrounded by gas. Situations may also arise in which the probe is surrounded by a combination of gas and liquid (case type II). In this case, it is often necessary to combine a rise time (termed T_r) to detect the passage of drops on the probe. The fine tuning of these different parameters is typically found using situation tests where bubbles having a controlled but variable size are successively injected into a vertically ascending or descending flow at a variable rate so as to vary the mean void fraction.

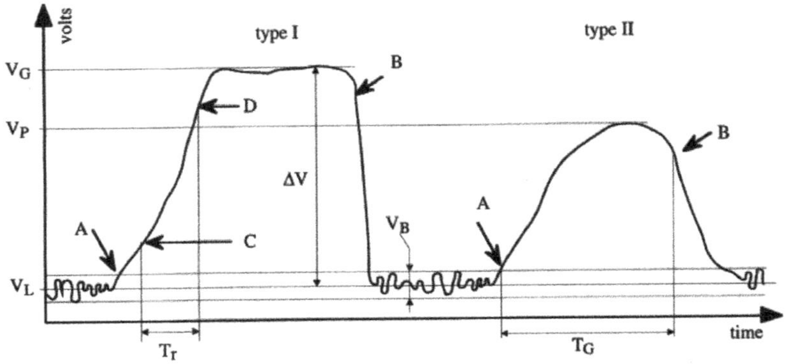

Figure 10.5. *Voltage measured at the terminals tips of one of the probes in a bi-probe when a gas bubble passes on it [BAR 99]*

Once the raw signals have been preprocessed to separate the sections corresponding to the presence of bubbles on the probe, the following characteristic variables are calculated:

– The mean bubbling frequency, obtained by accountancy of the number of bubbles having touched the first fiber of the bi-probe over a given time T.

– The mean void fraction, obtained directly from the crenellated signal by calculating, similarly to the hot film probe, the sum of the time durations for which a bubble was present on the first fiber of the bi-probe: $\overline{\alpha} = \dfrac{\sum \Delta t_{AB}}{T}$.

– The bubble speed, V_{bi}, by associating each detected bubble on the first probe with its detection on the second probe, taking into account the distance between the bi-probe's two elements and each bubble's time-of-flight (t_i). This procedure is quite delicate to implement as it has to be ensured that the bubbles detected on each of the two probes are actually the same. A specific validation procedure thus needs to be set up. This will not be discussed here due to lack of space. Different methods and their variations have been put forward by various authors. It is therefore possible to infer the mean velocity of bubbles, their variance, etc.

– The bubble size, for each bubble detected on both fibers, using chord length calculations, $y_i = V_{bi}.t_i$. The average of these chord lengths will give the mean chord length. Adjusting the chord length function to a log-normal distribution, and combining it with an assumption about the shape of the bubble (spherical or elliptical) allows the mean bubble size and their Sauter diameter d_{32} to be found. With this, it is possible to find the mean interfacial area per unit volume $\overline{\sigma_i}^s$ by the relationship: $\overline{\sigma_i}^s = \dfrac{6\overline{\alpha}}{d_{32}}$. As we have discussed in previous chapters, $\overline{\sigma_i}^s$ is a very important quantity for the characterization and modeling of two-phase flows.

Even if these probes are still in current use, a number of approximations are required in order to use them, especially with regard to the actual three-dimensional (3D) bubble movement. Some authors have therefore proposed to use triple or quadruple probes in order to remove these inaccuracies. Quadruple probes in particular (see [SHE 05] and Figure 10.6) enable the three velocity components to be found more precisely. Above all, the value of $\overline{\sigma_i}^s$ obtained is then much more accurate and is given by:

$$\overline{\sigma_i}^s = \frac{1}{T} \sum_{l} \frac{1}{\left| v_{sl}^{\alpha} n_{sl}^{\alpha} \right|} \qquad [10.3]$$

where v_{sl}^{α} represents the α velocity component of the lth interface passing at the point being considered, with n_{sl}^{α} then being the α component of the surface normal unit vector of this lth interface.

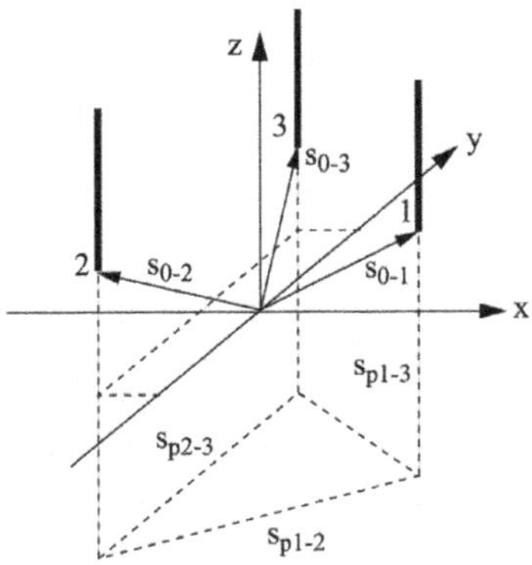

Figure 10.6. *Geometric arrangement of the probes in an optical quadri-probe [SHE 05]*

As we have mentioned earlier, single optical probe can also be used to characterize the properties of drops or droplets scattered in a gas flow. The response of these probes is similar to the example shown in Figure 10.5, the difference being that the reference level is the high voltage corresponding to the probe being completely immersed in gas, whereas the passage of droplets will be represented by fast drops in voltage to a low level corresponding to the passage of a droplet entirely wetting the sensitive part of the probe. Similarly, specific treatment is used in order to accurately determine the time taken for a droplet to pass over the probe (see, among others, [HON 04]) or the droplet's velocity. Variables such as the dispersed liquid phase's volume flux, the chords' probability distribution, the droplets' mean and Sauter diameters, the mean presence rate of the dispersed phase or the interfacial area per unit volume can thus be inferred.

Furthermore, conductive needle probes are also available and are based on similar operational principles [MAN 09, HOG 97, CAR 91]. These probes are made of a metal resistant to oxidation, usually steel, corundum or a platinum-based alloy. The probe is isolated along its longest length, the end is tapered (generally so as the sensitive part is under 1 mm) and the whole probe is mounted on a steel chassis. A voltage of 1 to several volts is applied to the probes and a potentiometer assesses the potential difference between the probe and the chassis that also serves as an electric grounding. When the probe is surrounded by liquid (in such a way that the circuit is closed, fictional switch in Figure 10.7), a voltage under that of the power source is obtained and a higher voltage in the opposite case. The liquid medium may not be conductive enough to obtain a clear reading and salt may therefore be added. This type of probe operates in assemblies with four needles, similarly to optical probes. A detailed comparative study performed by Le Corre *et al.* [LEC 03] showed that the perfomances of the two systems ware very close to each other and the small differences between the velocity, void fraction and interfacial area per unit volume measurements were actually caused by the varying geometry of the needles (mainly the size).

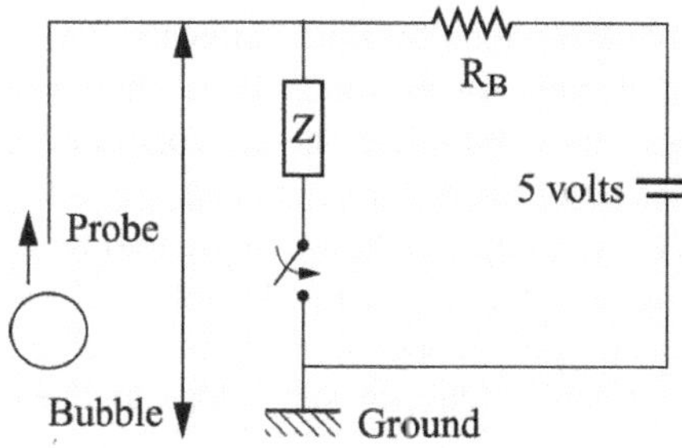

Figure 10.7. *Principle scheme of conductive needle probes*

The simple version of this system, sometimes called turbidimeter, is used for solid particles within a liquid. The system is based on the use of diodes that project infrared light in a fluid. The solid particles suspended in the fluid then backscatter the projected light, which is detected by the receiving diode. A specific electronic device then calculates the fluid's relative turbidity using the signal received in order to determine the flux of solid particles. Even though the technique is not new and is limited in the results that it provides, it is still the subject of improvements that, for example, aim to optimize the angle between the emitter and receiver, limiting the background noise and making the measurement principally sensitive to the first reflection from particles, which therefore reduces the errors caused by multiple reflections of the emitted light on particles.

10.2.4. *Wire networks*

Wire network methods are based on crossed wires set up in such a way that they occupy the space where the flow takes place (see Figure 10.8). There are usually two layers of orthogonal wires, although sometimes three layers are used. The principle of the method is based on the fact that the conductivity of the medium separating the two wire layers changes point-by-point according to the gas content within the liquid. The upstream layer consists of transmitting wires and the downstream layer of receiving wires. When electric voltage pulses are sent consecutively through all the transmitting wires, the position of the wire in the row being important, an induced current is created in the receiving wires that depends on the conductivity of the small volume surrounding the area adjacent to the intersection of receiving and transmitting wires. For every pulse in a transmitting wire, the response of all the receiving wires is recorded. A mutual influence matrix is therefore obtained between transmitting and receiving wires that allows the volume fraction of the gas in each elementary volume, α_{ijk}, to be found. A preliminary calibration in the liquid alone is therefore required (where i is the position of the volume in one of the space directions, j is for the perpendicular direction and k is the specific time of acquisition). To process the data, it is necessary to start by locating the areas in space where gas entirely surrounded by liquid appears. This enables the location of bubbles to be known at any given time. It is then possible to analyze the characteristic properties of these bubbles. The networks in each plane consist of 16–64 wires, approximately 100 µm in diameter, and are activated by pulses mostly ranging from 1 to 10 kHz. The distance between the two wire planes is kept as small as possible while making sure that the wires do not touch. The values obtained for variables such as the mean void fractions, the bubble velocity, the chord length or the interfacial volume per unit area, are very close to those obtained using other methods, for example needle probes or X-ray tomography (see, among others, [MAN 09]). Even though the bulkiness of the device in comparison to the flow can be problematic if the adequate precautions are not taken, the method is still of significance as it gives direct measurements for the entire plane, unlike the optical probes that only give information on particular points.

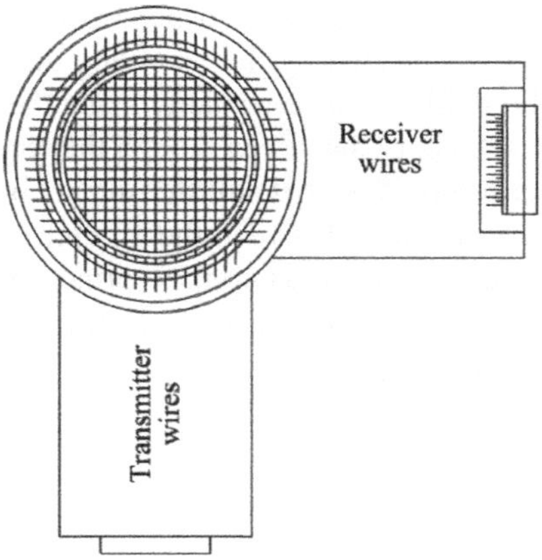

Figure 10.8. *Positioning of a two-layer wire network [MAN 09]*

10.3. Non-intrusive methods

Ideally, the flow should not be disturbed by the use of intrusive probes when taking measurements for flow dynamics. This can be achieved via various methods that will be discussed now. However, these methods are not necessarily well adapted for all multiphase flows and are often only applicable for relatively small α values due to multiple reflections of one or more laser beams on inclusions or because the laser beams are attenuated too significantly as they pass through the considered medium.

10.3.1. *Particle image velocimetry (PIV)*

Over the last three decades, PIV has become a standard tool in experimental fluid mechanics. It is an imaging technique that allows optical flow measurements in liquids and gases. It is based on injecting tracking particles in the flow. A high concentration of tracers is used to obtain more accurate measurements (the meaning of "high" will be discussed later) so that groups or packets of particles may be tracked rather than individual particles. The thickness of the volume being measured is reduced to a thin sheet of light, which lights up the particles that move in it. The principle of PIV consists of taking images (grouped into pairs) of the particles, using a digital camera, at two time instants that are very close to each other. The images are used to measure the displacement of the particles in the nodes of the light sheet mesh (positioned according to the direction of the main flow). The velocity is then calculated from the displacement as the time delay between the two images in a pair is known. The use of one camera alone enables only the measurement of the two transverse components of velocity (i.e. the components located in the direction of the light sheet). Synchronized PIV uses two cameras in stereoscopic configuration and allows the axial component to be measured as well (this technique is called "stereo-PIV" (SPIV)). In general, only the most common basic principles will be discussed hereafter. More specific or technical developments (that may, for example, use two or more cameras to determine the three velocity components simultaneously or vorticity components) will not be mentioned here and the reader should refer to more specialized works such as [BOU 12a] or [WES 13].

The light source is commonly a pulsed Nd:Yag laser with two cavities and the light sheet is generated by a cylindrical lens. Two-frame video cameras with charge-coupled device (CCD) sensors are used. The velocity is easily calculated by the relationship: velocity = displacement/time (where time, representing the interval between two images from the same pair and often symbolized by Δt, is very small and of the order of several microseconds, thus justifying the use of double-cavity lasers). On the other hand, the flash repetition rate for each cavity is rather small, usually only approximately 10–15 Hz, although laser-camera systems are now available, which can reach repetition rates of more than 1 kHz (this is referred to as "time-resolved-PIV" (TRPIV)). Each frame is then divided into small scanning areas to obtain intercorrelation using fast Fourier transform (FFT). The resulting correlation peak corresponds to the most probable particle displacement. An instantaneous velocity field is obtained by scanning all scanning areas. Statistical quantities associated with the measured velocity components, such as mean values, standard deviations or their correlation (at one or several points in space), can be found using the high number of images acquired.

PIV systems include the following elements:

– a particle generator (unless the flow naturally contains particles or bubbles that are the subject of the dynamics study);

– a lighting system (cavity pulsed lasers, specialized applications requiring more than one);

– a camera (one or more);

– a synchronization system that synchronizes laser flashes and image capture;

– a digitization and image processing system;

– tuning, calibration and related adjustment systems when a single camera is no longer being used.

Even if the basic principle of PIV (Figure 10.9) is very simple, there are nonetheless a number of complex problems relative to its implementation and optimization so that it is the subject of continuous improvements, the main subjects will be discussed hereafter. To obtain good quality velocity measurements, it is crucial for the user to have a good understanding of these problems and the precise quantification of associated errors and uncertainties.

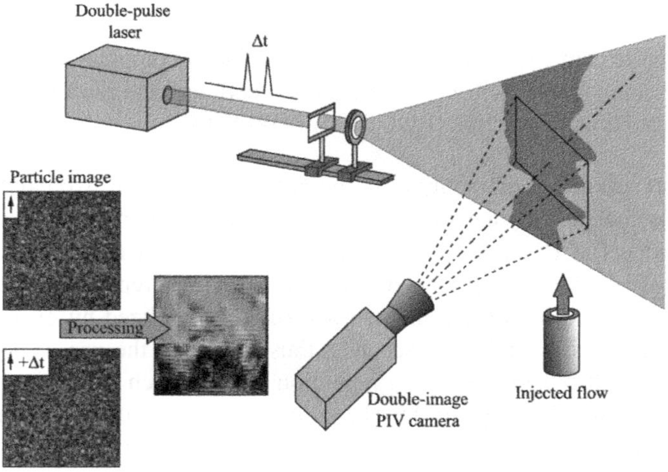

Figure 10.9. *Operating principle of PIV. Excerpt from documents by Dantec Dynamics*

There is indeed nothing trivial about particle generation and numerous solutions are available. The solution/must take into account a number of criteria:

– Chosen particles must be small enough to "follow" the flow well, yet large enough to be "seen". Their density must be as close as possible to that of the fluid being studied. Larger particles can be used in liquids as their density is generally in the same order of value as the density of the fluid. The criterion for particles to follow the flow well is generally that their characteristic time scale, equal to $d^2(2\rho_p + \rho)/36\mu$, must be small in comparison to the smallest characteristic time within the turbulent flow. This is equivalent to their falling velocity, given by $gd^2\rho_p/18\mu$, being small in comparison to the velocities in the flow.

– The particle concentration must be sufficiently high for the number of measurements to match the required resolution and sufficiently low for the lighting and the shooting not to be too disturbed by the particles (the projected surface of the particles in the flow on the horizontal plane must be small relative to the projected flow surface, thus guaranteeing a flow that is unmodified by the presence of particles).

– If the injection of particles is done in an area close to the measurement zone, care must be taken to perform the injection at the local velocity of the flow to minimize disturbances.

The number of particles contained in the laser sheet thickness (Δz_0) is given by:

$$N_s = \frac{C\Delta z_0 \pi D_{ei}^2}{4M^2} \qquad [10.4]$$

where C is the particle concentration in the flow, D_{ei} is the actual diameter of the particle's image and M is the optical system's magnification factor. When $Ns >> 1$, it is generally considered that the diffraction spots interfere with each other, causing the phenomenon of "speckling". On the other hand, when $Ns << 1$, the particle images are individually visible and this subject is relevant to this chapter. Two methods are possible and the choice between them depends on the particle image density Ni, calculated by applying the equation given above to the size of the analysis window. If $Ni \approx 1$, the number of particle images per unit area is very small and each image can be processed individually. This method is called "particle tracking velocimetry" (PTV) and will be discussed very briefly hereafter, even though the method is similar to the method that will be discussed later (DTV). If $Ni > 1$, the number of particle images is high and it is no longer possible to follow each one. The most common method is therefore PIV that allows the displacement of groups or packets of particle images in the analysis window to be measured via a correlation method. The image is subdivided (see Figure 10.10) into subimages (or windows), preferably square, such that their pixel side is a power of two (this allows faster calculations by using an FFT algorithm to obtain correlations). The particle displacement in the window is therefore measured over time Δt. The size of the windows determines the method's resolution as the measured velocity is the overall velocity in each window. The most commonly used sizes range from 16 × 16 to 128 × 128 pixels (for cameras that now have a minimum resolution of 1,000 × 1,000 pixels). Most of the time, an overlap between the windows is used to increase the number of velocity vectors calculated for each pair of images. With an overlap of 50%, for example, and a 16 × 16 scanning window, 255 × 255 vectors can be obtained from images that contain 2,048 × 2,048 pixels.

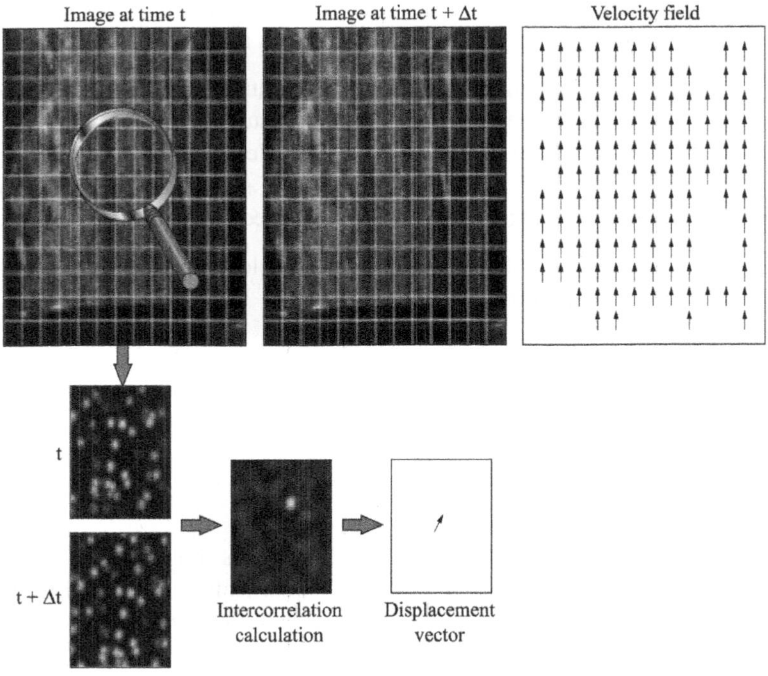

Figure 10.10. *Example of the assessment of a velocity field*

The intercorrelation obtained by FFT (Figure 10.11) shows a main peak caused by the correlation of particle images from the first analysis window recorded at time t with the same images in the second window recorded at time $t + \Delta t$. The displacement is given by the distance between the peaks from the center of the window. Figure 10.11 shows an example of analysis windows processed by intercorrelation. The shape of the peak is similar to the shape of the particle images and can be approximated using a Gaussian function.

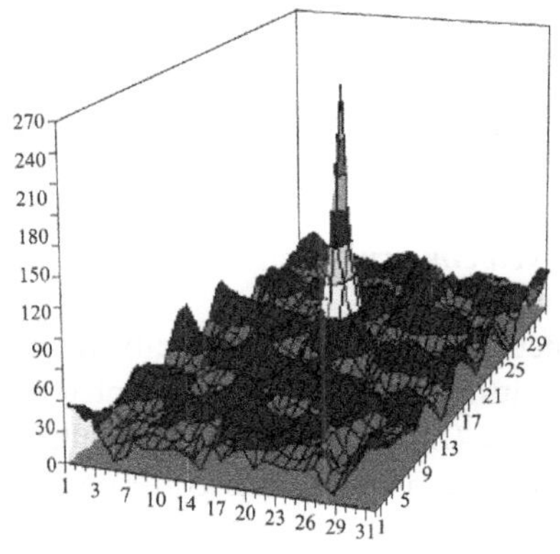

Figure 10.11. *Example of a correlogram*

In the case of CCD-type sensors, the light signal is integrated over the surface corresponding to the size of a pixel. It is possible to show that pixelization spreads the correlation peak as it is convoluted by a tetrahedral function that has a base twice the size of a pixel. It is often necessary to improve the measurement accuracy by implementing subpixel interpolation. This consists of determining the overall displacement value of the particle images within the correlation window with a step smaller than a pixel so as to improve the accuracy of the measurement. The correlation peak's subpixel position is therefore measured using a number of finite points at the peak's enclosure. The centroid can be determined, or a regression can be performed with the help of a function similar to the shape of the theoretical peak. Various continuous functions can be used for this regression and can be either Gaussian or parabolic.

Moreover, the detectability of the correlation peak and the accuracy of displacement measurements are closely related to the signal-to-noise ratio. This signal-to-noise ratio is proportional to the number of particle doublets and inversely proportional to isolated particles. Increasing the analysis window size may contribute to a better signal-to-noise ratio, at the expense of spatial resolution. To significantly increase the signal-to-noise ratio, the second analysis window must be staggered by an amount similar to the mean particle displacement. The particle displacement is therefore determined using a larger number of doublets. Its vector is affected mid-way between the two analysis windows in such a way that the accuracy is similar to that of the centered difference. This stagger is then implemented in successive steps in a multipass-, multigrid-type process. Fluid flows, in particular turbulent flows, contain non-negligible velocity gradients. It is not always possible to consider the entire particle group as a rigid pattern. The window deformation is therefore calculated from a predictor field obtained through iterative interpolation that uses vectors adjacent to the point being considered. The pictures are thus deformed so as to increase the signal-to-noise ratio and thus allow the correlation peak to be reinforced. For more details on these methods, as well as on their various practical uses, the reader may refer to [BOU 12a] or [WES 13].

10.3.2. *Droplet tracking velocimetry*

Although the PIV method provides the velocity field of the flow's continuous phase by using injected particles, the technique adapted from PTV and often called droplet tracking velocimetry (DTV) enables the simultaneous measurement of the velocity and characteristic size of inclusions corresponding to the dispersed phase (most commonly found in the form of droplets). The cornerstone of the method is the same and consists of using imaging as the flow is lit by a laser sheet. The images and movements of inclusions are recorded by a camera in the form of paired images, similarly to PIV. However, significant image processing has to be performed in order to determine the characteristic size of inclusions (usually their diameter if we assume that they are spherical). Their velocity components, in the plane of the laser sheet, can then be found by following their displacement in a pair of images taken one after the other. For example, the images shown in Figure 10.12 demonstrate what is obtained when a liquid spray is atomized; first as a raw image (Figure 10.12(a)), and then as an image that has had a circular mask applied so that all the pixels corresponding to the background lighting have the same maximum value (Figure 10.12(b)). The mask also deletes all peripheral areas where inhomogeneities in the laser's light intensity are especially important from the areas analyzed by computer processing. The highest grayscale value in the image therefore corresponds to a value 4,095, which depends on the camera's sensitivity (12 bits), and the lowest level is zero [STÉ 12].

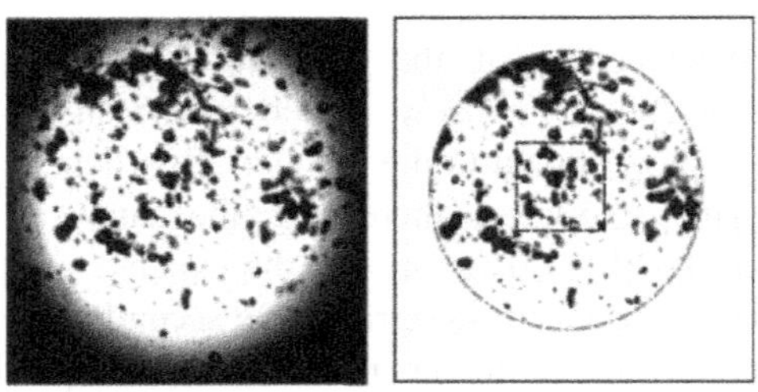

Figure 10.12. *DTV measurements [STÉ 12]*

When performing image processing, a primary thresholding is used to locate the liquid elements present in the image. The most widely used thresholding method consists of recognizing the pixels in the image that have an intensity below a certain threshold. However, by using this method, the drops that have a contrast below the threshold level will not be detected. This method therefore rejects an important number of drops due to lack of fine tuning. Moreover, the presence of a large drop having several different contrast zones is sometimes detected as several drops corresponding to the more important contrast zones.

An alternative method is to perform thresholdings according to the norm of the gradient of the pixel intensity in the image. This is calculated by convolution with a derivative filter. Sobel, Roberts and Deriche filters are a few examples. The image obtained highlights the areas where the pixel intensity varies significantly: the clearer a drop is, the higher gradient value it will have. However, drops that are not as clear will not be detected via this method. In addition, the method has a tendency to open contours and the use of advanced methods is required, such as calculating the wavelet transform of the image. This is obtained by image convolution with a two-dimensional (2D) filter, whose coefficients depend on the wavelet type being considered and its parameterization. A "Mexican hat" type wavelet is generally used as it is sensitive to grayscale concavities and thus detects the interfaces present in the image.

A histogram of wavelet transform values is computed and presented in Figure 10.13. This figure shows values ranging from –0.5 to 0.5, translated along 0.5 to obtain a display in the [0; 1] interval. The histogram's peak corresponds to the uniformly lit background. The value associated with the peak is zero (0.5 in the figure after translation). As we move away from the peak toward 0, the values for the image's wavelet transform are characteristic of detected objects that are of better quality. If the threshold is slightly below the median value, many unclear droplets will be detected. If the value is closer to the wavelet transform's minimum values, only very clear objects will be detected. In general, this thresholding technique permits a larger number of unclear droplets to be detected than the standard thresholding of the image's pixel intensities in grayscale (see Figure 10.14).

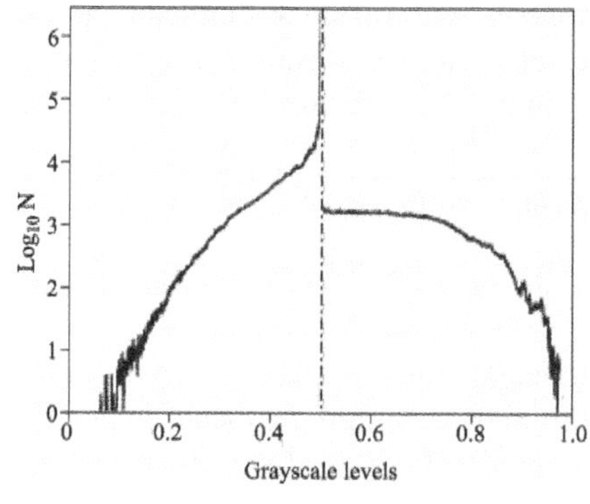

Figure 10.13. *Histogram example of the wavelet transform of the reference image above [STÉ 12]*

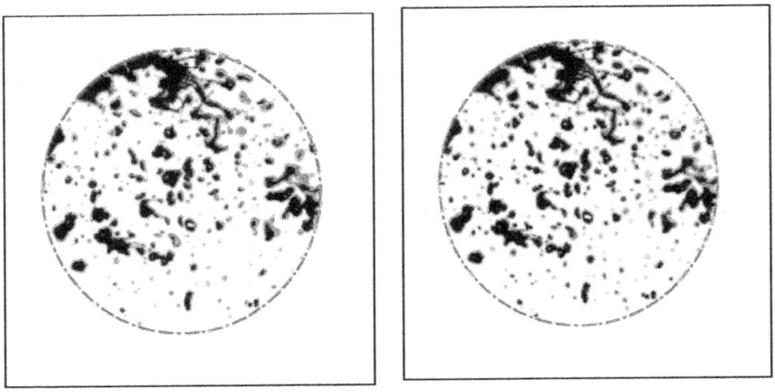

Figure 10.14. *Detection results for the reference image: by standard thresholding, 245 objects (left) and by wavelet transform, 305 objects (right) are detected in the image [STÉ 12]*

This first stage therefore enables the presence of drops to be located in the pictures. The second processing step consists of analyzing each contour referred to, and analyzing, contour-by-contour, the mapping of the grayscale level of each drop, enabling elements that form each drop to be recognized in addition to obtaining their shape's geometrical parameters (diameter, eccentricity of drops considered elliptical, other characteristics for drops requiring a more complex model). These criteria must be capable of evaluating whether the detected objects are on the camera's focal plane, in which case the characteristics found are correct, or whether, on the contrary, the objects are out of plane, in which case the measured characteristics must be corrected. Several methods can be used to determine the characteristic parameters of drops. The method developed by Stévenin [STÉ 12] seems to be one of the most accomplished methods. The basis of the method is to determine the apparent diameters of spheres with different diameters, placed consecutively in and out of the focal plane, at a known distance counted in terms of a number of diameters, and by combining the information relative to the maximum grayscale gradient G_{max} ($G_{max} = \dfrac{h}{K(D_{l=75\%} - D_{l=25\%})}$, where K is the number of pixels per unit length) and to the contrast C ($C = \dfrac{i_{max} - i_{min}}{i_{max} + i_{min}}$). A mask is first applied locally for each drop detected in order to isolate the drop from its environment. The data for contrast C, the object's maximum and minimum grayscale intensities, respectively i_{max} and i_{min}, and the grayscale gradient G_{max} along its edges are then collected. The equivalent diameter of the drop is assumed based on the relative level i_{ref} at 50% of the grayscale profile height $h = i_{max} - i_{min}$ and is called $D_{l=50\%}$. The calibration data are then used to improve the size estimate, taking into account object defocusing, approximated according to the grayscale contrast and gradient at the interface of the drops. When each drop has been detected and geometrically characterized, it is easy to obtain the two velocity components in plane with the image by following the movement of the drop from one of the images in the pair to the other.

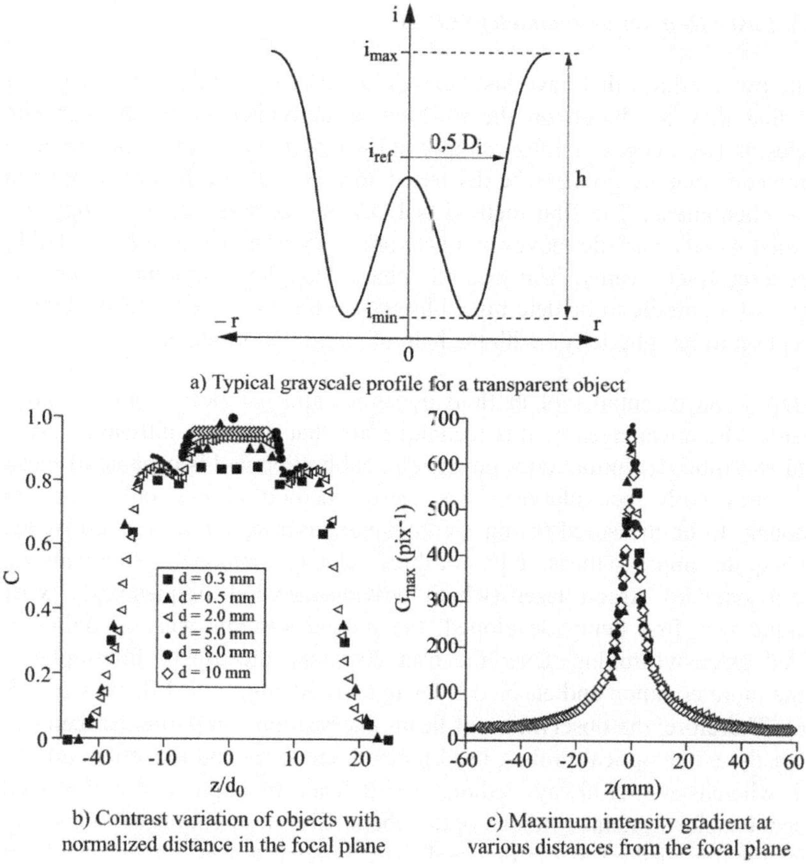

a) Typical grayscale profile for a transparent object

b) Contrast variation of objects with normalized distance in the focal plane

c) Maximum intensity gradient at various distances from the focal plane

Figure 10.15. *Determination of the apparent diameters for spheres having different diameters [STÉ 12]*

This DTV method is therefore very significant as it is capable of simultaneously obtaining the size and velocity of inclusions without having to inject particles. It is then possible to calculate either simple statistics for velocity or size, or joint statistics between size and velocity, such as the mean velocity or its standard deviation, according to the size class. It is important to note that fluorescent particles can be injected into the continuous phase. Two cameras can then be used, one equipped with a filter to suppress the wavelength of the fluorescent light, and the other with a filter that eliminates the frequency of the laser, to simultaneously provide the velocity field of the continuous phase by PIV and the inclusion characteristics by DTV.

10.3.3. *Laser Doppler anemometry (LDA)*

The two methods that have just been discussed can be related to imaging in the sense that they are based on the tracking of individual particles, or packets of particles, in two images of a flow simply lit by a laser sheet and taken consecutively. The methods that are going to be discussed now, in contrast, involve more complex optical phenomena. The first method is LDA and is based on the Doppler effect generated when a particle moves in a network of interference fringes created by two intersecting laser beams. The second, phase Doppler anemometry, enables the velocity of a particle to be determined by measuring its diameter (depending on the assumption of its sphericity), with the help of several photo-detectors.

LDA is an essential tool in fluid dynamics and has been used for more than 30 years. The advantages of this technique are that it is non-intrusive, has a high spatial and time resolution, does not require calibration and is capable of measuring flow zones with recirculations. The basic method allows only one velocity component to be measured (using a continuous laser source at a given frequency), yet it is quite simple to measure two or three velocity components simultaneously by using a standard Argon laser (which provides several frequencies). When the technique was first being developed, the optical systems were complicated and required extensive tuning. Over the past 20 years, the use of fiber-optic probes became more common and enabled LDA to be used simply and flexibly (see Figure 10.16). Therefore, the observation of beam intersections, in particular, can be made very easily as the optical probes, used for both emission and reception, are factory-tuned, whereas it was always tedious and delicate to improve the observation of intersections for systems without optic fibers. The observation of intersections is crucial for this method as the detected optical signal is sent directly to the photo-detector (also called photo-multiplier). As shown in the figure, the initial beam is split into two, and one is sent via a Bragg cell. This Bragg cell consists of a crystal block fixed to a piezoelectric crystal; the vibrations emitted by the piezoelectric crystals create sound waves that act as an optical grid, thus enabling the laser light to be diffracted and the frequency altered by sound waves. The frequency offset is very important as it allows its direction of the flow to be found (and therefore the sign of the velocity) by forcing the interference fringes, located at the intersection of two beams, to move continuously. Without this, we can only measure the velocity module.

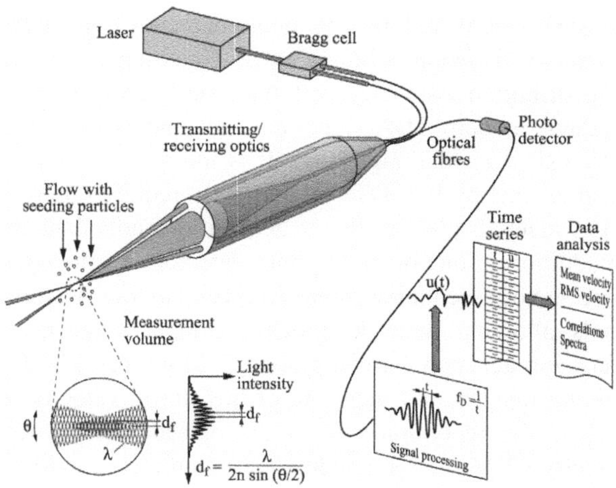

Figure 10.16. *Diagram of the principle behind LDA. Excerpt from documents by Dantec Dynamics*

The network of interference fringes created at the beam intersections (see Figure 10.17) is therefore the active part of the device that takes the measurements; the measurement volume is often considered as being elliptical. The interfringe, the basis of the particle velocity measurement, is linked to the characteristics of the two laser beams, namely the divergence angle θ and the wavelength λ, by $d_f = \frac{\lambda}{2n \sin(\theta/2)}$ (where n is the index of the medium in which the measurement is being taken, and θ is the divergence angle of the incident laser beams in the medium). It should be noted that the fringe network is located in the plane created by the two incident beams, and that it is perpendicular to their optical axis. The light diffused by the injected particles, when they pass through the interfringe network, is therefore offset by frequency f_D, directly linked to their velocity V by the equation $V = d_f . f_D$.

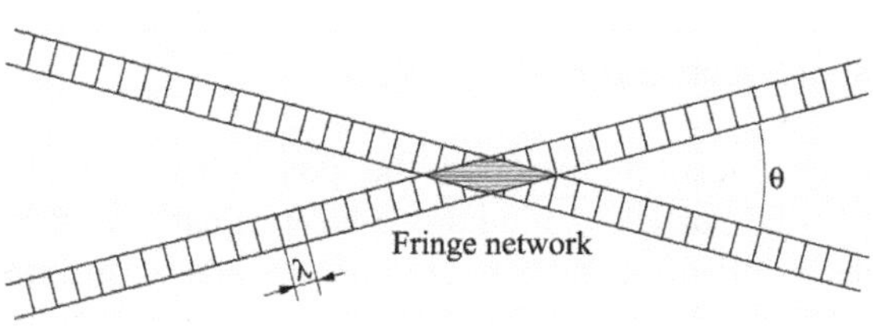

Figure 10.17. *Diagram of a* fringe *network*

The electric signal delivered by the photo-detector is a signal consisting of modulations (often called "Doppler bursts") with Gaussian envelopes. This is due to the light energy distribution associated with the laser beams, from where f_D must be extracted. Nowadays, modern LDA signal analysis and processing systems extract this frequency by using an FFT algorithm. It is important to note in passing that, contrary to measurements by PIV/DTV, which only require one or several cameras to be bought, LDA methods also require an acquisition and a data processing system to be bought in addition to one or more photo-detectors. In theory, even if the data processing system is not essential as some research laboratories prefer to internally develop the tools allowing them to process Dopper signals, in practice it is commonly preferred to get "commercial systems". Moreover, PIV/DTV methods are capable of obtaining images, and therefore of measuring velocity/size, at a regular frequency, which is the repetition rate of the pulsed laser, whereas LDA/phase Doppler anemometry (PDA) measurements are completely random as based solely on the passing of a particle in the measurement volume. In addition, the data acquisition and processing systems enable the implementation of criteria to select measurements that are valid and reject others. These criteria rely on the Doppler signal's envelope being sufficiently larger than the background noise, or each particle detected to pass through a sufficient number of fringes to ensure that its movement is mainly perpendicular to the fringe networks and not "skewed". Even if the basic principle is relatively simple, the method needs to be well understood for high-quality measurements to be obtained.

The choice of injected particles is also very important, in particular with respect to their size: they must be capable of diffusing light while following the velocity fluctuations in an often-turbulent flow. Solid particles are most widely used and can be injected into both gas and liquid flows. Of course, if the flow contains a dispersed phase, and the size and shape of the inclusions are compatible with the quality criteria for LDA measurements, it is possible to measure the velocity of inclusions directly instead of that of injected particles. Finally, the size of the measurement volume is directly linked to the lenses that are used at the output, and therefore to the angle θ, for the major axis of the ellipse, or to the diameter of the beams, for its minor axis. Depending on the ease with which the area of interest in the flow can be accessed, a long focal length may be required. However, this increases spatial integration in such a way that it is often necessary to add a device that separates the beams coming out of the optic fibers.

Finally, it is significant to mention the possibility, namely for fluid flows carrying dispersed solid particles, of adding a discrimination criterion that is used to determine whether a velocity measurement corresponds to an injected particle or a particle from the dispersed solid phase. The discrimination criterion is based on the voltage measured by an additional photo-detector. Depending on a predetermined threshold, the photo-detector separates measurements related to dispersed-phase particles (levels above the threshold) from those related to injected particles in the continuous phase (levels below the threshold). This separation is mainly feasible if a clear size difference between the two types of particles exists. It is then possible (see [MUS 96]) to obtain mean values and statistics associated with the two velocity fields, and therefore the mean concentration of the dispersed phase. In certain cases, where Gaussian distribution measurement of the energy in the laser beam can be corrected, it is possible to link the measured voltage in the discrimination system to the diameters of the solid particles without having to use the PDA method discussed hereafter.

Microprobes have been developed in order to bypass the difficulties encountered when the concentration of the dispersed phase is too high, or when the experimental device itself is too large, and are placed within the flow. They have a diameter around 15 mm and the measurement point is approximately 20 mm upstream of the probe. These optical microprobes are therefore able to take localized measurements, and are very useful in certain situations, even though their intrusive nature is a subject of debate.

10.3.4. *Phase Doppler anemometry (PDA)*

The basic principle of PDA is nearly identical to that of LDA, the difference being that the particle movement is that of inclusions transported by the continuous phase. The movement is then analyzed using the signals delivered by three photo-detectors that are located as shown in Figure 10.18.

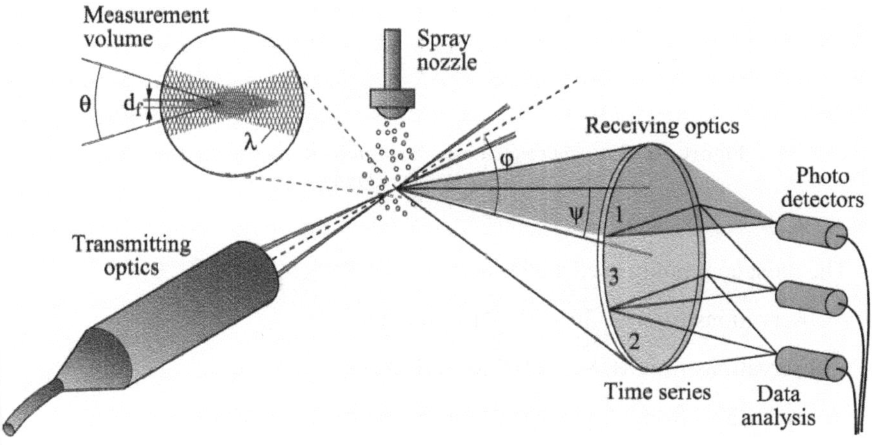

Figure 10.18. *Diagram of the principle of PDA. Excerpt from documents by Dantec Dynamics*

The device is capable of determining the particle size, because of additional detectors, by using the phase between the signals recorded by the various sensors. PDA can be used with light beams that are either reflected or refracted on the particles. The light intensity is therefore different in all directions as it decreases with each successive reflection within the sphere (Figure 10.19). The intensity is only important for the first three modes so the location of the photo-multipliers must be chosen wisely.

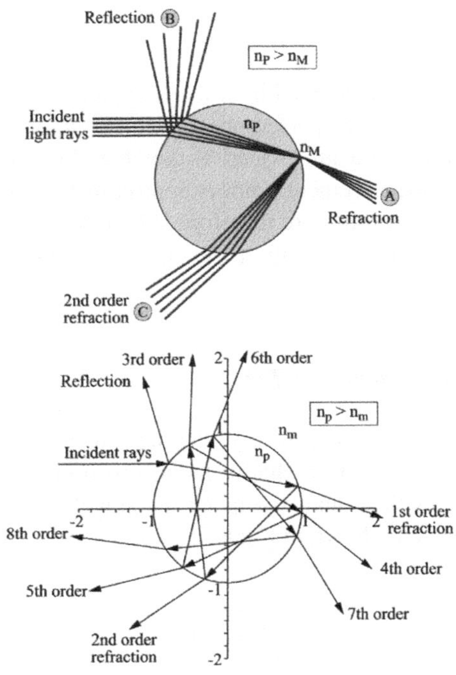

Figure 10.19. *Beam reflection/refraction on a spherical particle. Excerpt from documents by Dantec Dynamics*

The most commonly used angles are:

– observations from A: 30°–70° for refraction;

– observations from B: 80°–110° for reflection;

– observations from C: 135°–150° for second-order refraction.

The influence of size on phase shift between the sensor signals can be simply visualized as shown in Figure 10.20.

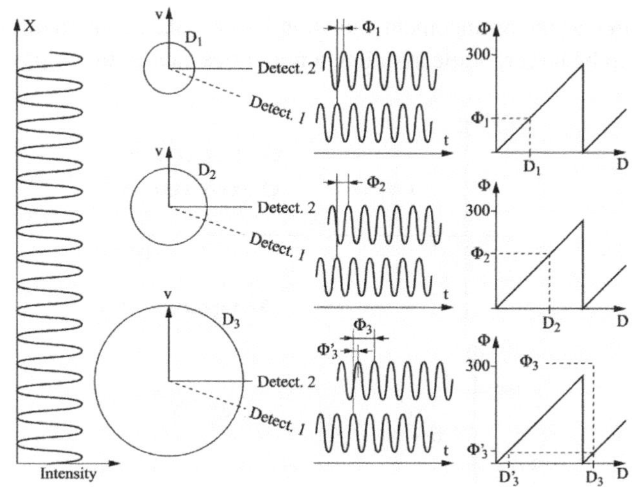

Figure 10.20. *Diagram of the phase shift between signals measured by PDA*

If we are observing in such a way that a dominating transmission mode exists, the relationship between the size of the particles and the phase shift is linear and shown in Figure 10.21, as the paths taken by the rays impacting each detector are slightly different from one another.

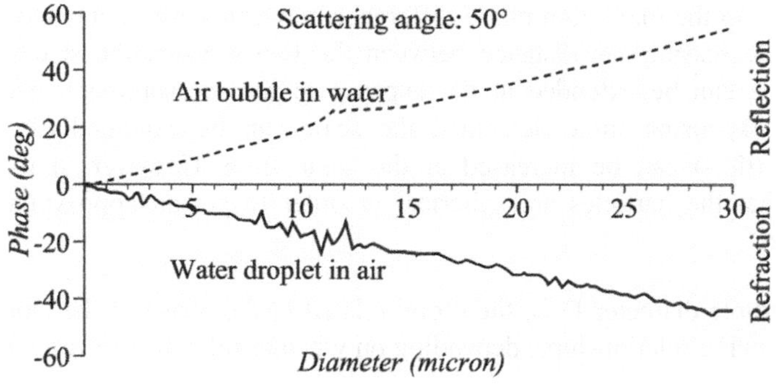

Figure 10.21. *Phase shift variation as a function of the diameter for bubbles and drops*

The three detectors are actually placed in the same optic fiber receptor head (LDA works mainly in retro-diffusion since the creation of optical fiber probes, whereas PDA works in forward diffusion). It is, therefore, necessary to add masks, ensuring that the detector will only see the light coming from the desired observation angle and not from parasitic light. The actual arrangement of detectors is often the same as the arrangement shown in Figure 10.22 as it enables the range of particle sizes to be larger, while avoiding ambiguities caused by 2π passings.

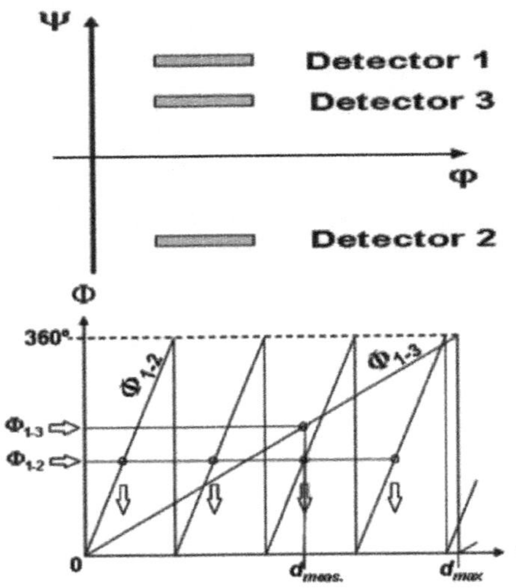

Figure 10.22. *Measuring principle of PDA using three detectors*

The maximum size of particles that can be measured by two detectors corresponds to the maximum phase difference between sensors one and two, $\Phi_{1\text{-}2} = 360°$. By decreasing the distance between the two sensors, the size scale of the particle size can be extended at the expense of the measurement resolution that decreases. By using three detectors, the scale can be extended ($\Phi_{1\text{-}3}$) and the resolution ($\Phi_{1\text{-}2}$) can be increased at the same time. However, it must still be assumed that the particles are spherical in order to get an approximation of the particle size.

The particle diameter D is, therefore, related to the phase of the Doppler signal by the following relationships, depending on whether reflective or refractive PDA is used.

For reflection:

$$\Phi = -\frac{2\pi m D}{\lambda}\frac{\sin\theta\sin\psi}{\sqrt{2(1+\cos\theta\cos\psi\cos\phi)[1+m^2-m\sqrt{1+\cos\theta\cos\psi\cos\phi})]}} \quad [10.5]$$

For refraction (first order):

$$\Phi = \frac{2\pi D}{\lambda} \frac{\sin\theta \sin\psi}{\sqrt{2(1+\cos\theta\cos\psi\cos\phi)}} \qquad [10.6]$$

with:

$$m = n_{rel} = \frac{n_{particle}}{n_{medium}} \qquad [10.7]$$

However, the choice of the precise optimal observation angle also depends on the fact that the transmitted light does not have a uniform intensity. This factor also depends on the polarization of the incident light, and must be taken into account when choosing the optimal set up. In general, errors are considered to be approximately 1% for velocity measurements and 1–2% for the size of perfectly spherical particles. In practice, however, the adjustment in real flows is very intricate and the signal-to-noise ratio is far from ideal. Such small error values are, therefore, difficult to obtain. PDA method is, nonetheless, the reference technique for the simultaneous characterization of particles sizes and velocities. The diameter size that can be measured varies from approximately 0.5 μm to 2–3 mm even though the maximum dynamics is only approximately 50 for a given set up. The dynamics for velocity measurements is approximately 30. To keep this discussion simple, further details will not be discussed here. The reader may refer to [BOU 12b] for a more comprehensive and detailed discussion.

10.3.5. *Ultrasonic Doppler Anemometry*

Ultrasonic Doppler anemometry (UDA) is a variation of LDA. This technique is widely used in studies related to environmental applications of fluid mechanics (rivers, coastal flows, etc.) and situations where the transport of sediments, or various other solid particles, is common. The method requires specific particles to be injected in the flow. It measures the velocity by scanning the flow with a high-frequency acoustic wave (ultrasounds: $f \sim 1$ MHz). In water, this gives a sound wave celerity $c \approx 1{,}500$ m/s, and a wavelength $\lambda \approx 1.5$ mm. This method only works in liquids (whether transparent or not) as the acoustic impedance in gases is infinite for these wavelengths.

Moreover, the method does not require calibration as it has a linear response and provides one or more of the velocity components. The method is advantageous as systems have been created, "profilers", that are capable of measuring the velocity along the whole distance that the sound waves are propagated along if sufficient power can be transmitted. A gel must be applied between the emitter and the wall that contains the liquid medium being studied, so as to prevent the impedance rupture caused by air if the system is placed out from the liquid. However, the system can also be placed directly within the flow depending on the situation being studied.

The generated wave is analyzed in three steps:

– the sound wave's outward path;

– the sound wave's return path;

– the measurement of the phase shift (proportional to the velocity of the particles).

For a system with a transceiver, usually placed on the wall that contains the flow (see Figure 10.23), the distance required for the first impulse to make a return journey between the emitter and the particle is given by $l_1 = \frac{c}{2} T_{s1}$. The distance required by the second impulse to make a return journey between the emitter and the particle is given by $l_2 = \frac{c}{2} T_{s2}$. In the time separating the two impulses, the particle has moved a distance *d* at a velocity U so that $l_2 - l_1 = UT_e \cos\theta = \frac{c}{2}(T_{s2} - T_{s1})$, with $\lambda = cT_e$.

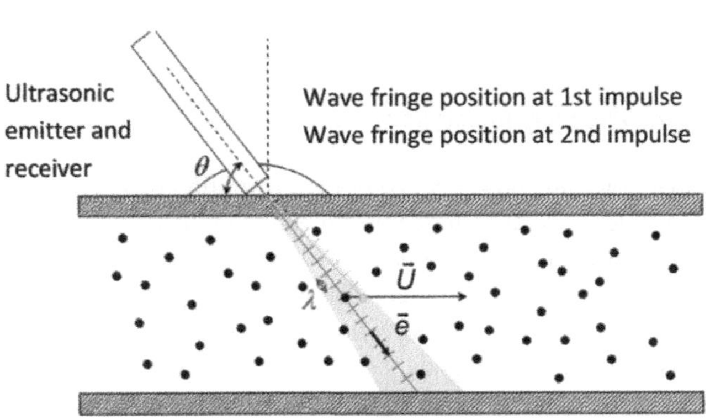

Figure 10.23. *Diagram of the principle of ultrasonic Doppler anemometry (transceiver head)*

As the time difference $T_{s2} - T_{s1}$ is very small (less than a microsecond), it is beneficial to replace this measurement by the measurement of the phase shift $\delta\phi$ between the two echoes, $\delta\phi = 2\pi f_e (T_{s2} - T_{s1})$, so that the particle velocity is given by

$U=\dfrac{c\delta\phi}{4\pi\cos\theta}$, causing a Doppler phase shift, equal to $f_D=\dfrac{\delta\phi}{2\pi T_e}$, to intervene and the velocity, therefore, becomes $U=\dfrac{c}{2f_e\cos\theta}f_D$. The spatial resolution is of the millimeter order, but can be improved by increasing the frequency of the wave emission rate and the signal power. However, the main drawback of ultrasonic Doppler velocimetry is that small scales cannot be resolved in turbulent flows. Two or three velocity components are measured by using a common emitter linked to two or three receptor heads, slanted relative to the emitter, so as to create a rigid probe, which is then placed within the flow.

10.3.6. *Densimetry by attenuation of gamma, X-ray or neutron radiation*

Densimetric methods by attenuation of gamma, X- or neutron rays have been highly developed in a number of fields, namely medicine, the metal industry and geotechnics. They are also used to measure the presence rate of the dispersed phase in two-phase flows, including two non-miscible liquids (the importance is that the phases have electrical conductivities that are sufficiently different). The main advantage of the methods is that they are non-intrusive and are very accurate (the accuracy is considered to be greater than 97%). The techniques can be used for laboratory studies and are also an important element in flowmeters routinely used in the oil and gas extraction industries, for example. With respect to geotechnical applications, compact systems have been developed and are easily transportable to the field. Gamma ray densimetry is the most interesting of the three techniques, as gamma rays can penetrate along a longer distance, making them particularly suitable for implementation at large-scale industrial sites. Only the method for gamma rays will be discussed here and the reader can refer to [FAL 09] and [CHA 97] for information on X-ray and neutron ray methods.

The source of the radiation is inserted into a shielded cavity, usually conical in shape, in such a way that the gamma rays can only leave the enclosure through a window placed in the side giving out onto the flow area being studied (in this case a circular tube, see Figure 10.24). Various elements can be used as a source of radiation, including americium, cesium, cobalt or barium. The source must be surrounded by a collimation system for the radiation to leave in the form of an axial ray and to have a well-defined size. The source type is chosen according to the element's half-life as well as the radiation energy that is emitted. The radiation from cesium is approximately 660 keV, whereas that from americium is approximately 60 keV. This has a considerable effect on the thickness of the shield required, as well as on the ease with which the two phases can be detected: the higher the emitted energy is, the more difficult it will be to discern the media.

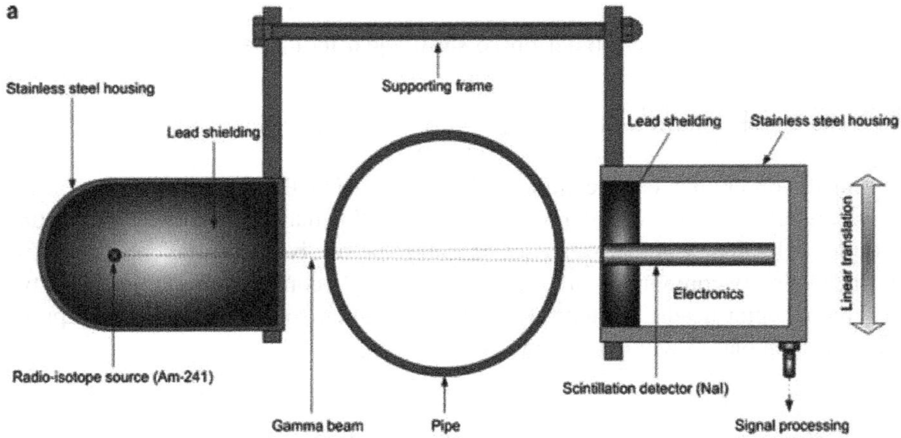

Figure 10.24. *Diagram of an experimental set up for gamma densimetric measurements [KUM 10]*

The radiation from the liquid medium is detected by a scintillation detector, sodium iodide doped with thallium being widely used. The ionizing rays lose their energy when they interact with the sodium iodide network. The energy then induces the scintillator to emit light which is proportional to the energy of the captured radiation. A photomultiplier then detects the photons and transforms them into an electric signal. Finally, the intensity of the energy transmitted in, first, the single phase (I1), and then the second phase (I2), is used to relate the measurement of transmitted energy in real situations (I) to the integral of the average content value along the volume of the beam going through the liquid, with average content of the dispersed phase in the volume ε_2. This is independent of the intensity of the emitted radiation and is given by:

$$\varepsilon_2 = \frac{\ln(\frac{I}{I_1})}{\ln(\frac{I_2}{I_1})} \qquad [10.8]$$

For more detail on the measurement restrictions, such as the selection of different parameters or electronic systems, or problems emerging when the area close to the wall is no longer negligible in comparison to the volume that the ray passes through, the reader can refer to [KUM 10].

10.4. Advanced optical methods

The diagnostic methods discussed in section 10.3.6 were mainly based on the tracking or detection of either particles injected into the flow or particles present as droplets forming the dispersed phase, with the aim of characterizing their properties. The optical methods discussed in this section are more elaborate methods and developments that bring into play the laws of light transmission in the medium, or the reaction of the medium to a given laser beam (or sheet) excitation. Laser-induced fluorescence (LIF) is a method that was developed a while ago and is therefore widely used in research laboratories or in industry. Other methods, in contrast, are still being developed or improved and their use still remains limited to research teams specialized in their development. Theoretical developments, on the propagation of light in the relevant media, underlying these methods will not be discussed in detail here. The reader may refer to [BOU 12b] or [ALB 03], in addition to [TRO 11]. It should be noted that these methods are still being perfected by specialized research teams. However, several methods are already being marketed by instrumentation companies in the field of instrumentation (not only the LIF method, but also Interferometric laser imaging for droplet sizing (ILIDS)/ interferometic particle imaging (IPI) or inline holography).

10.4.1. *Laser induced fluorescence*

10.4.1.1. *One-color LIF method*

LIF, most often used in its planar version, planar laser induced fluorescence (PLIF), is a powerful diagnostic method for the investigation of liquid and gas flows, both reactive and non-reactive. It is an instantaneous and non-intrusive technique that gives a high spatial and time resolution, and can be applied to determine various flow variables in the plane of the laser light sheet: concentration (mole fraction), density, velocity field and temperature can be measured using calibrated LIF images. In general, the method is based on spontaneous emission of rays by a molecule after absorption of a laser ray. It can equally be applied to flows whether the continuous phase consists of a liquid or a gas.

LIF consists of exciting a marker that induces a fluorescent signal. The marker is either previously diluted in the flow being studied, or appears spontaneously in the flow due to a chemical reaction. The intensity of the fluorescence signal depends on the temperature, among other things, as well as the tracer concentration in the measurement volume and the intensity of the local excitation. The most common version only uses one excitation at a given wavelength, and depends on the molecule being used as a tracer. Other methods also exist but are more complex as they combine excitation on two or three wavelengths (the method is then termed two- or three-color LIF).

For measurements carried out in liquids, the most commonly used molecules are rhodamine 6G for concentration, rhodamine B for temperature or fluorescein for concentration or temperature. Acetone tracers are the most common tracers for non-reactive gases and, in reacting flows, OH, NO and CH radicals are often used to track the progress of combustion reactions, although biacetyl or toluene injections are also sometimes used. As the excitation response, and therefore the received signal intensity, depends on a number of parameters besides concentration, for example temperature, pressure or pH, choosing the tracer is often an important step in the set up of LIF-based diagnostics. In particular, the phosphorescent emissions can, in some cases, be a source of disturbance when taking measurements, even if it takes place at scales completely different from the fluorescence. This is similar to quenching, which corresponds to a non-radiative relaxation of excited electrons to their fundamental state (referred to as "dynamic quenching" when the excited fluorophore is deactivated by contact with another "quencher" molecule, and "static quenching" when the fluorophore establishes a stable and non-fluorescent complex with a second molecule).

When the molecule is in its basic state and is excited by an energetic laser source, it reaches a higher and more unstable energy state: this is the phenomenon of absorption. To return to a stable state, the molecule must be de-energized through different processes as follows:

– spontaneous emission of photons (radiative process);

– collision with another molecule, thus transferring the excess energy to the second molecule without emitting a photon, which is therefore a form of quenching (a non-radiative process);

– collision with a photon, emitting two photons having the same wavelength as the laser. This is stimulated emission (radiative process).

Fluorescence is, therefore, characteristic of the first case, and brings about the natural relaxation of an excited molecule to a lower energy state through spontaneous emission of photons over short timescales. According to the implied electronic states, a second process called phosphorescence also emits photons when the molecule is de-energized, but the timescales are longer in this case. The luminescent process involves three electronic states, which are the base state singlet S_0, the first excitation S_1 and the first state triplet T_1, involved in the phosphorescent process. Transfers from one state to other state are shown in Figure 10.25.

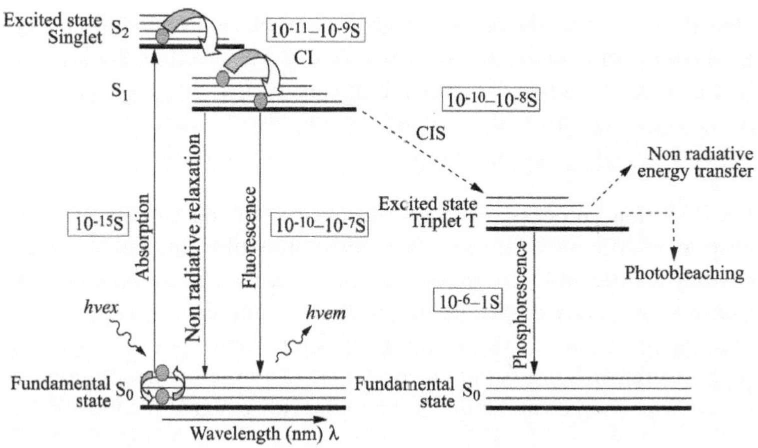

Figure 10.25. *Principle of laser-induced fluorescence (simplified version of the Jablonski diagram)*

The fluorescent signal recorded by the camera's charged-coupled device (CCD) sensor is linked to the fluorophore's mole fraction percentage in the mixture (χ_{fluo}). The relationship between the fluorescent signal S_{fluo} and other various carrier-flow parameters is given by equation [10.9] (the pressure can be important for gas flows, whereas the pH is important for liquid flows and must be taken into account):

$$S_{fluo} = \eta E n_{fluo} \sigma(\lambda, T) \phi(\lambda, T, p, \chi_i) \qquad [10.9]$$

where η is the optical efficiency, E is the energy of the laser pulse, $\sigma(\lambda,T)$ is the effective absorption cross-section as a function of temperature T and the exciting laser's wavelength λ, and ϕ is the quantum fluorescence yield that depends on T, pressure p and the composition of the different elements χ_i forming the carrier fluid. The variable n_{fluo} represents the concentration of fluorophore molecules that absorb the energy. The concentration of molecules is then converted to mass concentration. The experimental setup is similar to that used for PIV. The two methods are often combined to take simultaneous measurements, even though the wavelengths required for the laser by each method are different. PIV will, therefore, use a green light for the pulsed laser source – having a wavelength approximately 532 nm – and LIF will use a UV source – wavelengths equal to or approximately 266 nm. However, the same 532 nm source can be used by splitting the beam into two beams. One of the two beams goes through a device that doubles the frequency to obtain a wavelength of 266 nm. Moreover, the camera used to detect fluorescence has to be equipped with both a very narrow band-pass filter (so as to be sensitive to the emitted fluorescence's frequency), or at the very least a low-pass filter, and a light intensifier (as the fluorescent signal is weak in terms of energy). Laser emission and camera opening can be synchronized very well as the laser impulse is very short (a few nanoseconds), and the picture of the fluorescence must also be recorded very quickly after the end of the impulse to capture everything while avoiding parasitic light or phosphorescence from being recorded.

The calibration process is also necessary to determine the values of multiplication coefficients for n_{fluo} that come into play in equation [10.9]. The sensitivity to pressure and temperature must also be closely analyzed if temperature and/or pressure are likely to change (or fluctuate) significantly within the flow. It is important that the limit for the linearity of fluorescent response is not exceeded when the measurements are being taken. As the response is already quite weak, the power of the laser source can be increased to provide a signal having a better signal-to-noise ratio. However, if the laser excitation is too high, the response will no longer be linear and it is therefore essential to make sure that the saturation mode is not reached when measurements are being taken. Finally, the background noise signal and non-uniform energy distribution of the laser sheet have to be corrected for data analysis. Equation [10.10] is used to achieve this:

$$S(x,y)=\frac{S_{fluo}(x,y)-S_{noise}(x,y)}{L(x)} \qquad [10.10]$$

S(x,y) represents the signal values (level of grayscale for each image in the measurements' plane as a function of the pixel coordinates (x,y)) after correction, S_{fluo}(x,y) represents the raw values measured, S_{noise}(x,y) represents the background noise values (found when the flow is not active) and L(x) represents the energy distribution in the laser sheet (usually similar to a Gaussian distribution). It should be noted that the beam attenuation signal along the optical path (therefore along the *y*-axis in this case) may need to be corrected if the injection (or the content of various particles) is sufficiently high. The Beer–Lambert law is usually used in this case and is a simple decreasing exponential function of distance.

The technique of basic induced fluorescence (or one-color fluorescence) enables the study of very different situations corresponding to multiphase flows. Examples of these studies include the detection of aromatic polycyclic hydrocarbons (APH), often considered as precursors to the appearance of soot when combustion is not complete due to an important lack of oxygen, the quantification of nitrogen oxides (NOx) or OH radicals, or simplified academic versions of bubble flows aiming to understand the basic mechanisms involved in the interaction between bubbles and gas carrier flows. In practice, however, situations are often too complicated for LIF methods to be used on their own. The method is, therefore, associated with PIV methods (as mentioned previously) or PDA methods. In the former case, the association enables the velocity field of the continuous phase in a spray, for example, to be measured by intercorrelating two LIF images, taken a short time apart, of a tracer injected in the gas continuous phase; whereas the PDA method allows the size and velocity of droplets to be found. It is often necessary to find the temperature as well, as flows are subject to large temperature variations either caused by chemical reactions or because they are heated liquid-vapor equilibrium flows. Common practice is therefore to use two- or three-color LIF techniques. This will be discussed in section 10.4.1.2.

10.4.1.2. *Two- or three-color LIF methods*

These methods measure different spectral bands of the fluorescence sent back by one tracer. The influence of several variables that influence the measurement can be eliminated with the method that ensures a more precise temperature measurement. By using two-color LIF and a correlation between the measured signals in the two spectral bands λ_1 and λ_2, a signal can be obtained that is independent of the fluorophore concentration, the measurement volume and the excitation intensity. Using this variation to measure the temperature in two phase flows is of particular significance as it enables the ever-changing size of the measurement volume to be measured independently. The size of the measurement volume changes due to the passing of drops that carry fluorophore in the measurement volume. Three-color LIF has an additional advantage: it allows the absorption of the fluorescence signal by the tracer to be measured.

The equation given above for one-color LIF can be rewritten as follows for the two-color method:

$$S_{fluo}(T,\lambda)=\eta En_{fluo}K_{opt}(\lambda)K_{spect}(\lambda)e^{\beta(\lambda)/T} \tag{10.11}$$

where K_{opt} and K_{spect} are two constants, respectively, related to the optical system and the spectral properties of the selected products. These two constants are a function of the wavelength used. $\beta(\lambda)$ is the temperature sensitivity coefficient. By using suffixes 1 and 2 to denote the spectral bands being used by two-color PLIF, equation [10.11] can be written as:

$$R_{12}(T)=\frac{S_{fluo}(\lambda_1,T)}{S_{fluo}(\lambda_2,T)}=\frac{K_{opt}(\lambda_1)K_{spect}(\lambda_1)}{K_{opt}(\lambda_2)K_{spect}(\lambda_2)}e^{\Delta\beta/T} \tag{10.12}$$

where ΔB is equal to $\beta(\lambda_1) - \beta(\lambda_2)$. The posttreatment algorithm gives the temperature by using the following relationship:

$$T=\frac{T_0}{1-\frac{T_0\ln(G_{12})}{\Delta\beta}} \quad \text{and} \quad G_{12}=\frac{R_{12}(T_0)}{R_{12}(T)} \tag{10.13}$$

where $R_{12}(T_0)$ is the light intensity ratio relative to a reference for which the temperature T_0 is known, thus allowing K_{opt} and K_{spect} to be taken out of the equation. To retain only the information on temperature, an intensity ratio is calculated by measuring the fluorescent signal on two carefully selected spectral bands. For two-color PLIF, the fluid is lit by a plane laser (of wavelength λ), and the pictures are recorded by two cameras, both equipped with an interferometric filter that selects one of the two spectral bands (see Figure 10.26, taken from [SIM 12]).

This procedure is widely used and has been amply validated for individual monodispersed drops. However, phenomenology is more complex for sprays having a wide dispersion diameter and drop presence on the optic path as, for example, the ratio R_{12} then depends not only on temperature, but also on the pressure conditions at the injector's output, which namely changes the size distribution of drops and induces a bias in the temperatures that would be measured. The approach must, therefore, be altered and a three-color method must be set up (see Figure 10.27). As a result, the equations need to be reconsidered as follows by bringing in the function g_i that characterizes the drop size effect on the fluorescent signal. Equation [10.14] is, therefore, obtained:

$$R_{ij}(T)=\frac{K_{opt,i}K_{spect,i}g_i f_i(T)}{K_{opt,j}K_{spect,j}g_j f_j(T)} \qquad [10.14]$$

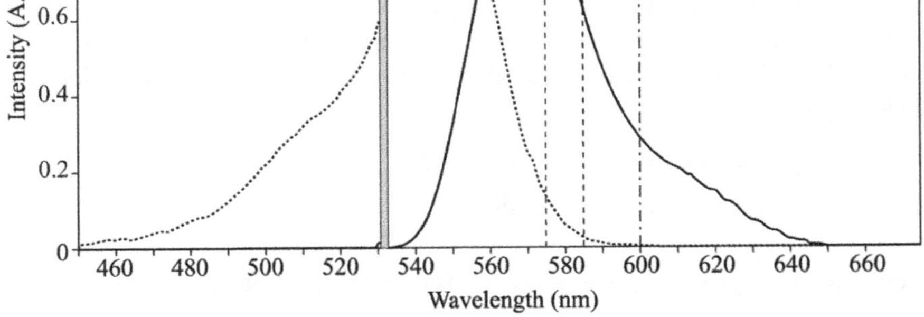

Figure 10.26. *Emission and absorption spectra as a function of wavelength. The two traced lines show the spectral bands and the thick line shows the laser's wavelength [SIM 12]*

The equation brings in a new variable that is the ratio of g_i functions. The adopted strategy is to measure the intensity of fluorescence on a third spectral band and to calculate the two ratios: R_{12} and R_{32}. The third band is chosen so that R_{32} will not be very sensitive to temperature but as sensitive as possible to the effect of the drop sizes (see Figure 10.27). The use of a reference obtained in a tank (R_{120}, R_{320} at T_0 and with $g_0 = 1$) enables the constants K_{opt} and K_{spect} to be taken out of the equation once again.

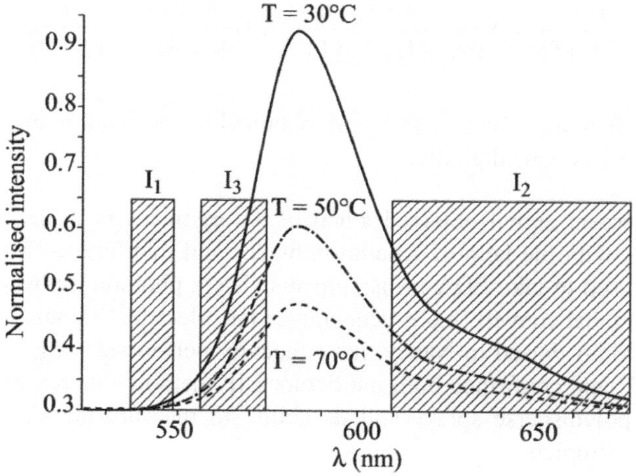

Figure 10.27. *Fluorescence spectra for three temperatures as a function of the wavelength and position of the three spectral bands chosen for LIF measurements [LAB 12]*

The following equation is obtained:

$$\frac{R_{ij}(T)}{R_{ij}(T_0)} = \frac{f_j(T_0)}{f_i(T_0)} \frac{f_i(T)}{f_j(T)} g_{ij} = \frac{f_{ij}(T)}{f_{ij}(T_0)} g_{ij} \qquad [10.15]$$

The influence of drop size distribution on the fluorescence ratio can be found once and for all by finding functions g_{12} and g_{32} for isothermic conditions ($T = T_0$). This is achieved by modifying the injection pressure and simultaneously measuring the R_{12} and R_{32} ratios near the injector at the temperature T_0. The latter is measured by a thermocouple placed inside the injector. The following relationship is given:

$$g_{12} = \frac{R_{12}(T)}{R_{12}(T_0)} \quad \text{and} \quad g_{32} = \frac{R_{32}(T)}{R_{32}(T_0)} \qquad [10.16]$$

It becomes apparent that an empirical equation relating g_{12} and g_{32} can be found by tracking the evolution of g_{12} as a function of g_{32} for each pressure value.

$$g_{12} = Ag_{32}^2 + Bg_{32} + C \qquad [10.17]$$

so that:

$$\frac{R_{12}(T)}{R_{12}(T_0)}\frac{f_{12}(T_0)}{f_{12}(T)} = A\left[\frac{R_{32}(T)}{R_{32}(T_0)}\frac{f_{32}(T_0)}{f_{32}(T)}\right]^2 + B\frac{R_{32}(T)}{R_{32}(T_0)}\frac{f_{32}(T_0)}{f_{32}(T)} + C \qquad [10.18]$$

The exact functions for f_{12} and f_{32} are obtained by calibration and the resolution of temperature T is done digitally.

The fluorescent signal obtained when using argon lasers is divided into three spectral bands with the help of separation blades and interference filters. The laser beam is also used to take PDA measurements. Three photomultipliers are equipped with amplifiers that convert fluorescent signals into electrical signals, which are then digitized. LEMTA, based in Nancy, have recently performed measurements in this way [LAB 12] and confirmed both multicolor LIF method's increased sensitivity to drop size in polydisperse sprays and its ability to measure the size, velocity and temperature of droplets.

10.4.2. *Interferometric methods (digital inline holography, Fourier interferometric imaging, ILIDS/IPI, rainbow)*

10.4.2.1. *Digital inline holography*

Digital inline holography (DIH) of particles exploits the diffraction patterns produced by interactions between laser beams and objects having sizes in the order of a few millimeters (see the operating diagram shown in Figure 10.28 [Gabor configuration]). Particles are generally injected into a fluid lit by a laser beam. The beam is enlarged in order to obtain interference pictures (or holograms) that occur between the reference wave emitted by the laser and the wave diffracted by the particles. The particles used can often be likened to spheres and the diffracted intensity patterns (quasi) concentric circles. Two pieces of information relative to the particle can be extracted from these diffraction patterns: its location along the three axes and its size (since particles are assumed to be spherical). The information on particle location is recorded in the rings' frequency modulations, whereas the size information is recorded using the ring contrast. The light intensity is recorded by a camera that allows images to be digitally processed. The process of capturing an image of the diffracting object (the particle) using the recorded light intensity is called "hologram restitution". This is performed by calculating the propagation of the laser beam in the direction opposite to the direction used for recording, and amounts to lighting the recorded hologram with the reference laser beam. The process can be performed digitally using mathematical diffraction models. The 3D image of particle clouds is, therefore, rebuilt exactly as if the screen had been moved along the optical restitution assembly's axis. The image becomes clear when the restitution distance is such that the screen is virtually identical to the particle plane. It is, therefore, possible to measure the third component to determine the object's position in space. Several digital methods can be used for this and are discussed in more detail in other literature. Examples of such methods include restitution by Fresnel transformation, with strong restrictions on the propagation distance, restitution by wavelet conversion, restitution by window Fourier transform and also restitution by fractional Fourier transform (see [BOU 12b] for more details).

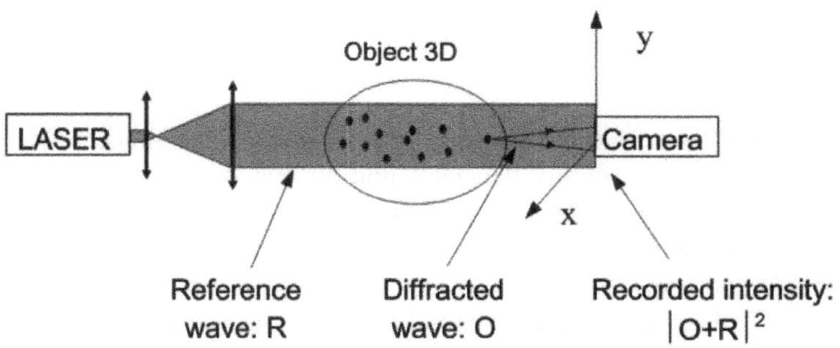

Figure 10.28. *Diagram depicting the principle behind assemblies for inline holography*

If we consider the transfer function in free space under "Fresnel's" approximation, as in [LAM 12], the image A(x,y,z), restored in each z-plane of the optical axis, is given by:

$$A(x,y,z) = FT^{-1}[FT(I(x,y))H_z(\mu,\nu)],$$

where I(x,y) is the intensity recorded by the sensor, $H_z(\mu,\nu) = \exp(i\kappa z).\exp(-i\pi\lambda z(\mu^2+\nu^2))$ is the transfer function at distance z, $\mu = x/\lambda z$ and $\nu = y/\lambda z$ are the spatial frequencies according to Gaussian approximation and $\kappa = 2\pi/\lambda$ is the wave number. The notation FT is used to symbolize the Fourier transform in two dimensions. The intensity measurement taken by the sensor causes a twin image to appear. Figure 10.29 shows typical results that are obtained for diffraction figures in a given z-plane (taken from [PU 05]).

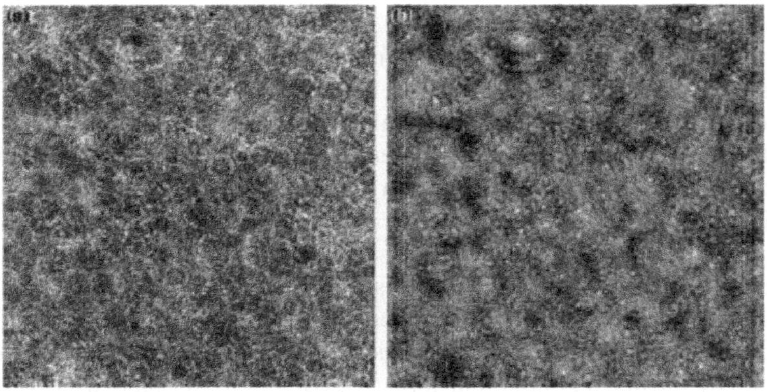

Figure 10.29. *Hologram obtained by inline holography. a) Raw intensity I(x,y); b) Restored image A(x,y) [PU 05]*

A limit does exist for the measurable drop density for a single hologram. It is intrinsic to the inline assembly that uses non-diffracted parts of the wave as the reference wave. The limit is often quantified through the use of a criterion that depends on several parameters (depth of field and mean drop diameter), and therefore only estimates the order of magnitude of the hologram's quality. With this approach, the hologram can still be exploited as long as this criterion is equal to or smaller than 10%. In the example given here, this is equivalent to 35 drops with a diameter of 1 mm in the field of survey. It is, therefore, important to always limit the number of drops when acquiring data in order to respect this criterion.

Finally, one of the key points of digital holography is the method that places drops according to the optical axis. It must be adapted to the experimental conditions and enables holograms to be automatically restored. Several methods are put forward in literature and can be arbitrarily categorized into three categories. The first category consists of methods that are based on local grayscale analysis of the restored image. The positioning of the objects in this method is generally undertaken by searching for intensity extrema. The methods in the second category use the restored image's characteristics. This can be the use of data contained in the module, the imaginary part, the phase or even the calculation of correlation coefficients between the restitutions. The methods in the third and final category depend on the direct use of restitution methods. These can be projection coefficients on the base of discrete wavelets that are used for restitution, and adjustment parameters, in the case of an inverse problem approach. Finally, it should be noted that this method is mainly used in a large number of fields other than the fields discussed here, at much smaller spatial scales (of micron order), and namely as microscopic systems in biology. It can also be used as a PIV system and provides the three velocity components for each particle located in the survey volume.

10.4.2.2. *Fourier interferometric imaging*

DIH discussed in section 10.4.2.1 is currently the only 3D technique actually used. However, it requires significant computations to be made and its longitudinal precision is inferior to its lateral precision. Fourier interferometric imaging (FII) is one of its derivatives and allows the particles to be located with micronic precision in all three space directions. Moreover, the process is relatively economic in terms of computational time. The number of 2D Fourier transforms is of the same order as the number of particles present in the volume being measured and FII can be extended to measure the diameter and the refraction index of diffusing particles.

FII was first introduced as an alternative method to holography. The method is based on recording interference fringes created by the light diffused by particles present in the measurement volume (there is no reference wave), and the analysis of fringes in the associated Fourier space. With this method, it is possible to transform the study of clouds into the study of all possible particle pairs and to measure the relative 3D positions of particles with submicronic precision in all three space directions.

Figure 10.30 shows the experimental device used for FII methods. The analysis is carried out in a setting where the diffusion goes forward: the main diffusion contributions are associated with the light reflected on particles' external surface and to the refracted light (see [MEU 12]).

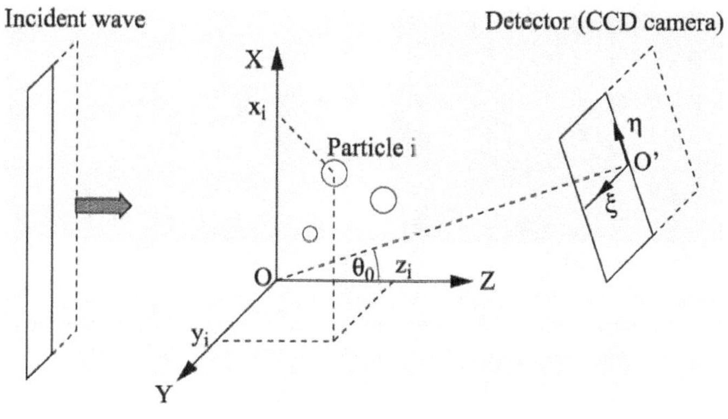

Figure 10.30. *Principle behind experimental devices used for FII [MEU 12]*

The total amplitude at a given point in space is given by the sum of light amplitudes scattered by each particle:

$$A_{total} = \sum_{i=1}^{n} A_i \qquad [10.19]$$

As the intensity is proportional to the square of the amplitude, under the far-field assumption, the following is obtained:

$$I=\sum_{i=1}^{n} A_i \sum_{j=1}^{n} A_j^* \qquad [10.20]$$

an equation that can also be written as:

$$I_t=\sum_{i=1}^{n} A_i A_i^* + \sum_{i=1}^{n} \sum_{j=1(j\neq i)}^{n} A_i A_j^* = \sum_{i=1}^{n} I_i + \sum_{i=1}^{n} \sum_{j=1(j\neq i)}^{n} A_i A_j^* \qquad [10.21]$$

where the term I_i includes the interference terms between different rays emitted by one drop, whereas the second term corresponds to interferences between "rays" emitted by different drops.

Figure 10.31(a) shows the interference system created by a group of five drops of water in a volume of 500 μm^3. The structures visible in the figure are mainly caused by the Moiré effect.

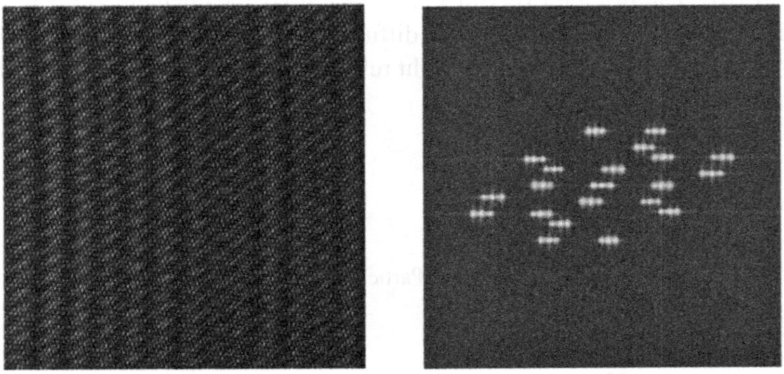

a) b)

Figure 10.31. *a) Interference field created by five drops for a diffusion between 30° and 35° and b) its Fourier transform using 2D FFT [BRI 12]*

The figure corresponds precisely to the intensity distribution predicted by the equation given previously, but is not easy to interpret directly. The fringe interferences in the associated Fourier space, however, are related to the spatial frequencies. Each pair of particles, therefore, has a couple of spots in the Fourier space associated with it. In other words, the central spot in the figure corresponds to the first term of the equation ($\sum_{i=1}^{n} I_i$), whereas each spot that is not centered corresponds to one of the $\sum_{i=1}^{n} \sum_{j=1(j\neq i)}^{n} A_i A_j^*$ terms. The distance to the center depends on the distance between particles, whereas the spots' angular positions depend on the particle position. Figure 10.31(b) is the 2D Fourier transform of Figure 10.31(a). The relationship between the 3D positions of a pair of particles (x,y,z), the collection angle (θ_0) and the spots' coordinates in the Fourier space (η, ξ) is, thus, given by the following equation:

$$\begin{bmatrix} \eta \\ \xi \end{bmatrix} = \pm C \begin{bmatrix} -(x_2 - x_1)\cos\theta_0 + (z_2 - z_1)\sin\theta_0 \\ (y_2 - y_1) \end{bmatrix}. \qquad [10.22]$$

Forward diffusion (typically between 20° and 80°) is dominated by the light reflected on the external surface of the drops, and the twice refracted light. The Fourier space position of each spot (corresponding to high-frequency datum) is dependent on the relative position between two particles, whereas the structure of each spot (corresponding to low-frequency datum) will depend on the diameters and refraction indices of the two particles that are the source of the spots. For a first approximation, this dependency can be neglected for forward diffusion. Finally, the method can be extended to measure particles of micrometric or nanometric size, which is currently also being developed. More details can be found in [BRI 12] and [MEU 12].

10.4.2.3. *Interferometric laser imaging for droplet sizing (ILIDS or IPI)*

This method is used to measure the size of particles (usually droplets in a spray) and has a relatively simple configuration as shown in Figure 10.32.

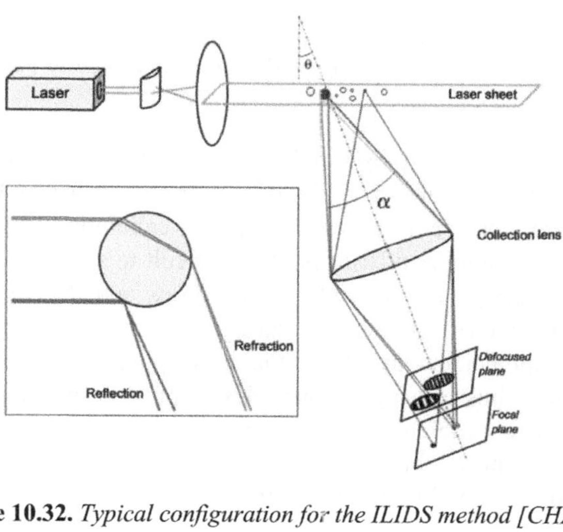

Figure 10.32. *Typical configuration for the ILIDS method [CHA 09b]*

The beams diffracted in the diffusion angle θ are collected by a spherical lens. The collection angle is denoted as α. Two "glare points" appear in the image plane, and their images in the defocused plane overlap. Interference fringes appear in the overlapping areas (see Figure 10.33). The CCD sensor for ILIDS is placed in the defocused plane, thus differentiating the method from holography. The size of the fringes depends on the degree of defocusing, and their number depends on the particle size and collection angle (see [CHA 09b]). An optical compression can be performed to increase the concentration of measurable particles. The equation for ILIDS obtained using the optical geometry approximation is written as follows:

$$\Delta\theta = \frac{2\pi}{\lambda}\left(\cos(\theta/2) + \frac{m\sin(\theta/2)}{\sqrt{m - 2m\cos(\theta/2) + 1}}\right)^{-1} \qquad [10.23]$$

where m is the drop's refractive index.

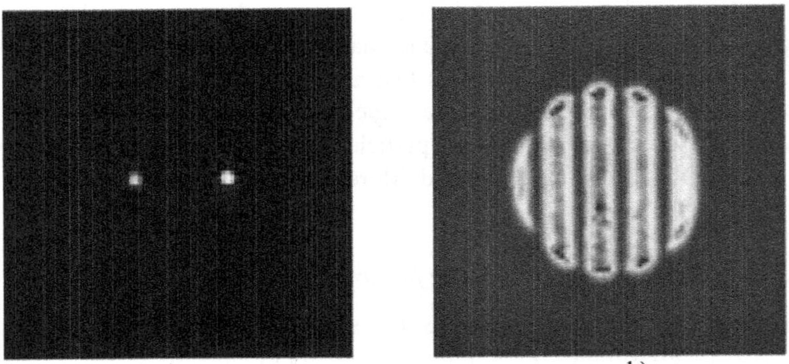

a) b)

Figure 10.33. *ILIDS method: a) glare points obtained when the camera is focused and b) interference fringes obtained when the camera is defocused [CHA 09b]*

The images are processed by finding the angular deviation of interfringes, obtained either from the number of fringes or from the fringe frequencies from fringe interference. The two methods require calibration and present some difficulties: given the number of fringes, it is difficult to see the collection angle α, whereas the fringe frequency depends on the drops and the various coefficients related to the geometry of the measuring system. Moreover, most ILIDS configurations use an incident laser beam that is at an angle to the optical axis and to its transverse plane, thus causing the particles to produce different spot sizes and aperture angles. The problem of spot size can be solved by a Scheimpflug configuration (also required for stereoscopic PIV, although it has not been discussed in this chapter, for more information see [BOU 12a]), but leaves the other problem unresolved.

It should also be noted that this method can also be used to find the evaporation rate of droplets from the instantaneous variation of the drop's diameter, obtained by measuring the diffusion diagram phase. Such an approach has been studied by Saengkaew *et al.* [SAE 12].

10.4.2.4. *Methods based on rainbows*

The rainbow phenomenon appears when an isolated droplet is lit by a laser sheet (see Figure 10.34(a)). The standard refractometric method is used to measure the size and temperature of the isolated droplets. The technique is based on the detection of the rainbows created by interference between reflected rays inside the drop. The angle denoted by θ_{rg} in the figure represents the rainbow's orientation angle and depends solely on the drop's refractive index m, thus allowing the temperature to be determined. In general, the method presents several problems linked to the temperature gradient inside the droplet, the non-sphericity of the droplet and, for small drops, the high-frequency oscillations of the signal (emitted by interference between internally and externally reflected drops) that overlap (see Figure 10.34(b)) with the basic scheme (called Airy fringes, which have a low frequency), and that are called "ripple structures". Global rainbow refractometry is usually used to remove these two difficulties (see, for example, [LEM 05]).

Global rainbow refractometry superimposes several rainbow responses generated by a sufficiently high number of drops assumed to represent the spray statistically so as to remove parasitic high-frequency oscillations. The same phenomenon occurs (linked to destructive interferences) when a single drop is subjected to small diameter variations when several non-spherical drops are present, with a randomly distributed orientation. Simple Airy theory can, therefore, be used in these different situations and only considers interferences between the rays internally reflected in one or more drops.

The following equation is obtained:

$$I_{grw} = \sum_{i=1}^{n} \Omega^2(d_i,\theta) d_i^{7/3} \quad [10.24]$$

where Ω is the intensity given by Airy's law:

$$\Omega^2(z) = \left(\int_0^\infty \cos\frac{\pi}{2}(z\eta - \eta^3)d\eta \right)^2 = \left(2(\frac{-12}{\pi^2})\mathrm{Ai}\left[(\frac{-12}{\pi^2})^{-1/3} z \right] \right) \quad [10.25]$$

with $z(\theta) = -(\theta - \theta_{rg})\left(\frac{16 \tan\tau_{rg} \sin^2\tau_{rg}}{\lambda^2} d^2 \right)^{1/3} \frac{180}{\pi}$, $\tan\tau_{rg} = \sqrt{\frac{m^2-1}{4-m^2}}$, Ai being the Airy function.

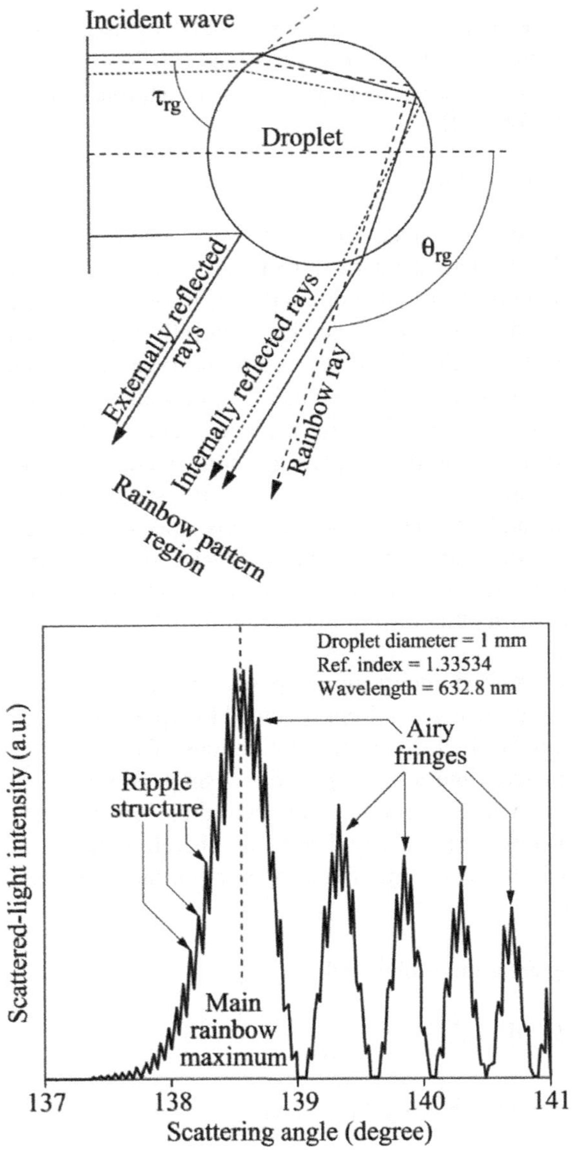

Figure 10.34. *Principle of rainbow refractometry a) with, parasitic waves appearing when isolated drops are considered b) [LEM 05]*

Analysis of the position of main rainbow peaks and secondary peaks enables the size of the diffusing particle and its refractive index to be measured accurately up to the fourth decimal point, thus permitting the drop temperature to be measured.

Figure 10.35 is an example of the angular response obtained for different temperatures with clouds of droplets having the same granulometry.

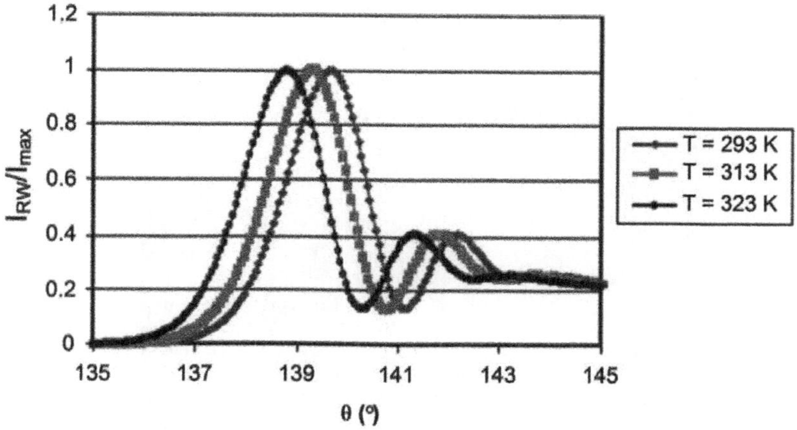

Figure 10.35. *Temperature sensitivity of the angular response for global rainbow refractometry [LEM 05]*

For more details on the technique and its difficulties, the reader can refer to [SAE 09] and [WAN 11].

It is also important to point out that the technique is derived from rainbow interferometry (see [BEE 97]), which has not been widely used due to its low sensitivity.

10.4.2.5. *Method of "Critical angle refractometry and sizing"*

This section will conclude with a brief discussion of the critical angle refractometry and sizing (CARS) method that was recently developed, even though it is obviously not possible to be exhaustive as the advanced optical methods being developed are very diverse and varied. This method is based on light diffused by a cloud of bubbles in a liquid, in the far field and close to the critical diffusion angle (see Figure 10.36 for one bubble only). The singular behavior of the diffusion from a bubble near the critical angle is caused by the semi-discontinuous character of reflection coefficients of waves p = 0 in the vicinity of the angle for total reflection, so that for an air bubble in water with a relative index $m = m_b/m_w < 1$, $\phi_c = \pi - 2\theta_c = \pi - 2\sin^{-1}m = 82°$. The diffusion properties of a spherical bubble can be rigorously modeled by Lorenz–Mie theory (LMT), the geometrical optics being unable of describing the properties of the diffused field in the immediate vicinity of the critical diffusion angle, where the wave character of the light is strongly "expressed" (namely by tunnel effect).

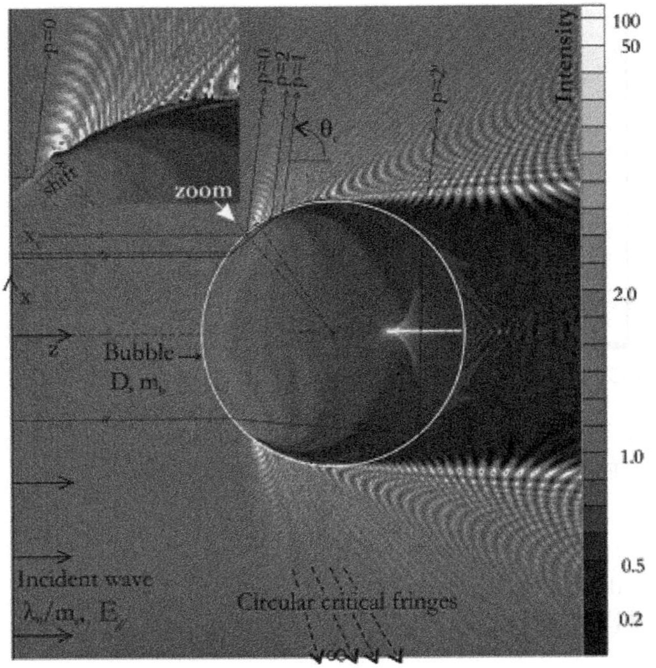

Figure 10.36. *Intensity of the electromagnetic field around and inside an air bubble in water (D = 100 μm), lit by a parallel polarized light beam of wavelength λ = 532 nm [ONO 11]*

Without going into the details of LMT theory, it seems important to mention that this new optical technique enables the size distribution to be reconstructed, and the average refractive index of clouds of spherical bubbles to be measured almost instantly and with very little computer resources. However, the method can also assess the size distribution when the bubbles are elliptical. Also, the wave fringe related to the critical diffusion seems to be quadratic, whereas it is cubic for the rainbow (see [ONO 12]).

Figure 10.37 shows an example of the results found for elliptical bubbles produced so as their major axis is in line with the flow's main axis. The LMT theory represents the measured data behavior very well. More details on the method and its implementation difficulties are discussed in [ONO 11].

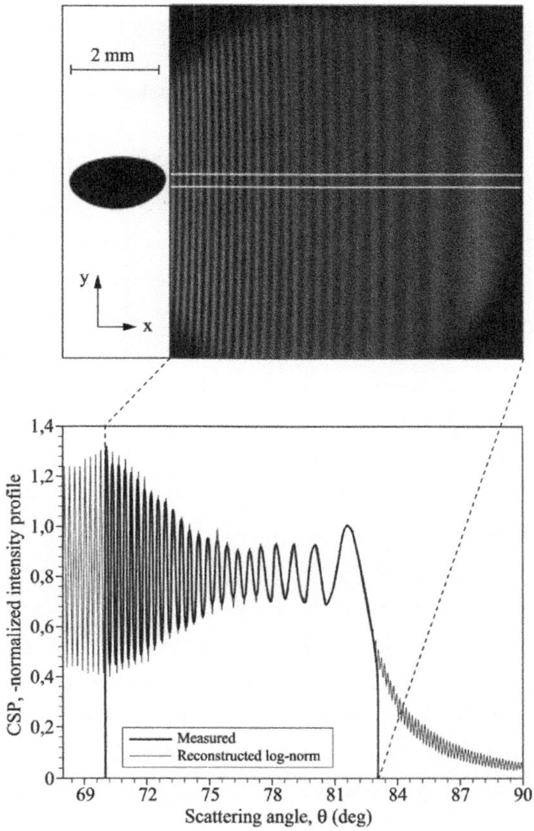

Figure 10.37. *Results obtained for elliptical air bubbles in water. Comparison of the data measured to the inversion model based on LMT theory [ONO 11]*

Chapter 11

Some Experimental Results Pertaining to Multiphase Flow Properties that Are Still Little Understood

11.1. Atomization/fragmentation of liquid jets

Fragmentation of jets or liquid sheets, often called atomization when the velocity is high, in surrounding gases at rest or in movement, is a situation encountered in a number of applications. For example, it concerns inkjet printers, diesel engine fuel injectors, propulsion systems in cryogenic rocket engines and applications in agriculture such as sprinkler irrigation or pesticide spraying. The variety and complexity of the mechanisms involved are partly due to the great diversity of possible situations, from a fluid mechanics point of view, in terms of the number of dimensionless characteristic parameters (Reynolds number, Weber number, velocity and density ratios between liquid and gas flows, etc.), and the field is therefore still the subject of many fundamental studies. Only typical results will be presented in this chapter, corresponding to cases where the surrounding gas is at rest. The reader may refer to the review papers [LAS 00, VIL 07, GOR 08] for additional details.

The problem with the atomization of liquid objects (jets, sheets, drops, etc.) has been studied over the past 50 years. It was first discovered that the Weber number – the ratio between the aerodynamic drag force and the forces caused by capillary action, $We = \rho_2(u_2 - u_1)^2 D/\sigma$, where ρ_2 is the density of the fluid that destabilizes the jet (commonly ambient air), $u_2 - u_1$ is the difference between the velocities of the destabilizing fluid and the liquid object, D is the object's characteristic length and

σ is the liquid's surface tension – is the main parameter that controls the process. The critical Weber number, above which the object disintegrates, is approximately 10, and sometimes smaller if the destabilizing flow is turbulent. However, the detailed analysis of the fragmentation process shows that the phenomenology involved is very complex. For example, when the Weber number is moderate, a drop will first distort itself into a pancake-like shape, slowly becoming flatter to create a bag-shaped membrane that then disintegrates into smaller (originating from the membrane itself) and larger drops (originating from the edge of the membrane). For higher Weber numbers, the drop no longer distorts entirely. Instead, the surface gradually deforms to form ligaments that then create droplets. The characteristic size of the droplets thus created is close to the diameter of the ligaments. This step corresponds to what is commonly termed primary atomization. The fragments rearrange themselves and their size either increases (by coalescing) or decreases (by collision or evaporation) during the secondary atomization stage. Throughout the fragmentation process, there is an exchange of momentum between the two phases, as the initial objects and the drops created by atomization will cause the air, initially at rest, to generate turbulence in the gas phase. At the same time, this gas turbulence will interact with dispersed drops to a greater or lesser extent, depending on their inertia.

We will now focus on the fragmentation of liquid jets. For these jets, the Reynolds number at the nozzle's exit is clearly another important parameter, and its value will control the phenomena that occur. Experimentally, different regimes can be distinguished (according to [ERO 91]) when the exit velocity of the liquid increases (Figure 11.1):

– For very low velocities, the drops form directly at the exit of the nozzle and there is no continuous liquid column. This is the drop-by-drop regime (section A).

– For low velocities, the jet is fragmented at a significant distance from the nozzle and produces drops having a diameter roughly twice the size of the nozzle diameter. This is known as the Rayleigh regime (section B).

– As the output velocity increases, the aerodynamic effects become more important and fragmentation occurs closer to the nozzle. The granulometric distribution is more widespread than in the Rayleigh regime, and the drops formed have a much smaller diameter than that of the nozzle. This regime is called "wind-induced break up" (and corresponds to sections C and D).

– For high ejection velocities, very fine drops form at the exit of the nozzle. This is the atomization regime (section E). Hereafter, primary and secondary atomizations will be considered as the primary and secondary fragmentations of this atomization regime.

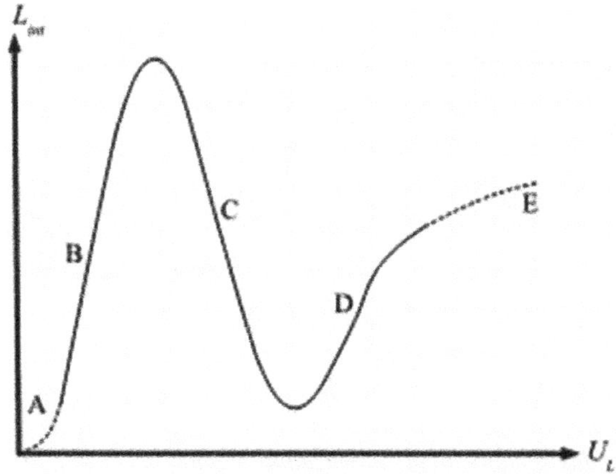

Figure 11.1. *Length of the liquid core L_{int} for a jet as a function of its ejection velocity U_L [ERO 91]*

The different regimes for the primary fragmentation described above are characterized by the length of the jet's intact area, or the rupture length, corresponding to an uninterrupted, continuous part of the liquid jet. According to the authors, the definition of this length varies. For example, it was proposed to define the length as the length of the jet hydraulically linked to the injection orifice. However, this definition can make the experimental determination of the length quite intricate, and is the reason why other definitions have been proposed, using a threshold value for the liquid's volume concentration, or even the length to the nozzle where the first drop separates from the liquid core. The range of definitions partly explains the difference in experimental results, which are effectively not related to the same item. The atomization regimes can be classified according to two dimensionless numbers, previously presented, to which the Ohnesorge number, given by $Oh = \frac{\mu_L}{\sqrt{\sigma \rho_L D}}$ (where μ_L is the liquid's molecular viscosity), is often added. For example, it has been proposed (see [FAE 90]) to classify the atomization regimes according to their Ohnesorge and gas Weber numbers (Figure 11.2(a)). However, another classification is possible (see [REI 82]) that takes into account the density ratios (Figure 11.2(b)). As the Oh number depends on We and Re, any two of the three can be used when the density ratio is given.

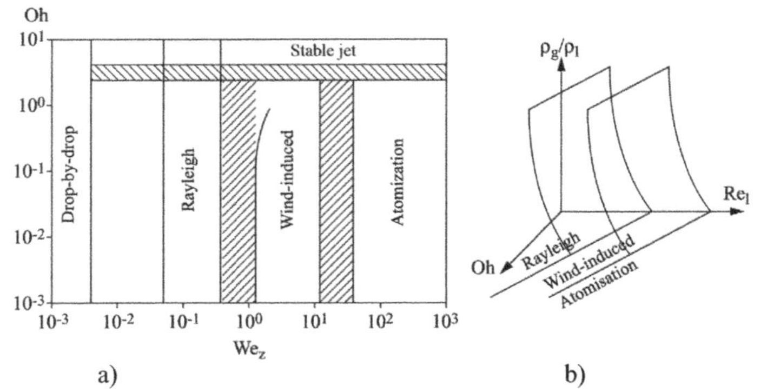

Figure 11.2. *Classification of the atomization regimes according to characteristic dimensionless number values; a): [FAE 90], b): [REI 82]*

Various mechanisms can be considered to explain the atomization of liquid jets having a high Weber number, which will be discussed hereafter. These mechanisms include aerodynamic effects, liquid turbulence, cavitation, ejection conditions (namely pressure fluctuations) and rearrangement of the velocity profile. As far as diesel injectors are concerned, which are the subject of numerous studies to date, the atomization of the liquid generally does not seem to be the result of one simple physical mechanism alone, but the combination of different phenomena linked together.

Aerodynamic effects originating from shearing of the jet's surface with air can lead to the development of Kelvin–Helmholtz-type instabilities. The jet's surface waves can therefore contribute to the jet fragmentation by amplifying it, or can facilitate the creation of ligaments and therefore of drops. This evidence can be tempered for heavy turbulent flows, having a density ratio greater than 500, as a number of authors have found the aerodynamic effects to be negligible in this case, as far as primary atomization is concerned. In this case, the creation of drops at the start of the jet is more likely to be linked to the turbulence of the liquid. However, as the liquid core is distorted by the turbulence, in other words the jet is more distorted downstream of the flow, the influence of aerodynamic effects progressively increases and becomes predominant: mechanisms corresponding to secondary fragmentation are then observed.

In addition, a number of studies conclude that atomization depends on the turbulence in the liquid flow created by the injector. It is hypothesized that the turbulence is influenced by the injection system's internal geometry. In particular, it became apparent that for experimental injectors, the spray characteristics changed radically depending on the liquid turbulence in injectors. For a given injector and the same exit Reynolds number, the spray angle increased with the nozzle's turbulence level.

It was also observed that cavitation, the appearance of gas pockets, inside the injection systems had a strong influence on the appearance of the jets emitted by diesel injectors. Some authors consider that this is an important phenomenon in liquid fragmentation observed in diesel jets. Cavitation results from partial liquid vaporization, occurring when the liquid pressure drops below its saturated vapor pressure, and is often obtained in high-velocity flows. For example, the presence of a bend or a neck can incur excessive pressure gradients in the flow, causing the boundary layer to detach and a recirculation zone to be created. If pressure in the recirculation zone decreases locally below the liquid's saturated vapor pressure, the liquid is vaporized and pockets of gas appear. This phenomenon can appear even in the core of the flow if the acceleration (or pressure gradient) is sufficiently large.

Other authors have noted that the pressure fluctuations in the liquid affect the liquid core's interface by provoking the appearance of a series of contractions and bulges downstream of the injection orifice. However, the frequency of these pressure modulations is low and is only of significant importance during the steady injection phase.

This brief presentation of the phenomena potentially involved in liquid jet atomization demonstrates the complexity, for the most part a consequence of the large number of parameters (velocity, density, turbulence level, surface tension, molecular viscosity, etc.) that can be varied from one practical situation to another. If the aim is to optimize the atomization process for a given situation, generally in terms of drop size or velocity at a given distance from the nozzle, it is important to understand the relative importance of all these phenomena. Ideally, a numerical modeling code should be used, capable of integrating all these phenomena and accurately simulating their relative importance. Unfortunately, this is not yet possible for the reasons mentioned in previous chapters (see, for instance, Chapter 7). This is why it is a common practice to resort to experiments in order to analyze the phenomena involved, as well as the specific effect of each parameter on them, and therefore to be able to model them as accurately as possible. The examples that follow will illustrate this.

First, the high-speed camera visualizations (Figure 11.3) performed by [MAR 04] demonstrate the formation of ligaments and the creation of droplets at their extremities. The case studied by the authors consisted of a liquid jet subjected to atomization as a result of a fast gas flow parallel to its surface. These experiments, carried out with different liquids, showed that the air flow destabilizes the liquid by a two-step mechanism. First, a shearing instability produces waves on the liquid. The liquid is subject to transient acceleration, suggesting that a

Rayleigh–Taylor-type instability is triggered at crest level of the waves, creating liquid ligaments that are stretched by the airflow until rupture, thus generating droplets. The primary wavelength, given by $\lambda \approx \delta(\rho_L/\rho_G)^{1/2}$, is a function of the vorticity thickness value δ and the density ratio between the liquid and the gas, ρ_L/ρ_G. The transverse corrugations of the crests have a characteristic size $\lambda_\perp \approx \delta We_\delta^{-1/3}(\rho_L/\rho_G)^{1/3}$ (where the dimensionless Weber number We_δ is defined by $We_\delta = \rho_G u_G^2 \delta/\sigma$ here). The ligament dynamics gives rise, after break-up, to a well-defined droplet size distribution of which the mean value is given by $\lambda_\perp$. This distribution curves exponentially.

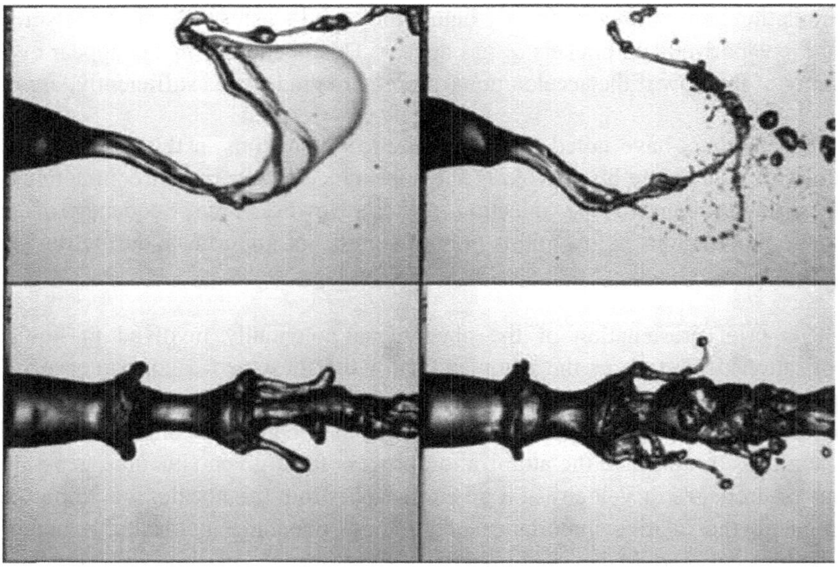

Figure 11.3. *Regimes for a slow, dense liquid jet atomized by a fast, light co-axial flow. Top: low gas velocity, the liquid jet meanders in the gas flow, leading to the possible creation of bags. Bottom: for high gas velocities, the surface of the jet is "peeled-off" and disintegrates after several surface instabilities [MAR 04]*

When a high-velocity horizontal liquid jet issues in ambient air at rest, similar mechanisms are observed, the variety of which is even larger than in the prior case. Complete disintegration of the liquid jet therefore only occurs at a very far distance from the nozzle, equal to approximately 900 times the nozzle's diameter d_{nozzle} (see Figure 11.4).

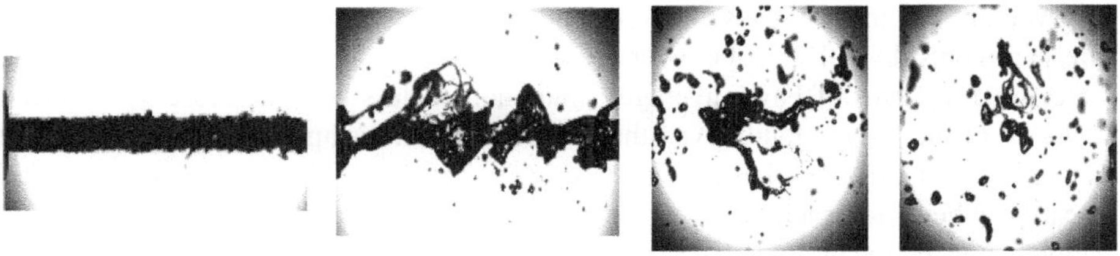

a) b) c) d)

Figure 11.4. *Atomization of a horizontal liquid jet issuing in ambient air at rest. Visualizations acquired with a camera used for PIV (with a very short shutter time that can be set by the investigator) for four sections: a) at the exit of the nozzle, b) at 70*d_{nozzle}, *c) at 460* d_{nozzle} *and d) at 900* d_{nozzle} *[STE 12]*

To quantify the properties of this two-phase flow at different positions relative to the nozzle, the images given above must first be processed by a specific shape-recognition, detection and contour-characterization algorithm (permitting the major and minor axes of each drop to be found), while taking into account the effect of the focal depth that skews the characteristic lengths obtained when the drop is not perfectly located in the focal plane. The algorithm and appropriate calibration methods are described in detail by Stévenin [STÉ 12]. By pairing the drops in two consecutive images obtained with the particle image velocimetry (PIV) camera, it is then possible to find the horizontal and vertical velocity components of each drop, and therefore to calculate the statistics for drop size and velocity. Figure 11.5 shows the results obtained for the distributions of drop sizes in a volume (in terms of equivalent diameters, using the ellipticity hypothesis). Many authors have observed that these overlap when the equivalent drop diameter is made dimensionless by the mass median diameter (MMD), calculated using the cumulative distribution function of the volume distribution function. As usual, the ratio between the MMD and the Sauter diameter D_{32} is more or less constant and equal to 1.2. This was found for both primary and secondary atomizations. The distribution itself is log-normal, which is the best suited to this case, even if other authors sometimes prescribe a Gamma law or a normal square root law. It is significant to note that the standard deviation $\sigma_{\ln}$ for a log-normal distribution is directly related to the ratio MMD/D_{32}: $\sigma_{\ln} = 2\ln(\text{MMD}/D_{32})$.

The axial evolution of the mean longitudinal velocity is shown in Figure 11.6(a). The results found using the particle-tracking method droplet tracking velocimetry (DTV) are compared with those found using laser Doppler anemometry (LDA). The comparison shows that the two measurement techniques correlate very well. However, it should be noted that these two measurement techniques only provide the velocity of the fluid, as the surrounding air was not seeded by particles for measurements by LDA. The overall mean velocity of the two-phase medium, calculated in terms of Favre averages, must nonetheless be quite close to the mean velocity of the liquid alone, given that the density ratio is approximately 1,000. The dimensionless radial profiles ($r_{1/2}$ is the velocity half-width, in other words the radial location where, for each section, the mean velocity is equal to half the centerline velocity) of the mean velocity (Figure 11.6), measured at four axial positions, show that the two-phase jet flow quickly becomes self-similar, even though its morphology changes considerably from one section to another, as shown in the images given in Figure 11.4. A minor asymmetry is also visible between the positive z regions (upper part of the jet) and negative z regions (lower part of the jet), a consequence of gravity-related effects. From these experimental observations, it is apparent that modeling methods described in the initial chapters of this book can realistically represent the overall flow properties, as long as the adapted closure assumptions allow all distribution details of the dispersed phase within the carrier liquid to be evened out. The ability to completely numerically simulate all these details is clearly still out of reach, at least in the case of flows with high Reynolds numbers.

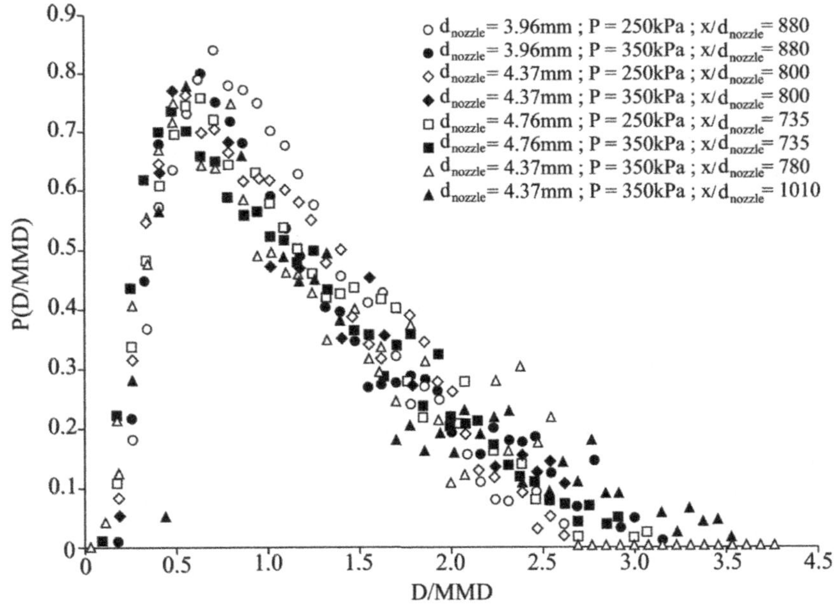

Figure 11.5. *Volume distribution of drop size for different Weber (We) and Reynolds (Re) numbers, and different abscissa x/d_{nozzle} [STE 12]*

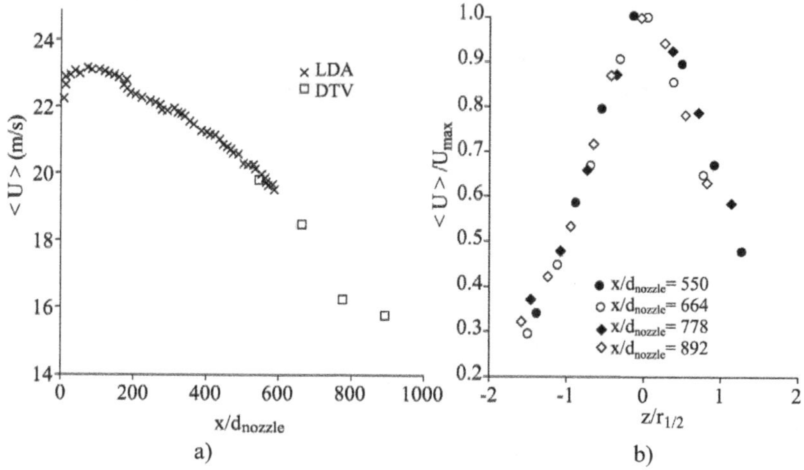

Figure 11.6. *a) Axial evolution of the mean velocity and b) normalized radial profiles for the mean axial velocity [STE 12]*

The results shown in Figure 11.7 enable the properties linked to the two-phase jet turbulence to be compared to the properties already established for standard, single-phase jets. Thus, it becomes apparent in Figure 11.7 that the ratio of velocity variances for axial and radial components is very different from the isotropic value currently used, equal to 1 (even though this ratio is in reality equal to ½ for jets, meaning a turbulent state that is more or less axisymmetric). In this case, very low values, less than 0.15, are found, and are even lower as the distance to the axis increases. This is original and unexpected. More advanced studies must therefore be undertaken in order to analyze this in more detail and to understand the physical reasons, as well as to propose adequate closure assumptions. On the other hand, the ratios between the Reynolds stress $<u'w'>$ and the turbulent kinetic energy k (Figure 11.7(b)) have values close to the standard ones (namely giving a value for coefficient C_μ equal to 0.09, which has an important role in turbulence modeling). It can be concluded that the link between k and the average velocity gradients remains unchanged for a two-phase jet, and that only the equilibrium between the three velocity variances is greatly affected. Finally, an analysis of the characteristics of velocity fields has been performed by diameter class. From this analysis, it has become apparent that the mean velocity of smaller drops is lower, but that the velocity variances are stronger, and anisotropy is more pronounced. If we wanted to implement a numerical model, based not only on overall results but also on results for each phase (the various phases being either the continuous phase, or several phases corresponding to several drop size classes), it would be necessary to include in closure assumptions all the ingredients permitting the full variety of behaviors, briefly discussed here, to be correctly simulated.

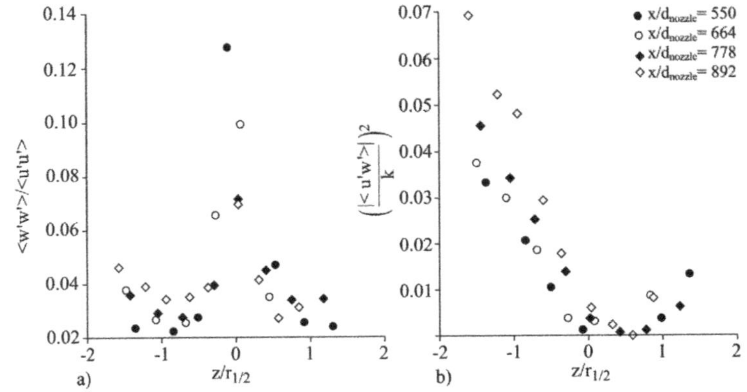

Figure 11.7. *a) Radial evolution of anisotropic factor $\langle w'w'\rangle/\langle u'u'\rangle$ and b) ratio $(\langle u'w'\rangle/k)^2$ that characterizes the value of C_μ [STE 12]*

11.2. Isolated bubbles, bubbles in swarm and their effects on carrier fluid

Practical situations where isolated particles, or relatively compact swarms of particles, move in a carrier fluid are numerous and varied. Common examples include solid particles in fluidized suspensions and granular media, discussed in Part 3 of this book. Other examples also exist, such as sedimentation of solid particles, both in natural media, such as rivers, lakes or oceans, and in industrial situations, such as wastewater treatment plants, rainwater reservoirs or droplet sprays in diesel engines.

This section will focus on situations where gas bubbles move in a carrier liquid. These cases are the subject of many laboratory studies, for which a variety of measurement techniques were implemented depending on the number of properties that needed to be found. The case of bubbles, considered to be isolated, will be discussed first, followed by cases where the gas content is larger, sometimes resulting in highly loaded media that are often called bubble swarms.

Bubbles and bubble flows play a very important role in a wide range of industrial and geophysical processes. Examples of these are the extraction and transport of oil and natural gas, mixtures in chemical reactors, manufacturing of alloys, cooling of nuclear reactors, two-phase heat exchangers, aeration processes, naval hydrodynamics or even ocean–atmosphere exchanges. All these applications have stimulated fundamental research on bubble flows with the aim of better understanding bubble hydrodynamics in flows with low-to-high Reynolds number values. One of the main questions relates to how the bubbles move in a flow, and how the continuous phase (in both laminar and turbulent flows) is affected by the dispersed phase. For classical results described in Chapter 5, the field remains largely open. More recently, computational codes used to predict two-phase flows have been developed. Their capacities are, to a large extent, limited by closure models used to represent the exchange of momentum between the dispersed and continuous phases. For this reason, the main goal of scientists in the field, over the last 20 years, has been to find rational expressions for the forces acting on the bubbles. Historically, the forces acting on rigid spheres (see [ODA 64]), whether they are moving steadily or not in a fluid at rest, have been determined through the use of strain gauges (fixed to the rods that hold the spheres and impose a specific displacement). Even if the measurement techniques were at first rather basic in comparison to the current techniques, these experiments enabled the drag coefficient C_D to be measured, as well as, more importantly, the added mass coefficient C_A that was subject of debate for a long time (see [MEI 96]). In particular, such experiments enabled an overall equation for the C_D value for uniform flows to be found, which is valid for a wide range of Reynolds numbers, based on the sphere diameter and its velocity, of up to several thousands:

$$C_D = \frac{16}{Re}\left\{1+\left[\frac{8}{Re}+\frac{1}{2}(1+3.315\,Re^{-1/2})\right]^{-1}\right\} \qquad [11.1]$$

where C_A = ½ for spheres. The added mass coefficient takes into account the force that the particle in non-steady movement (accelerated or decelerated) must provide to move the fluid in front of it. Of course, when spheres can no longer be considered as rigid, often the case for gas bubbles, additional phenomena need to be considered, such as the effect of the surface tension and the presence of surfactant products in the liquid phase. In these cases, the bubbles distort and cannot maintain their perfect sphericity due to the pressure gradients induced by flow features, namely on their downstream side. The C_D value then depends on the Weber number value, the bubble's aspect ratio, often considered as elliptical, and also Re. Characterizing these situations is hence much more complicated as the aim is often to characterize in detail the velocity disturbances induced by the bubble displacement in the carrier fluid, and to find the values of the drag and lift coefficients for non-uniform flows, such as flows characterized by imposed shearing or solid rotation (see [MAG 00]).

Among the most recent experiments, we can focus on those conducted by Rastello *et al.* [RAS 09], typical of the methods used nowadays to determine the values of C_D, C_L or C_A for specific flow cases. The basic principle is to consider that, for a real flow, there is an overlap between the regions where more or less important shearing dominates, and others where solid rotation dominates. An analysis will therefore have to be made of the corrections introduced by one or the other effect on the values found in uniform flow situations. Such an approach has been successfully implemented on the basis of computer simulations for flows having a constant shear (but more or less strong) by Legendre and Magnaudet [LEG 98b]. The study by Rastello *et al.* combines fast PIV and visualization measurements with the help of a high-speed camera for the detailed study of velocity fields near the bubble that is forced to move within a rotating flow, as well as for the quantitative study of the equilibrium positions of a bubble subjected to the vertical rise caused by buoyancy effects. However, it has been shown from equations of motion that when a solid rotation characterized by a rotational velocity ω occurs, a bubble of diameter R, inserted into the flow, reaches an equilibrium point defined by the angle θ_e in the plane perpendicular to the axis of rotation (horizontal) and by the distance r_e to this axis, such that:

$$C_D = \frac{4}{3}\frac{gd\cos\theta_e}{r_e^2\omega^2} \quad [11.2]$$

$$C_L = \frac{1}{2}(1 + C_A - \frac{g\sin\theta_e}{r_e\omega^2}) \quad [11.3]$$

$$C_A = \frac{\gamma_0}{2-\gamma_0}, \textit{where } \gamma_0 = \frac{2}{e^2}(1-\sqrt{1-e^2}\frac{\arcsin(e)}{e}), \textit{ with } e = \frac{\sqrt{\chi^2-1}}{\chi} \quad [11.4]$$

where χ is the ratio between the ellipse's major and minor axes if the bubble is deformed to an elliptical shape, and d is its equivalent diameter. The results obtained show not only that the method is very accurate, but also that it can differentiate between the behavior of clean bubbles (those that have not been contaminated by surfactants) and contaminated bubbles. The results show a systematic gap (see Figure 11.8) between the behavior of clean bubbles (supposed to correspond on their surface to a boundary condition corresponding to zero shear) that follow as evolution law for C_D, as a function of the Reynolds number (determined by the relative velocity of the bubble U_R, relative to the liquid, and the equivalent diameter d), much closer to that of solid spheres rather than close to that of very clean bubbles.

By analyzing the results in more detail, a clear dependency appears (Figure 11.9) for C_D on the Rossby number $Ro = U_R/\omega d = r_e/d$ (where C_{D0} is the value of C_D for uniform flows), which can be parameterized by $C_D = C_{D0}(1+0.55Ro^{-2})$. Similarly, the results obtained for the lift coefficient C_L show that the effects of shear and solid rotation are additive at the first order, so that the following can be written:

$$C_L(\text{shear}, \text{Magnus}) = C_L(\text{shear}, 0) + C_L(0, \text{Magnus}), \quad [11.5]$$

where Magnus represents the Magnus effect induced by the solid rotational movement of the fluid on the “free flight” bubbles. The rotation of the bubbles themselves is characterized by a velocity Ω_B, which is adjusted to experimental results with the following equation:

$$\frac{\Omega_B}{\omega} = 1.5 - \exp(-\frac{Re}{40}) \quad [11.6]$$

so that:

$$C_L(\text{shear,Magnus}) = \underbrace{\frac{1}{2}(1+C_A) - 0.25 + 1.2\,\text{Re}^{-1/3} - 6.5\,\text{Re}^{-1}}_{\text{shear}} + \underbrace{0.55\frac{3\Omega_B}{8\omega}}_{\text{Magnus}} \qquad [11.7]$$

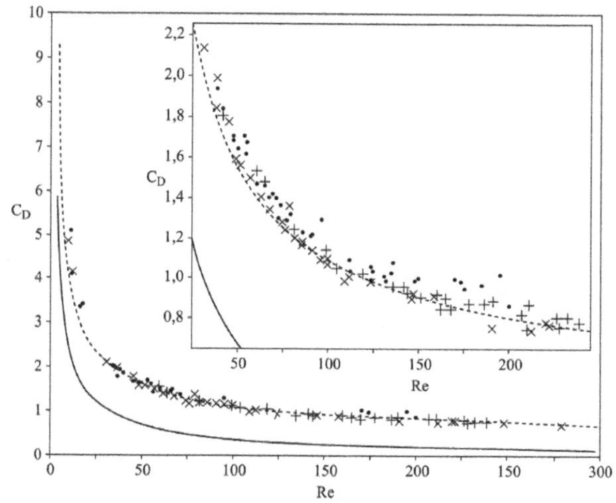

Figure 11.8. *Evolution of drag coefficient C_D as a function of Reynolds number Re for contaminated bubbles studied by Rastello et al. [RAS 09]. The insert is an enlargement of the results for small Re. The continuous line corresponds to the standard results for very clean bubbles, and the dotted line corresponds to those for solid spheres*

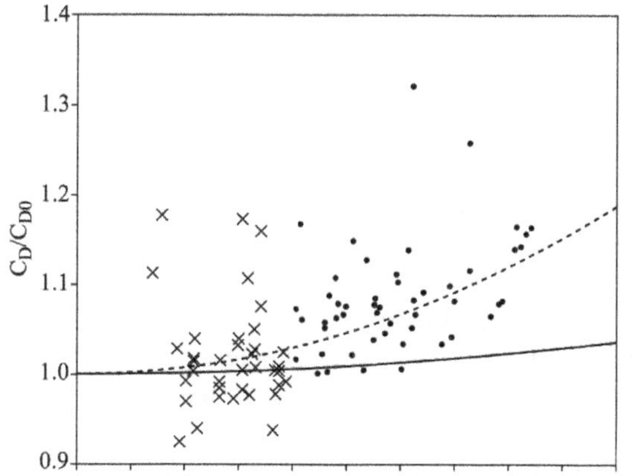

Figure 11.9. *Evolution of drag coefficient C_D as a function of the Rossby number for contaminated bubbles studied by Rastello et al. [RAS 09]*

It should be noted that Shew *et al.* [SHE 06] also measured the forces acting on isolated bubbles moving vertically upward, but combined bubble tracking using a high-speed camera to find their position in the horizontal plane, with an ultrasonic acoustic anemometer to measure their vertical velocity component. This experimental setup also allowed the detailed study of instabilities (zigzag and spiral instabilities) that often appear in the movement of rising bubbles because of the specific property of wake vortices, which have also been studied using more standard methods, although these will not be discussed here due to lack of space.

When bubbles can no longer be considered as isolated or independent of each other, the scenario becomes more complex as the wakes created by the bubbles are likely to interact very strongly with the flow of the carrier phase, which in turn affects the bubbles' evolution. The mean void fraction α must be taken into account in these cases. It is difficult to theoretically define this evolution as a function of α, as well as to measure the overall drag coefficient. It is therefore most often the case that experimental studies define overall properties, such as laws of evolution for the velocity of rising bubbles, or variances and spectra of velocity fluctuations, as a function of α. The equation that gives the C_D values will be given later (Figure 11.12); it is usually obtained by measuring the rise velocities of bubble swarms. A wide variety of such studies have been undertaken, and the most standard results found by Lance and Bataille [LAN 91] and Rensen *et al.* [REN 05a] will be discussed hereafter. The experiments by Lance and Bataille were performed in vertical flows, and the turbulence was generated by a grid. The flows studied had low velocities at the top of the water flow (0.3 m/s) used as the carrier phase. The bubbles were injected from the bottom through the use of small tubes, arranged in a compact configuration. To measure the turbulent velocity field, with or without bubbles, the basic techniques used were hot-film anemometry and LDA. In the presence of bubbles, an optical probe was used to determine the value of α accurately. For these experiments, the value of α remained less than 3%. The setup studied by Rensen *et al.* was essentially the same, although the carrier flow was directed toward the bottom (counterflow situation relative to the natural movement of bubbles) and the turbulence was generated by a grid, a device that enables the investigator to insert a much higher turbulent energy, so that the characteristic Reynolds number thus obtained would therefore be approximately five times larger than the Reynolds number obtained in the study by Lance and Bataille. The bubble diameters are similar in both studies, approximately 5 mm in the study by Lance and Bataille and ranging from 2 to 5 mm in that by Rensen *et al.*, roughly corresponding to the size of the Taylor microscale for both studies. However, given that two different grids were used, the intensity of the turbulence (equal to the ratio between the standard deviation u'_0 of vertical velocity fluctuations in the liquid without injected bubbles and their mean rise velocity relative to the flow of the carrier phase U_R) is higher in the study by Rensen *et al.*, ranging from 20% to 30%, whereas it was always less than 10% in the study by Lance and Bataille. The detailed measurements of the carrier and dispersed phase properties were determined by Rensen *et al.* using hot-film anemometry and an optical probe with four sensors. In both studies, standard signal processing methods were used, similar to the methods discussed in Chapter 10.

The figures below summarize the main results found. Figure 11.10 shows the evolution of the ratio between the variance of velocity fluctuations, induced by the presence of bubbles, u'^2_E, and U^2_R in terms of α, for different turbulence rate values u'_0/U_R. Even though this break down is a very simplified one, and often subject to debate, the effect of the velocity fluctuations measured in the liquid phase with bubbles (having a variance u'^2_L) is effectively considered as the superpositon of fluctuations in the bubble-free liquid and those generated by bubble movement, so that the following is obtained: $u'^2_E = u'^2_L - u'^2_0$. The figure clearly shows how a transition is established when α is larger than a critical value equal to approximately 0.012, and has a much steeper increase in u'^2_E / U^2_R when α is larger than the critical value. This increase is due to the excess turbulence generated in the carrier phase, which comes from interactions between bubble wakes. For values of α equal to or higher than 0.012, the interactions between bubble wakes, for bubble swarms, deeply alter the properties of the carrier phase. This important alteration of the turbulence properties in the liquid phase is also visible when the spectra of velocity fluctuations are studied. As shown in Figure 11.11, the spectrum in k^{-3} (here, $k^{-8/3}$ precisely) appears progressively in regions beyond the region in which Kolmogorov's law, having a turbulent energy spectrum that evolves as $k^{-5/3}$, is verified, which reflects the characteristics of the turbulence induced by the presence of wakes from bubbles (of diameter D). These wakes have a relatively small size and supply turbulent energy to a spectrum region that usually has a very fast decrease due to the dissipation of turbulent energy by viscous forces (case 1).

The results reported by Lance and Bataille [LAN 91], which can be qualified as "historic", have since been found again in a great number of studies. However, the results obtained by Rensen *et al.* partially disagreed. The reason behind this is not yet clear (a reasoned discussion is provided in the article by Rensen *et al.*, as well as a thorough bibliographic study of the results obtained on the subject over the past 20 years). It seems that the "bubblance parameter", given by $b = \frac{\frac{1}{2}\alpha U^2_R}{u'^2_0}$, could play an important part. When b < 1, the preexisting turbulence in the bubble-free flow is significant (explaining why the effect of bubbles is less in these situations, and may therefore limit the appearance of the region in k^{-3}, as was the case for the experiments by Rensen *et al.*), whereas when b > 1, the pre-existing turbulence is less important, such that the part played by bubble wakes is much more significant (with the marked appearance of a region in k^{-3} as in the experiments by Lance and Bataille). This must also be weighted by the bubbles' relative size in comparison to the characteristic scales of turbulence (and namely the Taylor and Kolmogorov scales, respectively, λ and η). The fact that bubbles can be considered as either very clean or, on the contrary, more or less contaminated by the presence of surfactants can equally play a non-negligible role in the results obtained, as was the case for the C_D value of isolated bubbles.

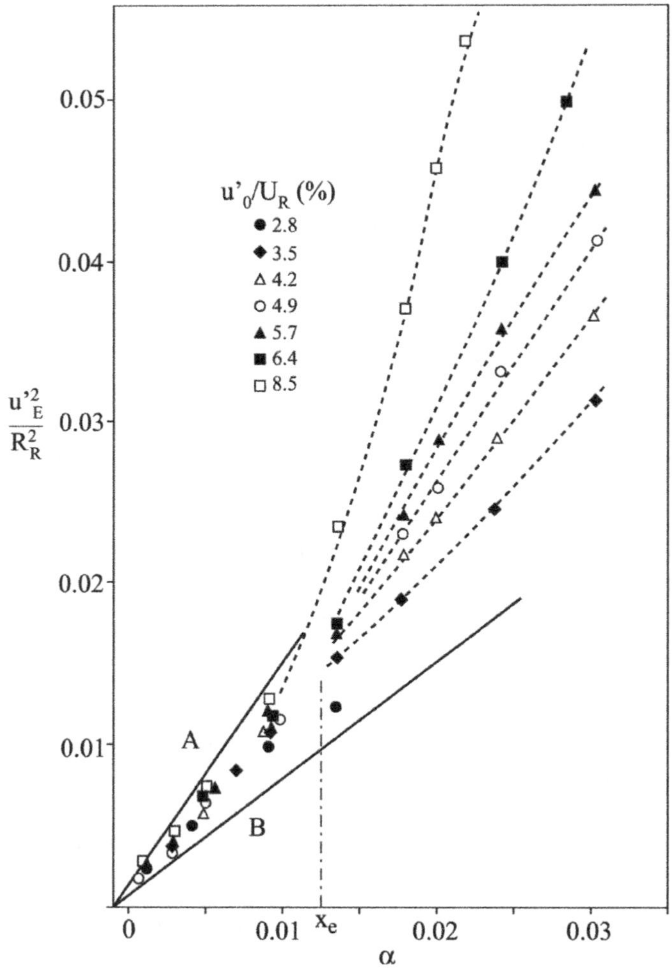

Figure 11.10. *Evolution of the normalized variance of velocity fluctuations induced by bubbles, in terms of α [LAN 91]*

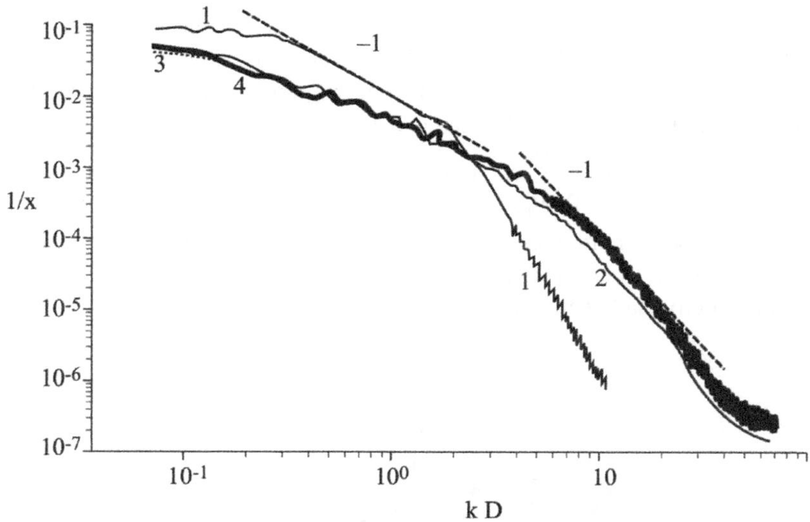

Figure 11.11. *Evolution of the turbulent energy spectrum in terms of α. The curves numbered 1, 2, 3 and 4 correspond to values of α = 0, 0.01, 0.025 and 0.04, respectively [LAN 91]*

Another highly significant aspect of bubble swarm flows concerns the transfer properties of gases contained in moving bubbles to the liquid. This occurs in many practical applications including industrial ozonation, oxygenation processes and aeration tanks in wastewater treatment plants, for example. To quantify the efficiency of these transfers between gas phases, usually consisting of a swarm of bubbles, and liquids to be processed, and namely to find numerical modeling methods based on the equations given in previous chapters, which are adapted to the case in question, specific modeling has to be introduced. This specific modeling mainly concerns the value of the drag coefficient, as well as the exchange coefficient, called Sherwood coefficient. The drag coefficient C_D is very complex to find experimentally. The most widely accepted equation is given by (e.g. [SIM 07]):

$$C_D = \frac{4}{3}\frac{\rho_l - \rho_g}{\rho_l} gD \frac{1-\alpha_{loc}}{U_R^2} \qquad [11.8]$$

where α_{loc} represents the local void ratio and U_R, as before, the relative velocity of the bubbles in the carrier liquid. The parameterization of the evolution of these two variables in terms of hydrodynamic conditions is complicated and is still the subject of numerous controversies. The experiments conducted by Simmonet *et al.* [SIM 07], in a vertical tank filled with water in which air bubbles were injected from the base, therefore combined mean and fluctuating velocity measurements in the liquid phase using LDA with a double optical probe to find the local void ratio α_{loc}, the bubble velocities and chord lengths. Moreover, as is often the case, the overall void ratio in the tank α_{glob} was estimated from the pressure difference ΔP between a section close to the entrance of the tank and a section close to its exit, separated by a height h_{pres}, by $\alpha_{glob} = 1 - \frac{\Delta P}{\rho_l g h_{pres}}$. Figure 11.12 therefore shows the overall results obtained by these authors once the influence of parameters, including the diameter of emitted bubbles, the void ratio α_{loc} (which generally has a value close to α_{glob}) or the liquid velocity, had been studied.

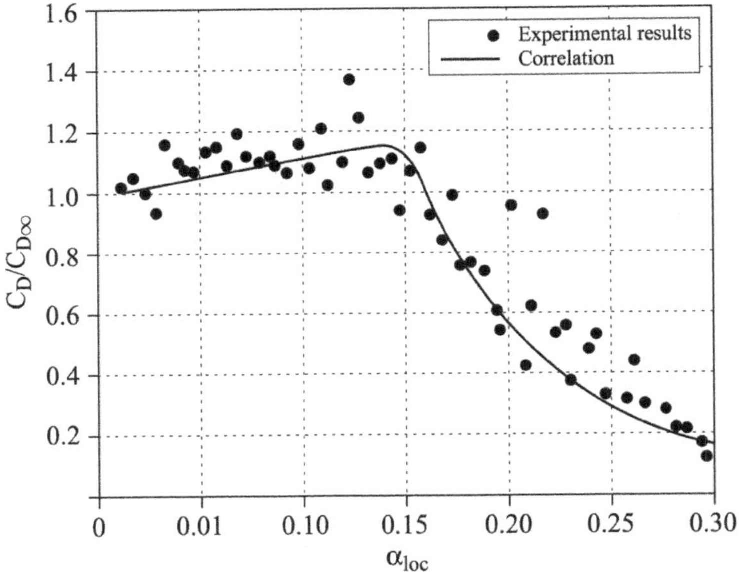

Figure 11.12. *Evolution of the drag coefficient, normalized by its value for individual bubbles, as a function of the number α_{loc} [SIM 07]*

For the range of bubbles studied, with diameters ranging from 2 to 7 mm, the transition observed in the region of $\alpha_{loc} = 0.15$ is due to a change in the hydrodynamic regime, which strongly influences the behavior of U_R. When $\alpha_{loc} < 0.15$, the relative bubble velocity is determined by the congestion effect of bubbles relative to each other, and implies that the relative velocity decreases as the void ratio increases. On the other hand, when $\alpha_{loc} > 0.15$, the suction effect induced by the wake of the upstream bubbles on the downstream bubbles leads to an important increase in the relative velocity with α_{loc}. Therefore, $\frac{C_D}{C_{D\infty}} \approx (1-\alpha_{loc})^{-1}$ when $\alpha_{loc} < 0.15$ and $\frac{C_D}{C_{D\infty}} \approx \frac{1}{23}\frac{(1-\alpha_{loc})^3}{\alpha_{loc}^2}$ when $\alpha_{loc} > 0.15$.

The transfer of gases to the liquid is usually characterized by using a transfer coefficient defined by $k_l = \frac{2}{\sqrt{\pi}}\sqrt{\frac{D_l}{t_c}}$, where D_l is the molecular diffusion coefficient, and t_c is a contact time, sometimes called exposure time. It represents the characteristic time taken to transfer to the moving liquid, sometimes interpreted as the residence time of fluid particles on bubble surfaces. Different models have been put forward for t_c. Sometimes, it is expressed as the ratio between the bubbles' diameter D and their rise velocity U, or as the ratio between the interfacial area and its rate of change, although it occasionally involves the variances of velocity fluctuations. However, the first proposal will be discussed, namely $t_c = D/U$, so that the Sherwood number is given by $Sh = k_lD/U$, as in the study by Colombet *et al.* [COL 11]. This equation is thus equivalent to the equation used by Boussinesq, who proposed an equation to find the value of Sh, based on the theory of potential flows, and takes into account transfer that occurs across a thin boundary layer surrounding the bubble, a valid assumption, and gives very good results for high Péclet numbers, Pe, and isolated bubbles, or at least for bubbles that are a long distance from each other: $Sh = \frac{2}{\sqrt{\pi}}(Re\,Sc)^{1/2} = \frac{2}{\sqrt{\pi}}Pe^{1/2}$ (with, Pe = ReSc, where Re is the Reynolds number based on D and U, and Sc is the Schmidt number). When bubbles can no longer be considered isolated, Colombet *et al.* carried out an experimental study to determine the evolution of the Sherwood number, the exact oxygen transfer rate, the density of the interfacial area and the relative bubble velocity for a wide range of void fractions ($0.45\% < \alpha < 16.5\%$). This was also performed using a vertical tank filled with water, in which air bubbles were injected from the base. High-speed camera and double optical probes were used to measure the void fraction α and the bubble velocity, in addition to particle tracking velocimetry (PTV). The latter technique, PTV, restricted the measurements to void fractions equal to 16.5%, as it is not possible to accurately follow the movement of each bubble for higher values of α. The interfacial area density a_i is calculated by assuming that bubbles are ellipsoids. The major and minor axes are measured so that $a_i = \alpha\frac{\sum S_b}{\sum V_b}$, where S_b and V_b are, respectively, the surface area and the volume of a given bubble. The concentration of oxygen dissolved in the liquid is found via Clark electrodes that deliver a microcurrent having intensities proportional to the concentration of oxygen dissolved in the liquid.

Assuming that the tank used is a "well-stirred reactor", due to the presence of bubbles and the turbulence generated by their movement within it, the evolution of the oxygen concentration C in the tank as a function of time can be described by the relationship:

$$\frac{\partial C}{\partial t}=\frac{k_l a_i}{(1-\alpha)}(C_s^*-C) \qquad [11.9]$$

where C_s^* is the saturation concentration, which is effectively that obtained at the top of the tank after the first few seconds of the experiment have elapsed. The evolution of C therefore appears to follow the law:

$$\frac{C}{C_s^*}=1-\frac{1}{\tau}(\tau e^{-t/\tau}) \qquad [11.10]$$

if, as is the case here, the time constant for the oxygen probe can be neglected. The transfer time τ is thus given by $\tau=\frac{1-\alpha}{k_l a_i}$. Figure 11.13 shows the results obtained by Colombet *et al.* [COL 11] for the evolution of C as a function of time, for $\alpha = 0.6\%$ and $\alpha = 15.2\%$, for which it is easy to find τ by adjusting the above law.

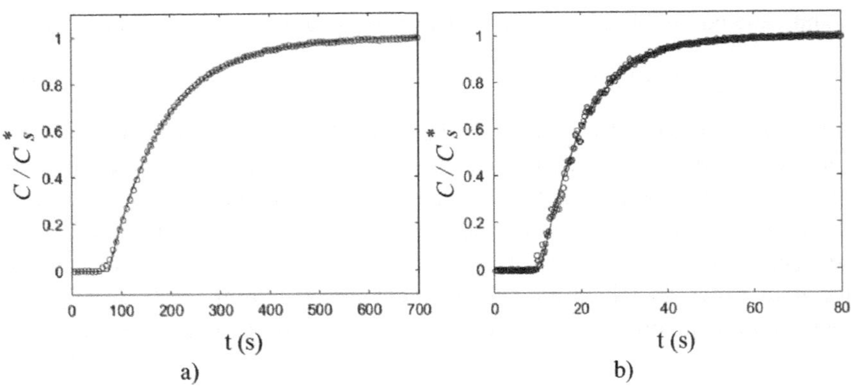

Figure 11.13. *Evolution of the oxygen concentration in the tank, for a) $\alpha = 0.6\%$ and b) $\alpha = 15.2\%$ [COL 11]*

The experimental equation for the variation of τ, $\tau \approx \alpha^{-0.74}$, is obtained by repeating the experiment for a whole range of α values. From this equation, the value of k_l can be calculated as $k_l = \frac{1-\alpha}{\tau a_i}$. The evolution of k_l is shown in Figure 11.14. It should be noted that even if k_l varies with α, a good approximation would, *a priori*, be to simply take $k_l = O(10^{-4})$ m/s. In other words, it does not require a complex experimental strategy to be developed. It must also be noted that the equation proposed by Boussinesq, for an isolated bubble, provides a remarkable adjustment of the experimental data, at least up to $\alpha = 16\%$. A possible explanation for this is that the liquid flow region located just upstream of a bubble is only a little affected by bubbles if they are present as an isolated bubble or as a surrounding swarm of bubbles, contrary to the wake area, which is highly dependent on the α value. However, while the value of Pe is close to 4×10^5 in this case, the thickness of boundary layers can be expected to be larger when cases have a smaller Pe value, so that the effect of the α value may then be more pronounced.

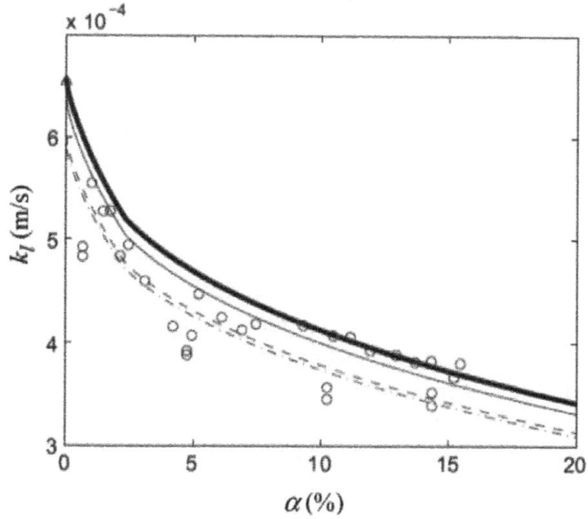

Figure 11.14. *Evolution of the transfer coefficient k_l as a function of α, and comparison with different laws in the literature. The continuous line corresponds to the relationship proposed by Boussinesq [COL 11]*

11.3. Boiling crisis

The boiling phenomenon has not yet been particularly mentioned here, even though it is very important in both domestic every-day life and in numerous industrial processes such as vapor cycles in thermal power plants and two-phase loops for satellite cooling. Boiling is sometimes desirable, as for the case of evaporators, and must sometimes be avoided, as in hydraulic circuits in solar panels. Over the past 50 years, the number of studies performed has been constantly increasing. However, heat transfer during boiling is not well understood and remains a subject open to debate: the basic mechanisms of heat transfer are still subject to discussion, and the correlations used to predict heat fluxes are based on very specific experiments and are difficult to extrapolate to other experimental conditions (fluid considered, pressure, geometry, etc.).

Among the numerous problems encountered, boiling crisis is one of the most important critical phenomena, as it may lead to the rupture of fuel cladding in nuclear power stations. It is therefore the subject of in-depth studies related to thermo-hydraulic design of pressurized water reactors (PWR) as one of the main issues concerns the safety of nuclear installations. This section will therefore describe the complexity of this phenomenon and will describe the way in which experimental techniques, discussed in Chapter 10, can be used with the aim, on one hand, of characterizing the various mechanisms involved and, on the other, of testing the validity of semi-empiric correlations that have been proposed by different authors, giving the base hypothesis for numerical modeling allowing the critical boiling flows to be determined in different industrial situations.

For permanent regimes, pool boiling phenomena, in other words the progressive evolution from nucleate to film boiling of heat transfer between a heated surface and the liquid in which it is immersed, have frequently been studied. One of the first studies was by Nukiyama [CAR 92] who analyzed the exchange through a metallic wire, heated electrically by Joule effect and placed horizontally in an infinitely large tank of saturated water. The Nukiyama curve (Figure 11.15), or "boiling curve", allows the exchange characteristics to be presented in a heat flux diagram as a function of the temperature difference between the wall and the saturation value. This curve (see also [VIS 12]) begins with a section where the heat is evacuated by natural conduction and convection up to point A in Figure 11.15. This section is then followed by a region of nucleate boiling. Point A, where nucleate boiling starts, corresponds to the appearance of small vapor bubbles at nucleation sites: the first phase of boiling is therefore characterized by isolated bubbles that rise in the shape of chains. The separation of the first vapor bubbles guarantees a good heat transfer, and therefore a strong increase in the heat flux, during which the temperature of the wall remains the same (see points A and B, Figure 11.15) and close to the saturation temperature. Once the bubble is detached, the amount of vapor increases and the bubbles coalesce to form pockets and columns (from point C in Figure 11.15). Heat transfer is still efficient. Critical heat flux (CHF) conditions occur when vapor production is such that it prevents the wall from being rewetted (point D in Figure 11.15); beyond this point, the behavior of the system differs depending on whether heating is controlled by temperature or heat flux. In the case of temperature control (black curve), the regime is that of transition boiling: an increasing number of bubbles coalesce and may lead to the appearance of a vapor layer that periodically isolates part of the hot surface. The increased overheating is no longer accompanied by an increase in the flux transmitted and, beyond the minimum film boiling (MFB) point – sometimes called the Leidenfrost point (point E) – the heating wire is covered by a stable vapor film from which bubbles regularly escape. If, on the other hand, the curve is controlled by heat flux, the heat exchange is significantly deteriorated and the wall temperature therefore increases quickly, resulting in a direct switch to film boiling (points D and F). Fusion of the heating element may occur and becomes a problem: this is termed "burn-out". It should be noted that the critical flux, for the maximum nucleate boiling regime, is in the order of 1 MW/m^2 for water. When the curve is followed in the opposite direction, there is a phenomenon of hysteresis: if the flow is decreased from the film regime, once the minimum flux has been reached, there is a direct change to nucleate boiling (EF′). However, it should be noted that more recent studies than that of Nukiyama have shown that this hysteresis does not always exist. For examples, authors have found that, in the case of very-wet fluid boiling, and when the heating surface is clean, the boiling curve remains exactly the same, independent of the increasing or decreasing heating direction.

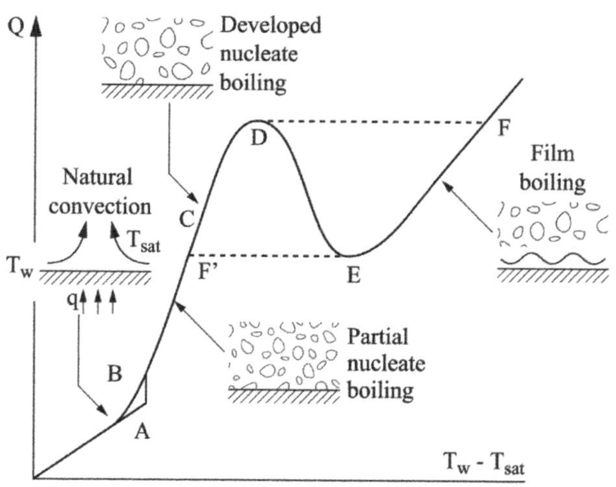

Figure 11.15. *Nukiyama boiling curve [VIS 12]*

In the presence of tube flows, the boiling regimes are different. Consider a subcooled liquid at the input of a heated vertical tube (Figure 11.16). When boiling initiates (see also [MON 09]), vapor bubbles separate from the wall and condense at the center of the tube, where the liquid has a temperature lower than (subcooled nucleate boiling) or equal to the saturation temperature (saturated nucleate boiling).

It is a “bubble flow” regime. When the amount of vapor increases, the bubbles coalesce. Pockets of vapor form and are separated by liquid plugs that may also be carrying small bubbles. This is the “slug flow” regime. The coalescence of vapor pockets can lead to the appearance of a destructured flow of “churn flow” type, or directly to the appearance of an “annular flow”. This consists of a liquid film around the wall and a vapor core. The liquid film becomes increasingly thin, waves appear at the liquid–gas interface and liquid droplets are torn away and dragged to the center of the flow. Finally, vapor may be present anywhere in the tube together with liquid droplets (“mist flow” regime). Wall dry-out occurs when the liquid layer in annular flow disappears completely due to the tearing away and evaporation of the liquid film.

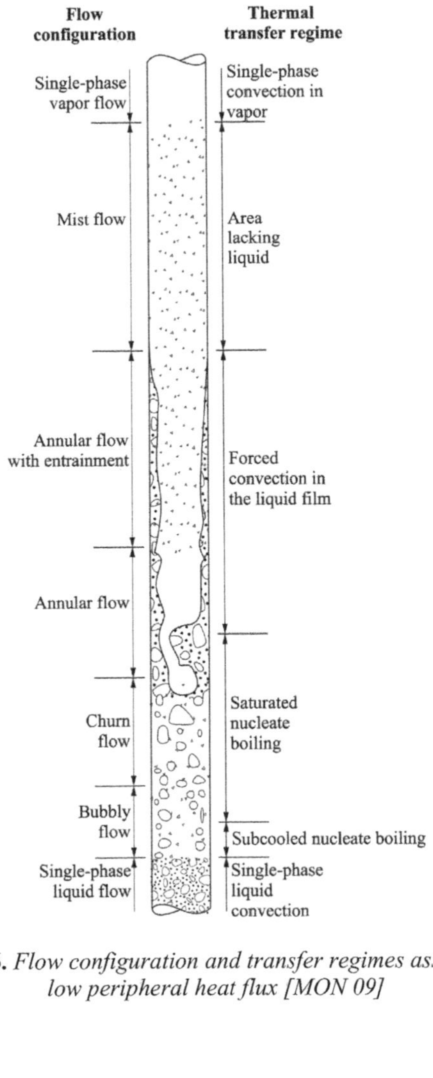

Figure 11.16. *Flow configuration and transfer regimes associated with low peripheral heat flux [MON 09]*

When the heat flux density increases, the dryness of the vapor also increases, and the different regimes migrate toward the entrance of the tube (Figure 11.17). In this case, a flux level can be achieved where nucleate boiling is reached at the surface and is immediately followed by a film boiling regime, in which a vapor layer spreads out on the hot surface. The heat exchange, improved by the appearance of vapor bubbles in comparison to the transfer in the liquid, then deteriorates as the vapor film separates the wall from the liquid (which is often called "wall dry-out").

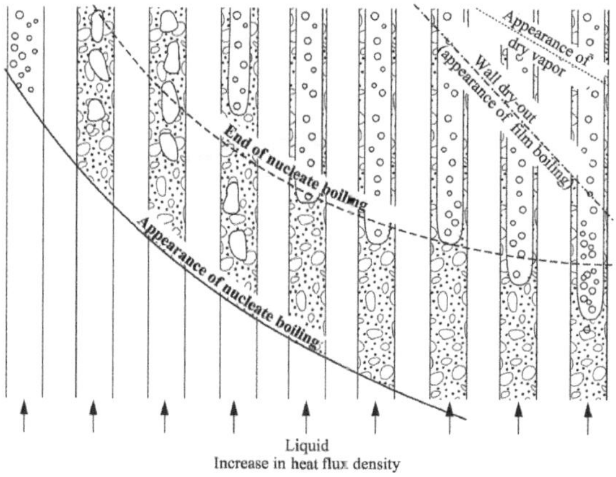

Figure 11.17. *Convective boiling regimes, influence of heat flux density [VIS 12]*

Historically, and up to 10 years ago, when the use of high-resolution, high-speed cameras appeared, experimental approaches to the problem have mainly been based on the use of optical-needle probes, such as those described in Chapter 10 (see [DIH 91] and [DIH 98]). By using these probes, the mean vapor content (or mean void fraction) can be found at different locations in the flow, or in the vicinity of heating walls, for different heat fluxes, in addition to the mean passing frequencies of interfaces that separate the liquid from gas bubbles (or pockets). These measurements also enable us to analyze, more finely, the hydrodynamics mechanisms that take place when departure from nucleate boiling occurs, and form the basis for semi-empirical relationships characterizing variables such as the mean bubble diameters when they detach themselves from the wall, or the mean number of nucleation sites per unit area on the wall. For example, Figure 11.18 (published in the article by Carrica *et al.* [CAR 95]) shows the time evolution for the response of Freon 11 when periodically heated (at a frequency of 0.05 Hz), in terms of mean void fraction and mean passing frequency of interfaces. Similar results are shown in Figure 11.19 for water. The appearance of the boiling crisis is well illustrated by this type of diagram (the point denoted by CHF), separating the initial nucleate boiling regime from the film boiling regime, which corresponds, for Freon, to the power that matches the maximum mean passing frequency of interfaces (in other words, to the situation in which the greatest number of bubbles are present in the flow when boiling is occurring) and has an associated relative maximum for the mean void fraction. Beyond this point, the transition to film boiling is established and the number of bubbles therefore decreases drastically to a constant until the heating power has returned to a level that is low enough for the flow to switch back to the nucleate boiling regime.

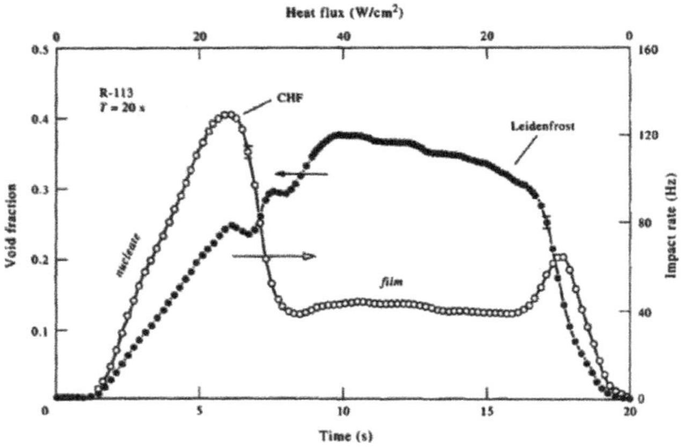

Figure 11.18. *Time evolution of the mean void fraction and mean passage frequency of interfaces obtained with Freon R11 in a periodic heating experiment by Carrica et al. [CAR 95]*

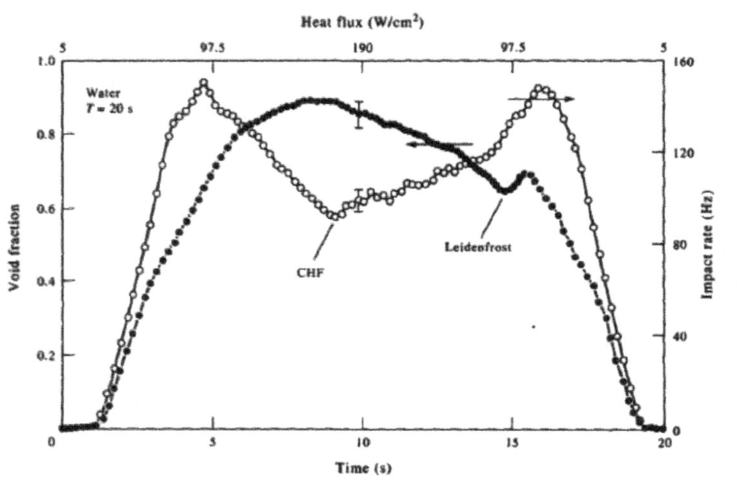

Figure 11.19. *Time evolution of the mean void fraction and mean passing frequency of interfaces obtained with water in a periodic heating experiment by Carrica et al. [CAR 96]*

In a different scenario, using water instead of Freon, this transition is much less sudden, and the maximum (at 80 W/cm^2) mean frequency of passing interfaces does not concur with the appearance of departure from nucleate boiling. Departure from nucleate boiling appears at a higher power level (of 180 W/cm^2), characterized by a minimum mean frequency of passing interfaces and a maximum mean void fraction. The two behaviors are clearly different and illustrate the fact that the hydrodynamic mechanisms associated with departure from nucleate boiling are not universal, contrary to what is assumed by historical semi-empirical models, such as those proposed by Kutateladze [KUT 50] and Zuber [ZUB 58] (see [THE 02]). Although a hydrodynamic instability mechanism for Freon, corresponding to the appearance of Helmholtz instabilities, seems to be the only one involved, a second transition occurs for water. When the heat flux increases, bubbles coalesce and form vapor columns. Then, the velocity of the vapor column, relative to the liquid, is such that the liquid–vapor interface oscillates and becomes unstable (Helmholtz instabilities). The liquid is no longer able to reach the wall and the surface is thus covered by a vapor film. For water, on the other hand, a second instability occurs as convection mushroom clouds appear afterward, through the grouping of vapor columns, before departure from nucleate boiling occurs. These different behaviors are linked to the different wettability properties of Freon and water, corresponding to different contact angles between the three elements involved, namely the liquid, its vapor and the solid wall. The contact angle for Freon is usually only around a few degrees, whereas the contact angles for water are higher by a factor equal to 10 or more. To analyze, interpret and model the phenomena involved, an overall approach is therefore not sufficient as the small-scale structure of contact between a liquid, its gas and the solid surface must be analyzed in detail.

To obtain more detailed information on the mechanisms involved, and in particular to better understand the gaps commonly observed with the semi-empirical laws such as that defining the value of the dimensionless Kutateladze number, Ku, by:

$$\mathrm{Ku} = U_{cr} \sqrt[4]{\frac{\rho_v^2}{\sigma g(\rho_l - \rho_v)}} \qquad [11.11]$$

(where σ is the surface tension, ρ_l and ρ_v are the liquid and gas densities and U_{cr} is the critical velocity for gas, generally considered as equal to $U_{cr} = q_{cr}/H_{lv}\rho_v$, with q_{cr} the CHF and H_{lv} the latent heat for vaporization), the measurements implemented over the last 10 years are principally based on the use of high-resolution, high-speed cameras. The historically accepted values for Ku were approximately equal to 0.145, but systematic studies, such as that by Carrica *et al.* [CAR 96], that vary the liquid, the pressure or the material used for the solid wall, showed a significant scattering of data relative to these values. The Ku parameter is indeed essential as its value can be used to calculate the value of q_{cr} with the following equation:

$$q_{cr} = Ku\rho_v H_{lv} \sqrt[4]{\frac{\sigma g(\rho_l - \rho_v)}{\rho_v^2}} \qquad [11.12]$$

The semi-empirical value above is easily comparable to experimental results such as those which have just been discussed.

Therefore, Theofanous *et al.* [THE 02] used an infrared camera capable of acquiring up to 1,000 images per second, in conjunction with a video camera capable of taking 20,000 images per second, to accurately measure the temperature on the surface of the heated wall with a resolution of 250 µm/pixel for a 128 × 128 pixels image. Different types of walls were also tested – new walls, walls already moderately used and already highly-used walls – in order to determine the influence of wall ageing on the results. The surface nanomorphology of these different walls was studied in detail by X-ray radiography and showed that the more used the walls were, the more their ageing changed the state of their surface by the appearance of titanium oxide, creating randomly distributed areas on the surface which have a different microcoarseness to that associated with the young material. The older the walls are, the higher the proportion of these areas is. These particular areas result in an increase in the surface density of nucleation sites and, as a result, decreases the level of overheating required to initiate departure from nucleate boiling (see Figures 11.20(a) and 11.20(b)). For new walls, the overheating required to reach departure from nucleate boiling lies between 22 and 32 K, whereas it varies only from 18 to 25 K for aged walls. It is also important to note that, as a consequence, the heat fluxes are significantly higher on aged walls. It should also be noted that despite all the care taken by the investigators, there is a large dispersion between results from the different tests, originally meant to be identical. The sensitivity of the results to micro- or nanophenomena, which cannot be controlled by the investigators, is therefore very high. Of course, this explains the large dispersion of results found in the past by various authors.

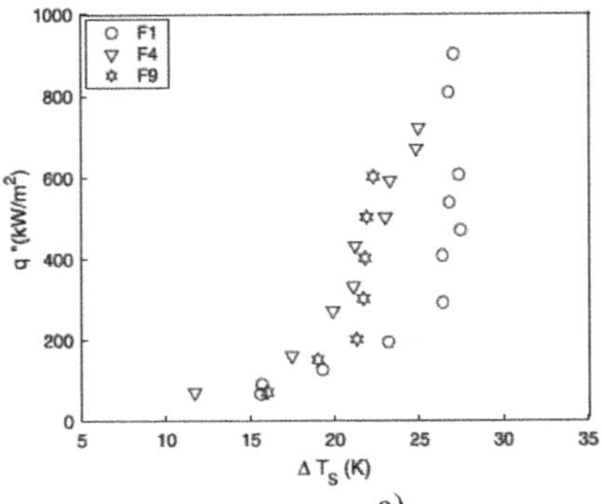

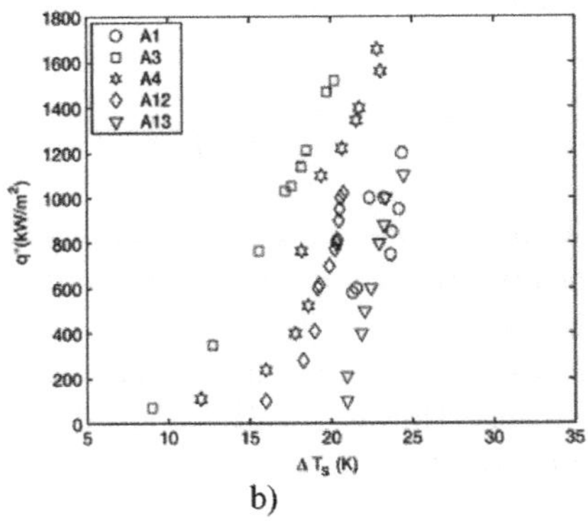

Figure 11.20. *Boiling curves obtained for a) new walls and b) aged walls [THE 02]*

A detailed analysis of temperature measurements of the surface shows the presence of hot and cool areas, the temperature level of which varies greatly with time (Figure 11.21). The hot areas are identified as dry hot points, which can be considered as precursors to the appearance of gas films on the wall, responsible for departure from nucleate boiling if the heat flux is sufficiently high. These hot areas are located at the base of gas bubbles and can evolve either reversibly (in this case rewetting of dry hot points occurs) or irreversibly (leading to dry-out of the heating wall).

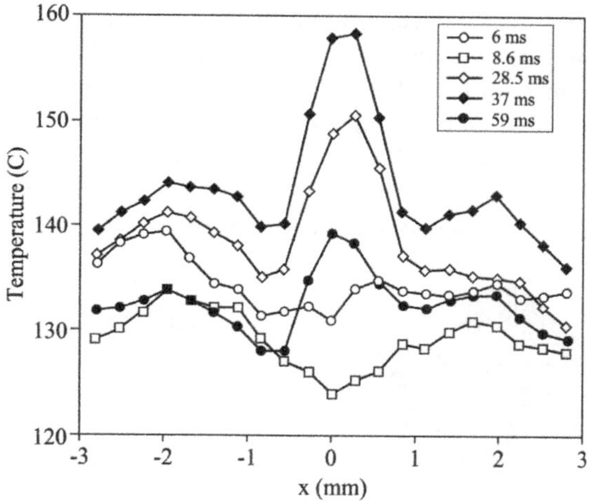

Figure 11.21. *Temperature distribution for dry hot points at five different time instants, for a heat flux of 900 kW/m² [THE 02]*

This type of highly detailed local information is very important as the characteristic elements of the concerned area's geometry are used for unsteady numerical modeling of departure from nucleate boiling phenomenon. Therefore, in the studies by Pezo and Stevanovic [PEZ 11], based on the two-fluid modeling approach discussed in Chapter 4, it is necessary to parametrically introduce the evolution of the surface density of nucleation sites, and the residence time of bubbles on the heating surface τ, as a function of heat flux and wall ageing. These data are used as boundary conditions that link the mechanisms associated with the thermal transfers in the heating wall, to the two-phase fluid medium that makes up the liquid and the vapor. For example, the mean diameter of the bubbles when they leave the wall, D_B, and the surface density of nucleation sites, n, are related by $n D_B^2 = 1$. Similarly, D_B and τ are often considered as being linked by the relationship $D_B = 2(\gamma Ja + \sqrt{\gamma^2 Ja^2 + 2\beta Ja})\sqrt{a\tau}$, where Ja is the Jacob number calculated by $Ja = \dfrac{c_{p,water}\,\Delta T_{sat}}{h_{vaporization}} \dfrac{\rho_{water}}{\rho_{vapor}}$, where $h_{vaporization}$ is the latent heat for vaporization and $c_{p,water}$ is the water's specific heat; parameter γ depends on the value of the contact angle, lying between 0.1 and 0.5 for angles ranging from 40° to 90°, whereas the constant β takes a value of 6 from semi-empirical adjustment. The transfer terms between the two phases, and namely for momentum equation, are modeled in the standard way (see Chapter 4 or [ISH 90]). The parameterization of the drag coefficient C_D originates from the results presented in the previous section, and brings in an α^3 dependence, relative to the void fraction α. As an example, Figure 11.22 shows the evolution of the void fraction at the wall (top), and 1 mm away from it (bottom), for heat fluxes of 200 kW/m^2 (left) and 1,000 kW/m^2 (right). Bubbles remain isolated and individualized for lower heat fluxes, whereas a near-continuous gas film forms at the higher heat flux, prohibiting the wall to be rewetted by the liquid, thus provoking departure from nucleate boiling. The complete computation domain, with horizontal dimensions of 40×40 mm^2, is presented here.

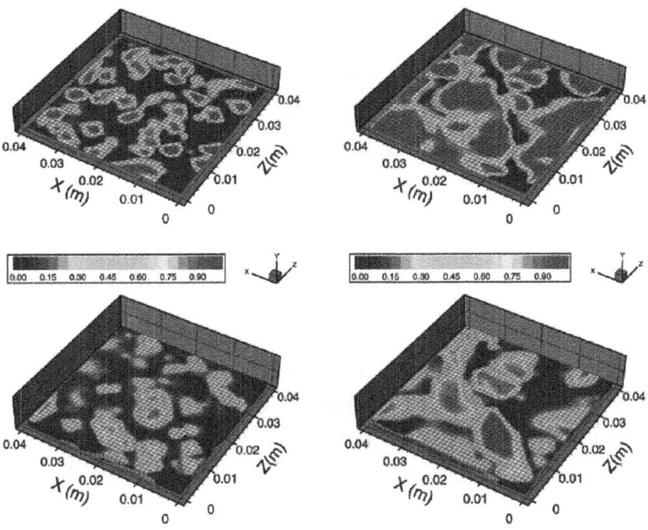

Figure 11.22. *Void fraction on the surface (top) and very close to the wall (bottom) for heat fluxes of 200 kW/m^2 (left) and 1,000 kW/m^2 (right) [PEZ 11]*

More quantitatively, Figure 11.23 compares the results from this model to the experimental data found by Theofanous *et al.* [THE 02], discussed previously, for the mean vertical distribution of the void fraction in the central vertical section of the cell. There is an exceptionally good correlation between the two, with density of nucleation sites equal to 2×10^5, for a heat flux of 1,000 kW/m^2, and the residence time of a bubble on the heating surface of 5 ms. The very high values of α close to the wall indicate the presence of a gas film.

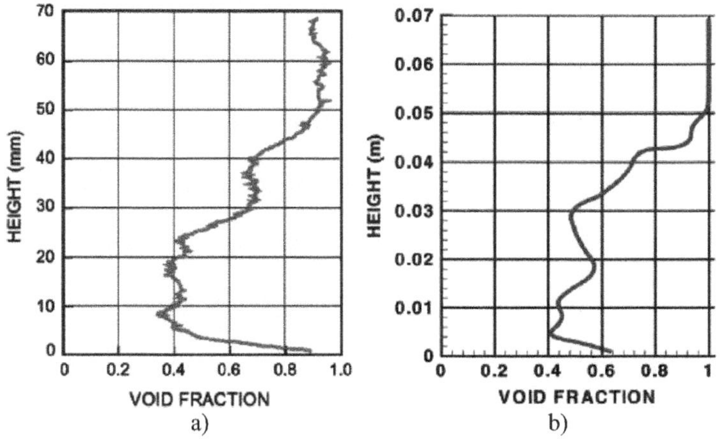

Figure 11.23. *Vertical distribution of the void fraction for a heat flux of 1,000 kW/m². a) Experimental data and b) results obtained by numerical modeling [PEZ 11]*

Finally, it is possible to deduce the critical boiling curve by a systematic study dependent on the wall heating, which is shown in Figure 11.24 for different nucleation site densities. The data correspond to a value of the Kutateladze constant approximately equal to 0.13.

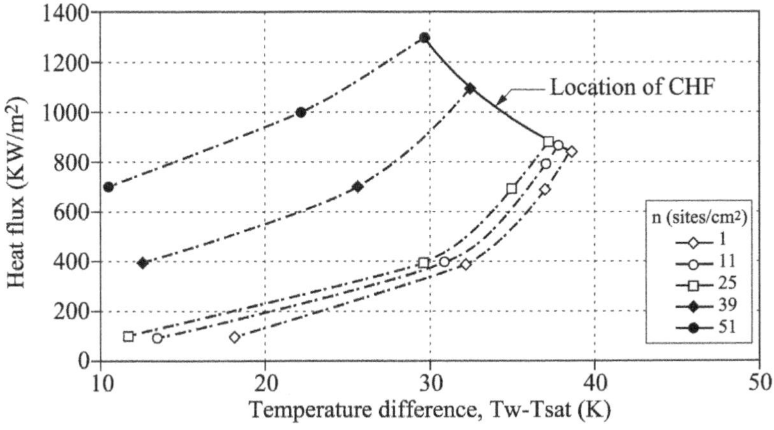

Figure 11.24. *Boiling curve numerically obtained for different nucleation site densities [PEZ 11]*

PART 3

From Fluidized Beds to Granular Media

Part 3 shows how the use of the knowledge described in the previous parts has its possibly most complex application in the modeling of fluidized beds, with all their aspects that are widely used in industry. In the first chapter of Part 3, these serve as an example to illustrate first of all the need for complex multiphase modeling, then to show how the existing modeling takes advantage of the developments previously presented, and finally to emphasize the profitable improvements but also the inherent difficulties with these particularly complex situations, where problems related to dynamics are associated with those of heat transfer and chemistry. In the next three chapters, the authors then study one of the necessary hydrodynamic improvements for very dense granular media in the neighborhood of the fluidization threshold, in situations where contact between particles or grains is more complicated than simply collisions, and may display lasting contacts with sliding. These situations are at the boundary of soil mechanics and the authors do not really go into this field. They limit the discussion to showing how the approach can lead to a model where the stress tensor of the medium is given by a constitutive law in the form of a differential balance equation, such as a type of viscoelastic medium with a threshold.

Chapter 12

Fluidized Beds

12.1. Introduction

Fluidized beds correspond to a two-phase medium where the dispersed phase is dense and consists of solid grains. They are found in a number of industrial devices. Detailed modeling of fluidized beds can be based on everything that has been discussed in this book, although it is also possible to consider a more global approach to modeling, often sufficient for some practical cases. Fluidized beds are, therefore, a good example to show that there are different levels of approximation and modeling, given the complexity of the phenomena involved, and that the implementation of one of these various models depends on the level of accuracy desired.

12.1.1. *Classification of different fluidization regimes*

When studying particle beds lying on a porous distributor, under which a gas, called fluidizing gas, is injected, the criterion for fluidization can generally be defined as follows (see Figure 12.1) based on the evolution of pressure drop Δp between the input and output of the fluidized bed system in terms of the gas's circulation velocity U inside the cell (refer to, for example, [LAG 88, YAT 83, JAC 00]). For relatively low velocities, the pressure drop is roughly proportional to U, and the solid medium behaves similar to a simple porous medium. As U increases, a threshold is reached at which particles start to vibrate, and the solid particle layer expands. The pressure drop then increases to a maximum before decreasing to a value roughly equal to the weight of the particles per unit area of the cell's cross section, P/A_c.

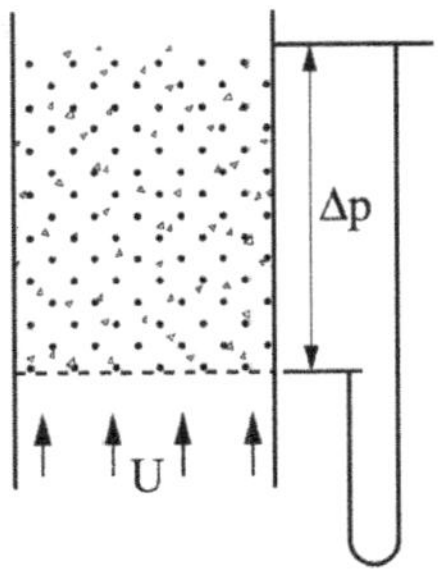

a) experimental set-up

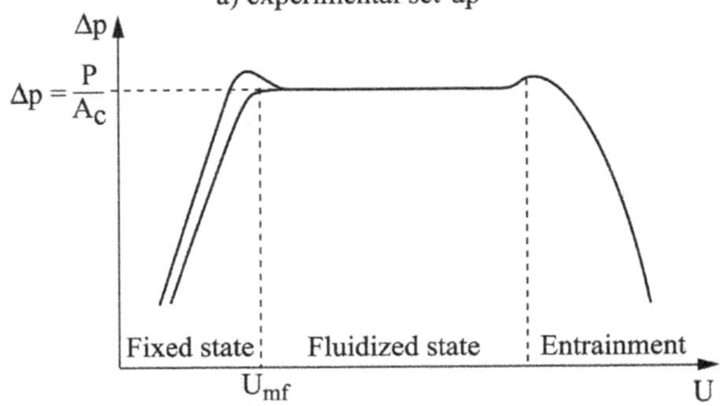

b) diagram representing the pressure drop through a layer of identical particles

Figure 12.1. *Diagram of the pressure drop Δp in a bed with respect to the velocity of the fluidizing gas. Excerpt from [LAG 88]*

Beyond this critical velocity, denoted by U_{mf}, the particles become independent of each other. They are, therefore, suspended and the bed behaves like a fluid: the layer is then said to be fluidized. Several fluidization regimes can be defined according to the velocity of the fluidizing gas (see Figure 12.2, where it is assumed that the gas is injected in the lower section of the cell by a device similar to a grid):

– First regime, $U_{mf} < U < U_{mb}$: above the minimum fluidization velocity, the fluidized bed expands homogeneously as the velocity of the gas increases. This regime is that of particulate fluidization. The bed's upper surface is perfectly defined and the particles are subjected to low-amplitude movements.

– Second regime, $U_{mb} < U < U_{ms}$: gas bubbles appear in the area of the gas distribution grid, located at the base of the cell. They then rise in the bed and grow larger, mainly by coalescing, before bursting on the upper surface. The surface between the bursting bubbles is quite well defined. This regime is called bubbling fluidization. Sensitive pressure sensors are used to measure significant amplitudes of pressure fluctuations.

– Third regime, $U_{ms} < U < U_{ft}$: the diameter of the bubbles reaches dimensions comparable to that of the cell. The passing of large bubbles and a dense bed phase alternate. This regime is called slugging regime. The surface of the bed rises and falls at a regular frequency. The pressure variations are considerable. In general, it is acceptable for both the analysis of experimental results and numerical simulations (no matter how sophisticated) to consider the medium as actually being a two-phase medium for these three regimes.

– Fourth regime, $U_{ft} < U < U_{fr}$: the bed consists of small elements from the gas and solid phases that come and go very quickly. The surface of the bed, therefore, becomes difficult to distinguish. The amplitude of the pressure variations is low. This regime is called turbulent fluidization.

– Fifth regime, $U_{fr} < U$: particles are entrained out of the cell by the gas current in a much diluted form and, therefore, need to be replaced. It can also be noted that small, falling packets of particles are present near the walls. The concept of the bed surface disappears entirely. The pressure drop in the gas becomes very low. This is the regime of fast fluidization. Contrary to the first three regimes, the latter two are generally considered as having only one phase because the residual bubbles of carrier fluid are either highly distorted and elongated, with very variable shapes (turbulent fluidization regime), or very small and difficult to distinguish from the solid pockets (fast fluidization regime).

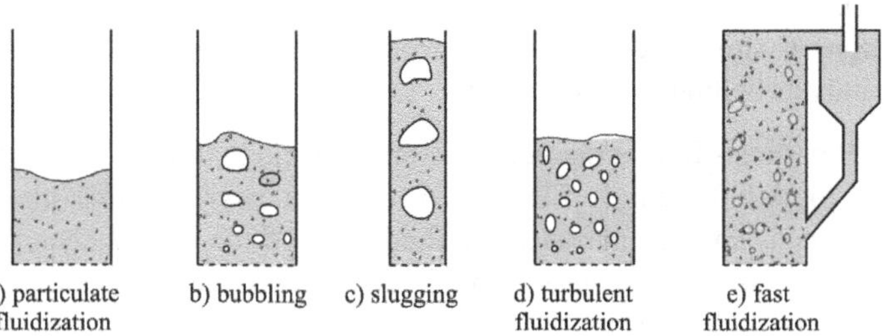

Figure 12.2. *Fluidization regimes Excerpt from [LAG 88]*

However, it is conventionally noted that all fluidization regimes presented above are not systematically observable when the velocity of the gas going through the cell is increased. The transitions between regimes effectively depend on other factors, besides the nature and properties of the gas and solid particles, such as the bed diameter, its height, the type of distributor used to inject gas or, also, the presence of baffles or transfer surfaces inside the cell.

For numerous systems, U_{mf} and U_{mb} are particularly hard to distinguish. There is, therefore, no actual particulate fluidization regime. Similarly, if the diameter of the cell is large, or if the height of the bed is low, it is impossible to identify the slugging threshold U_{ms}. The same is true for the turbulent fluidization threshold U_{mt}. As for the minimum velocity required for fast fluidization, it depends on the supply of particles to the cell. Furthermore, other transitions between regimes occur more or less gradually, with the exception of U_{mb} and U_{ft}, which are generally well defined.

Besides, the simple fluidization principles listed here cannot be simply or universally applied to all types of particle as fluidization depends directly on the compensation of the weight by the drag force acting on a particle, usually assumed spherical for simplicity, and this force acting from the moving fluid depends on the Reynolds number associated with the particle. To take into account the various particle specificities, Geldart developed a classification for powders, according to their size and density, which carries his name and enables particles to be classified according to their behavior in fluidized beds (see Figure 12.3). This classification consists of four distinct groups and depends on particle density and mean size. It also depends on the density of the fluid (according to the concept of submerged weight).

– Class A particles have a diameter between 20 and 100 µm and a density below 1,500 kg.m^{-3}. Fluidization of this class of particles is easy, of the particulate type (between U_{mf} and U_{mb}), and the fluidized bed expands considerably in comparison to the fixed state. It also sometimes has a tendency to form preferential channels (channeling). Beyond U_{mb}, few bubbles appear. When the supply of gas is stopped, defluidization is slow and this phenomenon is called air release. Hydrocarbon cracking catalysts belong to this category.

– Class B particles have a diameter between 80 and 800 µm. Their density ranges from 1,500 to 4,000 kg.m^{-3}. Fluidization is also easy for this class, although bubbling occurs as soon as particles are fluidized and the risk of channeling is low. Defluidization is fast for this class of particles. Sand is one of the particles that falls into this category. The boundary between classes A and B is represented by the following equation:

$$\rho_p - \rho_g = 44000 \frac{\rho_g^{0.1} \mu_g^{0.9}}{g d_p}$$

– Class C particles have a diameter below 30 µm. This class includes highly cohesive powders, such as flour and talcum powder, for which fluidization is very difficult. Fluidization must be assisted by exciting or vibrating the bed to avoid the occurrence of channeling phenomena. The layers do not expand significantly. The boundary between classes A and C is not well defined, depending namely on the gas humidity and the relative permittivity and resistivity of the particles.

– Class D particles have a diameter over 800 μm and/or a high density. Fluidization is possible for this type of particle but particular techniques, such as bed jets, are preferred. Cereals and lead shots are among the materials categorized into this class. The boundary between classes B and D corresponds to the following relationship:

$$\rho_p - \rho_g = d_p^{-2}.10^{-3}$$

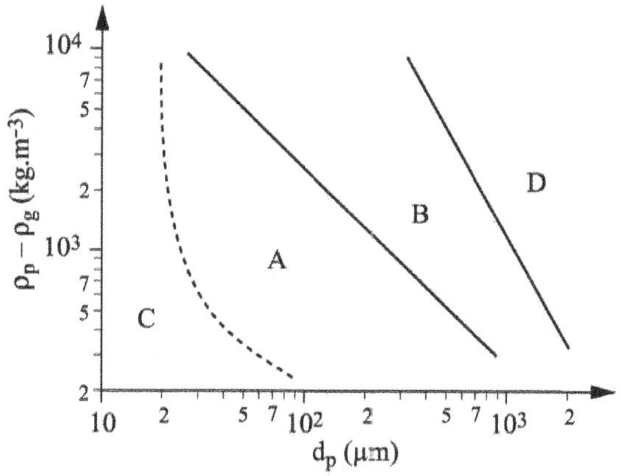

Figure 12.3. *Geldart classification. Excerpt from [LAG 88]*

Finally, Table 12.1 summarizes the different characteristics that differentiate the various fluidization regimes and their main properties (for a more detailed discussion, the reader may refer to [GRA 00]). The slip velocity given in the table is defined as equal to the difference $(U/\varepsilon) - G_s/(\rho_p(1-\varepsilon))$ (where ε is the mean void fraction and G_s is the net circulation mass flow rate of solid elements), and the slip rate given by $U\rho_p(1-\varepsilon)/(G_s\varepsilon)$. The regime called high-density fast fluidization was recently identified as having slightly different properties to fast fluidization regimes. However, as there is little information on this regime yet, and as the debate on its specificities is still far from being closed, the two regimes will not be differentiated from each other here.

The criteria for fluidization and the classification discussed above in terms of phenomenology, based on experimental observations, will now be re-examined in more detail by returning to the basic physical principles of these two-phase flows. The main modeling methods, mainly developed for bubbling, slugging, fast and turbulent fluidization, will then be presented as they are by far the most common in industry.

Regimes	Main applications	Most common particle class	Most common void fraction	Typical slip velocities (m/s)	Common slip factors
Bubbling fluidization	Polymers, Combustion	B, D	0.45–0.65	0.1–0.8	$\to \infty$
Slugging	–	B, D	0.55–0.7	0.3–0.9	$\to \infty$
Turbulent fluidization	Acrylonitrile, Oxichlorination, Fischer–Tropsch synthesis, Phthalic anhydride, Regenerators for catalytic cracking, Fluidized bed driers	A, B	0.65–0.8	0.5–1.5	>100
Fast fluidization	Combustion, calcination	B	0.9–0.97	2–6	1.6–5
High-density fast fluidization	Catalytic cracking, Maleic anhydride	A	0.75–0.93	3–10	5–10

Table 12.1. *Main characteristics of different fluidization regimes. Excerpt from [GRA 00]*

12.1.2. *Minimum fluidization and bubbling velocities*

12.1.2.1. *Reminder for head loss in porous media*

In general, if we consider a fluid flow in a medium cluttered with fixed grains, corresponding to a porous medium, the fluid velocity U is defined by Darcy's law as a function of the total pressure drop $\Delta\mathcal{P}$ between the input and output, where $\mathcal{P}$ is the motor pressure, $\mathcal{P} = P + \rho_g g x$ (with x, the height of the section being considered relative to a reference level, as x will indicate the longitudinal axis along which the gas flows in a cell assumed to be vertical):

$$U = \frac{k}{\mu_g}\frac{1}{L}\Delta P \qquad [12.1]$$

where k is the specific permeability of the medium and L is the thickness of the porous area. In this particular case, the medium's mean void fraction, ε, is often involved, as well as the differences from the linear relationship proposed by Darcy, when the velocities increase significantly. This leads to the following relationship, called Ergun's law:

$$\Delta P = UL\left[\frac{A\mu_g(1-\varepsilon)^2}{d_p^2\varepsilon^3}+\frac{B\rho_g U(1-\varepsilon)}{d_p\varepsilon^3}\right] \qquad [12.2]$$

for which the standard values of A and B are 150 and 1.8, respectively.

In reality, there is no need to use Darcy's law as the relationship above can be found directly (in terms of form) from the equation provided in Chapters 4 and 5, namely relationship [4.5] where some terms are neglected, which is associated with the law for friction on the solid phase [5.7b] by assuming that the solid phase is immobile. When a porous medium has a fluid flow going through it that is also a multiphase flow, the equations from Chapters 4 and 5 can then explain Darcy's law, representing the overall equilibrium created between the pressure drop, $\Delta\mathcal{P}$, between the input and output of the porous column (or, more accurately, the pressure gradient dP/dx in an elementary section, of infinitely small thickness dx) and the friction forces acting on the fluid caused by the fixed solid medium.

12.1.2.2. *Minimum fluidization velocity*

In order to determine the minimum fluidization velocity, U_{mf}, particles will now be assumed to be moving under the effect of friction generated by the fluid in motion. When the layer becomes fluidized, the absolute value of the friction force is equal to the submerged weight of the particles and leads to:

$$\Delta\mathcal{P}\, A_c = (A_c\, L_f)(1-\varepsilon_f)(\rho_p-\rho_g)g \qquad [12.3]$$

where ε_f, therefore, denotes the mean void fraction of the fluidized bed, and L_f, its height. Using the Ergun equation, the following is found for minimum fluidization:

$$\frac{1.8}{\varepsilon_{mf}^3}\left(\frac{d_p U_{mf}\rho_g}{\mu_g}\right)^2+\frac{150(1-\varepsilon_{mf})}{\varepsilon_{mf}^3}\frac{d_p U_{mf}\rho_g}{\mu_g}=\frac{d_p^3\rho_g(\rho_p-\rho_g)g}{\mu_g^2} \qquad [12.4]$$

For small particles with a particulate Reynolds number below 20, this defines the value of U_{mf} as being:

$$U_{mf}=\frac{d_p^2}{150}\frac{(\rho_p-\rho_g)g}{\mu_g}\frac{\varepsilon_{mf}^3}{(1-\varepsilon_{mf})} \qquad [12.5]$$

and on the contrary, the value of U_{mf} for particles with a Reynolds number between 20 and 1,000 is given by:

$$U_{mf} = \left[\frac{d_p}{1.8} \frac{(\rho_p - \rho_g) g}{\mu_g} \varepsilon_{mf}^3 \right]^{1/2} \quad [12.6]$$

The limiting gas velocity for which particles are fully entrained into the gas current is approximately equal to their free-fall velocity. For spherical particles, this is given by:

$$U_t = \left[\frac{4 g d_p}{3} \frac{(\rho_p - \rho_g)}{\rho_g C_D} \right]^{1/2} \quad [12.7]$$

where C_D is the particle's drag coefficient (dependent of the Reynolds number, according to standard laws that will not be repeated here).

12.1.2.3. *Minimum bubbling velocity*

For class B particles of the Geldart classification, the minimum bubbling velocity is equal to the minimum fluidization velocity. This is not the case for particles in class A, and the equation proposed by Abrahamsen and Geldart is often used to calculate it:

$$U_{mb} = 33 d_p \left(\frac{\mu_g}{\rho_g} \right)^{-0.1} \quad [12.8]$$

In the case where the equation gives a value of U_{mb} smaller than that of U_{mf}, it is necessary to take $U_{mb} = U_{mf}$. Finally, it should be noted that the difference between U_{mb} and U_{mf} increases as the size of the particles and their density decrease.

12.2. Complete models for the dynamics of fluidized beds

Unfortunately, given the complexity of the phenomena involved, no single and unique model has been universally acknowledged to date. However, several models exist that are based on equations, initially the same, which are then transformed into a number of different versions, depending on the way the terms linked to the interactions between the carrier fluid (composing one phase) and the solid particles (composing a second phase), including inter-particle interactions, are modeled. On the one hand, the initial equation can be written in two ways, depending on whether the reduced stress tensors are considered, as mentioned in Chapters 5 and 6.

Moreover, the best way to model interaction mechanisms between particles (friction, elastic or inelastic collisions, aggregation, etc.) and drag forces acting on the particles due to the fluid (which control the exchanges of momentum between the fluid and solid phases) is still subject to debate.

12.2.1. *Bubbling fluidization regime*

The equations associated with some of the models currently being used are presented and discussed hereafter. The indices g and s refer to the carrier fluid phase and the solid particle phase, respectively. Furthermore, to keep the presentation clear and tidy, the symbol ∇ will here represent the gradient operator; $\overline{\overline{\tau}}$, the stress tensor; u, the mean velocity vector (when it is not necessary to define which component u^i is being considered); g, the vector for gravitational acceleration and ".", the scalar product.

Continuity equations:

$$\begin{aligned}&\frac{\partial \varepsilon_g}{\partial t}+\nabla.(\varepsilon_g u_g)=0\\&\frac{\partial \varepsilon_s}{\partial t}+\nabla.(\varepsilon_s u_s)=0\end{aligned} \qquad [12.9]$$

Momentum equation:

Jackson's model [JAC 97].

$$\begin{aligned}&\rho_g\left[\frac{\partial u_g}{\partial t}+(u_g.\nabla)u_g\right]=\nabla.\overline{\overline{\tau_g}}-\nabla P-\frac{\beta}{\varepsilon_g}(u_g-u_s)+\rho_g g\\&\rho_s\varepsilon_s\left[\frac{\partial u_s}{\partial t}+(u_s.\nabla)u_s\right]-\rho_g\varepsilon_s\left[\frac{\partial u_g}{\partial t}+(u_g.\nabla)u_g\right]=\frac{\beta}{\varepsilon_g}(u_g-u_s)+\\&\qquad\varepsilon_s(\rho_s-\rho_g)g+\nabla.\overline{\overline{\tau_s}}-\nabla P_s\end{aligned} \qquad [12.10]$$

These equations can also be written differently for the two phases to play a more symmetrical part, as discussed in Chapter 4:

$$\begin{aligned}&\rho_g\varepsilon_g\left[\frac{\partial u_g}{\partial t}+(u_g.\nabla)u_g\right]=\varepsilon_g\nabla.\overline{\overline{\tau_g}}-\varepsilon_g\nabla P-\beta(u_g-u_s)+\varepsilon_g\rho_g g\\&\rho_s\varepsilon_s\left[\frac{\partial u_s}{\partial t}+(u_s.\nabla)u_s\right]=\varepsilon_s\nabla.\overline{\overline{\tau_g}}-\varepsilon_s\nabla P+\nabla.\overline{\overline{\tau_s}}-\nabla P_s+\beta(u_g-u_s)+\varepsilon_s\rho_s g\end{aligned} \qquad [12.11]$$

Jackson's model considers the reduced stress tensor for both phases, as well as the mean stress tensor in each phase $\bar{\Phi}_i(\overline{\tau}_i^{\alpha\beta *} - \overline{p}_i^* \delta_{\alpha\beta})$, as was discussed in section 5.3.1. This reduced tensor includes part of the forces in contact with the other phase. However, it also includes, in principal, the additional Reynolds-type contribution: in particular, the pressure defined here includes the kinetic energy of fluctuations in each phase, although Jackson does not outline this explicitly. The influence of this total tensor for the solid particle phase is denoted by $\nabla \overline{\overline{\tau}}_s - \nabla P_s$.

Moreover, the sum of the buoyancy forces acting on the particles is also included in the equations and is directly visible for the solid phase by the terms $\varepsilon_s \nabla \overline{\overline{\tau}}_g - \varepsilon_s \nabla P$, written in the form [5.1b] in section 5.3.1. However, it also includes the viscous tension effect in addition to pressure (an additional term that is negligible in practice). In the equation for the gas phase, the buoyancy forces must have the opposite sign. This becomes obvious as $\varepsilon_g \nabla \overline{\overline{\tau}}_g - \varepsilon_g \nabla P = \nabla \overline{\overline{\tau}}_g - \nabla P - \varepsilon_s \nabla \overline{\overline{\tau}}_g + \varepsilon_s \nabla P$, where it is found in addition to the total stress tensor of the gas. The other forces involved are friction forces between phases and gravitational forces acting on each phase.

Ishii's model [ISH 75]:

$$\begin{aligned} \rho_g \varepsilon_g \left[\frac{\partial u_g}{\partial t} + (u_g.\nabla) u_g \right] &= \nabla.\varepsilon_g \overline{\overline{\tau}}_g - \varepsilon_g \nabla P - \beta(u_g - u_s) + \varepsilon_g \rho_g g \\ \rho_s \varepsilon_s \left[\frac{\partial u_s}{\partial t} + (u_s.\nabla) u_s \right] &= \nabla.\varepsilon_s \overline{\overline{\tau}}_s - \varepsilon_s \nabla P + \beta(u_g - u_s) + \varepsilon_s \rho_s g \end{aligned} \qquad [12.12]$$

In the version rewritten by Enwald *et al.* [ENW 96], this also becomes:

$$\begin{aligned} \rho_g \varepsilon_g \left[\frac{\partial u_g}{\partial t} + (u_g.\nabla) u_g \right] &= \nabla.\varepsilon_g \overline{\overline{\tau}}_g - \varepsilon_g \nabla P - \beta(u_g - u_s) + \varepsilon_g \rho_g g \\ \rho_s \varepsilon_s \left[\frac{\partial u_s}{\partial t} + (u_s.\nabla) u_s \right] &= \nabla.\overline{\overline{\tau}}_s - \varepsilon_s \nabla P - \nabla P_s + \beta(u_g - u_s) + \varepsilon_s \rho_s g \end{aligned} \qquad [12.13]$$

In Ishii's model, the stress tensors are the mean tensors for each phase, and not the reduced ones. The Reynolds-type contributions are neglected, and it is assumed that the mean pressure in the fluid is equal to the pressure in the solid phase. The pressure terms, therefore, include buoyancy forces, but in the way it is written in Chapter 5, section 5.3.1, equation [5.1c], and taking into account that $P = \overline{p}_g^* = \overline{p}_g^s = \overline{p}_s^s = \overline{p}_s^*$. Enwald's model, on the other hand, adds a Reynolds-type term for solid particles that takes into account fluctuations and collisions between particles: therefore, $\overline{\overline{\tau_s}}$ has a different meaning here than in Ishii's model, including the Reynolds-type tensor deviator, and the diagonal part of the tensor is left as $-P_s$.

To calculate the pressure P_s, the equation proposed by Lun *et al.* [LUN 84] is considered and is based on granular gas theory:

$$P_s = \rho_s \varepsilon_s \Theta + 2 g_0 \rho_s \varepsilon_s^2 \Theta (1+e)$$

It is important to note here that when equations are established via a statistical average, as was done in Chapter 4, the equation for the velocity component u^i is found to have a Reynolds-like term in the form of $\frac{\partial}{\partial x^j}(\overline{-\rho u'^i u'^j})$. Taking the average is justified even if the flow is not turbulent in the classical sense, i.e. even if the fluid velocity, and probably that of the solid particles, is not very high. When a gas is injected between particles, this causes the particles to move suddenly and unpredictably, especially when the number of particles per unit volume allows them to collide into each other. The gas flow between the particles may stay laminar, although the particles can exhibit disorganized motion very early on as soon as there is a small gas flow. If the particles fluctuate, they push the gas between them and the gas itself, therefore, has a random movement: the characteristic randomness of the movement in both phases, therefore, exists early on, even at low mean velocities. In addition, even if the particle mean velocity is very low, collisions between particles, occurring as soon as the there is a sufficient number of particles, induce sudden accelerations and the particles' Reynolds-type terms are for mean velocities lower than those in the case of fluid flows. However, because the densities are different, and if the velocities are not too high, the Reynolds-like term in the gas balance equations for momentum can be neglected in the momentum balance equation for the fluid, especially when compared to the vertical gradient of the mean pressure and the gravitational force.

From this follow the concepts of solid pressure P_s and granular temperature Θ, which is defined as being equal to two-thirds of the mean kinetic energy of the solid particles at any point. In granular gas theory, Θ can be calculated at any point using a transport equation, derived from the equation for turbulent kinetic energy of solid particles:

$$\frac{3}{2}\left[\frac{\partial}{\partial t}(\varepsilon_s \rho_s \Theta) + \nabla.(\varepsilon_s \rho_s \Theta u_s)\right] = (-\nabla P_s \overline{\overline{I}} + \overline{\overline{\tau_s}}) : \nabla u_s - \nabla.(\kappa_s \nabla \Theta) - \gamma_s - J_s \quad [12.14]$$

Conductivity k_s adds the effects linked to the kinetics of the solid particles and their collisions. Different modeling approaches are possible, as for μ_s (which will soon appear below), but the relationships will not be given here as they are very similar. The article by Van Wachem *et al.* [WAC 01] presents the principles of the models. In general, the ratio μ_s/k_s is of the order of 1; yet, it remains less than 1, for any given model for μ_s or k_s.

The term γ_s represents the dissipation of Θ resulting from inelastic collisions between particles. In general, it is taken as equal to $\gamma_s = 12(1-e^2)\dfrac{\varepsilon_s^2 \rho_s g_0}{d_s\sqrt{\pi}}\Theta^{3/2}$.

Finally, the term J_s represents the dissipation of Θ resulting from the fluctuating forces caused by the fluid, $J_s = \beta(\overline{u'_s u'_s} - \overline{u'_g u'_s})$ and is often modeled by $J_s = \beta(3\Theta - \beta d_s \dfrac{(u_g - u_s)^2}{4\varepsilon_s \rho_s \sqrt{\pi\Theta}})$ when quantity $\overline{u'_g u'_s}$ is not negligible when compared to $\overline{u'_s u'_s} \equiv 3\Theta$ by definition.

It should also be noted that Θ is sometimes defined by a simple analytic relationship derived from the previous equation by approximating that the dominating terms are those for production and dissipation γ_s:

$$0 = (-\nabla P_s \overline{\overline{I}} + \overline{\overline{\tau_s}}) : \nabla u_s - \gamma_s \qquad [12.15]$$

As for g_0, which appears in a number of models, it represents the function for the radial distribution of particles (in the column's transverse section), and *e* is the energy restitution coefficient (for particle collisions). Different analytical formulas have been put forward for g_0 and they generally involve, at the very least, the value ε_s at the point being considered and its maximum value in the section $\varepsilon_{s,max}$, for example $g_0 = \dfrac{3}{5}\left[1 - (\varepsilon_s / \varepsilon_{s,max})^{1/3}\right]^{-1}$. For more details on this kinetic approach to the solid phase, the reader may refer to [LUN 84, PEI 98, GID 94].

The equation for granular temperature is no more than a turbulent kinetic energy equation for the particles, similar to the one previously derived in Chapter 4, section 4.2.1, the modeling of which is discussed in sections 4.2.2 and 4.2.3. The equation given above assumes that grains are spherical and non-deformable in a void, and are in a uniform, infinite medium, and that they collide and bounce inelastically (otherwise, the kinetic energy would not be dissipated), but do not stay in contact nor rub against each other. All these restrictions are not necessary to establish the equation in Chapter 4, section 4.2.1, but hypotheses of a different order must be used to close the equation, as discussed in sections 4.2.2 and 4.2.3. A crucial difference between granular gas theory and modeling from Chapter 4 is that the length scales that must be prescribed to close the granular temperature equation are only linked to the grain diameter, whereas for turbulence modeling they must take into account the geometry of the flow itself. Of course, this is due to the fact that granular gas theory considers gas in a uniform infinite medium; there is no geometric length scales of the flow. It is known that instabilities can develop in granular flows satisfying this equation, and the appearance of bubbles in such models is a manifestation of these instabilities. Therefore, it is possible to calculate these bubbles by using such models, provided that the mesh and time steps of numerical computations are sufficiently small. The choice of these scales should be changed to avoid this flaw as discussed in Chapter 6, but a large eddy simulation (LES)-type model is better suited to this unsteady situation than a model using statistical averages.

Granular gas theory also provides the stress tensor (written here for any phase, denoted by the index k) defined by $\overline{\overline{\tau_k}} = 2\mu_k \overline{\overline{D_k}} + (\lambda_k - \frac{2}{3}\mu_k)\mathrm{tr}(\overline{\overline{D_k}})\overline{\overline{I}}$, or $\overline{\overline{D_k}} = \frac{1}{2}(\nabla u_k + (\nabla u_k)^T)$ (where the exponent T refers to the transpose tensor of the tensor being considered) and I is the identity tensor. The viscosity coefficient of the solid phase μ_s can be calculated using different models, whereas λ_s is usually modeled according to the relationship proposed by Lun *et al.* [LUN 84]. $\lambda_s = \frac{4}{3}\varepsilon_s^2 \rho_s d_s g_0 (1+\varepsilon)\sqrt{\frac{\Theta}{\pi}}$.

Once again, the granular gas theory gives an expression for the stress tensor that is identical to a Boussinesq relationship used for turbulence models, see Chapter 6.

Main models for μ_s:

Model by Lun *et al.* [LUN 84]

$$\mu_s = \frac{5\sqrt{\Theta\pi}}{96}\rho_s d_s \left[\left(\frac{1}{\eta g_0} + \frac{8\varepsilon_s}{5}\right)\left(\frac{1+\frac{8}{5}\eta(3\eta-2)\varepsilon_s g_0}{2-\eta}\right) + \frac{768}{25\pi}\eta\varepsilon_s^2 g_0 \right]$$

$$= \frac{4}{5}\varepsilon_s^2 \rho_s d_s g_0 (1+e)\sqrt{\frac{\Theta}{\pi}} + \frac{1}{15}\sqrt{\Theta\pi}\,\frac{\rho_s d_s g_0 (1+e)(3/2e-1/2)\varepsilon_s^2}{(3/2-1/2e)} + \frac{1}{6}\sqrt{\Theta\pi}\,\frac{\rho_s d_s \varepsilon_s (3/4e+1/4)}{(3/2-e/2)} + \frac{10}{96}\sqrt{\Theta\pi}\,\frac{\rho_s d_s}{(1+e)(3/2-1/2e)g_0} \quad [12.16]$$

Model by Hrenya and Sinclair [HRE 97]

$$\mu_s = \frac{5\sqrt{\Theta\pi}}{96}\rho_s d_s \left[\left(\frac{1}{\eta g_0} \frac{1}{1+\frac{\lambda_{mfp}}{R}} + \frac{8\varepsilon_s}{5} \right) \left(\frac{1+\frac{8}{5}\eta(3\eta-2)\varepsilon_s g_0}{2-\eta} \right) + \frac{768}{25\pi}\eta\varepsilon_s^2 g_0 \right]$$

$$= \frac{4}{5}\varepsilon_s^2\rho_s d_s g_0(1+e)\sqrt{\frac{\Theta}{\pi}} + \frac{1}{15}\sqrt{\Theta\pi}\,\frac{\rho_s d_s g_0(1+e)(3/2e-1/2)\varepsilon_s^2}{(3/2-1/2e)} \quad [12.17]$$

$$+\frac{1}{6}\sqrt{\Theta\pi}\,\frac{\rho_s d_s \varepsilon_s (1/2(1+\frac{\lambda_{mfp}}{R})(3/4e-1/4)}{(3/2-1/2e)(1+\frac{\lambda_{mfp}}{R})} + \frac{10}{96}\sqrt{\Theta\pi}\,\frac{\rho_s d_s}{(1+e)(3/2-1/2e)g_0(1+\frac{\lambda_{mfp}}{R})}$$

Model by Syamlal *et al.* [SYA 93]

$$\mu_s = \frac{4}{5}\varepsilon_s^2\rho_s d_s g_0\sqrt{\frac{\Theta}{\pi}} + \frac{\varepsilon_s\rho_s d_s\sqrt{\Theta\pi}}{6(3-e)}\left[1+\frac{2}{5}(1+e)(3e-1)\varepsilon_s g_0\right]$$

$$= \frac{4}{5}\varepsilon_s^2\rho_s d_s g_0(1+e)\sqrt{\frac{\Theta}{\pi}} + \frac{1}{15}\sqrt{\Theta\pi}\,\frac{\rho_s d_s g_0(1+e)(3/2e-1/2)\varepsilon_s^2}{(3/2-e/2)} \quad [12.18]$$

$$+\frac{1}{12}\sqrt{\Theta\pi}\,\frac{\rho_s d_s\varepsilon_s}{(3/2-e/2)}$$

Model by Gidaspow [GID 94]

$$\mu_s = \frac{4}{5}\varepsilon_s^2\rho_s d_s g_0(1+e)\sqrt{\frac{\Theta}{\pi}} + \frac{\frac{10}{96}\rho_s d_s\sqrt{\Theta\pi}}{(1+e)g_0}\left[1+\frac{4}{5}(1+e)\varepsilon_s g_0\right]^2$$

$$= \frac{4}{5}\varepsilon_s^2\rho_s d_s g_0(1+e)\sqrt{\frac{\Theta}{\pi}} + \frac{1}{15}\sqrt{\Theta\pi}\,\rho_s d_s g_0(1+e)\varepsilon_s^2. \quad [12.19]$$

$$+\frac{1}{6}\sqrt{\Theta\pi}\,\rho_s d_s\varepsilon_s + \frac{10}{96}\sqrt{\Theta\pi}\,\frac{\rho_s d_s}{(1+e)g_0}$$

For all these equations, λ_{mfp} denotes the mean free path of a solid particle, i.e. the mean distance between two particles, and R is a characteristic dimension of the column forming the fluidized bed. It should be noted that for most models, λ_{mfp} does not appear, whereas in Hrenya and Sinclair's model, its magnitude is limited by the dimension R, characterizing the system forming the fluidized bed. In this last model, it is naturally found that μ_s tends toward zero as the volume ratio of the solid particles tends toward zero. The model proposed by Hrenya and Sinclair involves R, the radius of the tube through which the medium flows. This is the only attempt made to take into account the geometry of the flow and is suggested for reasons of confrontation with experiments.

Finally, only parameter β needs to be modeled to close the balance equations written at the start of this section. This parameter appears in the term for exchange of mechanical energy between the two phases. It is equal to $\beta(u_g - u_s)$ in all models considered here and can be modeled in several ways, the main ones being listed hereafter. They differ namely in the equation used for the drag coefficient C_D. It is simply taken from the Stokes equation, adapted to a dense particle medium, and particulate Reynolds numbers that are not small, as discussed in Chapter 5, sections 5.3.2.2 and 5.3.2.3.

Model by Wen and Yu [WEN 66]

$$\beta=\frac{3}{4}C_D\frac{(1-\varepsilon_s)\varepsilon_s\rho_g\left|u_g-u_s\right|}{d_s}(1-\varepsilon_s)^{-2.65}, with$$

$$C_D=\frac{24}{Re_p(1-\varepsilon_s)}\left[1+0.15((1-\varepsilon_s)Re_p)^{0.687}\right]\ if\ (1-\varepsilon_s)Re_p<1000. \quad [12.20]$$

$$=0.44\ if\ (1-\varepsilon_s)Re_p>1000,\ with\ Re_p=\frac{d_s\rho_g}{\mu_g}\left|u_g-u_s\right|$$

Ergun equation [ERG 52] used in the model proposed by Gidaspow [GID 94]:

$$\beta=150\frac{\varepsilon_s^2\mu_g}{(1-\varepsilon_s)d_s^2}+\frac{7}{4}\frac{\varepsilon_s\rho_g\left|u_g-u_s\right|}{d_s}\quad \text{if } \varepsilon_s>0.2$$

$$\frac{3}{4}C_D\frac{(1-\varepsilon_s)\varepsilon_s\rho_g\left|u_g-u_s\right|}{d_s}(1-\varepsilon_s)^{-2.65}\quad \text{if } \varepsilon_s\le 0.2 \quad [12.21]$$

Model by Syamlal *et al.* [SYA 93]

$$\beta=\frac{3}{4}C_D\frac{(1-\varepsilon_s)\varepsilon_s\rho_g\left|u_g-u_s\right|}{d_sV_r^2},$$

with

$$C_D=\left(0.63+4.8\sqrt{\frac{V_r}{Re}}\right)^2, \quad [12.22]$$

$$V_r=\frac{1}{2}\left[a-0.06\,Re+\sqrt{(0.06\,Re)^2+0.12\,Re(2b-a)+a^2}\right]$$

and
$$a = (1-\varepsilon_s)^{4.14}$$
$$b = 0.8(1-\varepsilon_s)^{1.28} \quad \text{if } \varepsilon_s \geq 0.15$$
$$(1-\varepsilon_s)^{2.65} \quad \text{if } \varepsilon_s < 0.15$$

The models meant for the "bubbling" regime are mainly interested in the identification of "bubbles", in other words, flow areas that are nearly particle-free and that are mobile.

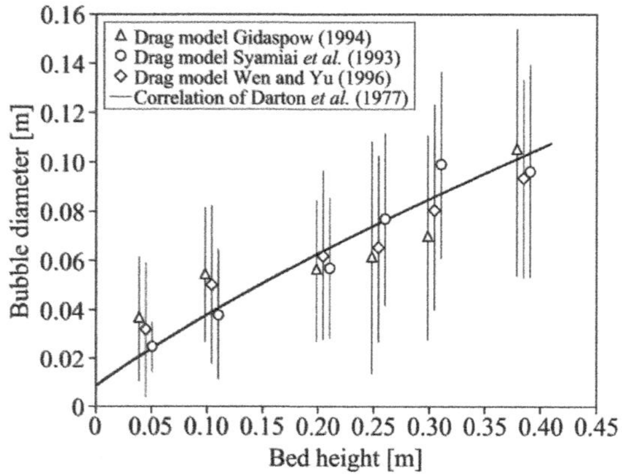

Figure 12.4. *Bubble diameter with respect to the fluidized bed height (for bubbling regime). Comparison of the predictions obtained from different drag coefficient models and synthesis of experimental data. Excerpt from [WAC 01]*

Figures 12.4 and 12.5 compare the results obtained via different models with experimental data. The characteristic bubble diameter is calculated from the velocity of the gas phase, using the relationship proposed by Darton *et al.* [DAR 77], based on experimental results (for solid particles belonging to class B of the Geldart classification), $D_b = 0.54(U - U_{mf})^{0.4}(x + 4\sqrt{A_0})^{0.8} g^{-0.2}$. The term $4\sqrt{A_0}$ represents some sort of virtual origin for the fluidized bed, equal to 0.03 m when a porous plate distributor is used at the input (from [DAR 77]). These results illustrate well how the chosen model for drag has little influence in bubbling and slugging regimes, associated with solid particle void fractions in the order of 0.4 to 0.5. The same is true for the friction model between solid particles, although this is not shown here. Similarly, the drag model does not greatly influence the prediction results for the evolution of a gas jet in a particle bed, even if this choice has a noticeable effect on the predicted shape of the bubbles that develop (the model underestimates the jet evolution and leads to bubbles that are more spherical than those observed in experiments). However, it is also important to note that, by not including any model for friction between solid particles, both the results (as there is less air penetrating between the solid particles, reflected by the bed having a lower expansion rate) and the computational time are penalized as the solution takes longer to converge.

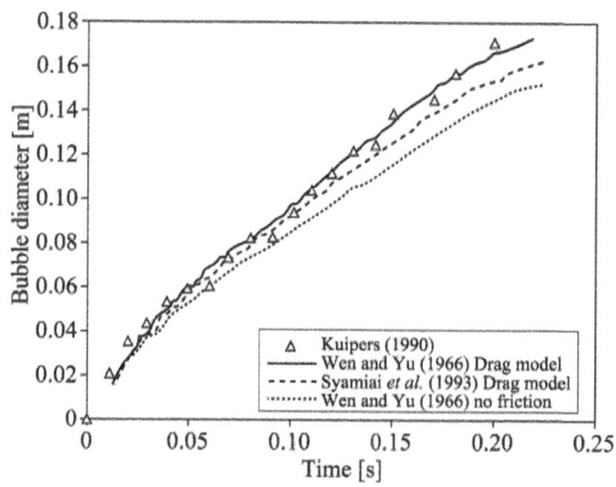

Figure 12.5. *Development of an air jet injected into a solid particle bed with respect to time. Comparison of the predictions obtained from different drag coefficient models and synthesis of experimental data. Excerpt from [WAC 01]*

12.2.2. *Turbulent fluidization regime*

Turbulent and fast fluidization regimes have large fluctuations comparable to those observed in turbulent, classic, single-phase flows (leading to a greater variety in the shape of gas bubbles, and much smaller sizes, contrary to regimes discussed in previous examples). For these regimes, it seems that there is even less consensus as to the model that should be adopted, the problems being rather similar to those encountered in classic, turbulent, single-phase flows (a greater number of equations need to be considered, as the equations for the kinetic energy of the turbulence and its dissipation rate in each phase must be taken into account in addition to the mean momentum equations; the greater number of phenomena that require modeling and the parameters that need being adjusted; the role and importance of explicitly taking into account the large turbulent scales of fluidization; the validity of taking mean overall statistics given the great diversity of events that can occur, etc.). This field is still subject to many advanced academic studies, namely using numerical simulations as a means of performing in-depth studies of phenomena occurring at microscopic levels. They should, therefore, enable reliable models to be put forward for macroscopic levels in the future.

However, an example of a turbulence model based on equations for turbulent kinetic energy k_g and its dissipation rate ε_g in the carrier fluid, as well as the kinetic energy k_s and covariance between the fluid velocity and the solid particle velocity k_{gs}, used by Peirano *et al.* [PEI 02] will be presented hereafter. Equations that are formally similar to the previous ones must be considered as far as the continuity equation and balance equation for momentum are concerned. When the influence of the turbulence is explicitly taken into account, the main difference is that these equations are written in terms of statistical averages of the velocity field (if the turbulent fluctuations in density are considered to be negligible) and the pressure. In addition to the viscous terms that were denoted by the tensor $\overline{\overline{\tau}}$, a supplementary term appears in the stress tensor due to turbulence. This is expressed as $\varepsilon_k \rho_k \overline{u_k^{'i} u_k^{'j}}$ (for a phase k, using tensor notations with similar indices to those in Chapters 2 and 3) and is commonly called "Reynolds tensor" (the notation $\bar{.}$ represents the operator for the statistical average with respect to turbulent fluctuations).

With the requirements of this section in mind, we will limit discussion to the simplest models that only involve four additional unknowns: the turbulent kinetic energy for solid and gas phases, respectively, denoted by k_s and k_g, the dissipation rate of the gaseous phase ε_g and the fluid-particle velocity covariance k_{gs}. A balance equation must, therefore, by derived analytically and modeled according to physical principles for each of the variables so as to then recalculate the Reynolds stresses that intervene in the averaged balance equations for momentum using the relationship $\overline{u_k^{'i} u_k^{'j}} = -\nu_{kt}\left(\frac{\partial \overline{u_k^i}}{\partial x^j} + \frac{\partial \overline{u_k^j}}{\partial x^i}\right) + \frac{2}{3} k_k \delta^{ij}$. The quantity ν_{kt} is often called the turbulence viscosity (for phase k), whereas the turbulent kinetic energy k_k, for phase k, is defined by $k_k = \frac{1}{2}\overline{u_k^{'i} u_k^{'i}}$.

According to the model used by Peirano *et al.* [PEI 02], the different terms are then modeled as follows [SIM 96]:

Balance equations for $\mathbf{k_g}$ *and* $<\varepsilon_g>$ (special attention must be paid to avoid confusion between ε_g, representing the mean volume fraction of the continuous phase (most often a gas), and $<\varepsilon_g>$, representing the mean dissipation for this phase):

$$\varepsilon_g \rho_g \left(\frac{\partial k_g}{\partial t} + \overline{u_g^i}\frac{\partial k_g}{\partial x^i}\right) = \frac{\partial}{\partial x^i}\left(\varepsilon_g \rho_g (\nu_g + \nu_{gt}/\sigma_{kg})\frac{\partial k_g}{\partial x^i}\right) - \varepsilon_g \rho_g \overline{u_g^{'i} u_g^{'j}} \frac{\partial \overline{u_g^i}}{\partial x^j} + \varepsilon_g \rho_g \Pi_{kg} - \varepsilon_g \rho_g < \varepsilon_g >, \quad [12.23]$$

$$\varepsilon_g \rho_g \left(\frac{\partial < \varepsilon_g >}{\partial t} + \overline{u_g^i} \frac{\partial < \varepsilon_g >}{\partial x^i}\right) = \frac{\partial}{\partial x^i}\left(\varepsilon_g \rho_g (\nu_g + \nu_{gt} / \sigma_{\varepsilon g}) \frac{\partial < \varepsilon_g >}{\partial x^i}\right)$$
$$-\varepsilon_g \rho_g \frac{< \varepsilon_g >}{k_g}\left(C_{\varepsilon 1} \overline{u_g^{\prime i} u_g^{\prime j}} \frac{\partial \overline{u_g^i}}{\partial x^j} + \quad . \qquad [12.24]$$
$$C_{\varepsilon 2} < \varepsilon_g >\right) + \Pi_{\varepsilon g} - \varepsilon_g \rho_g < \varepsilon_g >$$

with $\nu_{gt} = C_\mu^* k_g^2 / < \varepsilon_g >$, $C_\mu^* = C_\mu / (1 + C_{gs} X_{gs} / St(1 - k_{gs} / 2k_g))$,

$$\Pi_{kg} = (X_{gs} / \tau_{gs}^x)(-2k_g + k_{gs} + U_r^i U_d^i),\ \Pi_{\varepsilon g} = C_{\varepsilon 3} (< \varepsilon_g > / k_g) \Pi_{kg}$$

where U_r^i and U_d^i are, respectively, the mean relative velocity and the drift velocity components of index i, $U_r^i = \overline{u_s^i} - \overline{u_g^i} - U_d^i$.

The magnitude U_d^i is calculated from the relationship $\varepsilon_g \varepsilon_s U_d^i = -D_{gs}^t \partial \varepsilon_s / \partial x^i$, where the binary dispersion coefficient D_{gs}^t is given by $D_{gs}^t = k_{gs} \tau_{gs}^t / 3$. Coupling of the solid and gas phases is, therefore, partly performed via the variable k_{gs}, mentioned above, and also involves the interaction time between the turbulent fluid and the particles, τ_{gs}^t. The time τ_{gs}^t is itself a characteristic particle relaxation time. More standard, the constants that intervene in the previous terms have values of:

$$\sigma_{kg} = 1.0, \sigma_{\varepsilon g} = 1.2, C_\mu = 0.09, C_{\varepsilon 1} = 1.44, C_{\varepsilon 2} = 1.92, C_{\varepsilon 3} = 1.2 \text{ and } C_{gs} = 0.314 .$$

Finally, the magnitude of X_{gs} is given by $X_{gs} = \varepsilon_s \rho_s / \varepsilon_g \rho_g$, whereas the Stokes number is given by $St = \tau_g^t / \tau_{gs}^x$ where τ_g^t is what is generally called turbulence's turnover time (defined by $\tau_g^t = k_g / < \varepsilon_g >$, using the current notations). The details of how τ_{gs}^x and τ_{gs}^t are evaluated will not be given here to prevent the presentation from being too complicated. The interested reader will find them in [PEI 02, SIM 96].

This is in agreement with the previous Chapter 4, with the remark that the coefficient C_μ^* takes into account a specific particle effect by including their Stokes number and the energy k_{gs}. Indeed, $C_\mu^* = C_\mu / (1 + C_{gs} X_{gs} / St(1 - k_{gs} / 2k_g))$, where C_μ is the standard constant equal to 0.09.

Balance equations for $\mathbf{k_s}$ and $\mathbf{k_{gs}}$:

$$\varepsilon_s\rho_s(\frac{\partial k_s}{\partial t}+\overline{u_s^i}\frac{\partial k_s}{\partial x^i})= \frac{\partial}{\partial x^i}(\varepsilon_s\rho_s(D_s^t+D_s^c)\frac{\partial k_s}{\partial x^i})-\Sigma_s^{ij}\frac{\partial\overline{u_s^i}}{\partial x^j}-\frac{\varepsilon_s\rho_s}{\tau_{gs}^x}(2k_s-k_{gs})-\varepsilon_s\rho_s\frac{(1-e)^2}{3\tau_s^c}k_s, \quad [12.25]$$

$$\varepsilon_s\rho_s(\frac{\partial k_{gs}}{\partial t}+\overline{u_s^i}\frac{\partial k_{gs}}{\partial x^i})= \frac{\partial}{\partial x^i}(\varepsilon_s\rho_s v_{gs}^t/\sigma_{kg}\frac{\partial k_{gs}}{\partial x^i})-\varepsilon_s\rho_s M_{gs}^{ij}S_{gs}^{ij}+\Pi_{kgs}-\varepsilon_s\rho_s k_{gs}/\tau_{gs}^t, \quad [12.26]$$

where

$$M_{gs}^{ij}=\frac{1}{3}k_{gs}\delta^{ij} -\frac{\eta_r}{1+\eta_r}v_g^t\left[(\frac{\partial\overline{u_g^i}}{\partial x^j}+\frac{\partial\overline{u_g^j}}{\partial x^i})-\frac{2}{3}\frac{\partial\overline{u_g^m}}{\partial x^m}\delta^{ij}\right]-\frac{1}{1+\eta_r}v_{gs}^t\left[\frac{\partial\overline{u_g^i}}{\partial x^j}+\frac{\partial\overline{u_s^j}}{\partial x^i}-\frac{1}{3}(\frac{\partial\overline{u_g^m}}{\partial x^m}+\frac{\partial\overline{u_s^m}}{\partial x^m})\delta^{ij}\right],$$

with

$$\Sigma_s^{ij}=(P_s-\xi_s\frac{\partial\overline{u_s^m}}{\partial x^m})\delta^{ij}-\varepsilon_s\rho_s v_s(\frac{\partial\overline{u_s^i}}{\partial x^j}+\frac{\partial\overline{u_s^j}}{\partial x^i}-\frac{2}{3}\frac{\partial\overline{u_s^m}}{\partial x^m}\delta^{ij}) \text{ and } S_{gs}^{ij}=\frac{\partial\overline{u_g^i}}{\partial x^j}+\frac{\partial\overline{u_s^j}}{\partial x^i}.$$

The mean pressure P_s is still defined for the solid phase, as previously indicated, using the granular temperature Θ: $P_s=\varepsilon_s\rho_s(1+2\varepsilon_s g_0(1+e))\Theta$, and $\xi_s=4d_p\varepsilon_s^2\rho_s g_0\sqrt{\Theta/\pi}/3$. As for the the viscosity coefficients, v_s is calculated by $v_s=v_s^t+v_s^c$, with the colliding viscosity defined as $v_s^c=4\varepsilon_s g_0(1+e)(v_s^t+d_p\sqrt{\Theta/\pi})$ and the turbulent viscosity by $v_s^t=(2\eta_r k_{gs}/3+\Theta(1+\varepsilon_s g_0 A))/(2/\tau_{gs}^x+B/\tau_s^c)$. The coefficient η_r is the ratio given by τ_{gs}^t/τ_{gs}^x, g_0 is calculated here by $g_0=(1-\frac{\varepsilon_s}{\varepsilon_{s,max}})^{-2.5\varepsilon_{s,max}}$ and coefficients A and B by $A=2(1+3e)(3e-1)$ and $B=(1+e)(3-e)/5$. The coefficients for turbulent and colliding diffusion, D_s^t and D_s^c, are defined by

$$D_s^t=(3\eta_r k_{gs}/5+\Theta(1+\varepsilon_s g_0 C))/(9/5\tau_{gs}^x+D/\tau_s^c)$$

and

$$D_s^c=\varepsilon_s g_0(1+e)(6D_s^t/5+4d_p\sqrt{\Theta/\pi}/3),$$

with $C=3(1+e)^2(2e-1)/5$ and $D=(1+e)(49-33e)/100$.

The term for the interaction between the two phases, Π_{kgs}, is given by $\Pi_{kgs} = \varepsilon_s \rho_s ((1+X_{gs})k_{gs} - 2X_{gs}k_s - 2k_g)/\tau^x_{gs}$.

Finally, it must be noted that the dissipation of quantity k_s is not calculated by means of balance equations, as is the case for $<\varepsilon_g>$. However, the deletion of k_s (corresponding to ε_s) supposedly only happens when particles collide. It is standard to assume that there is no correlation for the respective grain positions before collision and, therefore, the collision time τ^c_s involves the mean distance between grains, proportional to the grain diameter, weighted by the volume fraction $\tau^c_s = (d_p / 24\varepsilon_s g_0)\sqrt{\pi / \Theta}$, and the only physical length scale that appears in the end is, therefore, the grain diameter d_p, as the fluctuating velocity gradients are neglected. This expression for τ^c_s is fairly standard in turbulence modeling, where ε and $k^{3/2}$ are usually linked by a length scale which is the turbulence's integral scale l, where d_p is used here as the length scale.

Again, these equations concur with the discussions in Chapter 4; yet, two important remarks need to be made. First, the length scale required to close the equation for k_s is directly set algebraically, relative to the particle diameter, while an equation for the turbulent energy dissipation rate is used for the gas phase. There is no theoretical justification for this choice, which prevents any geometrical length of the flow from playing a direct role in k_s and only an empirical one is possible (see [SIM 96]). The possibility of taking into account a new equation for ε, which would cover a wider range of possible situations, to calculate this length, is discussed in Chapter 4. Second, it must also be noted that the term $2k_s - k_{gs}$, appearing instead of a "standard dissipation" term for the solid phase, enables it to act either as a source term or as a sink term, depending on whether k_s is smaller or larger than k_{gs}.

It must also be noted that the equation for k_s is almost identical to the one for granular temperature, seen in the previous section, as no other scale besides the grain diameter appears in it. However, although the equation for Θ was previously approximated using simple algebraic formulae, the equation for k_s is here modeled with all the inherent complexities of the setup derived for turbulent flows as a solution to the partial differential equations.

The exchange of momentum occurring between the fluid and the particles requires additional terms to be introduced, in comparison to classic cases of turbulent single-phase flows (proportionality to the gradients of some variables, or characteristic time of certain mechanisms, for example), for which the proportionality constants must then be tuned as accurately as possible by comparing with experimental data. Figures 12.6 and 12.7 present the results obtained. Even though the pressure fluctuation spectrum in this case coincides very well with experiments carried out in parallel by the same authors, there is a noticeable shift in the probability distribution of the volume fractions occupied by the solid particle phase.

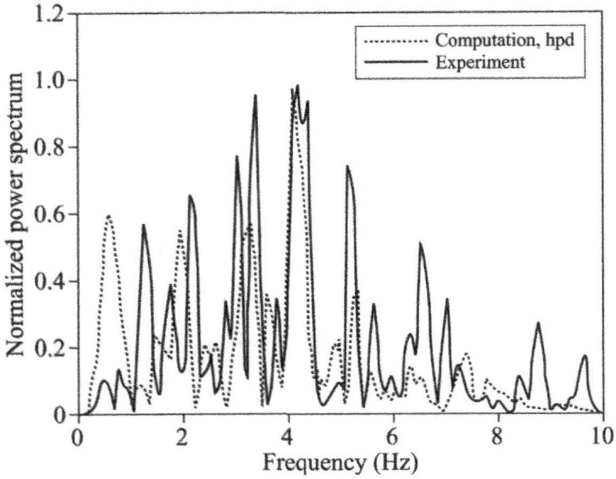

Figure 12.6. *Pressure fluctuation spectrum inside the fluidized bed (at about mid-height of the bed). Comparison of the predictions obtained by the model presented here with experimental data. Excerpt from [PEI 02]*

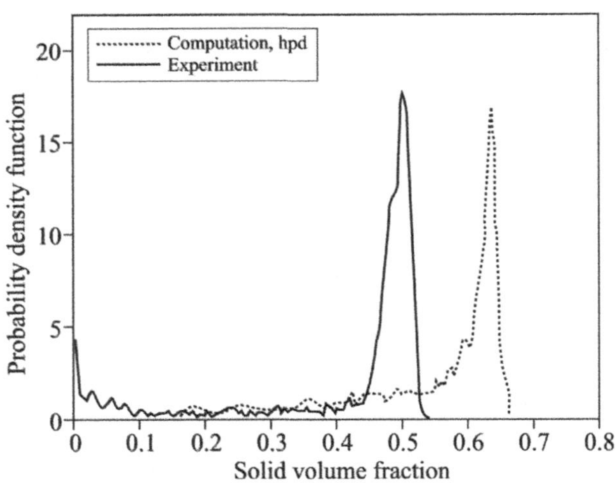

Figure 12.7. *Probability distribution of volume fractions occupied by the solid particle phase inside a fluidized bed (at about mid-height of the bed). Comparison of predictions obtained by the model presented here with experimental data. Excerpt from [PEI 02]*

12.3. Global models for chemical conversion in fluidized beds

When the main interest is modeling chemical conversions in a fluidized-bed reactor, the mechanical fluid phenomena are often simply represented by a few velocities such as U, U_{mf} and U_{mb} previously defined and assumed to be constant in some areas of the flow. Describing such modeling will be the topic of this section. Section 12.4 will focus on the thermal aspects, with an even more simplified presentation of fluid mechanics as only characteristic times or frequencies, determined empirically, are taken into account. Ideally, it is hoped that complete and detailed models, similar to those presented above, can be developed for these two types of situation. This still remains to be done given the complexity of the chemical and thermal processes that must be modeled. An example of the complexity is given in Part 4 of this book, where the modeling of flows using a probability density function (PDF) is discussed. This approach is currently the most efficient one for reactive turbulent flows, as the chemical production and consumption terms are highly nonlinear and cannot be treated satisfactorily in the context of models with second-order closures.

12.3.1. *Bubbling regime fluidizations*

In gas–solid systems, homogeneous fluidization is rarely observable. Bubbles form in the area above the fluidization grid as soon as the gas reaches the minimum fluidization velocity (U_{mf}). The bubbles grow in size and coalesce as they rise toward the surface of the bed and can also divide so that the overall movement induces vigorous agitation in the solid suspension and is in part responsible for the particular properties of fluidized beds.

In order to present global models, a bubbling fluidized bed is considered, consisting of global phases: (1) a bubble phase in which the solid is nearly entirely absent, (2) a cloud phase that accompanies the bubbles as they rise and (3) an emulsion phase in which the solid particles are in a fluidized state. Moreover, the bursting of bubbles at the surface of the bed results in particles being entrained into the area that will be called withdrawal area. If this particle withdrawal is significant, the chemical reaction may continue beyond the surface of the bubbling bed. The withdrawal area will not be specifically discussed here. Similarly, only two models will be presented here as they seem to be the ones used most widely and rated as the best in the literature on this field (at least when this book was prepared).

By using the expression given for U_{mf} (or another, as the one proposed above is not unique), global models can be found to calculate the conversion rate of the reactant in the bubbling part of the bed. The simplest models take into account two phases, namely bubbles and emulsion phases, with a slugging flow for the bubble phase and a perfectly stirred emulsion phase. The gas flow passing in the bubble phase corresponds to the total flow minus the flow required to achieve minimum fluidization, whereas the flow passing through the emulsion phase corresponds to the minimum fluidization flow. The balance for the bubble phase, for each section of height dx, can thus be written as follows:

$$(U-U_{mf})dC_{Ab}=K_g(C_{Ae}-C_{Ab})a_b\varepsilon_b dx\,, \qquad [12.27]$$

with C_{Ab}, the concentration of reactant A in the bubble phase; C_{Ae}, the reactant concentration in the emulsion; K_g, the transfer coefficient between the bubble and emulsion phases; a_b, the interfacial area of a bubble divided by its volume; and ε_b, the void fraction filled by the bubbles. The product a_b. ε_b corresponds to the bubble–emulsion interfacial area per unit volume of the bed. The global balance of matter in the emulsion phase, therefore, gives:

$$U_{mf}(C_{A0}-C_{Ae})+(C_{Ab}-C_{Ae})a_b\varepsilon_b dx=(1-\varepsilon)L_f K_R(C_{Ae})^n\,, \qquad [12.28]$$

where C_{A0} denotes the initial reactant concentration, K_R denotes the kinetic constant relative to the volume ratio of the catalyst and n denotes the order of the chemical reaction. The equations are solved simultaneously to define the concentration profiles of reactant A in both bubble and emulsion phases. The mean concentration at the surface of the bed, C_A, for $x = L_f$, can then be calculated by the relationship:

$$C_A=\beta C_{Ab}+(1-\beta)C_{Ae} \qquad [12.29]$$

where β is the fraction of gas that passes in the bubble phases. According to the order of the reaction being considered, an analytical relationship is found which gives the value of the conversion rate x_A defined by:

$$1- x_A = C_A/C_{A0}.$$

For models that take into account the three phases, the main hypothesis is that the entire fluid supplied goes through the bubble phase (the flux going through other phases is neglected). Cloud and emulsion phases still play an active part in the reaction by transferring matter between the three phases. The kinetic constant K_R is, therefore, defined relative to the volume of the catalyst. For a first-order reaction, the matter balance equations are written as:

– for the bubble phase, $dC_{Ab}/dt = \gamma_b K_R C_{Ab} + K_{bn}(C_{Ab} - C_{An})$ [12.30]

– for the cloud phase, $K_{bn}(C_{Ab} - C_{An}) = \gamma_n K_R C_{An} + K_{ne}(C_{An} - C_{Ae})$ [12.31]

– for the emulsion phase, $K_{ne}(C_{An} - C_{Ae}) = \gamma_e K_R C_{Ae}$ [12.32] with γ_b, the volume of the catalyst in the bubble phase per unit volume of bubbles; γ_e, the volume of catalyst in the emulsion phase per unit volume of bubbles; γ_n, the volume of catalyst in the cloud phase per unit volume of bubbles; K_{bn}, the material transfer coefficient in the cloud phase; and K_{ne}, the material coefficient in the emulsion phase. The following is thus obtained: $1 - x_A = C_A/C_{A0} = \exp(-k.L_f/U_b)$, where k represents the global conversion constant and includes the system parameters and mean rising velocity of bubbles in the fluid bed U_b (Davidson's semi-empirical law is often used, $U_b = 0.71\sqrt{gD_b} + U - U_{mf}$, where D_b is the bubble diameter, given by $D_b = 0.54(U - U_{mf})^{0.4}(x + 4\sqrt{S_0})^{0.8} g^{-0.2}$, with S_0, the bubble section divided by the number of bubbles in the gas-injection grid). k is expressed as:

$$k = \gamma_b K_R + \cfrac{1}{\cfrac{1}{K_{bn}} + \cfrac{1}{\gamma_n K_R + \cfrac{1}{\cfrac{1}{K_{ne}} + \cfrac{1}{\gamma_e K_R}}}} \qquad [12.33]$$

with $\gamma_b \approx 0.001$ to 0.01, $\gamma_n \approx (1 - \varepsilon_{mf})\left[\dfrac{3U_{mf}/\varepsilon_{mf}}{U_b - U_{mf}/\varepsilon_{mf}} + \alpha\right]$ and α is a constant that depends on the nature of the solid used ($\alpha = 0.25$ for sand, 0.4 for glass beads or 0.6 for Fee-type catalysts – with $\gamma_e = \dfrac{1-\varepsilon}{\varepsilon_b} - \gamma_n - \gamma_b$). The matter transfer coefficients are given by $K_{bn} = 4.5\dfrac{U_{mf}}{D_b} + 5.85\dfrac{D^{0.5} g^{0.25}}{(D_b)^{1.25}}$ and $K_{ne} \approx 6.78(\dfrac{\varepsilon_{mf} D U_b}{D_b^3})^{0.5}$ (where D denotes the diffusion coefficient of solid particles in the fluid).

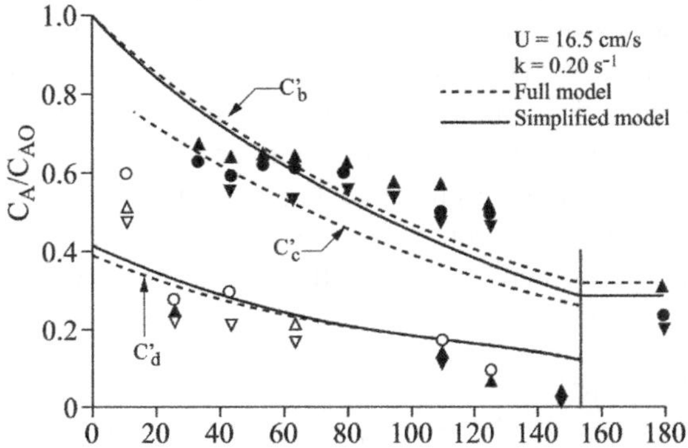

Figure 12.8. *Ozone decay rate. Excerpt from [CHA 75]*

Figure 12.8 shows an example of good results generally provided by a three-phase model and compares them to experimental data. The results are good even though a number of parameters need to be tuned.

These results are found experimentally (via catalytic cracking) and also via the global-phase model described above. The three series of points (identified by the indices b, c and d) correspond to three reaction kinetics resulting from the use of three different types of catalysts. The results were obtained by Chavarie and Grace [CHA 75] and presented and discussed in [YAT 83].

Similar models are used for slugging regimes, equally with the possibility of clearly defining different phases. The void fraction distribution is thus bimodal, with two identifiable peaks for $\varepsilon \approx 1$ for the bubble phase of the carrier fluid and $\varepsilon \approx \varepsilon_{mf}$ for the phase containing solid particles. On the other hand, turbulent fluidization and fast fluidization regimes have a more continuous void fraction distribution so the models, therefore, tend to consider only one single phase, with a particle concentration that varies almost continuously, thus becoming another variable in the model.

12.3.2. *Fast fluidization regime*

It is often the case that fast fluidization models are based on the fact that they experimentally appear to be characterized by two distinct regions, assumed to each consist of a single global phase. As has been mentioned before, residual fluid bubbles (given that stirring is very strong due to high-velocity injection of the carrier fluid and, therefore, that solid particles are withdrawn by this carrier fluid) are indeed very small and difficult to identify as a continuous phase as they are very well mixed within the phase consisting of solid particles. It is, therefore, impossible to make the distinction between a continuous phase transporting a dispersed phase and a single phase containing a mixture of the two original components. Besides, practical modeling of the regime is simplified by neglecting, as a first approximation, the longitudinal velocity gradient inside the column forming the fluidized bed (of radius R). The transverse gradient, however, cannot be neglected as observations show that the column can be divided into two distinct regions: a central core (of radius r_c), within which the carrier fluid and particles move at a velocity U and concentration C_c, and a peripheral annular region in which part of the particles recirculate to the bottom, without the gas moving (of concentration C_a). The mass balance, thus, can be written as follows (ε_c and ε_a represent the respective void fractions for each region):

In the core, $$U(\frac{R^2}{r_c^2})\frac{dC_c}{dx}+\frac{2K}{r_c}(C_c-C_a)+k_1(1-\varepsilon_c)C_c=0 \qquad [12.34]$$

in the annular region, $$\left(\frac{r_c^2}{R^2 - r_c^2}\right)\frac{2K}{r_c}(C_a - C_c) + k_1(1-\varepsilon_a)C_a = 0 \quad [12.35]$$

where K and k_1 are, respectively, the mass transfer coefficient and the chemical kinetic constant (assumed to be of first order). Using the initial condition $C_c = C_0$ at the base of the column, the following value of C_{Lf} at the top of the column (of height L_f) is given by the equation:

$$C_{L_f} = C_0 \exp\left(-k_1^* \left[\frac{(1-\varepsilon_a)(1-\phi^2)X}{k_1^*(1-\varepsilon_a)(1-\phi^2)+X} + (1-\varepsilon_c)\phi^2\right]\right), \quad [12.36]$$

where $\phi = r_c/R$, $k_1^* = \frac{k_1 L_f}{U}$ and $X = \frac{2KL_f r_c}{UR^2}$. Experiments usually show that K and ϕ range between 0.01 and 0.5 m/s, and 0.5 and 0.9. These two parameters are often found by tuning the data measured and depend namely on the velocity U, as well as the solid particle density.

Various variations of this simple model have been proposed and taken into consideration for a gas movement in the annular region and/or a longitudinal velocity gradient in the core. However, these will not be discussed here.

12.3.3. *Turbulent fluidization regime*

Historically, different approaches have been developed for turbulent fluidization regimes and were inspired by what has just been discussed. However, they will not be presented here as they are at present of little interest. On the other hand, the approach currently deemed to be more interesting is the one that considers turbulent fluidization regimes as an intermediate step between bubbling and fast fluidization regimes. This seems highly pertinent given that these two regimes have been discussed above. Therefore, only the equations used need to be studied to determine whether this is feasible.

In general, the mass balance equations can be written as follows for two-phase models for bubbling fluidization (one bubble phase and one solid or emulsion phase), assuming that a chemical component, indexed by i, can be either in the solid phase, at a concentration $C_{i,e}$, or in the bubble phase, thus having a concentration $C_{i,b}$):

$$C_i = q_b C_{i,b} + q_e C_{i,e} \quad [12.37]$$

$$u_b \frac{dC_{i,b}}{dx} = D_{x,b} \frac{d^2 C_{i,b}}{dx^2} + \rho_b R_{i,b}(C_b) - k_q a_I \varepsilon_b (C_{i,b} - C_{i,e}) \qquad [12.38]$$

$$u_e \frac{dC_{i,e}}{dx} = D_{x,e} \frac{d^2 C_{i,e}}{dx^2} + \rho_e R_{i,e}(C_e) + (\frac{\Psi_b}{\Psi_e}) k_q a_I \varepsilon_b (C_{i,b} - C_{i,e}) \qquad [12.39]$$

where Ψ_b and Ψ_e are, respectively, the volume fractions occupied by the bubble and emulsion phases, q_b and q_e are the output fractions for each phase, $D_{x,b}$ and $D_{x,e}$ are the diffusion coefficients for each phase, $R_{i,b}$ and $R_{i,e}$ are the chemical reaction terms for component i in each phase, k_q is the mass transfer coefficient between phases and a_I is the interfacial surface per unit volume of gas (for spherical bubbles with diameter d_b, $a_I = 6/d_b$). It is important to note here that the coefficients involved in the model are *a priori* dependent on the velocity U and that values generally considered are, therefore, only valid as long as the bubbling regime is maintained.

Assuming that the component is only present in the core and not in the annular part where solid particles can recirculate, the balance equation for C_i in fast fluidization regimes thus becomes (using the same notation, and adding a term for axial diffusion, in comparison to what was considered in section 5.3.2):

$$U \frac{dC_i}{dx} = D_x \frac{d^2 C_i}{dx^2} + \rho R_i(C). \qquad [12.40]$$

It, therefore, becomes apparent that the two approaches can be combined into a single one as long as the following relationship is satisfied:

$$\frac{dC_i}{dx} = q_b \frac{dC_{i,b}}{dx} + q_e \frac{dC_{i,e}}{dx} = \frac{1}{U}(\Psi_b u_b \frac{dC_{i,b}}{dx} + \Psi_e u_e \frac{dC_{i,e}}{dx}) \qquad [12.41]$$

which leads to:

$$U \frac{dC_i}{dx} = \alpha_D(x) D_x \frac{d^2 C_i}{dx^2} + \alpha_{R,i}(x) \rho R_i(C)$$

where $\alpha_D(x) = (\Psi_b D_{x,b} \frac{d^2 C_{i,b}}{dx^2} + \Psi_e D_{x,e} \frac{d^2 C_{i,e}}{dx^2}) / (D_x \frac{d^2 C_i}{dx^2})$

and $\alpha_{R,i}(x) = (\Psi_b \rho_b R_{i,b}(C_b) + \Psi_e \rho_e R_{i,e}(C_e)) / (\rho R_i(C))$. Therefore, the two models can be combined into a single one if α_D and $\alpha_{R,i}$ are equal to 1 for every point x. This only occurs if the phases converge to identical profiles $C_{i,b}(x)$ and $C_{i,e}(x)$ and if the properties D_x and ρ in each phase become identical. For a general model to reproduce the transition from a two-phase regime to a single-phase regime when the velocity U becomes very high (U tends toward $U_\infty >> U_c$), observed experimentally, the following boundary conditions must be met:

– for velocities in the gas, $\lim_{U\to\infty} u_b = \lim_{U\to\infty} u_e = U$;

– for diffusion coefficients, $\lim_{U\to\infty} D_{x,b} = \lim_{U\to\infty} D_{x,e} = D_x$;

– for particle concentrations, $\lim_{U\to\infty} \rho_b = \lim_{U\to\infty} \rho_e = \rho$.

These conditions are ensured by a statistical approach for velocities U, as the coefficients for two-phase models depend strongly on the velocity U, as mentioned before, and the stretch-deformation of gas bubbles observed for turbulent fluidization regimes can be interpreted as though the velocity U no longer locally takes a unique and uniform value, but can take values, spread over a wide range. Without going into technical details of the statistical process of the velocity U's random behavior, the authors of the model showed that it provided good quality results for turbulent fluidization regimes. This model (or variations of it) is, therefore, now the one most commonly used. The results obtained are indeed more accurate, and for a wider range of velocities, as shown in Figure 12.9.

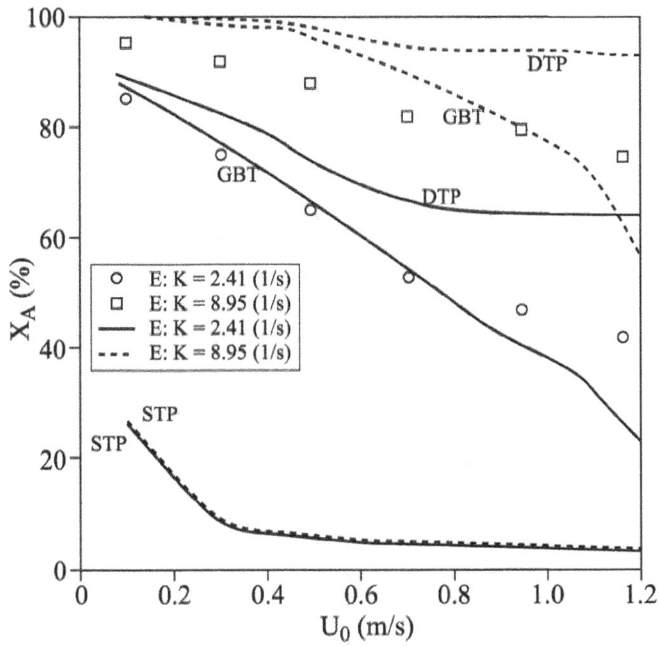

Figure 12.9. *Ozone decay rate. Excerpt from [JAF 04]*

These rates were obtained experimentally and from three standard models (including the one presented here, identified as GBT), for different fluidization velocities and two reaction kinetics, by Jafari *et al.* [JAF 04].

12.4. Global models for heat transfers in fluidized beds

The presentation on the main global models for fluidized beds will end by presenting the global models used to quantify heat transfers in fluidized beds. Indeed, many industrial processes that use fluidized beds bring into play phenomena that generate heat transfers. The most common processes are those that involve chemical reactions, both exothermic and endothermic. The most characteristic example is the combustion of coal. In all these processes, it is essential to have a good control of internal temperature of the fluidized bed in order to stay as close as possible to the optimum conditions for operation. It is, therefore, necessary to have reliable modeling tools capable of predicting heat transfers within the fluidized bed in order to estimate the quantities of heat released or removed at the walls or by cooling systems inserted inside the bed, for example. As before, experiments showed that the process involved depended on the fluidization regime being considered due to hydrodynamics playing an important role in heat transfers. Hereafter, bubbling regimes and fast fluidization regimes will be considered because the relative organization of phases in the carrier fluid and the solid particle phase is quite different in both cases, as discussed previously.

12.4.1. *Bubbling fluidization regime*

In the bubbling fluidization regime, the (global) phases of the carrier fluid and solid particles remain quite clearly separate in the bed. In this case, a cooling system is often used by placing tubes in the bed, either horizontally or vertically, which are filled with coolant. The bulk of heat transfers, therefore, take place on the walls of the tubes via both convection and conduction. Therefore, when a pocket of carrier fluid passes on part of the tube, convection occurs, and when a packet of solid particles passes, conduction and convection occur. In addition, radiation will also be involved if the temperature is high enough. By taking the usual decoupling hypothesis between these three phases, the heat transfer coefficient h can, therefore, be decomposed as:

$$h = f_d h_d + (1 - f_d) h_l + h_r , \qquad [12.42]$$

where h is the coefficient for the overall heat exchange, linked to the heat transfer rate q by:

$$q = hA(T_{bed} - T_{wall}), \qquad [12.43]$$

where A is the exchange surface, and T_{bed} and T_{wall} are, respectively, the temperatures inside the bed and on the wall of the cooler. These relationships can of course be considered globally, either for the entire height of the fluidized bed or, more commonly, as the local relationship for an intermediary transverse section of the bed. The term f_d thus represents the time proportion for which the solid particles are in contact with the wall being considered, and h_d, h_l and h_r are, respectively, the exchange coefficients associated with transfers linked to the solid particle phase, the carrier fluid phase and the radiation process. The problem now is to carry over these three terms to modeling, for which it is usually assumed that the surface element of interest is subjected to successive passing pockets of fluid that have nearly no particles or particle packets with very little fluid.

For solid particles, the following is often written:

$$h_d = 2\left[\frac{k_{pa}\rho_p c_{pp}(1-\varepsilon_{pa})}{\pi\tau_{pa}}\right]^{1/2} \qquad [12.44]$$

where k_{pa} is the heat conductive diffusivity of the particle packets, c_{pp} is the specific heat of the particle packets, ε_{pa} is the void ratio inside the packets and τ_{pa} is the mean contact time between packets and the wall surface. As is the case in most bubbling beds, the contribution of the carrier fluid phase is significantly inferior to that of the particles, and the overall contribution, excluding radiation, is deduced and will be denoted by h_c, as it results in fluid and particle convection on the wall elements. It is defined by:

$$h_c \approx f_d h_d = 2 f_d\left[\frac{k_{pa}\rho_p c_{pp}(1-\varepsilon_{pa})}{\pi\tau_{pa}}\right]^{1/2} \qquad [12.45]$$

such that modeling relies on the evaluation of the characteristic time τ_{pa} and the frequency f_d. Figures 12.10 and 12.11 show typical examples of results obtained experimentally for these quantities, with respect to the velocity (where U_{mf} represents the minimum fluidization velocity), when the obstacle is vertical.

Figure 12.12 shows a good correlation between this type of modeling and the experimental data for a vertical obstacle (for more details, the reader may refer to [CHE 05].

However, when the obstacle is horizontal, important variations for both τ_{pa} and f_d appear along the entire obstacle surface and generate important variations in the local heat transfer rate. This phenomenon is yet to be quantified experimentally for more effective models to be developed.

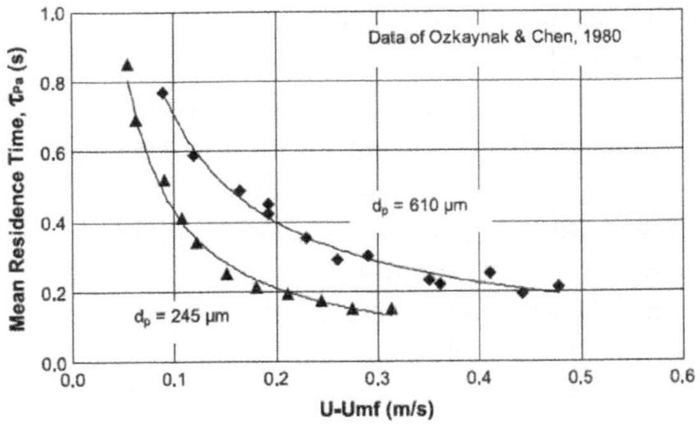

Figure 12.10. *Contact time for solids on a vertical obstacle, obtained from a bed of glass particles of diameter d_p. Excerpt from [CHE 05]*

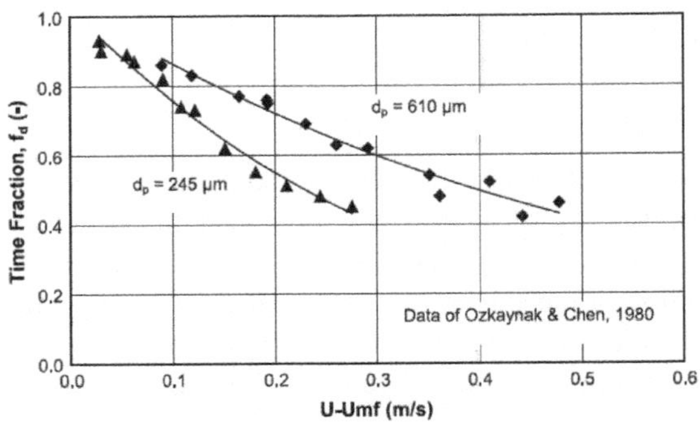

Figure 12.11. *Contact time fraction of solid particles on a vertical obstacle obtained from a bed of glass particles of diameter d_p. Excerpt from [CHE 05]*

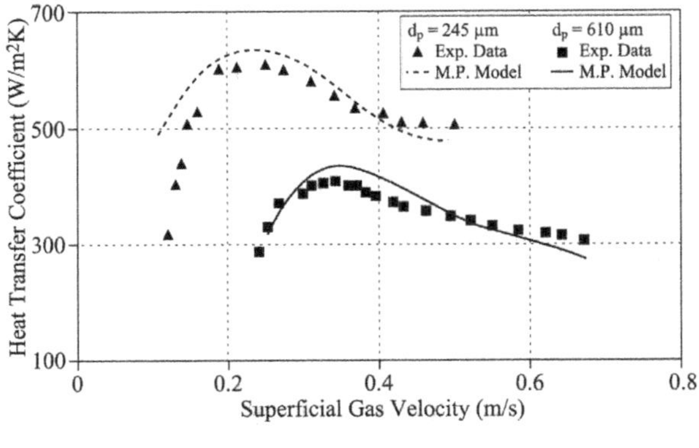

Figure 12.12. *Heat transfer coefficients. Comparison between prediction made by the model presented here and corresponding experimental data, for a bed of glass particles of diameter d_p, on a vertical obstacle. Excerpt from [CHE 05]*

As far as transfers via radiation are concerned, they are generally considered important when the temperature of the bed is over 500°C. These phenomena are most often modeled by considering a radiative exchange between "gray" solid objects, separated by the medium transparent to radiation, so that the following can be written:

$$q_r = \frac{\sigma(T_b^4 - T_w^4)}{\frac{A_w}{A_b}(\frac{1-e_b}{e_b}) + \frac{1}{F} + (\frac{1-e_w}{e_w})} \qquad [12.46]$$

where T_b and T_w are the temperatures of the bed and the surface of the obstacle wall of interest, respectively. The sight factor F is close to unity and A_w=A_b, except when the obstacle shape in the fluidized bed is very sinuous. The emissivity coefficient for the solid surface e_w depends on the material that it is made from (it usually has a value between 0.7 and 1 for the most common metallic materials). The parameter having the most uncertain value is the bed's emissivity coefficient, e_b. Some of the experimental results obtained for sand beds provided values for e_b ranging from 0.7 to 1. It is then deduced that:

$$h_r = \frac{q_r}{T_b - T_w} = \frac{\sigma(T_b^4 - T_w^4)}{(T_b - T_w)(\frac{1-e_b}{e_b} + 1 + \frac{1-e_w}{e_w})} \qquad [12.47]$$

12.4.2. *Fast fluidization regimes – circulating beds*

It has already been mentioned that the circulation of the carrier fluid and particles in the bed can be decomposed into two distinct regions for fast fluidization regimes. One region is the core of the device in which the carrier fluid phase can be considered as more or less continuous as the bubbles are very small and misshapen. The other, annular, region is close to the walls and large particle packets appear in it, which agglomerate and move in a fluid animated by relatively small velocities. This regime is very important from an industrial point of view as it is encountered in combustion systems and catalytic reactions. Most often, modeling of heat transfers for this type of fluidized bed is based on the breakdown into two superimposed contributions, one originating from transfers in the core and the other originating from contact between particles and the column walls. Given that large velocities are present in this regime, there is usually no exchange surface in the column so that heat transfers with the outside only occur on the walls of the setup. The contributions of transfers, both by convection in the core and radiation from the core to the walls, are considered and linked to the particles present on the walls or in their immediate vicinity. An electrical analogy is often used (see Figure 12.13), based on resistances in series and in parallel, according to the following diagram (by separating the contributions from the fraction of the wall without particles from that covered in particles):

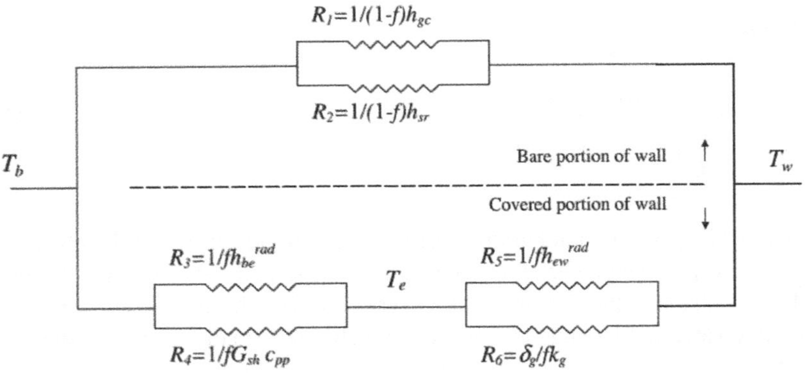

Figure 12.13. *Electrical analogy used to calculate heat transfers in fast fluidization regimes. Excerpt from [CHE 05]*

The mean height separating the clusters of particles from the wall is identified by δ_g, the mean fraction of the wall surface covered in particles by f, the mean exchange flux in the horizontal direction by G_{sh} and the mean void fraction in the core by ε_{sus}. The overall exchange coefficient can, therefore, be written as:

$$h=(h_{gc}+h_{sr})(1-f)+\frac{f}{\dfrac{1}{G_{sh}c_{pp}+h_{be}^{r}}+\dfrac{1}{(k_g/\delta_g)+h_{ew}^{r}}} \qquad [12.48]$$

The synthesis of experimental results leads to the following formulation for *f*:

$$f=1-exp\{-25000(1-\frac{2}{e^{0.5D}+e^{-0.5D}})(1-\varepsilon_{sus})\} \qquad [12.49]$$

where D is the hydraulic diameter of the column, whereas radiative transfer rates are given, as before, by:

$$h_{sr}=\frac{\sigma(T_b^4-T_w^4)}{(T_b-T_w)(\dfrac{1}{e_{sus}}+\dfrac{1}{e_w}-1)} \qquad [12.50]$$

$$h_{ew}^{r}=\frac{4\sigma(T_e^4-T_w^4)}{(T_e-T_w)(\frac{1}{e_{sus}}+\frac{1}{e_w}-1)} \quad [12.51]$$

and $$h_{be}^{r}=\frac{4\sigma(T_b^4-T_e^4)}{(T_b-T_e)(\frac{2}{e_{sus}}-1)} \quad [12.52]$$

where e_{sus} and e_w represent the emissivity of the suspension (bulk) and the walls, respectively, and T_e, the temperature of the suspension at position δ_g, is given by:

$$T_e=\frac{T_b\,R_{56}+T_w R_{34}}{R_{34}+R_{56}}\text{, with } R_{34}=\frac{R_3R_4}{R_3+R_4} \text{ and } R_{56}=\frac{R_5R_6}{R_5+R_6}$$

Finally, adjustment of experimental data has led to:

$$\delta_g=0.0282d_p(1-\varepsilon_{sus})^{-0.59} \text{ and } G_{sh}=0.0225\ln(\rho_{sus})+0.1093$$

Figure 12.14 shows the validity of this type of approach by comparing various experimental data. Despite the approximations, this type of model is currently considered as the one that provides the best results to size circulating fluidized beds of industrial size. While it is generally acknowledged that the rules of similarity that should enable heat transfers, associated with large setups, are to be found using pilot laboratory trials, they are not yet very robust, no matter what fluidization regime is being considered.

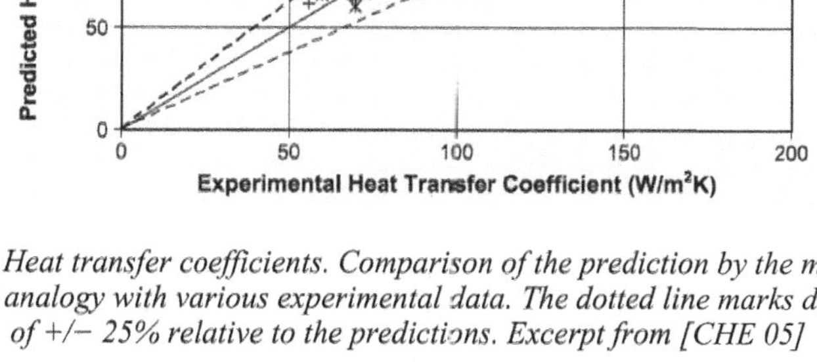

Figure 12.14. *Heat transfer coefficients. Comparison of the prediction by the model based on electrical analogy with various experimental data. The dotted line marks differences of +/– 25% relative to the predictions. Excerpt from [CHE 05]*

12.5. Conclusion

As has just been discussed, fluidized beds are a particularly complex field of multiphase media mechanics. The first research and developmental studies were mostly founded on empirical bases with significant contributions from experiments. Over the last few years, developments have been made on more fundamental bases and profit from the experience obtained in basic research on multiphase flows (see Chapters 2–4). However, they have not led to the definition of a Eulerian model that is consensual, even for a single regime. The most recent developments on the subject tend to turn toward numerical Euler–Lagrange simulations which should enable the different interaction processes between the carrier fluid and the solid particle phases, as well as interparticle interactions, to be analyzed in more detail. These models will be introduced in the last chapter of this book. It can be expected that the current level of knowledge, relative to the hydrodynamics of fluidized beds alone, and therefore of the physical and numerical models resulting from it, can become almost equal to that of single-phase fluids. However, the complex couplings with phenomena caused by chemical reactions and heat transfer still require more complicated studies, the main issues of which are discussed in the last part of the book. They make it a field in which experimentation with industrial prototypes still remains crucial.

Chapter 13

Generalizations for Granular Media

13.1. Introduction

Granular media occur in different types of practical situations: handling of powders, geophysical environments, landslides and even the denser industrial fluidized beds are granular media. These are two- or multiphase media where the proportion of the solid phase is high, and for which there is at least one fluid phase: in the case of dry granular media, the fluid is either air or gas, whereas it is a liquid, most often water, in "immersed" granular media. However, the fluid phase in dry media does not play a mechanical role as its density is very low in comparison to that of the grains, and the contact forces between the fluid and the grains are completely negligible. It simply prevents the grains' volume fraction from becoming unity, and thus justifies a maximum value that is not well known (of the order of 0.6). Depending on the conditions, granular media can have properties similar to those of gases, liquids or solids. Hutter and Rajagopal [HUT 94] present a synthetic overview of their main characteristics and of modeling attempts made since the study of the subject first started. The recent book by Andréotti *et al.* [AND 11] goes into more detail and presents a wider range of thoughts on the subject.

Various *ad hoc* models, based on global theories, have been proposed to represent the different aspects of granular media, with the exception of the limit case termed "granular gas", for which models are directly linked to, and extend the kinetic theory for gases. Numerous experiments on this type of medium have been performed to support modeling and attempt to validate them. This is not an easy task to do, as these experiments are difficult to perform because the number of overlapping phenomena has to be set to a minimum, thus simplifying their interpretation. A representative sample of recent models will be presented in the last section of this chapter. The main goal here is to show how the multiphase approach developed can provide safe bases for modeling, within the entire range of the medium's behavior.

Grains, even small ones, are continuous solid media in a granular medium, and the fluid phase between the grains is a continuous fluid medium (sometimes two, water and air). It therefore consists of a medium having several phases, and it is piecewise continuous at any point in time. The approach described in the first chapters of this book can also be applied to it. However, in these chapters, it was assumed that the solid phase always locally consisted of single grains, which could of course meet others by colliding with them but without being linked together in pairs or groups for any significant length of time.

This assumption is no longer realistic for granular media because the number of grains is too high (per unit volume) and their movement can be too slow. Moreover, it is clear that the grains as a whole cannot be considered as one single phase, as the phase's constitutive relations would remain unknown. Each grain must therefore be considered as a single phase, and its constitutive relations are then perfectly well known. Some grains may even behave differently from others, having different constitutive relations. However, the description would then be too rich to represent the media, and the grains will finally have to be represented as a whole. This approach will be adopted here, and this chapter will set several questions that are associated with this, which will be discussed hereafter. The exchange of mass between phases will not be considered here (an aspect that can be taken into account in the case of the fluid porosity of specific grains, for example).

13.2. *Balance equations for mean granular media*

First, let us recall the balance equations for an instantaneous piecewise continuous medium, the mass balance, the momentum and energy balances. The phase's indicator for each grain and for the fluid between them is Φ_i, unity in phase i and zero elsewhere. Therefore:

$$\frac{\partial}{\partial t}(\rho_i \Phi_i) + \frac{\partial}{\partial x_\alpha}(\rho_i \Phi_i v_i^\alpha) = 0 \qquad [13.1]$$

$$\frac{\partial}{\partial t}(\rho_i \Phi_i v_i^\beta) + \frac{\partial}{\partial x_\alpha}(\rho_i \Phi_i v_i^\alpha v_i^\beta) = \frac{\partial}{\partial x_\alpha}(\Phi_i \sigma_i^{\alpha\beta}) + \rho_i \Phi_i g_\beta + \sigma_i^{\alpha\beta} n_i^\alpha \sigma_i^s \qquad [13.2]$$

By defining the total mass energy in grain i as $e_{ti}=e_i+\frac{1}{2}v_i^\beta v_i^\beta$, the following equation is obtained:

$$\frac{\partial}{\partial t}(\rho_i\Phi_i e_{ti})+\frac{\partial}{\partial x_\alpha}(\rho_i\Phi_i v_i^\alpha e_{ti})=\frac{\partial}{\partial x_\alpha}(\Phi_i(-j_i^{Q\alpha}+\sigma_i^{\alpha\beta}v_i^\beta))+\rho_i\Phi_i v_i^\beta g_\beta$$
$$+\sigma_i^{\alpha\beta}n_i^\alpha\sigma_i^s v_i^\beta+q_{is}\sigma_i^s \qquad [13.3]$$

Remembering that the indicator of the phase i follows a simple convection equation:

$$\frac{\partial}{\partial t}(\Phi_i)+v_i^\alpha\frac{\partial}{\partial x_\alpha}(\Phi_i)=0 \qquad [13.4]$$

Even if there is no mass exchanged between the phases, there is an exchange in momentum, in other words, contact forces between the phases, included in the last term of [13.2], proportional to the contact surface per unit volume between phase i and other grains or the fluid. Energy is also exchanged by the power of contact forces and heat transfer, included in the last two terms of [13.3].

It is recalled that:

$$-n_i^\alpha\frac{\partial\Phi_i}{\partial x_\alpha}=\sigma_i^s\text{, where }-n_i^\alpha\sigma_i^s=\frac{\partial\Phi_i}{\partial x_\alpha} \qquad [13.5]$$

Here, the vector perpendicular to the interface n_i^α (with a positive orientation toward the exterior of phase i) is again assumed to always be defined; in other words, the total area of "triple" lines or points, where two grains and the fluid are in contact together, is negligible in comparison to the total interface area. The contact area is assumed to consist only of parts where two grains are in contact or where a grain is in contact with the fluid.

The tensor $\sigma_i^{\alpha\beta}$ is no more than the Cauchy stress tensor of the phase being considered. For example, if the phase is solid, it is given by an elastic or viscoelastic constitutive relation, which is known if the grain characteristics are known. The internal elasticity of grains permits the existence of contact surfaces between two grains not necessarily limited to a single point, depending on the forces acting between the grains, of course. If the phase is fluid, as is the case in between grains, the tensor will include the diagonal fluid pressure (negative) and the viscous stress tensor.

The velocity is continuous along the contact surface between a grain and the fluid, contrary to the density that is discontinuous. The tangential velocity along the contact surface between grains may only be discontinuous if a grain slips relative to another, as required by friction conditions. The perpendicular velocity may only be discontinuous at the beginning or at the end of contact, which can be called "a collision".

In compliance with Chapter 3, the granular medium being modeled will be a statistical average of an infinite number of instantaneous media, each having their own initial, slightly different, conditions, in particular concerning the location and the size of each grain, each having (slightly) different boundary conditions. Then, a particular interest will be taken in the mean mass of grains per unit of total volume and the mean momentum of grains, the grains' volume fraction, and, eventually, the fluid's mean density and momentum. It is assumed that there are N solid grains and that the fluid phase corresponds to $i = N + 1$.

By definition, the mean density of the medium is $\bar{\rho} = \sum_{i=1,N+1} \overline{\rho_i \Phi_i} = \sum_{i=1,N+1} \bar{\rho}_i^* \bar{\Phi}_i$, where $\bar{\rho}_i^*$ is the mean intrinsic density of phase i. For the fluid, we note $\bar{\rho}_f = \bar{\rho}_{N+1}^*$, the mean intrinsic density of the fluid (without an asterisk for simplicity). The volume fraction of grains is given by $\bar{\Phi}_s = \sum_{i=1,N} \overline{\Phi_i} = 1 - \bar{\Phi}_{N+1} = 1 - \bar{\Phi}_f$. For grains, $\sum_{i=1,N} \overline{\rho_i \Phi_i} = \bar{\rho}_s \bar{\Phi}_s$, where $\bar{\rho}_s$ is the mean intrinsic density, and $\bar{\rho} = \bar{\rho}_s \bar{\Phi}_s + \bar{\rho}_f (1 - \bar{\Phi}_s)$. This formula takes into account the fact that grains possibly present in the medium being considered are not necessarily made of the same material and can be compressed to different degrees. If the change in volume of each grain is neglected (even if it can change shape, at most), and if all the grains are made of the same material, then $\bar{\rho}_s$ is known and identical everywhere in the medium. The same is true if all grains are made of different materials (and do not change in volume) only as long as the distribution of the different grains in the medium is statistically homogeneous.

The mean momentum in the medium volume is $\bar{\rho}\tilde{v}^\alpha = \sum_{i=1,N+1} \overline{\Phi_i \rho_i v_i^\alpha}$ and using the same hypotheses, $\bar{\rho}\tilde{v}^\alpha = \bar{\rho}_s \bar{\Phi}_s \tilde{v}_s^\alpha + \bar{\rho}_f \bar{\Phi}_f \tilde{v}_f^\alpha$, where $\bar{\rho}_s \tilde{v}_s^\alpha \bar{\Phi}_s = \sum_{i=1,N} \overline{\rho_i \Phi_i v_i^\alpha}$.

The mean balance equations for the grain phase are obtained by adding equations [13.1] and [13.2], for $i = 1,\dots,N$, therefore:

$$\frac{\partial}{\partial t}(\bar{\rho}_s \bar{\Phi}_s) + \frac{\partial}{\partial x_\alpha}(\bar{\rho}_s \bar{\Phi}_s \tilde{v}_s^\alpha) = 0 \qquad [13.6]$$

$$\frac{\partial}{\partial t}(\bar{\rho}_s \bar{\Phi}_s \tilde{v}_s^\beta) + \frac{\partial}{\partial x_\alpha}(\bar{\rho}_s \bar{\Phi}_s \tilde{v}_s^\alpha \tilde{v}_s^\beta) = \frac{\partial}{\partial x_\alpha}\Big(\sum_{i=1,N} \overline{\Phi_i \sigma_i^{\alpha\beta}} - \sum_{i=1,N} \overline{\rho_i \Phi_i v''^\alpha_i v''^\beta_i}\Big)$$
$$+\bar{\rho}_s \bar{\Phi}_s g_\beta + \sum_{i=1,N} \overline{\sigma_i^{\alpha\beta} n_i^\alpha \sigma_i^s}$$

First, even though the grain density $\bar{\rho}_s$ is a known constant, the grains' phase does not have an overall constant density (thus, the mean grain velocity has a non-zero divergence) as the volume fraction can vary. Second, the granular medium's mean Cauchy tensor consists of two parts, one linked to the stresses on the grains and the contacts between grains, and the other linked to the fluctuating movement of grains. The fluctuating velocity that plays in the term is $v''^\alpha_i = v_i^\alpha - \tilde{v}_s^\alpha$, the difference between the instantaneous velocity of grain *i*, at a given point in space and time, and the mean velocity of all grains at that point.

The last term of the equation is related to the contact forces with the fluid and, sometimes, to the surface tension. If the surface tension effect is neglected, this term is nothing more than the opposite of the force exerted by the grains on the fluid. This force can be written in two different ways that have been described in Chapter 5: either, as in section 5.3.3,

$$\sum_{i=1,N} \overline{\sigma_i^{\alpha\beta} n_i^\alpha \sigma_i^s} = -\overline{\sigma_f^{\alpha\beta} n_f^\alpha \sigma_f^s} = -p_f \overline{\frac{\partial \Phi_f}{\partial x_\beta}} - \overline{\tau_f^{\alpha\beta} n_f^\alpha \sigma_f^s} = -\bar{p}_f^s \frac{\partial \bar{\Phi}_f}{\partial x_\beta} - \overline{\tau_{+f}^{\alpha\beta} n_f^\alpha \sigma_f^s}$$

or $$\sum_{i=1,N} \overline{\sigma_i^{\alpha\beta} n_i^\alpha \sigma_i^s} = -\bar{\Phi}_s \frac{\partial}{\partial x_\beta} \bar{p}_f^* + \frac{\partial}{\partial x_\beta}(\bar{p}_f^s \bar{\Phi}_s) + \bar{F}_{\tau f}^\beta,$$

as in section 5.3.2.

The first two terms are linked to the buoyancy acting on the grains, and the last term is linked to the viscous friction between the fluid and the grains (the extra term for non-homogeneous media is negligible for viscous forces in this case, in comparison to the one linked to buoyancy, the second term, see the discussion in section 5.3.2 of Chapter 5). When it is assumed that, in a fluid (moving slowly),

$\overline{p}_f^s = \overline{p}_f^*$, these two formulations are identified and $\overline{F}_{\tau f}^{\beta}$ is nothing more than an approximation of $-\overline{\tau_{+f}^{\alpha\beta} n_f^{\alpha} \sigma_f^s}$, and $\overline{\Phi}_s + \overline{\Phi}_f = 1$.

Then the following is finally obtained:

$$\frac{\partial}{\partial t}\left(\overline{\rho}_s \overline{\Phi}_s \tilde{v}_s^{\beta}\right) + \frac{\partial}{\partial x_\alpha}(\overline{\rho}_s \overline{\Phi}_s \tilde{v}_s^{\alpha} \tilde{v}_s^{\beta}) = \frac{\partial}{\partial x_\alpha}\Big(\sum_{i=1,N} \overline{\Phi_i \sigma_i^{\alpha\beta}} - \sum_{i=1,N} \overline{\rho_i \Phi_i v''^{\alpha}_i v''^{\beta}_i}\Big)$$
$$+ \overline{\rho}_s \overline{\Phi}_s g_\beta - \overline{p}_f^* \frac{\partial \overline{\Phi}_f}{\partial x_\beta} - \overline{\tau_{+f}^{\alpha\beta} n_f^{\alpha} \sigma_f^s} \qquad [13.7]$$

The mean balance equations for the fluid phase are:

$$\frac{\partial}{\partial t}\left(\overline{\rho}_f \overline{\Phi}_f\right) + \frac{\partial}{\partial x_\alpha}(\overline{\rho}_f \overline{\Phi}_f \tilde{v}_f^{\alpha}) = 0 \qquad [13.8]$$

$$\frac{\partial}{\partial t}\left(\overline{\rho}_f \overline{\Phi}_f \tilde{v}_f^{\beta}\right) + \frac{\partial}{\partial x_\alpha}(\overline{\rho}_f \overline{\Phi}_f \tilde{v}_f^{\alpha} \tilde{v}_f^{\beta}) = \frac{\partial}{\partial x_\alpha}\Big(\overline{\Phi}_f(-\overline{p}_f^* \delta_{\alpha\beta} + \overline{\tau}_f^{*\alpha\beta}) - \overline{\rho_f \Phi_f v''^{\alpha}_f v''^{\beta}_f}\Big)$$
$$+ \overline{\rho}_f \overline{\Phi}_f g_\beta + \overline{p}_f^* \frac{\partial \overline{\Phi}_f}{\partial x_\beta} + \overline{\tau_{+f}^{\alpha\beta} n_f^{\alpha} \sigma_f^s} \qquad [13.9]$$

The recent book by Andréotti *et al.* [AND 11, Chapter 7] gives the following equations:

$$\overline{\rho}_s \overline{\Phi}_s \Big(\frac{\partial}{\partial t}\left(\tilde{v}_s^{\beta}\right) + \tilde{v}_s^{\alpha} \frac{\partial}{\partial x_\alpha}(\tilde{v}_s^{\beta})\Big) = \frac{\partial}{\partial x_\alpha}(\sigma_P^{\alpha\beta})$$

$$+ \overline{\rho}_s \overline{\Phi}_s g_\beta + \overline{\Phi}_s \frac{\partial \sigma_f^{\alpha\beta}}{\partial x_\alpha} + \overline{F}_{\tau f}^{\beta}$$

and $$\overline{\rho}_f \overline{\Phi}_f \Big(\frac{\partial}{\partial t}\left(\tilde{v}_f^{\beta}\right) + \tilde{v}_f^{\alpha} \frac{\partial}{\partial x_\alpha}(\tilde{v}_f^{\beta})\Big) = \frac{\partial}{\partial x_\alpha}(\sigma_f^{\alpha\beta})$$

$$+ \overline{\rho}_f \overline{\Phi}_f g_\beta - \overline{\Phi}_s \frac{\partial \sigma_f^{\alpha\beta}}{\partial x_\alpha} - \overline{F}_{\tau f}^{\beta}$$

If we note that $\bar{\Phi}_s \dfrac{\partial \sigma_f^{\alpha\beta}}{\partial x_\alpha} = +\dfrac{\partial \bar{\Phi}_s \sigma_f^{\alpha\beta}}{\partial x_\alpha} - p_f \dfrac{\partial \bar{\Phi}_f}{\partial x_\beta} + \tau_f^{\alpha\beta} \dfrac{\partial \bar{\Phi}_f}{\partial x_\alpha}$, it becomes apparent that these formulas are comparable with [13.7] and [13.9] by writing $\sigma_P^{\alpha\beta} = \sum_{i=1,N} \overline{\Phi_i \sigma_i^{\alpha\beta}} - \sum_{i=1,N} \overline{\rho_i \Phi_i v''^{\alpha}_i v''^{\beta}_i} + \bar{\Phi}_s \bar{\sigma}_f^{*\alpha\beta}$ and $\sigma_f^{\alpha\beta} = \bar{\sigma}_f^{*\alpha\beta} = -\bar{p}_f^* \delta_{\alpha\beta} + \bar{\tau}_f^{*\alpha\beta}$. However, buoyancy then takes into account viscous friction in the liquid, which has not been considered here, and the velocity fluctuations in the liquid are neglected when they are considered here.

The equivalent Cauchy stress tensor for the grain phase, as defined by Andréotti *et al.*, is the sum of the so-called reduced tensor, mentioned in section 5.3.2, Chapter 5 (and not the mean sum of the internal Cauchy stress tensors of the grains), and the dynamic (Reynolds) tensor. On the other hand, their equivalent Cauchy stress tensor for the fluid phase is indeed the mean Cauchy tensor for the fluid in the fluid alone, and the dynamic term is neglected because the flow between grains is realistically considered to be very slow. This is a standard technique in the *a priori* modeling of granular media and is also adopted in [JAC 00], as mentioned previously (section 5.3.2, Chapter 5).

In cases where the fluid is air and the solid phase is sand, for example, then $\bar{\rho}_s \gg \bar{\rho}_f$. Grain and air velocities are of the same order of magnitude in most cases and therefore $\bar{\rho}_s \bar{\Phi}_s \tilde{v}_s^\alpha \approx \bar{\rho} \tilde{v}^\alpha \approx \bar{\rho}_s \bar{\Phi}_s \tilde{v}^\alpha$, thus $\tilde{v}_s^\alpha \approx \tilde{v}^\alpha, \bar{\rho} \approx \bar{\rho}_s \bar{\Phi}_s$. The movement of the grain phase can therefore be calculated using the balance equations for the mean medium given by (capillary action still being neglected):

$$\frac{\partial}{\partial t}(\bar{\rho}) + \frac{\partial}{\partial x_\alpha}(\bar{\rho}\tilde{v}^\alpha) = 0 \qquad [13.10]$$

$$\frac{\partial}{\partial t}(\bar{\rho}\tilde{v}^\beta) + \frac{\partial}{\partial x_\alpha}(\bar{\rho}\tilde{v}^\alpha \tilde{v}^\beta) = \frac{\partial}{\partial x_\alpha}\Big(\sum_{i=1,N} \overline{\Phi_i \sigma_i^{\alpha\beta}} - \bar{p}_f^* \bar{\Phi}_f \delta_{\alpha\beta} + \bar{\Phi}_f \bar{\tau}_f^{*\alpha\beta} - \overline{\rho v'^\alpha v'^\beta}\Big) + \bar{\rho} g_\beta \qquad [13.11]$$

The velocity fluctuation is given here by $v'^\alpha = v^\alpha - \tilde{v}^\alpha$, the difference between the velocity at a given point and a given instant within the medium, whether in the fluid or in grains, and the mean velocity. Equations [13.8] and [13.9] can be rewritten to give [13.10] and [13.11] as, in this case, the force of the fluid acting on the grains can be neglected, and the mean pressure gradient in the fluid is in the order of $\bar{\rho}_f g_\beta \ll \bar{\rho}_s g_\beta$ (see [13.9]). When the presence of the fluid can be completely neglected, equation [13.11] can be used by neglecting $-\bar{p}_f^* \bar{\Phi}_f \delta_{\alpha\beta} + \bar{\Phi}_f \bar{\tau}_f^{*\alpha\beta}$.

The mean total energy equation for the grains can also be found starting with [13.3], using the same approach. Then, by removing the mean and fluctuating kinetic energies of the grains, the balance equation for the internal energy is found. When the fluid is light, and has a temperature close to the grain temperature, the following equation is identical to the equation that gives the mean internal energy for the entire medium:

$$\frac{\partial}{\partial t}\overline{\rho}\tilde{e} + \frac{\partial}{\partial x_\gamma}(\overline{\rho}\tilde{v}_\gamma \tilde{e}) = \frac{\partial}{\partial x_\gamma}(-\overline{\varphi}_{Qt}^{\gamma})$$

$$+(\sum_{i=1,N} \overline{\Phi}_i \overline{\sigma}_i^{*\alpha\beta} - \overline{\Phi}_f \overline{p}_f^{*} \delta^{\alpha\beta} - \sum_{i=1,N} \overline{\rho_i \Phi_i v''^{\alpha}_i v''^{\beta}_i})\frac{\partial \tilde{v}^{\beta}}{\partial x_\alpha} \quad [13.12]$$

The form of [13.12] is standard, and viscous friction in the fluid phase has been neglected. The term $-\varphi_{Qt}^{\gamma}$ is linked to the total heat transfer, through conduction, radiation or even the additional effect due to random grain or fluid particle movement with different temperature. If the fluid and the grain phase have fairly different mean temperature, to be studied separately as in the case of fluidized beds, it is necessary to maintain two separate equations for the mean internal energies, one for the grains and one for the fluid, obtained from equation [13.3], with significant terms for the heat exchange between the two phases,

If $\overline{\rho}_s \gg \overline{\rho}_f$, or if it possible to assume that the mean fluid velocity is nearly equal to that of the grains, we must consider [13.10], [13.11] and [13.12], which are simpler, and even go so far as to neglect all contribution from the fluid. This case is applicable to granular media in air or for numerically simulated granular media.

13.3. Necessary closure approximations

After the averaging operation has been performed, the granular medium is represented by variables for mass and momentum $\overline{\rho}_f, \overline{\rho}_s, \overline{\Phi}_s, \tilde{v}_f^{\alpha}, \tilde{v}_s^{\alpha}$. Four equations have been established above, for which some terms need to be (and can be) modeled further. In compliance with the discussions in Chapter 4, they must be completed by another relationship. As mentioned in Chapter 4, it has clearly been assumed that the mean state equations for the fluid and the mean constitutive laws inside the grains are known as well as a model term for additional "turbulent fluxes" and models for exchanges between the grain phase and the interstitial fluid phase. For all cases in general, where the grains and the fluid phase may both exhibit fast compressible evolutions, the balance equation for the solid's phase mean volume must be used and modeled similarly to the equation by Baer and Nunziato, although adapted to this case for lasting contacts between grains.

For cases where phase evolutions are slow and incompressible, the hypothesis of equal mean pressures in each phase, mentioned in Chapter 4, is no longer realistic for granular media: effectively, the pressures at the interface between fluid and grains are always identical (if the surface tension effect is neglected) and thus $\overline{p}_s^s = \overline{p}_f^s$, and for low velocities $\overline{p}_f^* = \overline{p}_f^s$, but $\overline{p}_s^* \neq \overline{p}_s^s$ because it is evident that, if numerous contacts between grains are such that the stresses at the medium's boundaries can be balanced without involving the liquid phase, the mean stresses in the solid phase, at a given point, particularly the pressure, are not linked to the stresses imposed by the neighboring fluid alone. However, only four equations are needed as $\overline{\rho}_s$ is now a constant (later written ρ_s, the density of the solid that each grain is made of). As the grains' phase is not a standard solid phase, the fact that ρ_s is constant does not result in one of the constitutive relations for the granular media to disappear. The density of the grains' phase is not constant even though ρ_s is, as it is given by $\rho_s\overline{\Phi}_s$, involved in equation [13.6], and this can therefore replace the Baer and Nunziato-type equation. The phenomenon of "Reynolds dilatancy", emblematic of granular media, should then be allowed and can easily be represented as part of these equations.

The balance equations [13.6], [13.7] and [13.8], [13.9] also involve other averaged variables. First, the fluid's mean intrinsic pressure, related to the mean density (also intrinsic) by an averaged state equation, for example $\overline{p}_f^* = \overline{\rho}_f R_f \tilde{T}_f$, and the mean fluid temperature must be known. When the fluid can be assumed to be "incompressible", $\overline{\rho}_f$ is a given constant and the fluid pressure is calculated by the mass balance equation instead of the previous equation. The fluid's mean viscous stress tensor also plays a role in [13.9] and is calculated via an irreversible process law, which is easily applied using the fluid's mean velocity for a Newtonian fluid, for example. The Reynolds stress tensor is added to this and is familiar. It can be calculated via a turbulence model, in case it is non-negligible. Finally, the last two terms are linked to contact and friction due to solid phase and are also found, with the opposite sign, in the equation for the solid phase momentum. The topics discussed in Chapter 5 enable these terms to be realistically approximated. These terms are entirely negligible in the case of dry granular media. Only the two parts of the mean Cauchy tensor for the granular phase remain in [13.7], and they still need to be modeled.

The first part of the mean Cauchy tensor, $\sum_{i=1,N}\overline{\Phi_i\sigma_i^{\alpha\beta}}$, includes the effects of stress in the grains, and especially the effect of contact between grains. This will hereafter be called "sliding contact Cauchy tensor". The definition is simple, as shown by equation [13.6], yet does not provide an expression for the different components of the tensor in terms of the mean characteristics of the medium alone: only the $\sigma_i^{\alpha\beta}$ components of different grains, at any grain point, appear. Even in the simplest situation, for which grains are assumed to be simply elastic and to be static relative to each other, the internal stress distribution must be calculated in the same way as for a random heterogeneous material with "holes". When the possibility of grains moving relative to each other is taken into account, the bare minimum for granular media, the main difficulty immediately becomes the complex and changing distribution of collisions and lasting contacts, whether sliding or non-sliding, which change the internal stresses of the grains (to varying degrees depending on the grain stiffness). Although it is not immediately apparent in the definition, it is still implicitly present. The following chapter will describe a new approach, still based on instant equations for piecewise continuous media, thus giving an approximation for the problem.

The second part of the Cauchy tensor is more familiar, as it is similar to the one that dominates in multiphase fluid flows. Here, it will be called "dynamic Cauchy tensor", and Hutter and Rajagopal write "dynamic stress contribution". Its modeling will be discussed later by adapting it to granular media. In general, the flow velocities of granular media are low, and there is a certain temptation to neglect the term as it seems to vary with the square of the velocity. However, even when the flow of the medium is slow, the velocities, local and instant, may exhibit very sudden changes, but may, on the contrary, remain equal to zero for long periods of time. For example, if we assume that the velocity at a given point only takes values between 0 and 1, the latter value having a low probability, ε, of occurring, the average velocity is therefore ε, and not its square ε^2, and the average of the velocity squared is still ε, and the fluctuation squared $\varepsilon - \varepsilon^2$, therefore much higher than the mean velocity squared.

It is often said that experiments for granular media exhibit three different behaviors, depending on the shearing rate and the medium's mean density: a "quasi-static" regime (or "frictional"), a "dense liquid" regime and a "gas collision" regime. The volume ratio of the grains remains high for the first regime and decreases for the others. It is considered as difficult to build a model that covers all three regimes continuously; the reader may refer to [AND 11]. The equations given here are valid for the three regimes even if they still need to be closed for modeling. They show good potential for general modeling. It already becomes apparent that the Cauchy tensor for sliding contacts will dominate in the first regime, whereas the kinetic Cauchy tensor will do so in the gas collision regime. Both parts will influence the middle liquid regime to varying degrees.

13.4. Some already proposed methods

The models concerning "granular gases" will not be detailed here, we have given already some details in Chapters 6 and 12, even though they are very interesting (the reader may refer to [AND 11]). The problem related to modeling of the three regimes, relative to dry granular media, will be considered here.

According to Hutter and Rajagopal, the first model types, studied in 1985, defined a "quasi-static" and a "colliding" contribution in the medium's stress tensor by writing $\sigma_P^{\alpha\beta} = \sigma_S^{\alpha\beta} + \sigma_C^{\alpha\beta}$, thus already implicitly acknowledging the breakdown shown previously.

Acknowledging that the quasi-static part could not be of the viscous Newtonian type, but independent of the strains themselves, the following relationship was already proposed at that time [NOR 87]:

$$\sigma_S^{\alpha\beta} = -p_s \delta_{\alpha\beta} + a p_s \frac{S^{\alpha\beta}}{\|S\|}$$

where $S^{\alpha\beta}$ is the tensor deviator for the gradient of the medium's mean velocities, and $\|S\| = \sqrt{(\frac{1}{2} S^{\alpha\beta} S^{\beta\alpha})}$. In this way, the strain plays with its direction and not with its value. If the granular medium is considered as globally "incompressible", the divergence of the medium's mean velocity is zero, and this condition enables the pressure p_s to be found, which thus no longer needs to be prescribed by constitutive law. On the other hand, if the medium is considered to be "compressible", an additional law must be prescribed for pressure and the law $p_s = p_r / (\Phi_m - \bar{\Phi}_s)^n$, where p_r, Φ_m, n are constants, has been proposed, for example. The use of tensor $S^{\alpha\beta} / \|S\|$ seems to have been initiated by Goddard [GOD 86]. The factor ap_s represents the experimental occurrence of medium moving similarly to a solid that follows Coulomb's law, and a is linked to a "friction angle".

The colliding contribution $\sigma_C^{\alpha\beta}$ was identified as that implemented in granular gas models, already well developed at the time. With various hypotheses on the distribution function of two particles, they showed that it followed the relationship:

$$\sigma_C^{\alpha\beta} = -\rho_s f(\bar{\Phi}_s) \Theta \delta_{\alpha\beta} + 2\mu_{GG}(\bar{\Phi}_s, \Theta) S^{\alpha\beta}$$

where Θ is the "granular temperature" equal to 2/3 of the grain's shaking kinetic energy and satisfies a balance equation similar to k_p, defined in Chapter 6. As has already been mentioned in Chapter 6, section 6.2.3, it is important to note that the length scale involved in these equations (here implicitly in $\mu_{GG}(\bar{\Phi}_s, \Theta)$) is nothing other than the mean grain diameter, weighted by a function of the volume ratio when the distance between grains has to be taken into account.

More recently, an analogous model has been mentioned by Frenette *et al.* [FRE 02] to be used for flow computations. The medium, for this model, is assumed to be "incompressible" (the mean velocity field has zero divergence), and the average stress tensor is broken down into two separate contributions:

$$\sigma_S^{\alpha\beta} = -p_s \delta_{\alpha\beta} + (ap_s - b)\frac{S^{\alpha\beta}}{\|S\|}$$

The expression is similar, but a Druker–Prager criterion is used instead, with constant b depending on the possible cohesion between grains. This law can be adjusted to allow computations to be performed more easily at the start of the movement, only with numerical justification. The dynamic contribution is also simplified by approximating a balance equation for the granular temperature, thus explicitly providing this temperature in terms of the mean velocity gradients. This approximation is similar to the one called "mixing length", which has been discussed in Chapter 6. It is also often used for granular gases and corresponds to the formulation adopted by Bagnold to explain his well-known experimental results, giving the following:

$$\sigma_C^{\alpha\beta} = -\rho_s f_F(\bar{\Phi}_s)\|S\|^2 \delta_{\alpha\beta} + c\rho_s f_F(\bar{\Phi}_s)\|S\| S^{\alpha\beta}$$

where c is a constant. It can be seen that in this second contribution, in simple shear situations, it is obtained: $p_C^n = \rho_s f_F(\bar{\Phi}_s)\|S\|^2$, and $\tau_C = c\rho_s f_F(\bar{\Phi}_s)\|S\|^2$, and the constant c is thus linked to a "dynamic" friction coefficient, which is not the same as the first.

More recently, Jop *et al.* [JOP 06] proposed a model based on numerous experiments and computer simulations on sheared granular flows. They also assumed the medium to be "incompressible". Without dividing the stress tensor into two parts, they proposed:

$$\sigma^{\alpha\beta} = -p_s \delta_{\alpha\beta} + \eta\left(\frac{\|S\| d}{\sqrt{p_s / \rho_s}}\right) p_s \frac{S^{\alpha\beta}}{\|S\|}$$

The relationship is similar in form to that of the first models, but the function $\eta(\frac{\|S\|d}{\sqrt{p_s/\rho_s}})$ is based on experimental results (where d is the grain diameter), and uses the hypothesis that these results can lead to a law valid locally. It is (approximately) given by $\eta=\eta_1+\frac{\eta_2-\eta_1}{1+0,3\sqrt{p_s/\rho_s}/(\|S\|d)}$. As can be seen, this gives $\eta=\eta_1$ if $\frac{\|S\|d}{\sqrt{p_s/\rho_s}}$ is equal to zero, and $\eta=\eta_2$ if $\frac{\|S\|d}{\sqrt{p_s/\rho_s}}$ is infinite. It is therefore able to provide the same friction coefficients as the model by Frenette *et al.* for quasi-static regimes if $a=\eta_1$ and for colliding regimes if $c=\eta_2$.

In addition to the friction coefficients, the model by Jop *et al.* [JOP 06] proposes to use a general relationship between the grains' volume fraction and the "inertial" parameter $\frac{\|S\|d}{\sqrt{p_s/\rho_s}}$, identified by experiments in a number of situations. This relationship allows the volume ratio at each point in the medium to be found once its movement has been calculated, while maintaining incompressible modeling (in other words, neglecting the effects of the local variations of the medium's mean volume fraction). The relationship is simplified and given by the linear equation: $\overline{\Phi}_s=\Phi_{\max}-(\Phi_{\max}-\Phi_{\min})\frac{\|S\|d}{\sqrt{p_s/\rho_s}}$.

Although the model by Frenette involves the grains' volume fraction, it originally meant to prescribe it as data. The model by Jop *et al.* allows the volume fraction of the grains to be calculated thanks to the additional local relationship. The necessity of the relationship is due to the fact that the medium is assumed to be incompressible. The pressure p_s is therefore calculated directly by the zero velocity divergence condition, and it is not linked to the medium's mean density. On the contrary, this is the case for the older method quoted by Hutter and Rajagopal, in which the relationship $p_s=p_r/(\Phi_m-\overline{\Phi}_s)^n$ is used, allowing the grains' volume fraction and the pressure to be found locally at the same time.

The achievements and the deficiencies of these models are highlighted by Hutter and Rajagopal [HUT 94] and Andréotti *et al.* [AND 11] . The following chapters will discuss how the multiphase approach can explain some of these models and even provide an outlook for improvements and generalizations.

Chapter 14

Modeling of Cauchy Tensor of Sliding Contacts

14.1. Hypotheses and basic equations

This chapter will discuss how the tensor $\sum_{i=1,N} \overline{\Phi_i \sigma_i^{\alpha\beta}} = \overline{\sigma}_s$ can be calculated as a function of the average (or mean) variables previously defined, $\rho_s \overline{\Phi}_s, \tilde{v}_f^{\alpha}, \tilde{v}_s^{\alpha}$, and occasionally $\overline{p}_f^*$, taking into account the characteristics of the solid material or materials that constitute the grains. This is a core problem and probably has no exact solution as the variables $\rho_s \overline{\Phi}_s, \tilde{v}_f^{\alpha}, \tilde{v}_s^{\alpha}$, representing the composition and the motion of the media, only consider some of the characteristics of its motion and composition, and all the details are not known. It is only possible to find an approximate solution, nonetheless having a well-known validity range, or else the approximation would be problematic for practical applications. The range of validity has to be as wide as possible.

With this aim in mind, it is common practice to propose algebraic formulas that are as simple as possible, similar to formulas used for classical solids and fluids, and validated by some experiments.

By using the approach discussed in this book where the granular medium is nothing but an averaged instantaneous piecewise continuous medium, it will be seen that it is more natural to propose a differential constitutive to law. The article by Borghi and Bonnelli on this subject [BOR 07] will be referred to here.

Each grain is assumed to be homogeneous and isotropic and behaves elastically, the elastic strain remaining small, even if each grain can move and rotate. In this case, the velocity field in each grain *i* can be written as:

$$\vec{v}_i = \dot{\vec{c}}_i + Q_i.(\vec{x} - \vec{c}_i) + \vec{v}_i^e \qquad [14.1]$$

thus obtaining $\quad grad\vec{v}_i = Q_i + \frac{D^{Qi}}{dt}\varepsilon_i^e \qquad [14.2]$

where Q_i is the second-order antisymmetric tensor corresponding to the rotation of the grain, $\vec{c}_i$ is the position of the grain's center of gravity and $\dot{\vec{c}}_i$ is its total derivative, i.e. the displacement velocity of the grain's center. The point denotes the product contracted once, here the product of a second-order tensor and a vector. The tensor for small elastic strain $\boldsymbol{\varepsilon}_i^e$, which is symmetrical, satisfies the equation $\frac{D^{Qi}}{dt}\boldsymbol{\varepsilon}_i^e = \text{grad}\,\vec{v}_i^e$, where D^{Qi}/dt is the objective rotational derivative defined by:

$$\frac{D^{Qi}}{dt}(a_{\alpha\beta}) = \frac{\partial}{\partial t}\left(a_{\alpha\beta}\right) + v_i^\gamma \frac{\partial}{\partial x_\gamma}(a_{\alpha\beta}) + 2\left(a_{\alpha\gamma}Q_{\gamma\beta}\right)_{sym} \qquad [14.3]$$

In the next steps, it makes sense to consider the Kirchhoff stress tensor τ_ι^K, directly linked to the Cauchy stress tensor as $\boldsymbol{\tau}_i^K = \frac{\rho_i^0}{\rho_i}\boldsymbol{\sigma}_i$ (the density of the grains here can therefore vary relative to reference ρ_i^0, although it varies weakly if strains are small), and the linear law for elastic behavior is written as follows:

$$\frac{D^{Q_i}}{dt}\boldsymbol{\varepsilon}_i^{e\alpha\beta} = (\underset{\sim}{E}^{-1})^{\alpha\beta\gamma\delta}\frac{D^{Q_i}}{dt}\boldsymbol{\tau}_\iota^{\mathbf{K}\delta\gamma} \qquad [14.4]$$

where $\underset{\sim}{E}^{-1}$ is the tensor for elastic stiffness, of the fourth order:

$$\underset{\sim}{E}^{-1} = \frac{1+\nu}{E}\mathbf{I}\otimes\mathbf{I} - \frac{\nu}{E}\mathbf{I}\times\mathbf{I} \qquad [14.5]$$

E is the Young's modulus and ν is the Poisson coefficient of the grain material. Relationship [14.4] can be considered as a "general incremental elastic law" instead of the standard form $\boldsymbol{\varepsilon}_{i\alpha\beta}^e = (\underset{\sim}{E}^{-1})^{\alpha\beta\gamma\delta}\boldsymbol{\tau}_{\iota\delta\gamma}^{\mathbf{K}}$.

14.2. Unclosed balance equation for Cauchy tensor of sliding contact

Starting with the incremental law [14.4], incremental laws, first $\sum_{i=1,N} \Phi_i \tau_i^K$ and then $\bar{\sigma}_s^{\alpha\beta} = \sum_{i=1,N} \overline{\Phi_i \sigma_i^{\alpha\beta}}$, will be searched for.

It will be recalled that $grad \Phi_i = -\vec{n}_i \sigma_i^s$, taking a non zero value only at the surface of phase i (grain i), where $\vec{n}_i$ is normal to the interface separating phase i from others, directed toward the outside. Then, it can be written as:

$$\sum_{i=1,N+1} \Phi_i \, \mathrm{grad}\, \vec{v}_i = \mathrm{grad}\Big(\sum_{i=1,N+1} \Phi_i \vec{v}_i \Big) + \sum_{i=1,N+1} \sigma_i^s \vec{v}_i \otimes \vec{n}_i \qquad [14.6]$$

It is standard in fluid mechanics to assume that the fluid velocity and the solid's surface velocity are identical when the fluid and the grain are in contact. However, sliding or disconnection between two touching grains may occur, and a discontinuity may arise in the velocity there. For each grains couple in the direction (m,n) among all the couples K, the following notation will be used: $\vec{\Delta}_{mn} = \vec{v}_m - \vec{v}_n$, the difference (also called jump) in velocity on the contact surface between grains, and $\vec{n}_{mn}$ normal to grain m on the contact surface (n,m). Quantity σ_{nm}^s is σ_n^s restricted to the case where grain m is in contact with n (or vice versa). Of course, it is clear that $\vec{n}_{mn} = -\vec{n}_{nm}$. Therefore, only the contribution of contact surfaces between grains remains in the last term of [14.6], which can be rewritten as:

$$\sum_{i=1,N} \Phi_i \, \mathrm{grad}\, \vec{v}_i = \mathrm{grad}(\Phi_s \vec{v}_s) - \vec{v}_f \otimes \vec{n}_f \sigma_f^s - \sum_{(m,n)\in K} \sigma_{nm}^s \vec{\Delta}_{nm} \otimes \vec{n}_{mn} \qquad [14.7]$$

Therefore, by using relationship [14.2], the following is obtained:

$$\sum_{i=1,N} \Phi_i Q_i + \sum_{i=1,N} \Phi_i \frac{D^{Qi}}{dt} \varepsilon_i^e = \mathrm{grad}(\Phi_s \vec{v}_s) - \vec{v}_f \otimes \vec{n}_f \sigma_f^s - \sum_{(m,n)\in K} \sigma_{nm}^s \vec{\Delta}_{nm} \otimes \vec{n}_{mn}$$

It is known that $\frac{\partial}{\partial t}(\Phi_i) + v_i^\alpha \frac{\partial}{\partial x_\alpha}(\Phi_i) = 0$, in other terms $\frac{D^{Qi}}{dt}(\Phi_i) = \frac{\partial}{\partial t}(\Phi_i) + v_i^\alpha \frac{\partial}{\partial x_\alpha}(\Phi_i) = 0$, and a formula is obtained for the entire granular phase, analogous to [14.2]:

$$\mathrm{grad}(\Phi_s \vec{v}_s) = \sum_{i=1,N} \Phi_i Q_i + \sum_{i=1,N} \frac{D^{Qi}}{dt} \Phi_i \varepsilon_i^e + \vec{v}_f \otimes \vec{n}_f \sigma_f^s + \sum_{(m,n)\in K} \sigma_{nm}^s \vec{\Delta}_{nm} \otimes \vec{n}_{mn} \qquad [14.8]$$

It will be noted that $\sum_{i=1,N} \Phi_i Q_i = Q_s$, where Q_s is an antisymmetric tensor for grain rotations, and it can also be highlighted that $\vec{v}_f = \vec{v}_s$ at the surface of the grains.

$\text{grad}(\Phi_s \vec{v}_s) = \mathbf{D} + \mathbf{W}$ is split into $\mathbf{D} = (\text{grad}\,\Phi_s \vec{v}_s)_{\text{sym}}$ and $\mathbf{W} = (\text{grad}\,\Phi_s \vec{v}_s)_{asym}$, thus:

$$D = \sum_{i=1,N} \frac{D^{Qi}}{dt} \Phi_i \varepsilon_i^e + (\vec{v}_s \otimes \vec{n}_f \sigma_f^s)_{sym} + \sum_{(m,n)\in K} (\sigma_{nm}^s \vec{\Delta}_{nm} \otimes \vec{n}_{mn})_{sym} \qquad [14.9a]$$

$$W = Q_s + (\vec{v}_s \otimes \vec{n}_f \sigma_f^s)_{asym} + \sum_{(m,n)\in K} (\sigma_{nm}^s \vec{\Delta}_{nm} \otimes \vec{n}_{mn})_{asym} \qquad [14.9b]$$

The last terms in [14.9a and 14.9b] incorporate the effect of grain movements relative to their contact surface. These movements are caused by different stresses acting on the contact surface, and by the way in which the surface reacts according to its friction characteristics. Reversible and irreversible parts can be recognized in $\vec{\Delta}_{nm}$, $\vec{\Delta}_{nm} = \vec{\Delta}_{nm}^e + \vec{\Delta}_{nm}^p$, when the contact is represented using the approach given by Emeriault and Cambou [EME 96]. All components of the grains' Cauchy or Kirchhoff tensors, for the contact surface between two grains of identical composition, are discontinuous, although a well-defined value for $\boldsymbol{\tau}$ at contact can be defined as $\langle \boldsymbol{\tau} \rangle_{mn} = \frac{1}{2}(\boldsymbol{\tau}_m + \boldsymbol{\tau}_n)$. The same holds true for Q_s. Therefore, the contact force in all cases is $f_{mn}^{\alpha} = S_{mn} \langle \boldsymbol{\tau} \rangle_{mn}^{\alpha\beta} \cdot n_{mn}^{\beta}$ where S_{mn} is the contact surface, which is small in comparison to the total grain surface. The approach used by Emeriault and Cambou can be formally generalized:

$$\left(\vec{\Delta}_{mn}^e \otimes \vec{n}_{mn} \right)_{sym}^{\alpha\beta} = S_{mn} (\underset{\sim}{K}_{mn}^{-1})^{\alpha\beta\gamma\delta} \frac{D^{Q_{mn}}}{dt} \langle \boldsymbol{\tau} \rangle_{mn}^{\delta\gamma} \qquad [14.10]$$

The exact definition of K and the pertinent objective rotational derivative of the contact depend on the modeling of this contact given by Emeriault and Cambou. The details are presented in [EME 96], as well as in [BOR 07].

Starting from equations [14.9a] and [14.4], $\vec{\Delta}_{nm} = \vec{\Delta}_{nm}^e + \vec{\Delta}_{nm}^p$ is now introduced, and after a few calculations, the following is obtained:

$$\sum_{i=1,N} \frac{D^{Q_i}}{dt} \Phi_\iota \boldsymbol{\tau}_i^{K\alpha\beta} = (\underset{\sim}{c}^e)^{\alpha\beta\gamma\delta} \left[\mathbf{D} - \sigma_f^s (\vec{v}_s \otimes \vec{n}_f)_{sym} - \sum_{(m,n)\in K} \sigma_{nm} (\vec{\Delta}_{mn}^p \otimes \vec{n}_{mn})_{sym} \right]^{\delta\gamma}$$

$$-2\left(\sum_{i=1,N} \Phi_\iota \boldsymbol{\tau}_i^{K\alpha\gamma} \boldsymbol{\omega}_c^{\gamma\beta} \right)_{sym} \qquad [14.11]$$

where

$$\underset{\sim}{c}^e = \left[\underset{\sim}{E}^{-1} + \sum_{(m,n)\in K} \sigma_{nm} S_{mn} \underset{\sim}{K}_{mn}^{-1} \right]^{-1} \qquad [14.12]$$

$$\boldsymbol{\omega}_c = \sum_{(m,n)\in K} |n_{nm}| (\mathbf{Q}_{mn}^c - \langle \mathbf{Q} \rangle_{mn}) \qquad [14.13]$$

Equation [14.11] must then be translated in terms of the Cauchy stress tensor. Using $\boldsymbol{\sigma}_i = \frac{\rho_i}{\rho_i^0} \boldsymbol{\tau}_i^K$, it can be written that:

$$\frac{d_i}{dt} \boldsymbol{\sigma}_i = \frac{\rho_i}{\rho_i^0} \frac{d_i}{dt} \tau_i^K + \frac{\tau_i^K}{\rho_i^0} \left(\frac{\partial \rho_i}{\partial t} + v_i^\alpha \frac{\partial \rho_i}{\partial x_\alpha} \right) = \frac{\rho_i}{\rho_i^0} \frac{d_i}{dt} \tau_i^K - \frac{\tau_i^K}{\rho_i^0} \rho_i \frac{\partial v_i^\alpha}{\partial x_\alpha} \qquad \text{therefore}$$

$\frac{d_i}{dt} \Phi_i \tau_i^K = \frac{\rho_i^0}{\rho_i} \frac{d_i}{dt} \Phi_i \boldsymbol{\sigma}_i + \Phi_i \tau_i^K \frac{\partial v_i^\alpha}{\partial x_\alpha}$ and, because grain deformations are small, $\rho_i \approx \rho_i^0$, $\frac{d_i}{dt} \Phi_i \tau_i^K = \frac{d_i}{dt} \Phi_i \boldsymbol{\sigma}_i + \Phi_i \tau_i^K \frac{\partial v_i^\alpha}{\partial x_\alpha}$, $\sum_{i=1,N} \frac{d_i}{dt} \Phi_i \tau_i^K = \sum_{i=1,N} \frac{d_i}{dt} \Phi_i \boldsymbol{\sigma}_i + \sum_{i=1,N} \Phi_i \tau_i^K \frac{\partial v_i^\alpha}{\partial x_\alpha}$.

It is therefore obtained that:

$$\sum_{i=1,N} \frac{D^{Qi}}{dt} \Phi_i \tau_i^K = \sum_{i=1,N} \frac{d_i}{dt} \Phi_i \boldsymbol{\sigma}_i + \sum_{i=1,N} \Phi_i \tau_i^K \frac{\partial v_i^\alpha}{\partial x_\alpha} + 2\left(\sum_{i=1,N} \Phi_i \tau_i^K . Q_i\right)_{sym}$$

or

$$\sum_{i=1,N} \frac{D^{Qi}}{dt} \Phi_i \tau_i^K - 2(\tau_s^K . Q_s)_{sym} = \frac{\partial}{\partial t} \boldsymbol{\sigma}_s + v_s^\alpha \frac{\partial}{\partial x_\alpha} \boldsymbol{\sigma}_s + \sum_{i=1,N} \Phi_i \tau_i^K \frac{\partial v_i^\alpha}{\partial x_\alpha} \qquad [14.14]$$

with $\tau_s^K = \sum_{i=1,N} \Phi_i \tau_i^K$, $\sigma_s = \sum_{i=1,N} \Phi_i \sigma_i$, $v_s^\alpha = \sum_{i=1,N} \Phi_i v_i^\alpha$, in other words, the quantities indexed with *s* are those indexed with *i* in grain *i*, no matter which grain is being considered. The last term in [14.14] can also be written as

$\sum_{i=1,N} \Phi_i \tau_i^K \frac{\partial v_i^\alpha}{\partial x_\alpha} = \sum_{i=1,N} \Phi_i \frac{\rho_i^0}{\rho_i} \sigma_i \frac{\partial v_i^\alpha}{\partial x_\alpha} \approx \sigma_s \frac{\partial v_s^\alpha}{\partial x_\alpha}$ because each grain has a density that is nearly constant. Therefore:

$$\frac{\partial \boldsymbol{\sigma}_s^{\alpha\beta}}{\partial t} + \frac{\partial}{\partial x_\gamma}(\boldsymbol{\sigma}_s^{\alpha\beta} v_s^\gamma) = (\underset{\sim}{c}^e)^{\alpha\beta\gamma\delta} \left[\mathbf{D} - \sigma_f^s (\vec{v}_s \otimes \vec{n}_f)_{sym} - \sum_{mn} \sigma_{nm} (\vec{\Delta}_{nm}^p \otimes \vec{n}_{mn})_{sym} \right]^{\delta\gamma}$$
$$-2\left(\sigma_s^{\alpha\gamma} \cdot \boldsymbol{\omega}_c^{\gamma\beta}\right)_{\text{sym}} - 2(\sigma_s^{\alpha\gamma} . \mathbf{Q}_s^{\gamma\beta})_{sym} \qquad [14.15]$$

This equation is well defined for any point in space even though it is relative to the granular phase, with each term being zero in the fluid. The statistical average can be taken, and a balance equation is then found for the mean Cauchy stress tensor in the granular medium:

$$\frac{\partial}{\partial t}\bar{\boldsymbol{\sigma}}_s + \frac{\partial}{\partial x_\gamma}(\bar{\boldsymbol{\sigma}}_s \tilde{v}_s^\gamma) = -\frac{\partial}{\partial x_\gamma}\overline{(\boldsymbol{\sigma}_s v_s^{\gamma}{}')} + \bar{\underset{\sim}{c}}^e : \left(\bar{\Phi}_s (grad\vec{\tilde{v}}_s)_{sym} - \bar{\mathbf{D}}^p - \bar{\mathbf{D}}^f\right)$$
$$\underbrace{-2\left(\overline{\sigma_s \cdot \boldsymbol{\omega}_c}\right)_{\text{sym}}}_{\text{rotation of contact normal}} \underbrace{-2\left(\overline{\sigma_s \cdot \mathbf{Q}_s{}'}\right)_{\text{sym}}}_{\text{fluctuation of grain rotation}} \underbrace{-2\left(\bar{\sigma}_s \cdot \bar{\underset{\sim}{Q}}_s\right)_{\text{sym}}}_{\text{Jaumann spin (objectivity)}} \qquad [14.16]$$

where $\bar{D}^p = \overline{\sum_{mn} \sigma_{nm} (\vec{\Delta}_{mn}^p \otimes \vec{n}_{mn})_{sym}}$ [14.17]

and $\bar{D}^f = \overline{\sigma_f^s (\vec{v}\,'_s \otimes \vec{n}_f)_{sym}}$ [14.18]

This equation is interesting to discuss and has the classical appearance of balance equations. It should also be noted that it cannot be simply written as the total Lagrangian derivative with respect to the mean velocity because $\bar{\sigma}_s$ is a second-order tensor, and an objective derivative is in fact involved. It is also established that the objective rotational derivative, Jaumann's derivative, appears naturally, as shown in the last term. The mean rotation of the medium can be found using equation [14.9b], giving:

$$\overline{\underset{\sim}{Q}_s} = grad(\bar{\Phi}_s \vec{\tilde{v}}_s)_{asym} - \overline{(\vec{v}_s \otimes \vec{n}_f \sigma_f^s)}_{asym} - \overline{\sum_{(m,n)\in K} (\sigma_{nm}^s \vec{\Delta}_{nm} \otimes \vec{n}_{mn})}_{asym} \qquad [14.19]$$

According to Cambou [CAM 98], the first term in [14.19] is the most important term.

In this equation, production and destruction terms appear, as well as a term for "turbulent dispersion", due to velocity variations between grains. This is the first term in the right-hand side part. The left factor in the second term is found to be the mean elastic stiffness tensor of the medium, taking into account both grain and contact stiffness, according to the approach by Emeriault and Cambou. This tensor is formally defined, although its definition does not enable it to be easily calculated without more complex calculations that use additional hypotheses. For more details, the reader may refer to [EME 96]. As a matter of fact, simply its dependency on $\bar{\Phi}_s$ is of interest, and cannot unfortunately be found theoretically. The correlation between fluctuations of stiffness and fluctuations of the right-hand side multiplying factor has here been neglected. The first term of the right-hand side factor comes from:

$$\bar{D} - \overline{\sigma_f^s(\vec{v}_s \otimes \vec{n}_f)_{sym}} = grad(\bar{\Phi}_s \vec{\bar{v}}_s)_{sym} - (\vec{\bar{v}}_s \otimes grad\bar{\Phi}_s)_{sym} = \bar{\Phi}_s grad(\vec{\bar{v}}_s)_{sym}$$

The second term $\bar{D}^p$ is caused by slips or disconnections between grains, and the third $\bar{D}^f$ takes into account dilation and contraction effects, or variations in the volume of the fluid between grains. However, if the grains consist of the same material that cannot be compressed much, the density of each grain will be very similar and $\tilde{v}_s^\alpha \approx \bar{v}_s^\alpha$.

If it is assumed that the granular medium is statistically isotropic, a simple first hypothesis, a more accurate calculation of the tensor for elastic stiffness will enable it to be calculated simply as a function of two scalars: an actual compression modulus and an actual shearing modulus, χ^e, G^e, both functions of $\bar{\Phi}_s$, at least. Equation [14.16] can then be rewritten in terms of the trace and the deviator of the actual Cauchy tensor. The following is thus obtained:

$$\frac{\partial}{\partial t} tr\bar{\sigma}_s + \frac{\partial}{\partial x_\alpha}(tr(\bar{\sigma}_s)\tilde{v}_s^\alpha) = -\frac{\partial}{\partial x_\alpha}\overline{(tr(\sigma_s)v'^\alpha_s)} + 3\chi^e\bar{\Phi}_s\frac{\partial \tilde{v}_s^\alpha}{\partial x_\alpha} - tr(\bar{c}^e : (\bar{D}^p + \bar{D}^f)) \qquad [14.20]$$

$$\frac{\partial \bar{\sigma}_{sd}^{\alpha\beta}}{\partial t} + \frac{\partial}{\partial x_\gamma}(\bar{\sigma}_{sd}^{\alpha\beta}\tilde{v}_s^\gamma) = -\frac{\partial}{\partial x_\gamma}\overline{(\sigma_{sd}^{\alpha\beta}v'^\gamma_s)} + 2G^e\bar{\Phi}_s\left(\frac{\partial \bar{v}_s^\beta}{\partial x_\alpha}\right)^d_{sym} - (\bar{c}^e : (\bar{D}^p + \bar{D}^f))^{d\alpha\beta}$$
$$-2(\bar{\sigma}_{sd}^{\alpha\gamma}.\bar{\mathcal{Q}}_s^{\gamma\beta})_{skew} \qquad [14.21]$$

It will be recalled that the mean Cauchy tensor for the grains can be written in such a way to make the mean intrinsic tensor appear: $\sum_{i=1,N} \overline{\Phi_i \sigma_i^{\alpha\beta}} = \bar{\sigma}_s = \bar{\Phi}_s \bar{\sigma}_s^*$. It is no longer per unit volume of the medium, but per unit volume of the grains only. The balance equations for the trace and deviator of the mean intrinsic tensor are:

$$\bar{\Phi}_s \frac{d}{dt} tr\bar{\sigma}^*_s = -\frac{\partial}{\partial x_\alpha} (\overline{tr(\sigma^*_s)\Phi_s v'^{\alpha}_s}) + 3\chi^e \bar{\Phi}_s \frac{\partial \tilde{v}^\alpha_s}{\partial x_\alpha} - tr(\bar{c}^e : (\bar{D}^p + \bar{D}^f)) \qquad [14.22]$$

$$\bar{\Phi}_s \frac{d}{dt} \bar{\sigma}^{*\alpha\beta}_{sd} = -\frac{\partial}{\partial x_\gamma} (\overline{\sigma^{*\alpha\beta}_{sd} \Phi_s v'^{\gamma}_s}) + 2G^e \bar{\Phi}_s (\frac{\partial \tilde{v}^\beta_s}{\partial x_\alpha})^d_{sym} - (\bar{c}^e : (\bar{D}^p + \bar{D}^f))^{d\alpha\beta}$$
$$-2\bar{\Phi}_s (\bar{\sigma}^{*\alpha\gamma}_{sd} . \bar{Q}^{\gamma\beta}_s)_{skew} \qquad [14.23]$$

It is noted that the structure of equation [14.23] is similar to that corresponding to the one used in Maxwell's viscoelastic behavior model, for classical materials with a constant volume. This is written as:

$$\frac{\partial}{\partial t} \bar{\sigma}^{\alpha\beta}_d + \frac{\partial}{\partial x_\gamma} (v_\gamma \bar{\sigma}^{\alpha\beta}_d) = +2G^e (\frac{\partial v_\alpha}{\partial x_\beta} + \frac{\partial v_\beta}{\partial x_\alpha}) - \frac{2G^e}{\eta} \bar{\sigma}^{\alpha\beta}_d$$

In the context of extended irreversible thermodynamics (see [JOU 01, Chapter 2]), this equation has been shown to be in any case in agreement with the second thermodynamic principle, using a linear approximation, and extends Newton's standard viscous model.

In this case, the two equations mentioned above can no longer be used as they are, because the terms including irreversible effects, $\bar{D}^p$, $\bar{D}^f$, are not yet known in terms of the mean variables involved in modeling. The theoretical approach presented here has simply identified them by their physical meaning. Numerical experiments, using computer simulation by discrete elements (e.g. see [CAM 01]), could enable the value of these terms to be estimated, for certain reference experimental conditions, which could be used to infer their algebraic expressions. For example, Kruyt [KRU 03] investigates a two-dimensional granular medium subjected to quasi-static biaxial deformation, and [MID 04] investigates "shearing cells" in which a stable flow is either obtained or not, depending on the magnitude of the shearing stress applied to the boundaries. However, it is first necessary to have thoughts on the shape of such laws to obtain explicit relationships as a function of the mean variables defined above. This problem will be discussed in the next section, using "Extended Irreversible Thermodynamics". Hereafter, it will be noted that $\overline{I\Phi_s} = tr(\bar{c}^e : (\bar{D}^p + \bar{D}^f))$ and $\overline{I^{\alpha\beta}_d}\bar{\Phi}_s = (\bar{c}^e : (\bar{D}^p + \bar{D}^f))^{d\alpha\beta}$.

Before tackling this problem, it is noted that if these irreversible effects are negligible, as well as the velocity fluctuations, equation [14.21] becomes: $\frac{d}{dt} tr\bar{\sigma}^*_{sr} = 3\chi^e \frac{\partial \tilde{v}^\alpha_s}{\partial x_\alpha}$. Starting with equation [13.11], it is found that

$\frac{d}{dt}(\bar{\rho}) = -\bar{\rho}\frac{\partial \tilde{v}^{\alpha}}{\partial x_{\alpha}} \approx -\bar{\rho}\frac{\partial \tilde{v}_{s}^{\alpha}}{\partial x_{\alpha}}$, and thus $\frac{d}{dt}tr\bar{\sigma}_{sr}^{*} = -\frac{3\chi^{e}}{\bar{\rho}}\frac{d\bar{\rho}}{dt}$. The actual compression modulus is not necessarily a constant, but is at least a function of $\bar{\Phi}_{s}$, therefore $\bar{\rho}$ is a function of the compression modulus from $\bar{\rho} = \rho_{s}\bar{\Phi}_{s}$. If it is a function of this variable alone, there is therefore a direct relationship, an algebraic dependency, between the pressure and the density, as for a fluid:

$$tr(\bar{\sigma}_{sr}^{*}) - cste = -\int_{\rho_0}^{\bar{\rho}} 3\frac{\chi^{e}(\rho')}{\rho'}d\rho' \qquad [14.24a]$$

The value of ρ_0 from which one integrates is not very well known, and neither is the constant that is the trace magnitude when the mean density is ρ_0. The density corresponding to the volume fraction of the grains equal to unity cannot be taken as ρ_0, because this can only be reached if the stresses are infinite and the grains strongly deformed. The value corresponding to the limit of elastic behavior for the side that is less dense must be taken, and this has been done. This limit will be defined as $-tr(\bar{\sigma}_{s}^{*}) = 3\pi_{c} > 0$ later in the discussion and will therefore set the value of the constant, but not the value of the limit density that still needs to be found via experimentation. The grains' volume fraction at this limit may be considered to be roughly at "random loose packing" level (given by $\bar{\Phi}_{\min} \approx 0.55$ for equal-sized, spherical grains), corresponding to the grain size distribution of the medium of interest. In Chapter 13, it was seen that the law $\bar{\pi}_{s}^{*} = -tr(\bar{\sigma}_{sr}^{*})/3 = \bar{\pi}_{s}^{*}(\bar{\Phi}_{s})$ was used as a quasi-static contribution to the normal stress in a sheared granular flow. It did not reach zero for minimum volume fractions, but decreased significantly with the volume fraction, and was very large for a maximum volume fraction of 0.63.

In these reversible conditions, equation [14.23] gives:

$$\frac{d}{dt}\bar{\sigma}_{srd}^{*\alpha\beta} = 2G^{e}\left(\frac{\partial \bar{v}_{s}^{\beta}}{\partial x_{\alpha}}\right)_{sym}^{d} - 2(\bar{\sigma}_{srd}^{*\alpha\gamma}.\bar{\Omega}_{s}^{\gamma\beta})_{skew} \qquad [14.24b]$$

This is the incremental version, with large strains, of the following standard relationship for an elastic solid, where u^{β} is the displacement vector: $\bar{\sigma}_{srd}^{*\alpha\beta} = 2G^{e}\left(\frac{\partial u^{\beta}}{\partial x_{\alpha}}\right)_{sym}^{d}$

14.3. Closure approximations for irreversible terms

As mentioned previously, equations [14.22] and [14.23] cannot be used as they are. On the one hand, the terms for turbulent dispersion, $\overline{tr(\sigma_s^*)\Phi_s v'^{\alpha}_s}$ and $\overline{\sigma_{sd}^{*\alpha\beta}\Phi_s v'^{\gamma}_s}$, must be modeled. On the other hand, the $\overline{I}$ terms need to be modeled with respect to the mean variables $\overline{\sigma}_s, \tilde{v}_s, \overline{\Phi}_s$, and their gradients, or else with respect to other variables that can be calculated locally via balance equations, depending on what they are. Let us recall that $\overline{Q}_s^{\alpha\beta}$ can be approximated by the first term of [14.9]. The thermodynamics of irreversible processes can provide pointers to model such terms by following the approach that has already been used in Chapter 9.

It is proposed here to describe granular media with nine variables $\tilde{e}, \overline{\rho}, \overline{\pi}_s^* = -tr(\overline{\sigma}_s^{*\alpha\beta})/3, \overline{\sigma}_{sd}^{*\alpha\beta}, \tilde{v}^{\alpha}$, for which balance equations are available even if some terms still need to be "closed" by adequate approximations. A quasi-static granular medium will be considered here, and the kinetic Cauchy stress tensor is therefore negligible. In reality, it would be possible, and strictly preferable, to treat this question completely with the interaction of both parts of the Cauchy tensor. However, this first approach has its definite uses.

In these conditions, entropy is written as $\varsigma = \varsigma(\tilde{e}, 1/\overline{\rho}, \overline{\pi}_s^*, \overline{\sigma}_{sd}^{*\alpha\beta})$, and the Euler-Gibbs relationship is given in the form $d\varsigma = \frac{1}{T}d\tilde{e} + \frac{p}{T}d(\frac{1}{\overline{\rho}}) - \frac{y_p}{T}d\overline{\pi}_s^* - \frac{y_d^{\alpha\beta}}{T}d\overline{\sigma}_{sd}^{*\alpha\beta}$.

The quantities $T, p, y_p, y_d^{\alpha\beta}$ are functions (zero-order homogeneous) of variables mentioned above, which are the equations of state for the medium.

The balance equation for entropy can also be found using the variables' balance equations, i.e. [13.10–13.12] and [14.22] and [14.23]. After having performed several standard calculations, the following is found:

$$\overline{\rho}\frac{d}{dt}\varsigma = -\frac{\partial}{\partial x_\gamma}(\frac{\overline{\varphi}_{Qt}^{\gamma}}{T}) + \overline{\varphi}_{Qt}^{\gamma}\frac{\partial}{\partial x_\gamma}(\frac{1}{T}) + \frac{\overline{\rho}}{T}\frac{\overline{\sigma}_{sd}^{*\alpha\beta}}{\rho_s}(\frac{\partial \tilde{v}^{\beta}}{\partial x_\alpha})_{sym}^{d} - \frac{\overline{\rho}}{T}\frac{\overline{\pi}_s^*}{\rho_s}\frac{\partial \tilde{v}^{\alpha}}{\partial x_\alpha} - \frac{1}{T}(\overline{\Phi}_f\overline{p}_f^* - p)\frac{\partial \tilde{v}^{\alpha}}{\partial x_\alpha}$$

$$+\frac{\partial}{\partial x_\alpha}(\frac{y_p}{T}\overline{\pi_s^*\Phi_s v'^{\alpha}_s}) - \overline{\pi_s^*\Phi_s v'^{\alpha}_s}\frac{\partial}{\partial x_\alpha}(\frac{y_p}{T}) - \frac{y_p}{T}(-\chi^e\overline{\rho}\frac{\partial \tilde{v}^{\alpha}}{\partial x_\alpha} + \frac{\overline{\rho}}{3}\overline{I})$$

$$+\frac{\partial}{\partial x_\gamma}(\frac{y_d^{\alpha\beta}}{T}\overline{\sigma_{sd}^{*\alpha\beta}\Phi_s v'^{\gamma}_s}) - \overline{\sigma_{sd}^{*\alpha\beta}\Phi_s v'^{\gamma}_s}\frac{\partial}{\partial x_\gamma}(\frac{y_d^{\alpha\beta}}{T}) - \frac{y_d^{\alpha\beta}}{T}(2G^e\overline{\rho}(\frac{\partial \overline{v}^{\beta}}{\partial x_\alpha})_{sym}^{d} - \overline{\rho}\overline{I}_d^{\alpha\beta}) \quad [14.25]$$

The terms linked to reversible processes must disappear from the equation. This involves five relationships that prescribe five partial derivatives for entropy, thus providing five of the six equations of state required:

$$p = \overline{\Phi}_f \overline{p}_f^*,\ 2G^e y_d^{\alpha\beta} = \frac{\overline{\sigma}_{sd}^{*\alpha\beta}}{\rho_s},\ \chi^e y_p = \frac{\overline{\pi}_s^*}{\rho_s}$$

Maxwell's relationships then give the shape of the sixth equation of state:

$$\tilde{e} = \frac{(\overline{\pi}_s^*)^2}{6\chi^e \rho_s} + \frac{\overline{\sigma}_{sd}^{*\alpha\beta}\overline{\sigma}_{sd}^{*\beta\alpha}}{4G^e \rho_s} + fct(T)$$

Naturally, it is found that $\dfrac{\partial(p)}{\partial\overline{\pi}_s^*} = -\dfrac{\partial(y_p)}{\partial(1/\overline{\rho})} = 0$, $\dfrac{\partial(p)}{\partial\overline{\sigma}_{sd}^{*\alpha\beta}} = -\dfrac{\partial(y_d^{\alpha\beta})}{\partial(1/\overline{\rho})} = 0$, $\dfrac{\partial(y_p)}{\partial(\overline{\sigma}_{sd}^{*\alpha\beta})} = \dfrac{\partial(y_d^{\alpha\beta})}{\partial(\overline{\pi}_s^*)} = 0$.

The entropy production rate in [14.24a] is therefore written as:

$$\overline{\rho}P_\varsigma = \overline{\varphi}_Q^\gamma \frac{\partial}{\partial x_\gamma}\left(\frac{1}{T}\right) - \overline{\pi_s^*\Phi_s v'^\alpha_s}\frac{\partial}{\partial x_\alpha}\left(\frac{y_p}{T}\right) - \overline{\sigma_{sd}^{*\alpha\beta}\Phi_s v'^\gamma_s}\frac{\partial}{\partial x_\gamma}\left(\frac{y_d^{\alpha\beta}}{T}\right) - \frac{\overline{\rho\pi_s^*I}}{3\rho_s T\chi^e} + \frac{\overline{\rho\sigma}_{sd}^{*\alpha\beta}\overline{I}_d^{\alpha\beta}}{2\rho_s TG^e}$$

According to Curie's theorem, each separate term must be positive to satisfy the second principle. In the linear context, without aiming to create couplings, this means that the total heat flux can be represented by a Fourier law, although this is not the main interest here. Moreover, this justifies the fluxes to be expressed by gradient laws, with positive dispersion coefficients D_{es}, D_{esd}:

$$-\overline{\pi_s^*\Phi_s v'^\alpha_s} = \rho_s D_{es}\frac{\partial}{\partial x_\alpha}\left(\frac{y_p}{T}\right) = D_{es}\frac{\partial}{\partial x_\alpha}\left(\frac{\overline{\pi}_s^*}{\chi^e T}\right) \text{ and } -\overline{\sigma_{sd}^{*\alpha\beta}\Phi_s v'^\gamma_s} = D_{esd}\frac{\partial}{\partial x_\gamma}\left(\frac{\overline{\sigma}_{sd}^{*\alpha\beta}}{2G^e T}\right) \quad [14.26]$$

Dispersion is caused by fluctuations of the grain velocities, and the dispersion coefficients must therefore be linked to them. They can be found via a kinetic tensor $-\sum_{i=1,N} \overline{\rho_i \Phi_i v''^\alpha_i v''^\beta_i}$, which will be studied in the following chapter. The coefficients will therefore be mentioned again there.

Still in the context of linearity, the irreversible terms in the Cauchy stress tensor for sliding contacts must satisfy:

$$\overline{I} = -\frac{\chi^e}{\eta_v}\overline{\pi}_s^*,\ \overline{I}_d^{\alpha\beta} = \frac{G^e}{\eta}\overline{\boldsymbol{\sigma}}_{sd}^{*\alpha\beta}$$

Coefficients η, η_v are positive phenomenological coefficients for viscosity. A model is thus found similar to the Maxwell model, for a viscoelastic, yet dilatable and compressible, material.

Highly nonlinear modeling, with a "threshold", is also compatible with the second principle. It may be more realistic for some granular media for which flow thresholds have been observed.

For example, by noting H as the Heaviside step function, it can be proposed:

$$\overline{I} = -\frac{3\chi^e}{\alpha_v} H(\pi_c - \overline{\pi}_s^*) \frac{\left|\partial \overline{v}^\alpha / \partial x_\alpha\right|}{\pi_c} \overline{\pi}_s^* \qquad [14.27a]$$

$$\overline{I}_d^{\alpha\beta} = \frac{2G^e}{\alpha} H(\left\|\overline{\sigma}_{s\alpha\beta}^{*d}\right\| - (\tau_c + \left|\overline{\pi}_s^*\right| \sin\phi)) \frac{\left\|(\partial \overline{v}^\beta / \partial x_\alpha)_{sym}^d\right\|}{(\tau_c + \left|\overline{\pi}_s^*\right| \sin\phi)} \overline{\sigma}_{s\alpha\beta}^{*d} \qquad [14.27b]$$

The physical meaning of the threshold involved in [14.27a] is that the medium behaves identically to an elastic solid when it is compressed (then $\overline{\pi}_s^*$ is positive) above the positive value π_c, whereas if the medium is compressed less, or expanded, the normal contact forces weaken and the contact pressure then decreases exponentially because of the term $\overline{I}$.

With [14.27b], the medium remains near-elastic prior to the limit defined by the critical tension τ_c and the angle ϕ , corresponding to a Drucker–Prager criterion for plastic materials. The existence of such a limit demonstrates a "solid-fluid" transition and is also suggested by results from discrete elements computer simulation. For example, [MID 04] mentions a threshold T = 0.1 N, which then gives $\sin\phi = 0.1$, $\tau_c = 0$. The appearance of the solid fraction in $\overline{I}, \overline{I}_d^{\alpha\beta}$ is expected as these terms are, by definition, proportional to the mean interface area between grains per unit volume, a quantity that increases with the grains' volume fraction, but for which the linear dependency is probably a very rough approximation. Quantities α and α_v are therefore dimensionless, constant for first approximations, or functions of the volume fraction, to be specified by experiments as τ_c and ϕ.

If we consider equation [14.23] alongside approximation [14.27b], for conditions in which the last two terms predominate and counterbalance each other (when $\left\|(\partial \overline{v}^\beta / \partial x_\alpha)_{sym}^d\right\|$ is very big), then [14.23] is simplified to an algebraic form, therefore local, and gives:

$$\overline{\boldsymbol{\sigma}}_{sd}^{*\alpha\beta} = \alpha(\tau_c + \sin\phi\left|\overline{\pi}_s^*\right|) \frac{(\partial \overline{v}^\beta / \partial x_\alpha)_{sym}^d}{\left\|(\partial \overline{v}^\beta / \partial x_\alpha)_{sym}^d\right\|} \qquad [14.28a]$$

Similarly, for [14.22] and [14.27a], a simple relationship is obtained beyond the limit and when expansion $\left|\partial\overline{v}^{\alpha}/\partial x_{\alpha}\right|$ is very large:

$$\overline{\pi}_s^* = -\alpha_v \pi_c \frac{(\partial\overline{v}^{\alpha}/\partial x_{\alpha})}{\left|\partial\overline{v}^{\alpha}/\partial x_{\alpha}\right|} = \alpha_v \pi_c \qquad [14.28b]$$

On the one hand, the critical value π_c must surely depend on the pressure in the fluid between the grains when the granular medium loses nearly all contacts. If no grains are touching, the interface balance relation at grain level must provide $\overline{\pi}_s^* = \overline{p}_f$, therefore π_c must be larger than $\overline{p}_f$. It can also be written that $\pi_c = \pi_c' + \overline{p}_f$, and this would provide a limit linked to the "effective" pressure, in the same way as Terzaghi. This also implies that the plasticity limit depends on $\overline{p}_f$, as it involves $\overline{\pi}_s^*$. On the other hand, the pressure in the grains when they are all separated is that of the surrounding fluid plus the mean effect of the collisions with other grains. This would imply that the value of α_v must depend on the kinetic energy of grain fluctuations and always be larger than one.

In Chapter 13, a relationship similar to [14.28a] was proposed as early as 1986 by Goddard [GOD 86]. The existence of a flow threshold for granular media is sometimes accompanied by hysteresis phenomena (see the article by GDR MIDI, [MID 04]). It is thought that the use of complete differential equations, similar to [14.23] with closure [14.27b], will enable this phenomenon to be better represented, whereas the algebraic version [14.28a] is obviously only valid once it has been decided that a threshold is known and that only certain terms in the equations dominate. The use of a function more continuous than the Heaviside step function, although still stiff, may prove to be better suited, possibly depending on grain shape.

However, no matter which modeling is chosen, it becomes apparent that reversible and irreversible parts can be defined for stress tensors: $\overline{\pi}_s^* = \overline{\pi}_e^* + \overline{\pi}_i^*, \overline{\sigma}_{sd}^{*\alpha\beta} = \overline{\sigma}_{sed}^{*\alpha\beta} + \overline{\sigma}_{sid}^{*\alpha\beta}$. Equations [14.24a] and [14.24b] give the elastic parts and the total is given by differential equations [14.22] and [14.23] with closures [14.27a] and [14.27b].

Chapter 15

Modeling the Kinetic Cauchy Stress Tensor

In this chapter, we will discuss the contribution of $-\sum_{i=1,N} \overline{\rho_i \Phi_i v''^{\alpha}_i \, v''^{\beta}_i}$, caused by (sudden) fluctuations in grain velocities, relative to their mean velocity. For the flows of granular media, this term is, indeed, equivalent to the Reynolds tensor for turbulent fluid flows, and the approach taken for multiphase media, previously outlined, enables us to propose models similar to classic turbulence models. If the influence of the interstitial fluid is negligible, then the tensor is nearly identical to the overall Reynolds tensor, defined by fluctuations in the grain or fluid velocity at a given point, and is involved in equation [13.11]: $-\rho \, \overline{v'^{\alpha} \, v'^{\beta}}$. This modeling will now be studied.

Experiments on turbulence models first suggest a model of the "Prandtl mixing length" type, which has already proved to be efficient in Chapter 6 for an overall representation of turbulent two-phase flows. In section 15.2, modeling using a balance equation for the shaking kinetic energy of grains will be discussed, and it will be noted that it generalizes already proposed "granular gas"-type models. For both these approaches, the emphasis will be put on the assessment of the characteristic length scale. Finally, general modeling of granular flows will be discussed, jointly using the two Cauchy stress tensor (CST) contributions: the one caused by sliding contacts and the kinetic one. There is no reason why these two contributions should be uncoupled and coupling them will have to be considered. Questions related to the wall conditions will then also be identified.

15.1. Prandtl–Bagnold modeling

In Chapter 6, we discussed the overall modeling of a turbulent two-phase flow using the Prandtl mixing length (PML) theory. Discussions in Chapter 9 also showed that this theory agreed with the linear thermodynamics of irreversible processes. Modeling the kinetic stress tensor for a granular medium can use the same bases.

If the effects of the interstitial fluid are negligible (equation [13.11] is therefore used), the results found can be used directly and, in particular, it can be proposed:

$$-\overline{\rho v'^{\alpha} v'^{\beta}} + \bar{\rho}\frac{2}{3}\delta^{\alpha\beta}k = \bar{\rho}\nu_e(\frac{\partial \tilde{v}^{\alpha}}{\partial x_{\beta}} + \frac{\partial \tilde{v}^{\beta}}{\partial x_{\alpha}} - \frac{2}{3}\delta^{\alpha\beta}\frac{\partial \tilde{v}^{\gamma}}{\partial x_{\gamma}}) \qquad [15.1]$$

where coefficient ν_e is the "effective viscosity coefficient", which is positive, and k is the kinetic energy of the fluctuating grains and fluid motions (mainly the movement of grains when the fluid is air or gas), defined by $\bar{\rho}k = \frac{1}{2}\overline{\rho v'^{\alpha} v'^{\alpha}}$, to be discussed in the next section, which is linked to "granular temperature".

The velocity fluctuations are mainly a result of the mean velocity gradients in the medium, whereas the kinetic energy of fluctuating grain motions can be approximated by:

$$k = C_B l_M^2 (\frac{\partial \tilde{v}^{\alpha}}{\partial x_{\beta}} + \frac{\partial \tilde{v}^{\beta}}{\partial x_{\alpha}})(\frac{\partial \tilde{v}^{\alpha}}{\partial x_{\beta}} + \frac{\partial \tilde{v}^{\beta}}{\partial x_{\alpha}}) \qquad [15.2]$$

where C_B is a positive constant, the value of C_B must be provided by experiments and is ideally as general as possible. The effective viscosity coefficient can then be estimated by a formula dependent on the "mixing length" l_M, and it is written as follows:

$$\nu_e = l_M^2 ((\frac{\partial \tilde{v}^{\alpha}}{\partial x_{\beta}} + \frac{\partial \tilde{v}^{\beta}}{\partial x_{\alpha}})(\frac{\partial \tilde{v}^{\alpha}}{\partial x_{\beta}} + \frac{\partial \tilde{v}^{\beta}}{\partial x_{\alpha}}))^{1/2} \qquad [15.3]$$

This length l_M is characteristic of the correlation length of fluctuating motions. It has a minimum value equal to the mean grain size, although larger if the grains move in a correlated way. Prandtl theory for turbulence does not provide this, but states that it is firstly affected by the flow geometry, exclusively so in the case of turbulence with a high Reynolds number.

In 1954, Bagnold [BAG 54] used the same formulas (simplified for the case of a simple sheared flow) to describe his experiments as seen in Chapter 6. However, he considered this mixing length to be directly linked to the grain diameter *d*, increasing significantly if the volume fraction of the grains increased. According to his measurements, $l_M = 0{,}16\lambda d$, where $\lambda = ((\bar{\Phi}_{Max} / \bar{\Phi}_p)^{1/3} - 1)^{-1}$, which is in direct agreement with granular gas theory, also mentioned in Chapter 6. However, these experiments were recently subject to debate (see [HUN 02]), but this does not affect the identification of the influence of volume fraction, nor the involvement of grain diameters. Moreover, even if the experiments did not precisely exhibit the dependence of tangential and normal stresses on the square of the transverse velocity gradient, the granular gas theory does show this dependency, as well as the PML theory above.

The way in which the length scale l_M may be specified for two-phase flows in general was discussed in Chapter 6, and emphasis was put on the need to take into account the joint influence of the flow's geometry and a high volume fraction of grains, therefore capable of reconciling Bagnold and Prandtl. By transposing to make the grain diameter appear directly, the following can be proposed:

$$l_M = c \max\{\lambda d F_G(a\Delta/(\lambda d)), b y_p + d\} \qquad [15.4]$$

Quantity Δ is the "thickness of the velocity gradient layer", which is equal to half the space between cylinders in Bagnold's case. Length l_M is also limited by the distance to the nearest wall y_p. For highly diluted situations, *a* is 0.09, *b* is 0.435, $F_G = a\Delta/(\lambda d)$ and *c* is 1, although they are still unknown for denser two-phase flows.

For the case of granular media discussed here, having a high mass fraction of particles, the influence of Δ decreases and even disappears, and the influence of the correcting function F_G must therefore be weak. However, the influence of the presence of walls has to be maintained. The choice of this correlation is now the subject of an increasing number of debates, for example [MID 04] . This article provides numerous experimental results for a certain number of simple situations, as well as results from computer simulations using discreet elements, without macroscopic modeling, which can be considered as "numerical experiments". In particular, the experiments show the importance of this correlation or mixing length. The comparison between modeling results, which take into account a relationship similar to equation [15.4], with this wall influence, and the numerous and detailed results from numerical experiments, will determine to what extent the mixing length theory is valid, and will specify a function F_G and coefficients *a*, *b* and *c*.

15.2. K-l_t or "turbulent granular gas" modeling

Following this example of turbulence models, we can expect that a model can be found in which the shaking kinetic energy of grains (the mean motion being subtracted) is calculated by a balance equation and will be more widely applicable, to cases where fluctuations are not entirely defined by the local mean velocity gradient.

Indeed, the shaking kinetic energy in the grains is linked to what is called "granular temperature" in "granular gas" theories, studying cases where the behavior of the granular medium is solely controlled by shaking and collisions between grains, therefore where sliding contacts are negligible. This temperature is defined as $\Theta = 2k_s / 3$.

The general method developed in this book enables a balance equation to be found for k_s, and is not restricted to the regime dominated by short collisions and grain shaking. This equation was already discussed in Chapter 6, for either an overall two-phase medium, or one of the medium's phases.

Starting with equation [6.19], the mass transfer between phases is eliminated as it does not exist here, and the relationship becomes:

$$\frac{\partial}{\partial t}(\rho_s \bar{\Phi}_s k_s) + \frac{\partial}{\partial x_\alpha}(\rho_s \bar{\Phi}_s k_s \tilde{v}_s^\alpha) = \frac{\partial}{\partial x_\alpha}(-\overline{\rho_s \Phi_s v'^\beta_s v'^\beta_s v'^\alpha / 2} + \overline{\Phi_s v'^\beta_s \sigma_s^{\alpha\beta}}) - \overline{\rho_s \Phi_s v'^\alpha_s v'^\beta_s} \frac{\partial \tilde{v}_s^\beta}{\partial x_\alpha}$$

$$+ \overline{p_s \Phi_s \frac{\partial v'^\alpha_s}{\partial x_\alpha}} - \rho_s \bar{\Phi}_s \varepsilon_s + \overline{\sigma_s^{\alpha\beta} n_s^\beta v'^\alpha_s \sigma_p^s}$$

This can be applied to the solid phase of a granular medium as long as the CST $\sigma_s^{\alpha\beta}$ for the grain phase is identified to the one for sliding contacts, and pressure p_s to the opposite third of the trace for the elastic part of the sliding contact tensor. The term $-\sum_{i=1,N} \overline{\rho_i \Phi_i v''^\alpha_i v''^\beta_i}$ is renamed $-\overline{\rho_s \Phi_s v'^\alpha_s v'^\beta_s}$ for simplicity. The irreversible part of the tensor $\sigma_s^{\alpha\beta}$ must also be recognized in order to identify the term for dissipation of kinetic energy (with ε_s). As has been done above, a realistic simplification of the forces acting on the fluid phase, contained in the last term, is simply achieved by assimilating them to the interstitial fluid pressure. If the instant CST for sliding contact, the tensor in equation [14.15] in Chapter 14, is written as $\sigma_s^{\alpha\beta} = \sigma_{se}^{\alpha\beta} + \sigma_{si}^{\alpha\beta}$, then $\pi_s = \pi_e + \pi_i$ and $\sigma_{sd}^{\alpha\beta} = \sigma_{sed}^{\alpha\beta} + \sigma_{sid}^{\alpha\beta}$, and the following is obtained:

$$\frac{\partial}{\partial t}(\bar{\rho}k_s)+\frac{\partial}{\partial x_\alpha}(\bar{\rho}\tilde{v}_s^\alpha k_s)=\frac{\partial}{\partial x_\alpha}(\overline{v'^\beta_s\,\Phi_s\sigma_s^{\alpha\beta}}-\frac{1}{2}\overline{\rho_s\Phi_s v'^\alpha_s\, v'^\beta_s\, v'^\beta_s})$$

$$-\overline{\rho_s\Phi_s v'^\alpha_s\, v'^\beta_s}\frac{\partial\tilde{v}_s^\beta}{\partial x_\alpha}+\overline{\pi_e\Phi_s\frac{\partial v'^\alpha_s}{\partial x_\alpha}}-\overline{\sigma_{sed}^{\alpha\beta}\Phi_s\frac{\partial v'^\beta_s}{\partial x_\alpha}}-\bar{\rho}\varepsilon_s+\overline{p_f n_f^\alpha v'^\alpha_s\,\sigma_s} \qquad [15.5]$$

The first term on the right-hand side is a standard dispersion term due to fluctuations in the grains' velocities and the second term is a production term due to the mean velocity gradient in the solid phase, almost equal to the medium's mean velocity gradient, also standard. Standard modeling [15.1] leads to:

$$-\overline{\rho_s\Phi_s v'^\alpha_s\, v'^\beta_s}+\frac{2}{3}\bar{\rho}k_s\delta_{\alpha\beta}=\mu_{es}(\frac{\partial\tilde{v}_s^\alpha}{\partial x_\beta}+\frac{\partial\tilde{v}_s^\beta}{\partial x_\alpha}-\frac{2}{3}\frac{\partial\tilde{v}_s^\gamma}{\partial x_\gamma}\delta_{\alpha\beta}) \text{ (Boussinesq relationship)}$$

and also $\overline{v'^\beta_s\,\Phi_s\sigma_s^{\alpha\beta}}-\frac{1}{2}\overline{\rho_s\Phi_s v'^\alpha_s\, v'^\beta_s\, v'^\beta_s}=\frac{\mu_{es}}{Sc_k}\frac{\partial k_s}{\partial x_\alpha}$, with $\mu_{es}=\rho_s\bar{\Phi}_s C_{\mu s}k_s^{1/2}l_t$ [15.6a]

The other terms are responsible for reversible transfers of k_s with the kinetic energy of the mean motion, because of the elasticity of the solid medium (elasticity of grains and contacts), of transfers to thermal energy, thus "dissipation" of kinetic energy, and of transfers of k_s to the fluid phase. They are not all in their closed form as they still depend on fluctuations. However, this problem is resolved easily for the two elasticity terms of the medium with the hypothesis that the grain density is constant. It can, therefore, be written as:

$$\overline{\pi_e\Phi_s\frac{\partial v'^\alpha_s}{\partial x_\alpha}}=\overline{\pi_e\Phi_s\frac{\partial \tilde{v}^\alpha_s}{\partial x_\alpha}}+\overline{\pi_e\Phi_s\frac{\partial v^\alpha_s}{\partial x_\alpha}}=\bar{\pi}_e^*\bar{\Phi}_s\frac{\partial\tilde{v}_s^\alpha}{\partial x_\alpha},$$

and similarly $-\overline{\sigma_{sed}^{\alpha\beta}\Phi_s\frac{\partial v'^\beta_s}{\partial x_\alpha}}= -\bar{\sigma}_{sed}^{*\alpha\beta}\bar{\Phi}_s\frac{\partial\tilde{v}_s^\beta}{\partial x_\alpha}$. Modeling the term for exchange with the fluid phase can be achieved by referring to the discussions in Chapters 6 and 9. More simply, it can be written as $\overline{p_f n_f^\alpha v'^\alpha_s\,\sigma_s}=\overline{p_f n_f^\alpha v^\alpha_s\sigma_s}-\overline{p_f n_f^\alpha\sigma_s}\tilde{v}_s^\alpha$. The fluctuations of the fluid pressure are neglected to give $\overline{p_f n_f^\alpha v'^\alpha_s\,\sigma_s}=\bar{p}_f^*\frac{\partial\bar{\Phi}_f}{\partial x_\alpha}(\tilde{v}_s^\alpha-\bar{v}_{sf}^\alpha)$, where $\bar{v}_{sf}^\alpha$ is the mean velocity of the interfaces.

Also by assuming $\bar{v}_{sf}^\alpha=a\tilde{v}_s^\alpha+(1-a)\tilde{v}_f^\alpha$, $\overline{p_f n_f^\alpha v'^\alpha_s\,\sigma_s}=\bar{p}_f^*\frac{\partial\bar{\Phi}_f}{\partial x_\alpha}(1-a)(\tilde{v}_s^\alpha-\tilde{v}_f^\alpha)$

Finally, modeling of the dissipation rate can be done by bringing in the integral scale l_t:

$$\overline{\rho\varepsilon}_s = \rho_s \overline{\Phi}_s C_{Ds} k_s^{3/2} / l_t \qquad [15.6b]$$

When the fluid density is very small in comparison to that of the grains, the kinetic energy of grains' fluctuations is almost equal to the kinetic energy of the fluctuations in the entire medium, and, in this case, its equation can also be inferred from the equation for the entire medium, without involving a term for the exchange with the fluid phase. Equation [6.5] is thus used as a starting point:

$$\frac{\partial}{\partial t}(\bar{\rho}k) + \frac{\partial}{\partial x_\alpha}(\bar{\rho}k\tilde{v}^\alpha) = \frac{\partial}{\partial x_\alpha}(-\overline{\rho v'^\beta v'^\beta v'^\alpha}/2 - \overline{p v'^\alpha} + \overline{v'^\beta \tau^{\alpha\beta}}) - \overline{\rho v'^\alpha v'^\beta}\frac{\partial \tilde{v}^\beta}{\partial x_\alpha}$$
$$+\overline{p\frac{\partial v'^\alpha}{\partial x_\alpha}} - \overline{\tau^{\alpha\beta}\frac{\partial v'^\beta}{\partial x_\alpha}}$$

where $p = \Phi_f p_f + \Phi_s \pi_s$ and $\tau^{\alpha\beta} = \Phi_f \tau_f^{\alpha\beta} + \Phi_s \sigma_{sd}^{\alpha\beta}$.

Therefore:

$$\frac{\partial}{\partial t}(\bar{\rho}k) + \frac{\partial}{\partial x_\alpha}(\bar{\rho}k\tilde{v}^\alpha) = \frac{\partial}{\partial x_\alpha}(-\overline{\rho v'^\beta v'^\beta v'^\alpha/2} + \overline{\Phi_s \sigma_s^{\alpha\beta} v'^\alpha} - \overline{\Phi_f p_f v'^\alpha} + \overline{v'^\beta \Phi_f \tau_f^{\alpha\beta}}) - \overline{\rho v'^\alpha v'^\beta}\frac{\partial \tilde{v}^\beta}{\partial x_\alpha}$$
$$+\overline{\Phi_f p_f \frac{\partial v'^\alpha}{\partial x_\alpha}} + \overline{\Phi_s \pi_e \frac{\partial v'^\alpha}{\partial x_\alpha}} - \overline{\Phi_s \sigma_{sed}^{\alpha\beta}\frac{\partial v'^\beta}{\partial x_\alpha}} - \overline{\Phi_s \sigma_{si}^{\alpha\beta}\frac{\partial v'^\beta}{\partial x_\alpha}} - \overline{\Phi_f \tau_f^{\alpha\beta}\frac{\partial v'^\beta}{\partial x_\alpha}}$$

In this equation, k can be identified to k_s, and velocity $\tilde{v}$ can be identified to $\tilde{v}_s$, and the terms involving τ_f can be neglected in comparison to their analogs relative to the solid phase. However, the terms involving the fluid pressure must be kept. The equation is, therefore, rewritten as:

$$\frac{\partial}{\partial t}(\bar{\rho}k_s) + \frac{\partial}{\partial x_\alpha}(\bar{\rho}k_s\tilde{v}_s^\alpha) = \frac{\partial}{\partial x_\alpha}(-\overline{\rho v_s'^\beta v_s'^\beta v_s'^\alpha/2} + \overline{\Phi_s \sigma_s^{\alpha\beta} v_s'^\alpha}) - \overline{\rho v_s'^\alpha v_s'^\beta}\frac{\partial \tilde{v}_s^\beta}{\partial x_\alpha}$$
$$-\overline{v_f'^\alpha p_f \frac{\partial \Phi_f}{\partial x_\alpha}} - \overline{v_f'^\alpha \Phi_f \frac{\partial p_f}{\partial x_\alpha}} + \overline{\Phi_s \pi_e \frac{\partial v_s'^\alpha}{\partial x_\alpha}} - \overline{\Phi_s \sigma_{sed}^{\alpha\beta}\frac{\partial v_s'^\beta}{\partial x_\alpha}} - \overline{\Phi_s \sigma_{si}^{\alpha\beta}\frac{\partial v_s'^\beta}{\partial x_\alpha}} \qquad [15.7]$$

Comparison with [15.5] shows that they have the same form, term by term. Of course, the dissipation rate here is given by $\rho\varepsilon_s = \overline{\Phi_s \sigma_{si}^{\alpha\beta}\frac{\partial v_s'^\beta}{\partial x_\alpha}}$, and the effect of the fluid pressure is expressed in a slightly different way, although it is equivalent in this case.

It seems that the standard relationship between the kinetic energy of fluctuations and the mean velocity gradient, used in [15.3] for the Prandtl–Bagnold model, cannot correspond to a local equilibrium between production and dissipation of kinetic energy, unless the terms $+\pi_e \Phi_s \overline{\frac{\partial v'^{\alpha}_s}{\partial x_\alpha}} - \overline{\sigma^{\alpha\beta}_{ed} \Phi_s \frac{\partial v'^{\beta}_s}{\partial x_\alpha}} + \overline{\bar{p}^*_f n^{\alpha}_f v'^{\alpha}_s \sigma_s}$ are negligible in [15.5], equally for [15.7]. These terms are admittedly not crucial for balance equations, and are probably decreasing in importance, when the medium is increasingly shaken, but the role they play needs to be studied in more detail.

Second, it is interesting to compare this equation with the one used for granular temperature in "granular gas theory". Granular gas theory is an extension of the kinetic theory of gases that gives such an equation, of which there are several slightly different versions (the reader may refer to [AND 11]). For example, the equation used by [LUN 84] is written as:

$$\bar{\rho}\frac{\partial}{\partial t}(\frac{3}{2}\Theta) + \bar{\rho}\tilde{v}^{\alpha}_s \frac{\partial}{\partial x_\alpha}(\frac{3}{2}\Theta) = \frac{\partial}{\partial x_\alpha}(\bar{\rho}K\frac{\partial \Theta}{\partial x_\alpha} + \bar{\rho}D\frac{\partial \bar{\rho}}{\partial x_\alpha}) + \Sigma_{\alpha\beta}\frac{\partial \tilde{v}^{\beta}_s}{\partial x_\alpha} - \bar{\rho}\Theta\Gamma \qquad [15.8]$$

Tensor $\Sigma_{\alpha\beta}$ is the kinetic Cauchy stress tensor, $\Sigma_{\alpha\beta} = -\sum_{i=1,N} \overline{\rho_i \Phi_i v'^{\alpha}_i v'^{\beta}_i}$, and the theory expresses the following constitutive relationship:

$$\Sigma_{\alpha\beta} = -(\bar{\rho}\Theta(1 + 4\frac{1+e}{2}\bar{\Phi}_s g_0(\bar{\Phi}_s))\delta_{\alpha\beta} + \bar{\rho}\eta(\frac{\partial \tilde{v}^{\beta}_s}{\partial x_\alpha} + \frac{\partial \tilde{v}^{\alpha}_s}{\partial x_\beta}) - \bar{\rho}\eta_v \delta_{\alpha\beta}\frac{\partial \tilde{v}^{\gamma}_s}{\partial x_\gamma}) \qquad [15.9]$$

The first, diagonal, contribution in this tensor is simply proportional to the granular temperature, as for a gas, although here it is a non-ideal gas (e is the coefficient for collision restitution). The other two contributions are linked to the mean velocity gradient and involve coefficients for effective viscosity η, η_v, the theory of which relates to variables $T_G, d, \bar{\Phi}_s$. Coefficients K, D in equation [15.8] are diffusion coefficients, which are also proportional to η, η_v. The last term in [15.8] is a term for dissipation caused by non-elastic grain collisions, and theory provides: $\Gamma = \frac{12}{\sqrt{\pi}}(1-e^2)\bar{\Phi}_s g_0(\bar{\Phi}_s)\frac{\Theta^{1/2}}{d}$. Function g_0 is already mentioned in Chapter 6, $g_0 = 1/(1-\frac{\bar{\Phi}_s}{\bar{\Phi}_{Max}})^{5\bar{\Phi}_{Max}/2}$.

A strong analogy can be made between [15.8] and the $\bar{\rho}k_s$ equation here, [15.5]. The dispersion flux, the term for production due to the mean velocity gradient, and, finally, the dissipation rate are recovered. Equation [15.9] is similar in appearance to the Boussinesq relationship. The dissipation is conventionally related to $\frac{\Theta^{3/2}}{l_t}$, with the length scale $l_t = d/(\frac{12}{\sqrt{\pi}}(1-e^2)\bar{\Phi}_s g_0(\bar{\Phi}_s))$ corresponding to formula [15.6b]. Within the terms for diffusion fluxes, coefficients η, K, D also involve a characteristic length, which is proportional to *d*. However, it should be noted that the kinetic energy diffusion flux (also called pseudo-heat flux) is expressed in granular gas theory via two terms, one of which is proportional to the mass fraction gradient. Only the term proportional to the kinetic energy gradient has been considered in [15.5], but it is also possible to take into account the second term without creating any contradictions.

Equation [15.5], therefore, concurs with the basis provided by granular gas theory, without coinciding completely with the equation resulting from this theory. This is a generalization: on the one hand, the presence of interstitial fluid can be taken into account; on the other hand, the elastic properties of the grain phase intervene in ways different to how a gas would, and, finally, the effect of irreversible sliding contacts is not *a priori* neglected.

The length scale linked to the dissipation of "granular temperature", and the one linked to the momentum flux, as well as the "pseudo-heat flux", are both linked to the physical scale given by the grain size, although the weighting depends on the volume fraction, and is different for the two scales. Effectively, the length scale linked to the flux is related to a spatial correlation of the velocity fluctuations in grain movement; it is larger than the grain size and increases when the grains' volume fraction increases and it could be said that the grains "push" each other. The dissipation scale is linked to the distance between the grains because the dissipation occurs due to collisions between them. This distance decreases when the grains' volume fraction increases and it can become smaller than the grain size in dense media. It has never been considered for a granular gas that these length scales may be influenced by the flow geometry of the granular gas. The reason for this is that, in this theory, the size of the device containing the flow is always assumed to be very large in comparison to the particle size, and the flows considered are very simple. Even if a granular gas exhibits chaotic motions, it is not meant to be a flow subjected to saturated instabilities, which may bring about the appearance of length scales smaller than the device scale, and therefore competing with the ones produced by the grain size. Granular gas flows can be unstable, and this has been studied, but the saturated fluctuating state may lead to a different kind of granular flow that needs to be modeled differently.

Changing the focus to these granular flows, considered with saturated instabilities, what approach is used for their modeling? The approach discussed in this book shows that it is only necessary to link the equation already modeled for $\bar{\rho}k_s$ to an adequate formula for length l_t. Given all the previous discussions, the former mixing length formula is also suitable, with adaptation of the function F_G and the constants. Therefore:

$$l_t = C_l \max\{\lambda d F_G(a\Delta/(\lambda d)), by_p + d\} \qquad [15.10]$$

As before, the validity of this model needs to be tested. The comparison of modeling results with numerous, detailed results corresponding to the numerical experiments reported in [MID 04] will allow us to specify the extent to which the $\bar{\rho}k_s$ equation is capable of recovering or improving the mixing length approach, and specifying the numerical coefficients and functions involved.

15.3. Toward a general model for all regimes

Developing unified modeling, valid for all flow regimes of granular media, ranging from quasi-static to kinetic regimes, is directly possible in the approach presented here by adding both CST contributions, and by solving their equations simultaneously. Both equation modelings for the sliding-contact tensor and the kinetic tensor have been studied, using various models for the latter. The equations obtained will now be reviewed and their coupling will be discussed. The stress tensor for sliding contacts is given by equations [14.22] and [14.23], to be completed by closure hypotheses similar to [14.27a] and [14.27b], in addition to [14.26] for dispersion flux terms. The reversible part of the tensor is given by equations [14.24a] and [14.24b]. The kinetic stress tensor is defined by the algebraic relationship [15.1], the effective viscosity coefficient by [15.3] or else by [15.6], and the kinetic energy of velocity fluctuations is given by [15.5], modeled as specified, by using the length scale [15.10].

However, the two contributions in the overall CST are not completely decoupled. This is apparent from the presence of a dispersion coefficient in [14.26], D_{es}, D_{esd}. In the context of the current modeling, it now appears that they can only be proportional to the effective viscosity coefficient. They must therefore be defined as: $D_e = \frac{\mu_{es}}{\bar{\rho}S_{cs}}$ and $D_{es} = \frac{\mu_{es}}{\bar{\rho}S_{csd}}$. The Schmidt numbers S_{cs}, S_{csd} can be constant and must be found by experimentation, although they can also be functions of numbers, which represent the relative importance of the contributions from sliding contacts and kinetics. Likewise, the Reynolds number for turbulence can intervene in the C_μ coefficient for turbulence models. If shearing phenomena are the most important, the number $\left\|\bar{\sigma}_{sd}^{*\alpha\beta}\right\| / \rho_s k_s$ will be meaningful, but if dilation is also important, the number $\bar{\pi}_s^* / \rho_s k_s$ may possibly also interfere here, both playing a similar role as the Reynolds number in turbulence.

Similarly, coefficients $C_{\mu s}$, C_{Ds}, which occur in the equation for k_s, and C_l, or c, in the definition for the length scale, can also depend on these numbers. All this is yet to be studied and the comparisons between the results from calculations via the models, and detailed "measurements" provided by well-chosen experiments or direct numerical simulations, will enable these studies to be undertaken in the best possible conditions.

The number $\bar{\pi}_s^* / \bar{\rho} k_s$ was recently used in the model developed by Savage [SAV 98]. The dissipation rate for shaking kinetic energy is expressed as:

$$\varepsilon_s = \frac{\varphi(\bar{\Phi}_s)}{d} k_s^{3/2} \left(1 + \frac{C}{R(\bar{\Phi}_s)} \frac{\bar{\pi}_s^*}{\rho_s k_s}\right) \qquad [15.11]$$

resulting from a tentative physical model for a very dense granular medium in slow motion, where $\bar{\pi}_s^*$ is defined by an algebraic formula as a function of the grains' volume fraction, similar to a reversible model obtained in Chapter 14.

One possible important question is on the separate calculation of the three diagonal components in the kinetic Cauchy tensor $\overline{\Phi_s v'^x_s v'^x_s}$, $\overline{\Phi_s v'^y_s v'^y_s}$ and $\overline{\Phi_s v'^z_s v'^z_s}$, instead of their sum $2\bar{\Phi}_s k_s$. The tensor's anisotropy can indeed be directly influenced by the flow's anisotropy, and it is possible that its influence will not be cancelled so quickly by collisions between grains.

It is possible to have balance equations for all components of the tensor. To achieve this, the relationships obtained in Chapter 6, section 6.3.4, simply need to be transposed. Starting with [6.21], the following is obtained for $\bar{\rho}_s \bar{\Phi}_s R_s^{\beta\gamma} = \overline{\rho_s \Phi_s v'^\beta_s v'^\gamma_s}$:

$$\begin{aligned}
&\frac{\partial}{\partial t}(\rho_s \bar{\Phi}_s R_s^{\beta\gamma}) + \frac{\partial}{\partial x_\alpha}(\rho_s \bar{\Phi}_s \tilde{v}_s^\alpha R_s^{\beta\gamma}) = \\
&\frac{\partial}{\partial x_\alpha}\left(-\overline{\rho_s \Phi_s v'^\beta_s v'^\gamma_s v'^\alpha_s} - \overline{\pi_s \Phi_s v'^\beta_s}\delta^{\alpha\gamma} - \overline{\pi_s \Phi_s v'^\gamma_s}\delta^{\alpha\beta} + \overline{\Phi_s v'^\beta_s \sigma_{sd}^{\alpha\gamma}} + \overline{\Phi_s v'^\gamma_s \sigma_{sd}^{\alpha\beta}}\right) \\
&-\rho_s \bar{\Phi}_s \left(R_s^{\gamma\delta} \frac{\partial \tilde{v}_s^\beta}{\partial x_\delta} + R_s^{\beta\delta} \frac{\partial \tilde{v}_s^\gamma}{\partial x_\delta}\right) + \left(\overline{\Phi_s \pi_e \frac{\partial v'^\beta_s}{\partial x_\gamma}} + \overline{\Phi_s \pi_e \frac{\partial v'^\gamma_s}{\partial x_\beta}}\right) \\
&+\left(\overline{\Phi_s \sigma_{sed}^{\alpha\gamma} \frac{\partial v'^\beta_s}{\partial x_\alpha}} + \overline{\Phi_s \sigma_{sed}^{\alpha\beta} \frac{\partial v'^\gamma_s}{\partial x_\alpha}}\right) - \rho_s \bar{\Phi}_s \varepsilon_s^{\beta\gamma} + \left(\overline{\sigma_s^{\alpha\gamma} n_s^\alpha v'^\beta_s}^s + \overline{\sigma_s^{\alpha\beta} n_s^\alpha v'^\gamma_s}^s\right)\bar{\sigma}_i^s
\end{aligned} \qquad [15.12]$$

Similarly to the equation for $2\bar{\Phi}_s k_s$, the correlation terms need to be modeled. The first term on the right-hand side is a diffusion term that can be modeled using a gradient law. The terms for contact with the interstitial fluid phase can be simplified if the contact force is assumed to decrease to the mean fluid pressure, and a model for the mean velocity of interfaces can be brought in, as has been done for k_s. This leads to:

$$\begin{aligned}
&\frac{\partial}{\partial t}(\rho_s \bar{\Phi}_s R_s^{\beta\gamma}) + \frac{\partial}{\partial x_\alpha}(\rho_s \bar{\Phi}_s \tilde{v}_s^\alpha R_s^{\beta\gamma}) = \frac{\partial}{\partial x_\alpha}(\frac{\mu_{es}}{S_{sR}} \frac{\partial}{\partial x_\alpha} R_s^{\beta\gamma}) - \rho_s \bar{\Phi}_s (R_s^{\gamma\delta} \frac{\partial \tilde{v}_s^\beta}{\partial x_\delta} + R_s^{\beta\delta} \frac{\partial \tilde{v}_s^\gamma}{\partial x_\delta}) \\
&+ (\overline{\Phi_s \pi_e \frac{\partial v'^\beta_s}{\partial x_\gamma}} + \overline{\Phi_s \pi_e \frac{\partial v'^\gamma_s}{\partial x_\beta}}) + (\overline{\Phi_s \sigma_{sed}^{\alpha\gamma} \frac{\partial v'^\beta_s}{\partial x_\alpha}} + \overline{\Phi_s \sigma_{sed}^{\alpha\beta} \frac{\partial v'^\gamma_s}{\partial x_\alpha}}) - \rho_s \bar{\Phi}_s \varepsilon_s^{\beta\gamma} \\
&+ \bar{p}_f^* \frac{\partial \bar{\Phi}_f}{\partial x_\gamma}(1-a)(\tilde{v}_s^\beta - \tilde{v}_f^\beta) + \bar{p}_f^* \frac{\partial \bar{\Phi}_f}{\partial x_\beta}(1-a)(\tilde{v}_s^\gamma - \tilde{v}_f^\gamma)
\end{aligned} \quad [15.13]$$

The important terms left to model are the elastic terms and the dissipation term. For turbulent flow cases, experimentation shows that the pressure terms, analogous to the elastic terms here, bring about isotropization in the Reynolds tensor: the non-diagonal terms decrease and their contribution is carried over to the diagonal terms in the same way. Moreover, experimentation shows that the term for the dissipation rate tends toward a diagonal structure with a high Reynolds number, where $\varepsilon^{\alpha\beta} = \varepsilon\delta_{\alpha\beta}$. Are these properties also in operation for granular flows? This question needs to be studied via experimentation or by direct numerical simulations, and then it will be possible to establish a model. No matter what, this modeling still has to be completed by prescribing length scales. This can be done algebraically by taking a law similar to the one mentioned above for the mixing length approach, or by a balance equation for ε_s, following the example discussed in Chapter 6, in the general context of any multiphase flow.

15.4. Boundary conditions at walls

Choosing boundary conditions for a granular medium is also a delicate problem, as was mentioned at the end of Chapter 6. However, it is also in this case that the situation can be studied in more detail.

The balance relationship for momentum at an interface, written for a "median" impermeable wall (plotted in the middle of small-scale asperities on the wall), dictates that the force imposed by the medium on the wall is equal to and opposite to the force imposed by the wall on the medium. Therefore:

$$(\sigma_{tf}^{ij})_w N_w^j = F_w^i \quad [15.14]$$

where $(\sigma_{tf}^{ij})_w$ is the total CST for the mean medium, after modeling, N_w^j is the vector normal to the (median) wall and F_w^i is the force imposed by the wall on the fluid, which is either provided or else calculated, given the motion of the wall. The relationship [15.14] is still a boundary condition for the balance of momentum, no matter how the CST for the medium has been or will be modeled.

In granular media consisting of disorganized, moving solid grains, F_w^i is also balanced by the mean flux of momentum loss in grains, M_w^i, because of collisions with and friction against the wall, everything per unit area of the median wall. Therefore:

$$(\sigma_{tf}^{ij})_w N_w^j = M_w^i \qquad [15.15]$$

This relationship defines the link between macroscopic modeling and the microscopic structure (here at the same scale as grains and asperities) of both the medium and the wall.

It is possible to calculate all the quantities involved, for the modeling of $(\sigma_{tf}^{ij})_w$ and of M_w^i, with approximate, but realistic, approaches for granular gases and a wall. For example, all these quantities are calculated in [CAO 96] for spherical grains of a given size using a restitution coefficient for collisions and a friction coefficient, and a wall on which the asperities are semi-spheres with a given radius (to be compared with the radius of the particles in the medium), separated by plane areas of given dimension, in a perfectly regular manner. If interest is taken in the projection of M_w^i, tangential to the wall, a term appears which includes a possible slip velocity. This slip velocity at the wall, in the case of a uniform sheared flow, is related to the normal gradient of the mean tangential velocity of the flow, and this gives the boundary condition at the wall:

$$(\overline{u}_w^t - U_w) = a \left.\frac{\partial \overline{u}^t}{\partial n}\right|_w \qquad [15.16]$$

where U_w is the velocity of the wall, and coefficient a (with the same dimension as a length, here related to the mean grain and asperity radii) depends on the characteristics of the wall and of the granular gas. The value of a becomes zero when the asperity is large in comparison to the grain size, thus justifying an adherence condition at the median wall position. The velocity normal to the wall (and relative to the wall) is, of course, zero as no components of the medium pass through it. In the projection normal to the wall, the pressure (linked to the trace of σ_{tf}^{ij}) is found to be proportional to the kinetic energy of grain velocity fluctuations, and is already included in models for σ_{tf}^{ij}.

As far as the kinetic energy of velocity fluctuations is concerned, the mean balance condition at the wall must use the balance equation for the mean kinetic energy of fluctuations, studied in Chapter 6. Formally, it is written as follows:

$$(\varphi_k^n)_w = M_w^t(\overline{u}_w^t - U_w) + D_w \qquad [15.17]$$

which is interpreted as the diffusion flux of k toward this wall in this medium, as modeled, being equal to the energy dissipated when collisions occur on the wall, one part of which is the power of the mean kinetic energy loss in the mean motion and another part is the dissipation of the kinetic energy of fluctuations. The kinetic energy flux for velocity fluctuations depends on the model used for the multiphase medium, and magnitudes D_w and M_w^i, previously given, characterizing the interaction of particles with the wall. In the same paper, all these terms are accurately calculated for the same rough wall, and the following form is finally found:

$$c\left.\frac{\partial k}{\partial n}\right|_w = \frac{b}{k_w}(u_w^t - U_w)\left.\frac{\partial u^t}{\partial n}\right|_w - k_w \qquad [15.18]$$

where coefficients b and c are of the same dimension as a (a length), and they again depend on the wall and the granular gas, but not on the particle shaking that is included in k.

The question must be reconsidered in the case of the particulate phase of a multiphase flow that is not a granular gas, with non-spherical grains of various sizes, a high volume fraction of grains to allow lasting sliding contacts with the wall, and a more or less rough wall with irregular asperities. It is tempting to keep as boundary conditions equations [15.16] and [15.18], but by finding the coefficients empirically depending on the roughness of the wall relative to the grain size, and becoming nil at the limit case of a very rough wall. It would be best to reconsider the basic relationships [15.15] and [15.17], to keep the models for the wall effects in [CAO 96], presented in more detail (taking friction into account) and, on the contrary, to introduce a new model for σ_{tf}^{ij} and flux φ_k^n.

PART 4

Studying Fluctuations and Probability Densities

Part 4 concerns the improvement of the representation of the multiphase media by adding a better understanding of the fluctuations around mean values, by the introduction of probability density functions (PDF). This approach makes it possible to characterize any sort of fluctuation, which provides a huge improvement to the modeling. The fact that we have defined mean variables from the beginning does not actually mean that it is impossible to understand the fluctuations around them. On the contrary, by starting with the exact equations from Part 1, this improvement is allowed. Even if the PDFs are more complicated, it is possible to improve the modeling of thermal and chemical aspects, and to find less global models than those described in the chapter on fluidized beds, for example. The authors study, first of all, the modeling of the PDF of compositions for the gas phase of a two-phase medium in combustion, then discuss the possibility of better understanding the temperature fluctuations of the liquid phase or of drops or solid particles, which can have significant consequences on phase changes. Finally, in the final chapter, the authors study the possibility of better understanding the distribution of velocity fluctuations and parcel sizes of a phase. To develop a better understanding of the distribution of sizes of a phase, we must define different categories of parcels in this phase, and the treatment can thus become quite unwieldy. For situations in which the phases are always highly dispersed, the classically used stochastic Lagrangian-Eulerian simulation method is more appropriate. The authors show how the approach developed in this book is connected to this method.

Chapter 16

Fluctuations of the Gas Phase in Reactive Two-Phase Media

16.1. Specificities of reactive two-phase media

If we consider mist and "spray" flames, for example, the multiphase medium concerned mainly consists of droplets that are more or less dispersed in a composite gas phase. The latter consists of several species that mix and chemically react together, with at least one originating from droplet vaporization.

The species mass fractions in the gas phase are directly linked to the progression of chemical reactions, and the mean composition primarily needs to be known. Using the balance equations that were given in Chapter 4, section 4.4 (equation [4.11]), it can be found that:

$$\frac{\partial}{\partial t}\left(\bar{\rho}^*_{jg}\bar{\Phi}_g\right)+\frac{\partial}{\partial x_\alpha}(\bar{\rho}^*_{jg}\bar{\Phi}_g\tilde{v}^\alpha_g)=\frac{\partial}{\partial x_\alpha}(-\bar{\Phi}_g\overline{j^{Dj}_\alpha}^* - \overline{\rho_{jg}\Phi_g v'^\alpha_g}) - \overline{\rho_{jg}v^\alpha_{rjg}n^\alpha_g\sigma^s_{lg}} + \bar{\rho}^*_g\bar{\Phi}_g\tilde{w}^{ch}_j \quad [16.1]$$

$\bar{\rho}^*_{jg}$ is the mean mass of the gas species j contained in a unit volume of the gas phase. The mean mass fraction of species j in phase i, $\tilde{Y}_{ji}$, can be used such that $\overline{\rho_{ji}\Phi_i} = \bar{\rho}^*_{ji}\bar{\Phi}_i = \bar{\rho}^*_g\tilde{Y}_{ji}\bar{\Phi}_i$.

The presence of chemical reactions creates a very delicate problem. The calculation of the last term in equation [16.1], the mean reaction rate of species j, often strongly dependent on the mass fraction of the different species present, and on the temperature of the gas phase, requires us to know the fluctuations of these variables and not only their mean values. By looking closely at the detailed chemical and physical phenomena occurring at the scale of and between the droplets, it seems that the chemical reaction and the heat exchanges in gas media often enable small "flamelets", diffusion flamelets, premixing flamelets, triple flamelets, etc., to appear. These structures have already been discussed for turbulent flames (see [BOR 00, Chapter 9] and [REV 05]) and they must be taken into account even if an interest is only taken in the mean values. The way in which they can alter the vaporized flow rate in local exchanges has already been discussed in Chapter 5 (section 5.6.3). It is now necessary to find a way to obtain these fluctuations in order to better calculate the mean reaction rate in equation [16.1].

Another interesting and important question concerns the creation of soot in flames, typically relevant to reactive multiphase media. It is important to know whether soot is present for two reasons: first, it is a pollutant (in addition to inefficient combustion, although this effect is often very weak), and second, it plays an important role in heat transfer from the reactive medium through radiation. It is necessary to consider soot as one or several distinct phases, especially when soot particles agglomerate, while being oxidized by surrounding gases. Sufficiently small particle sizes often make it possible to assume that they have the same mean velocity and temperature as the gas phase, yet it is important that the size distribution must be known in order to predict the way in which they can be oxidized, as well as the temperature differences between particles to accurately predict their radiation. Temperature fluctuations in the gas phase or condensed phases will be discussed in Chapters 17 and 18. In this case, the determination of the particle size distribution depends directly on the ability to model particle growth and agglomeration phenomena. The complete study of flames producing soot is, nonetheless, one of the most difficult problems in the field of reactive multiphase media.

16.2. Probability density of composition fluctuations of the gas phase

The easiest problem to tackle for situations mentioned above is the knowledge of the fluctuation of the chemical composition of the gas phase where the combustion takes place. To calculate the mean reaction rate locally in the gas phase in the most general possible way, it is necessary to know the joint probability density function (PDF) of the fluctuations of the random variables corresponding to physical variables that are involved in this rate. Indeed, by definition:

$$\overline{\rho}_g^* \tilde{\dot{w}}_j^{ch} = \iint_{\hat{Y},\hat{T}_g} \rho_g \dot{w}_j^{ch}(\hat{Y},\hat{T}_g) P_g(\hat{Y},\hat{T}_g) d\hat{Y} d\hat{T}_g \qquad [16.2]$$

This PDF has been studied and widely used for flame modeling in the gas phase. $\hat{Y}$ is a vector of independent random variables corresponding to the mass fractions involved in the reaction rate of the gas phase (the number of its components is equal to the number of species present minus one, as they must be independent variables), and a pertinent reaction mechanism must be identified to find it. The PDF not only depends on random variables, but also on the location that is being considered.

To find how a variable fluctuates about its mean value, it was first considered to study the standard deviation of the fluctuations at any point in the mean continuous medium. This has been done, more or less, in particular, by studying the kinetic energy of fluctuations (either in the overall medium or in each phase) because the latter is directly linked to the standard deviation of velocity component fluctuations. The way in which a balance equation can be obtained for this quantity has been previously discussed and the method can be used again to study the standard deviation of temperature fluctuations and other quantities. However, experience shows that standard deviations do not provide enough information on fluctuations in the case of media where combustion occurs. The reaction rate has strong nonlinear properties, which explains that important errors can be obtained by knowing the standard deviation alone. This property is the one that has motivated studies directly interested in PDFs, and it is pertinent to do the same for the multiphase media that are of interest here.

Starting from the instantaneous equations for the medium, several techniques can provide an equation for the PDF's partial differental equation. The first attempt to obtain such an equation seems to have been undertaken by Frost in the 1960s [FRO 66]. Since then, the use of PDF's for turbulent combustion has been widely studied and significant progress has been made, as much so in Russia as in Western countries (for example [LIB 80], [KUZ 90] and [BOR 00]).

Two different techniques have been used to find the equation for the PDF. The first technique [KUZ 90] uses the "PDF characteristic function" and is possibly more elegant and safer. The reader may refer to [KUZ 90] for the definition of the function and the description of the technique. In the following, an attempt will be made to transpose the other technique, proposed by Lundgren, to the type of two-phase flow relevant here ([BOR 00] or [BOR 08b]). The reason for using this second technique is that the physical meaning of the "fine-grained PDF" used in the approach is deemed to be more direct.

Before going into more detail, it can already be said that the balance equation obtained for the PDF used in equation [16.2], which is relative to a single point in the medium, will not be closed directly. As for each equation for the mean values, including some terms involving fluctuations, in addition to means, the PDF equation will also involve fluctuations under forms other than the PDF concerned. Approximations for these "other forms" will therefore be needed, based on experimental results and theoretical considerations, in order for the equation to be used in practice. This method is strictly identical to the method applied to equations for mean values and for the kinetic energy of fluctuations. It is therefore simply the general version of the method already applied in Part 3 of this book.

16.2.1. *Instant basic equations of the gas medium*

It will be recalled that in Chapter 4, section 4.5.1, an instant balance equation was written to study the composition of a fluid phase that consisted of a mixture. The mass fraction of species j in phase i Y_{ji}, will be used here to characterize this mixture. The instantaneous balance equation for this quantity was established in Chapter 2. For species in the gas phase of the two-phase medium considered here, it is written as:

$$\frac{\partial}{\partial t}\left(\rho_g Y_{jg}\Phi_g\right)+\frac{\partial}{\partial x_\alpha}(\rho_g Y_{jg}\Phi_g v^\alpha)=\frac{\partial}{\partial x_\alpha}(-\Phi_g j_\alpha^{Dj})-\rho_{gj}v_{rjg}^\alpha n_g^\alpha\sigma_{lg}^s+ \rho_g\Phi_g\dot{w}_j^{ch} \quad [16.3]$$

To make things more clear, only one gas species is assumed here to originate from vaporization of the liquid phase: it is simply the vapor from the liquid parcels, here assumed to consist of a simple liquid. Starting with [16.3], the equation for species V, the vapor from the liquid phase, is written as:

$$\frac{\partial}{\partial t}\left(\rho_g Y_V\Phi_g\right)+\frac{\partial}{\partial x_\alpha}(\rho_g Y_V\Phi_g v^\alpha)=\frac{\partial}{\partial x_\alpha}(-\Phi_g j_\alpha^{DV})-\rho_V v_{rV}^\alpha n_g^\alpha\sigma_{lg}^s+\rho_g\Phi_g\dot{w}_V^{ch}$$

where the term for mass transfer by vaporization-condensation is the mass flow rate lost by the liquid phase, $-\rho_V v_{rV}^\alpha n_g^\alpha\sigma_{lg}^s=-\rho_l v_{rl}^\alpha n_g^\alpha\sigma_{lg}^s$. This will be detailed again further on.

Using the total mass balance in the gas phase, $\frac{\partial}{\partial t}\left(\rho_g\Phi_g\right)+\frac{\partial}{\partial x_\alpha}(\rho_g\Phi_g v^\alpha)=-\rho_l v_{rl}^\alpha n_g^\alpha\sigma_{lg}^s$, the mass balance equation for the liquid vapor can be rewritten as:

$$\rho_g\Phi_g\frac{\partial}{\partial t}(Y_V)+\rho_g\Phi_g v^\alpha\frac{\partial}{\partial x_\alpha}(Y_V)=\frac{\partial}{\partial x_\alpha}(-\Phi_g j_\alpha^{DV})-\rho_l v_{rl}^\alpha n_g^\alpha\sigma_{lg}^s(1-Y_V)+ \rho_g\Phi_g\dot{w}_V^{ch} \quad [16.4a]$$

It should be noted here that equation [16.4a] can also be written differently: on the one hand, the interface relationships for species V and for the total mass (refer to Chapter 4, section 4.5.1) are written as:

$$[Y_V \rho_g v_{rg}^{\alpha} + j_{\alpha}^{DV}] n_g^{\alpha} \sigma_{lg}^s = \rho_l v_{rl}^{\alpha} n_g^{\alpha} \sigma_{lg}^s \text{ and } \rho_g v_{rg}^{\alpha} n_g^{\alpha} \sigma_{lg}^s = \rho_l v_{rl}^{\alpha} n_g^{\alpha} \sigma_{lg}^s$$

Therefore $j_{\alpha}^{DV} n_g^{\alpha} \sigma_{lg}^s = \rho_l v_{rl}^{\alpha} n_g^{\alpha} \sigma_{lg}^s (1 - Y_V)$, and:

$$\frac{\partial}{\partial x_{\alpha}}(-\Phi_g j_{\alpha}^{DV}) = \Phi_g \frac{\partial}{\partial x_{\alpha}}(-j_{\alpha}^{DV}) - j_{\alpha}^{DV} \frac{\partial}{\partial x_{\alpha}}(\Phi_g) = \Phi_g \frac{\partial}{\partial x_{\alpha}}(-j_{\alpha}^{DV}) + j_{\alpha}^{DV} n_g^{\alpha} \sigma_{lg}^s$$

The following is, therefore, also found:

$$\rho_g \Phi_g \frac{\partial}{\partial t}(Y_V) + \rho_g \Phi_g v^{\alpha} \frac{\partial}{\partial x_{\alpha}}(Y_V) = \Phi_g \frac{\partial}{\partial x_{\alpha}}(-j_{\alpha}^{DV}) + \rho_g \Phi_g \dot{w}_V^{ch} \qquad [16.4b]$$

This equation could have been written directly, as it simply means that the classic balance equation for species V is valid in the gas phase but not in the liquid phase, nor at the interface between the two. It therefore becomes clear that the general differential balance equation [16.3] plus the balance equation at the interface also provide this balance equation, valid only in the gas.

Fick's law, a simple law valid for a binary mixture or a minor species in a mixture, will be used for the diffusion term in the gas phase: $j_{\alpha}^{DV} = -\rho_g D_V \dfrac{\partial Y_V}{\partial x_{\alpha}}$.

The latter term in [16.4b] is a term for chemical reactions in the gas phases, possibly equal to zero (if V is not reactive) and highly nonlinear otherwise. In principle, it can depend on many other species, as well as on the temperature: $\rho_g \dot{w}_V^{ch} = \rho_g \dot{w}_V^{ch}(Y_g, T_g)$, although it is hoped that only species V and one or two other species appear within the vector Y_g, once the involved chemical mechanism has been conveniently simplified.

The other species in the gas phase involved in chemical reactions also satisfy an equation which has the same shape. In the form of [16.3], this equation does not have a term for mass exchange between phases, but by using a total mass balance for the gas phase, the following is found for these species j that are different from V:

$$\rho_g \Phi_g \frac{\partial}{\partial t}(Y_{jg}) + \rho_g \Phi_g v^{\alpha} \frac{\partial}{\partial x_{\alpha}}(Y_{jg}) = \frac{\partial}{\partial x_{\alpha}}(-\Phi_g j_{\alpha}^{Dj}) + \rho_l v_{rl}^{\alpha} n_g^{\alpha} \sigma_{lg}^s Y_j + \rho_g \Phi_g \dot{w}_j^{ch} \qquad [16.5]$$

Because of equilibrium at the interface, different to the equilibrium of the vaporizing species, the same equation [16.4b] is, finally, found again.

Remember that the various terms for chemical reactions between gas species are linked to each other (see [BOR 08b] or [BOR 00]).

The temperature of the gas phase can be found from the phase's enthalpy, using the thermal equation of state. This has been described in Chapter 2, in a general form given by [2.12a]. It is possible to apply the equation to the gas phase and to rewrite it more simply by neglecting the kinetic energy, the power of friction forces and gravitational forces, which is realistic in the situations being considered here. In the context of negligible kinetic energy, pressure fluctuations are weak relative to the mean pressure and the corresponding term is simplified. The following is therefore obtained:

$$\frac{\partial}{\partial t}\left(\rho_g \Phi_g h_g\right)+\frac{\partial}{\partial x_\alpha}(\rho_g \Phi_g v_g^\alpha h_g)=\frac{\partial}{\partial x_\alpha}(-\Phi_g j_g^{Q\alpha})+\Phi_g \frac{\partial}{\partial t}(\overline{p}) + \rho_l v_{rl}^\alpha n_g^\alpha \sigma_{lg}^s h_g - j_g^{Q\alpha} n_g^\alpha \sigma_{gl}^s$$

By using the total mass balance for the gas phase, this equation can be rewritten in the following form:

$$\rho_g \Phi_g \frac{\partial}{\partial t}\left(h_g\right)+\rho_g \Phi_g v_g^\alpha \frac{\partial}{\partial x_\alpha}(h_g)=\frac{\partial}{\partial x_\alpha}(-\Phi_g j_g^{Q\alpha})+\Phi_g \frac{\partial}{\partial t}(\overline{p}) - j_g^{Q\alpha} n_g^\alpha \sigma_{gl}^s \qquad [16.6]$$

The first right-hand side term is the heat flux in the gas phase. Hereafter, an approximate law relating the heat flux to the enthalpy gradient will be used, which assumes that all the mass diffusivities and the thermal diffusivity are almost equal, therefore $j_g^{Q\alpha} = -\rho_g D_V \frac{\partial h_g}{\partial x_\alpha}$ and $j_\alpha^{DV} = -\rho_g D_V \frac{\partial Y_V}{\partial x_\alpha}$

Finally, it can be formally written that the physical variables of interest for the reaction are a vector of variables $c_{i,i=1,\ldots,N} = \left\{Y_{1g},\ldots,Y_V,h_g\right\}$ defined in the gas phase by a balance equation of type:

$$\rho_g \Phi_g \frac{\partial}{\partial t}\left(c_i\right)+\rho_g \Phi_g v_g^\alpha \frac{\partial}{\partial x_\alpha}(c_i)=\frac{\partial}{\partial x_\alpha}(\rho_g D_V \Phi_g \frac{\partial c_i}{\partial x_\alpha}) - J_{is}^\alpha n_g^\alpha \sigma_{lg}^s + \rho_g \Phi_g \dot{w}_i^{vol} \qquad [16.7]$$

Terms $\dot{w}_i^{vol}$ represent the source terms due to volume processes (chemistry or pressure effect), and terms J_{is}^α represent terms for mass or heat exchange between gas and liquid phases, defined in [16.4a], [16.5] and [16.6].

The density of the gas phase, ρ_g, is defined by the equation of state as a function of variables Y_{gj}, h_g and pressure. The relationship $\rho_g = \rho_g(Y_{jg}, h_g, \overline{p})$ is obtained by neglecting pressure fluctuations in comparison to the mean pressure.

16.2.2. *PDF equation*

Lundgren's method defines a "fine-grained PDF" that is effectively the probability, for a single experiment and not a large number of these, that all physical variables $c_i(x,t)$ are equal to the corresponding components of the random variable $\hat{Y}$, differing by up to $d\hat{Y}$, at point x at time t. This probability is then simply defined by the product $\prod_i (\delta(\hat{Y}_i - c_i(x,t))d\hat{Y}_i)$. The "fine-grained PDF" is nothing other than a Dirac peak in the multidimensional space of random variables. The probability density for many experiments is by definition the statistical average of the "fine-grained PDF":

$$P(\hat{Y};x,t) = \overline{\prod_i \delta(\hat{Y}_i - c_i(x,t))}$$

and the conditioned average in the gas phase, of interest here, is such that:

$$\overline{\Phi}_g P_g(\hat{Y};x,t) = \overline{\prod_i \Phi_g \delta(\hat{Y}_i - c_i(x,t))}$$

To simplify writing, it will be noted that $\prod_i \delta(\hat{Y}_i - c_i(x,t)) = \delta(\hat{Y} - c(x,t))$. In this expression, and for the expressions that follow, the absence of indices for Y or c indicates that all the components of these vectors are involved.

It is, therefore, possible to find an equation for the PDF from balance equations for c in the form of [16.7]. Therefore:

$\dfrac{\partial}{\partial t}(\delta(\hat{Y} - c(x,t))) = \sum_i \dfrac{\partial}{\partial \hat{Y}_i}(\delta(\hat{Y} - c)).(-\dfrac{\partial c_i}{\partial t})$, and similarly for the partial derivatives with respect to all space coordinates:

$$\frac{\partial}{\partial x_\alpha}(\delta(\hat{Y} - c(x,t))) = \sum_i \frac{\partial}{\partial \hat{Y}_i}(\delta(\hat{Y} - c)).(-\frac{\partial c_i}{\partial x_\alpha})$$

Therefore, using the balance equation [16.7], the following is found:

$$\Phi_g \frac{\partial}{\partial t}(\delta(\hat{Y}-c(x,t)))+\Phi_g v_g^\alpha \frac{\partial}{\partial x_\alpha}(\delta(\hat{Y}-c(x,t)))=$$

$$-\sum_i \frac{\partial}{\partial \hat{Y}_i}(\delta(\hat{Y}-c))(\frac{1}{\rho_g}\frac{\partial}{\partial x_\alpha}(\rho_g D_V \Phi_g \frac{\partial c_i}{\partial x_\alpha})-\frac{1}{\rho_g}J_{is}^\alpha n_g^\alpha \sigma_{lg}^s+\Phi_g \dot{w}_i^{vol})$$

After multiplying by the gas density, and using the continuity equation for the gas phase, the equation becomes:

$$\frac{\partial}{\partial t}(\rho_g \Phi_g \delta(\hat{Y}-c)+\frac{\partial}{\partial x_\alpha}(\rho_g \Phi_g v_g^\alpha \delta(\hat{Y}-c))-(\rho_l v_{rl}^\alpha n_g^\alpha \sigma_{lg}^s)\delta(\hat{Y}-c)=$$

$$-\sum_i \frac{\partial}{\partial \hat{Y}_i}(\delta(\hat{Y}-c)(\frac{\partial}{\partial x_\alpha}(\rho_g D_V \Phi_g \frac{\partial c_i}{\partial x_\alpha})-J_{is}^\alpha n_g^\alpha \sigma_{lg}^s+\rho_g \Phi_g \dot{w}_i^{vol}))$$

The average of the relationship must now be taken.

As defined above, the density is a function of the variables (and of the non-random parameter $\overline{p}$). The average of the product $\rho_g(c)\Phi_g \delta(\hat{Y}-c)$ is nothing other than $\rho_g(\hat{Y})\overline{\Phi_g \delta(\hat{Y}-c)}=\rho_g(\hat{Y})\overline{\Phi}_g P_g(\hat{Y})$ as $\rho_g(\hat{Y})$ is the average of $\rho_g(c)$ with the condition that its variables are equal to the components of $\hat{Y}$. Likewise, the average of the product $\rho_g(c)\Phi_g \dot{w}_i^{vol}(c)\delta(\hat{Y}-c)$ is $\rho_g(\hat{Y})\dot{w}_i^{vol}(\hat{Y})\overline{\Phi}_g P_g(\hat{Y})$. In addition, as the velocity fluctuation is defined as the difference with the mean value, the following is obtained by taking the average of the above equation:

$$\frac{\partial}{\partial t}\left(\rho_g(\hat{Y})\overline{\Phi}_g P_g(\hat{Y})\right)+\frac{\partial}{\partial x_\alpha}(\rho_g(\hat{Y})\overline{\Phi}_g P_g(\hat{Y})\tilde{v}_g^\alpha)=$$

$$\frac{\partial}{\partial x_\alpha}(-\overline{\rho_g \Phi_g v'^\alpha_g \delta(\hat{Y}-c)})+\overline{\rho_l v_{rl}^\alpha n_g^\alpha \sigma_{lg}^s \delta(\hat{Y}-c)}$$

$$-\sum_i \frac{\partial}{\partial \hat{Y}_i}(\overline{\frac{\partial}{\partial x_\alpha}(\rho_g D_V \Phi_g \frac{\partial c_i}{\partial x_\alpha})\delta(\hat{Y}-c)}+\rho_g(\hat{Y})\dot{w}_i^{vol}(\hat{Y})\overline{\Phi}_g P_g(\hat{Y})-\overline{J_{is}^\alpha n_g^\alpha \sigma_{lg}^s \delta(\hat{Y}-c)})$$

The second term on the right-hand side can be broken down:

$$\sum_i \frac{\partial}{\partial \hat{Y}_i}\overline{(\frac{\partial}{\partial x_\alpha}(\rho_g D_V \Phi_g \frac{\partial c_i}{\partial x_\alpha})\delta(\hat{Y}-c))} = \sum_i \overline{\frac{\partial}{\partial x_\alpha}(\rho_g D_V \Phi_g \frac{\partial c_i}{\partial x_\alpha})\frac{\partial \delta(\hat{Y}-c(x))}{\partial \hat{Y}_i}}$$

$$= \sum_i \overline{\frac{\partial}{\partial x_\alpha}(\rho_g D_V \Phi_g \frac{\partial c_i}{\partial x_\alpha}\frac{\partial \delta}{\partial \hat{Y}_i})} - \sum_i \overline{\rho_g D_V \Phi_g \frac{\partial c_i}{\partial x_\alpha}\frac{\partial}{\partial x_\alpha}(\frac{\partial \delta}{\partial \hat{Y}_i})} =$$

$$\frac{\partial}{\partial x_\alpha}\overline{(\rho_g D_V \Phi_g \sum_i \frac{\partial c_i}{\partial x_\alpha}\frac{\partial \delta}{\partial \hat{Y}_i})} - \sum_i \frac{\partial}{\partial \hat{Y}_i}\overline{(\rho_g D_V \Phi_g \frac{\partial c_i}{\partial x_\alpha}(\frac{\partial \delta}{\partial x_\alpha}))} =$$

$$\frac{\partial}{\partial x_\alpha}\overline{(-\rho_g D_V \Phi_g \frac{\partial \delta}{\partial x_\alpha})} + \sum_i \frac{\partial}{\partial \hat{Y}_i}\overline{(\rho_g D_V \Phi_g \frac{\partial c_i}{\partial x_\alpha}\sum_j \frac{\partial c_j}{\partial x_\alpha}\frac{\partial \delta}{\partial \hat{Y}_j})} =$$

$$\frac{\partial}{\partial x_\alpha}(-\overline{\rho_g D_V}\frac{\partial(P_g \bar{\Phi}_g)}{\partial x_\alpha} + \overline{\rho_g D_V n_g^\alpha \sigma_{lg}^s \delta})) + \sum_i \sum_j \frac{\partial^2}{\partial \hat{Y}_i \partial \hat{Y}_j}\overline{(\rho_g D_V \frac{\partial c_i}{\partial x_\alpha}\frac{\partial c_j}{\partial x_\alpha}\Phi_g \delta)}.$$

Fluctuations in coefficient $\rho_g D_V$, or rather its correlation with fluctuations in factor $\frac{\partial(\Phi_g \delta)}{\partial x_\alpha}$, have been neglected at the end of the above calculations. Fluctuations in thermal diffusivity are not expected to play an important role in the problem because the term in question is often negligible in comparison to the term due to turbulent dispersion.

Finally:

$$\frac{\partial}{\partial t}\left(\rho_g(\hat{Y})\bar{\Phi}_g P_g(\hat{Y})\right) + \frac{\partial}{\partial x_\alpha}(\rho_g(\hat{Y})\bar{\Phi}_g P_g(\hat{Y})\tilde{v}_g^\alpha) =$$

$$+\frac{\partial}{\partial x_\alpha}(\overline{\rho_g D_V}\bar{\Phi}_g \frac{\partial P_g}{\partial x_\alpha} - \overline{\rho_g \Phi_g v'^\alpha_g \delta(\hat{Y}-c)} + \overline{\rho_g D_V}(P_g \frac{\partial \bar{\Phi}_g}{\partial x_\alpha} - \overline{n_g^\alpha \sigma_{lg}^s \delta(\hat{Y}-c)}))$$

$$-\sum_i \frac{\partial}{\partial \hat{Y}_i}(\sum_j \frac{\partial}{\partial \hat{Y}_j}\overline{(\rho_g D_V \Phi_g \frac{\partial c_j}{\partial x_\alpha}\frac{\partial c_i}{\partial x_\alpha}\delta(\hat{Y}-c))} + \rho_g(\hat{Y})\dot{w}_i^{vol}(\hat{Y})\bar{\Phi}_g P_g(\hat{Y}) -$$

$$\overline{J_{is}^\alpha n_g^\alpha \sigma_{lg}^s \delta(\hat{Y}-c)}) - \overline{\rho_l v_{rl}^\alpha n_g^\alpha \sigma_{lg}^s \delta(\hat{Y}-c)} \qquad [16.8]$$

This equation cannot be used as it is now because it includes terms that are yet unknown averages. Closure assumptions need to be discussed which may provide an estimate of these terms at the lowest possible cost, while still being realistic. However, the term linked to chemical reactions in the gas phase appears in a form that does not require any additional hypothesis.

The first term on the right-hand side can be interpreted as the dispersion flux of the PDF in a physical space, with a contribution linked to molecular diffusion, one linked to turbulence (velocity fluctuations), and a new one linked to the presence of interfaces. The term for molecular diffusion is often negligible in comparison to the turbulent contribution. The latter cannot be calculated without a model for [16.8] because the velocity fluctuations are not taken into account by a random (vector) variable in the PDF considered. This has been studied ever since PDFs were first used for reactive, single-phase fluid media and will be discussed further in the next section. The third term is linked to the presence of interfaces and is also still to be modeled. However, as it is proportional to the molecular diffusion coefficient, it should also be negligible in comparison to turbulent dispersion. Because of the presence of the interface area per unit volume, the PDF for the composition of the gas phase, involved in this term, deals with the close proximity of the interfaces: $\overline{n_g^\alpha \sigma_{lg}^s \delta(\hat{Y}-c)} = \overline{n_g^\alpha \sigma_{lg}^s} P_s(\hat{Y}) = \frac{\partial \bar{\Phi}_g}{\partial x_\alpha} P_s(\hat{Y})$, and both terms with molecular diffusion can therefore be written as $+\overline{\rho_g D_V}(P_g(\hat{Y}) - P_s(\hat{Y}))\frac{\partial \bar{\Phi}_g}{\partial x_\alpha}$. The PDF for the gas phase, conditioned on the interface $P_s(\hat{Y})$, is not necessarily equal to the PDF for the gas phase volume as the interface plays a particular role concerning the chemical composition. This will be discussed further when other exchange terms at the interface are considered.

The second term on the right-hand side, a sum of partial derivatives with respect to all random variables, represents a probability transport in the random variables' space. It consists of three parts. The first is the term called "micromixing", already known in studies of PDFs in single-phase flows. It cannot be calculated without knowing the gradients of c_i, in the gas phase, which do not correspond to the PDF's random variables taken into account here, and must therefore be modeled. The micromixing term has also been studied in detail for fluid turbulent media; various models have been developed and tested. These models will be discussed in more detail in section 16.4. The second term is the contribution linked to chemical reactions, in its closed form, as for single-phase reactive media. The last term is a term analogous to the reaction term but is linked to exchanges between phases and will be discussed later. It can already be noted that it consists of an average conditioned on the interface, and then only the PDF $P_s(\hat{Y})$ is needed.

Similarly, the last term of equation [16.8] is due to the total mass exchange between phases and has never been studied to date. It is of significance to note that the equation obtained does not concern directly the multidimensional PDF $P_g(\hat{Y})$, for which the integral over space $\hat{Y}$ is always equal to 1, but $\bar{\Phi}_g P_g(\hat{Y})$, for which the integral is the gas volume fraction, which varies because of the source term for vaporization–condensation included in the last term.

As for all PDFs, it is known that $\iint_{\hat{Y}} P_g(\hat{Y})d\hat{Y} = 1$. The physical variables correspond to random variables and can vary from negative to positive infinity, although some of them, for example the mass fraction of a species, can only vary from zero to one, or less because of the restrictions imposed by chemistry. The space of corresponding random variables encompasses this area of physical variable variation, including boundaries (therefore for certain variables this will be 0– and 1+), and by definition the PDF has a value of zero at the boundaries of the random variable variations. In addition, any transport term in the random variable space cancels itself out at the space's boundaries because the integral of the PDF is constant and equal to one. Once equation [16.8] has been integrated over the random variable space, the probability transport term disappears and the following is therefore obtained:

$$\frac{\partial}{\partial t}\left(\bar{\rho}_g^*\bar{\Phi}_g\right) + \frac{\partial}{\partial x_\alpha}(\bar{\rho}_g^*\bar{\Phi}_g\tilde{v}_g^\alpha) =$$

$$+\frac{\partial}{\partial x_\alpha}\left(-\iint_{\hat{Y}} \overline{\rho_g(\hat{Y})\Phi_g v''^\alpha_g\,\delta(\hat{Y}-c)d\hat{Y}} + \overline{\rho_g D_V}\left(\frac{\partial\bar{\Phi}_g}{\partial x_\alpha} - \iint_{\hat{Y}} \overline{n_g^\alpha \sigma_{lg}^s \delta(\hat{Y}-c)d\hat{Y}}\right)\right)$$

$$-\iint_{\hat{Y}} \overline{\rho_l v_{rl}^\alpha n_g^\alpha \sigma_{lg}^s \delta(\hat{Y}-c)d\hat{Y}}$$

By definition, the appearance of the δ function under the mean line indicates that the statistical average has the condition that the c_i involved are equal to corresponding $\hat{Y}_i$. It is therefore obtained that:

$$\iint_{\hat{Y}} \overline{\rho_g(\hat{Y})\Phi_g v''^\alpha_g\,\delta(\hat{Y}-c)d\hat{Y}} = \overline{\rho_g(c)\Phi_g v''^\alpha_g} = 0$$

and $\iint_{\hat{Y}} \overline{n_g^\alpha \sigma_{lg}^s \delta(\hat{Y}-c)d\hat{Y}} = \overline{n_g^\alpha \sigma_{lg}^s} = \dfrac{\partial\bar{\Phi}_g}{\partial x_\alpha}$

and finally, $-\iint_{\hat{Y}} \overline{\rho_l v_{rl}^\alpha n_g^\alpha \sigma_{lg}^s \delta(\hat{Y}-c)d\hat{Y}} = -\overline{\rho_l v_{rl}^\alpha n_g^\alpha \sigma_{lg}^s}$

For the terms on the right-hand side, the overbars represent the averages for which the physical variables c_i, the gas velocity, Φ_g and all others that may be involved, fluctuate. The simple mean mass balance of the gas phase, established in Chapter 4, section 4.2, is thus recovered. This balance equation allows the mean gas volume fraction to be found because the mean (intrinsic) density is $\bar{\rho}_g^* = \iint_{\hat{Y}} \rho_g(\hat{Y}) P_g(\hat{Y}) d\hat{Y}$ and can be calculated if the PDF is known.

16.3. Modeling the terms due to exchanges between phases

In previous discussions, five types of terms for exchanges between phases were defined: one due to total mass exchange, given by $-\overline{\rho_l v_{rl}^\alpha n_g^\alpha \sigma_{lg}^s \delta(\hat{Y}-c)}$, one due to the exchange of species that vaporize, given by $-\overline{\rho_l v_{rl}^\alpha n_g^\alpha (1-\hat{Y}_V) \sigma_{lg}^s \delta(\hat{Y}-c)}$, one due to the mass exchange for the other species, given by: $\overline{\rho_l v_{rl}^\alpha \hat{Y}_i n_g^\alpha \sigma_{lg}^s \delta(\hat{Y}-c)}$, one due to heat exchange, given by $-\overline{j_g^{Q\alpha} n_g^\alpha \sigma_{lg}^s \delta(\hat{Y}-c)}$ and, finally, one due to enthalpy exchange per mass exchange, given by $-\overline{\rho_l v_{rl}^\alpha n_g^\alpha h_g \sigma_{lg}^s \delta(\hat{Y}-c)}$. All bring in to play the PDF of gas composition in its restricted form at the interface $P_s(\hat{Y})$. Moreover, they involve the vaporized, or condensed, mass flow rate. They all need to be modeled as they bring in variables that the considered PDF does not include. The way in which these terms can be modeled in the above PDF equation will now successively be examined.

16.3.1. *Total mass exchange*

The local and direct relationship that enables the total mass exchange to be calculated accurately is based on the interface balance relation for vapor of the liquid phase. It is written as $\left[Y_V \rho_g v_{rg}^\alpha + j_\alpha^{DV} \right] n_g^\alpha \sigma_{lg}^s = \rho_l v_{rl}^\alpha n_g^\alpha \sigma_{lg}^s$ (see Chapter 4, section 4.5.1) and the interface relationship for the total mass is given by $\rho_g v_{rg}^\alpha n_g^\alpha \sigma_{lg}^s = \rho_l v_{rl}^\alpha n_g^\alpha \sigma_{lg}^s$; therefore:

$$\rho_l v_{rl}^\alpha n_g^\alpha \sigma_{lg}^s = j_\alpha^{DV} n_g^\alpha \sigma_{lg}^s / (1 - Y_V)$$

and, using Fick's law, the following is finally written:

$$-\rho_l v_{rl}^\alpha n_g^\alpha \sigma_{lg}^s = \rho_g D_V \frac{\partial Y_V}{\partial x_\alpha} n_g^\alpha \sigma_{lg}^s / (1 - Y_V)$$

At the same time, prescribing the equilibrium for vaporization at the interface, resulting in an algebraic relation $Y_V\big|_s = Y_{Vsat}(T_s, p)$, is imposed at any time. This equilibrium is valid for temperature and pressure conditions such that the gas medium is not rarified, except when the interface starts to form during "nucleation" (which will not be discussed here). In the context of subsonic flow velocities considered here, pressure fluctuations can be neglected and the pressure inside is then the mean local pressure. The term for the total mass exchange in the PDF is therefore:

$$-\overline{\rho_l v_{rl}^\alpha n_g^\alpha \sigma_{lg}^s \delta(\hat{Y}-c)} = \overline{\rho_g D_V \frac{\partial Y_V}{\partial x_\alpha} n_g^\alpha \sigma_{lg}^s \delta(\hat{Y}-c)/(1-Y_V)}$$

As the PDF considered does not include the normal gradient of the vapor's mass fraction, nor the interface area per unit volume, this mean value cannot be calculated explicitly.

However, the vaporized mass flow rate per unit area can be modeled by considering local exchange situations, as in Chapter 5, in particular in the case of droplet vaporization, which is particularly useful here. The instant mass flow rate per unit area entering the gas phase, or leaving the liquid phase, can be found and is here simply noted as $\rho_l v_{rl}^\alpha n_g^\alpha = \dot{m}_V$ (therefore positive for vaporization and negative for condensation). For a droplet at rest of radius r_d in a quasi-static vaporization regime (see Chapter 5), the following is obtained:

$$\dot{m}_V = \frac{\rho_g D_V}{r_d} Ln\left(\frac{1-Y_{V\infty}}{1-Y_{Vs}}\right)$$

where $Y_{Vs} = Y_{Vs}(T_s)$ is given by the vaporization equilibrium at the surface of the drop. The variable $Y_{V\infty}$ is the mass fraction of the liquid's vapor in the gas phase at a distance from the drop which is large relative to its radius. In Chapter 5, this was replaced by the mean value $\tilde{Y}_V$ to calculate the mean exchange in mean balance equations. However, it is now possible to find the fluctuations in Y_V, and the fluctuations in $Y_{V\infty}$ need to be taken into account. They are also known in a PDF that can be called $P_{g\infty}(\hat{Y};x,t)$: this is $P_g(\hat{Y};x,t)$, but in which the realizations associated with the presence of the nearby interface are not taken into account. At this level, two approximations are possible: $P_{g\infty}(\hat{Y};x,t) = \delta(Y_V - \tilde{Y}_V)$, corresponding to neglecting fluctuations, or $P_{g\infty}(\hat{Y};x,t) = P_g(\hat{Y};x,t)$. The latter is opted for here, and this option will be re-examined later. In addition, the radius of the drop is calculated as a function of the mean interface area per unit volume and of the volume fraction of drops, assuming that all drops are identical. Using these two hypotheses, it is therefore obtained that $\dot{m}_V = \dot{m}_V(Y_V, T_s, ...)$, where it is explicitly written that the variable can fluctuate. This expression can be corrected if the motion of the drop, relative to the continuous fluid, is taken into account by involving other non-fluctuating variables, for example the mean relative velocity, or the turbulent kinetic energy in the fluid, as was discussed in Chapter 5.

The term provided for total mass exchange is therefore:

$$\overline{-\rho_l v_{rl}^\alpha n_g^\alpha \sigma_{lg}^s \delta(\hat{Y}-c)} = \overline{\dot{m}_V(Y_V, T_s)\bar{\sigma}_{lg}^s \delta(\hat{Y}-c)}$$

$$\text{where: } \overline{-\rho_l v_{rl}^\alpha n_g^\alpha \sigma_{lg}^s \delta(\hat{Y}-c)} = \overline{\dot{m}_{V|Y}}\ (\hat{Y}_V) P_g(\hat{Y}).\bar{\sigma}_{lg}^s \quad [16.9]$$

Function $\overline{\dot{m}_{V|Y}}\ (\hat{Y}_V)$ is the conditional average for $\dot{m}_V$ when the physical variable is set to the given value $\hat{Y}_V$. As the only variable that can still fluctuate in the model is T_s, the following is obtained:

$$\overline{\dot{m}_{V|Y}}\ (\hat{Y}_V) = \frac{\rho_g D_V}{3\bar{\Phi}_g} \bar{\sigma}_{lg}^s \int_{\hat{T}} P_{Ls}(\hat{T}) Ln\left(\frac{1-\hat{Y}_V}{1-Y_{Vs}(\hat{T})}\right) d\hat{T}$$

where $P_{Ls}(\hat{T})$ is the PDF for temperature fluctuations on the drop surfaces. This PDF can be studied and will be discussed in Chapter 17. For a first estimate, it can be assumed that all the drops have a homogeneous temperature, are at the same temperature and $\dot{m}_V$ can therefore be calculated via a mean energy balance equation for the liquid phase. Therefore:

$$\overline{\dot{m}_{V|Y}}\ (\hat{Y}_V) = \frac{\rho_g D_V}{3\bar{\Phi}_g} \bar{\sigma}_{lg}^s Ln\left(\frac{1-\hat{Y}_V}{1-Y_{Vs}(\tilde{T}_L)}\right)$$

16.3.2. *Mass exchange for species*

The problem is identical for exchange terms related to species, which are proportional to the total mass exchange term. Using the same approach, and the same hypotheses that must be checked later, the following is obtained:

$$\overline{-\rho_l v_{rl}^\alpha n_g^\alpha (1-\hat{Y}_V)\sigma_{lg}^s \delta(\hat{Y}-c)} = \overline{\dot{m}_{V|Y}}\ (\hat{Y}_V)(1-\hat{Y}_V) P_g(\hat{Y}).\bar{\sigma}_{lg}^s \quad [16.10]$$

$$\overline{\rho_l v_{rl}^\alpha \hat{Y}_i n_g^\alpha \sigma_{lg}^s \delta(\hat{Y}-c)} = \overline{\dot{m}_{V|Y}}\ (\hat{Y}_V)\hat{Y}_i P_g(\hat{Y}).\bar{\sigma}_{lg}^s \quad [16.11]$$

$$\overline{-\rho_l v_{rl}^{\alpha} n_g^{\alpha} h_g \sigma_{lg}^{s} \delta(\hat{Y}-c)} = \overline{\dot{m}_{V|Y}}(\hat{Y}_V)\hat{h}_g P_g(\hat{Y}).\bar{\sigma}_{lg}^{s} \qquad [16.12]$$

The discussion concerning $P_{Ls}(\hat{T})$ above is also relevant for these terms.

16.3.3. *Heat exchange*

The situation is analogous in the case of the term for heat exchange between phases. The heat flux received by the gas phase is given by $\overline{-j_g^{Q\alpha} n_g^{\alpha} \sigma_{lg}^{s} \delta(\hat{Y}-c)} = \rho_g D_V \dfrac{\partial h_g}{\partial x_\alpha} n_g^{\alpha} \sigma_{lg}^{s} \delta(\hat{Y}-c)$, but neither enthalpy's normal gradient nor the interfacial area per unit volume are taken into account by random variables. Modeling this term can nonetheless be achieved by using the formula provided by the study of heat transfer during drop vaporization. Indeed, it has been found in Chapter 5 that:

$$\left(\rho_g D_V \frac{dh_g}{dr}\right)_s = \dot{m}_V \left(\frac{C_g(T_{g\infty}-T_s)}{Y_{Vsat}(T_s)-Y_{V\infty}} - L\right)(1-Y_{Vsat}(T_s))$$

In the same way as before, $T_{g\infty}$ and $Y_{V\infty}$ can be incorporated into T_g and Y_V of the gas phase, the temperature can be calculated from the enthalpy and these two variables are therefore part of the phase's vector of variables c_i. It can therefore formally be written as:

$$-j_g^{Q\alpha} n_g^{\alpha} = \dot{m}_V(Y_V)\left(\frac{C_p(T_g-T_s)}{Y_{Vs}(T_s)-Y_V} - L\right)(1-Y_{Vs}(T_s)) = q(Y_V, T_s, h_g, \ldots)$$

by not writing the other variables that are not liable to fluctuating (in this model). Therefore, it can also be written as:

$$\overline{-j_g^{Q\alpha} n_g^{\alpha} \sigma_{lg}^{s} \delta(\hat{Y}-c)} = \overline{q_{|Y}}(\hat{Y}_V, \hat{h}_g) P_g(\hat{Y}).\bar{\sigma}_{lg}^{s} \qquad [16.13]$$

where the conditioned average of heat flux also involves $P_{Ls}(\hat{T})$:

$$\overline{q_{|Y}}(\hat{Y}_V, \hat{h}_g) = \int_{\hat{T}} P_{Ls}(\hat{T}) q(\hat{Y}_V, \hat{T}, \hat{h}_g) d\hat{T}$$

which could ultimately be replaced by the liquid's mean temperature.

Vaporization alters the shape of the PDF because the source term is localized in a particular area of the field for variables $\hat{Y}$. For example, in the case of vaporization of drops having the same temperature, the source term for vaporization is precisely located at $\hat{Y}_V = Y_{Vs}(T_s)$, the saturation value for the liquid's temperature, and the PDF will remain limited to lower values as the other terms in the equation, to be seen later, can only push the probability to lower values.

Starting from the PDF equation [16.8], balance equations can be recovered for each mean value of the physical variables associated with the random variables considered in the PDF. For example, the equation needs to be multiplied by the random variable $\hat{Y}_V$ (which is not a function of space or time) and integrated over the whole random variables' space to find the balance equation for $\tilde{Y}_V$. It will then be necessary to verify, without too many calculations, that the equation is of the same type as [16.1], and the term for mass exchange in this equation for $\tilde{Y}_V$ is written as:

$$-\iint_{\hat{Y}} \overline{\rho_l v_{rl}^\alpha n_g^\alpha (1-\hat{Y}_V)\delta(\hat{Y}-c)\sigma_{lg}^s} d\hat{Y} = \iint_{\hat{Y}} \overline{\dot{m}_{V|Y}}(\hat{Y}_V)(1-\hat{Y}_V)P_g(\hat{Y})d\hat{Y}.\overline{\sigma}_{lg}^s$$

A model of this mass exchange term was given directly in Chapter 5, and the PDF defined above was not looked for, simply as $-\overline{\rho_l v_{rl}^\alpha n_g^\alpha (1-\hat{Y}_V)\sigma_{lg}^s} = \overline{\dot{m}_V}(\tilde{Y}_V)(1-Y_{Vsat}(\tilde{T}_L)).\overline{\sigma}_{lg}^s$. Neither the fluctuations of Y_V, nor those of Y_{Vs}, due to the surface temperature of the drops, were taken into account. It is more convenient to consider the PDF for Y_V fluctuations as done here, and $P_{Ls}(\hat{T}_L)$ must also be studied more closely as will be done in Chapter 17.

In any case, the reader will probably realize that the solution to the PDF equation does not actually require the equations for mean values [16.1], given at the beginning of this chapter, to be considered. In fact, if $P_g(\hat{Y})$ is known, we can easily calculate all the mean values for all the physical quantities corresponding to random variables taken into account, and this is done by simple integration:

$$\overline{c}_i = \iint_Y \hat{Y}_i P_g(\hat{Y})d\hat{Y}\text{, and } \overline{\rho}_g\tilde{c}_i = \iint_Y \rho_g(\hat{Y})\hat{Y}_i P_g(\hat{Y})d\hat{Y}$$

16.4. Modeling micromixing and turbulent dispersion

16.4.1. *The "micromixing" term in PDF equations*

This term is in equation [16.8], defined by:

$$-\sum_i \frac{\partial}{\partial \hat{Y}_i}\left(\sum_j \frac{\partial}{\partial \hat{Y}_j}\left(\overline{\rho_g D_V \Phi_g \frac{\partial c_j}{\partial x_\alpha}\frac{\partial c_i}{\partial x_\alpha}\delta(\hat{Y}-c)}\right)\right)$$

It is the term that decreases fluctuations and makes the PDF tend toward a Dirac peak located at the mean values in the multidimensional space of random variables. The study of this equation is standard for turbulent gas media, whether reactive or not, and is at the heart of its modeling. These studies will not be discussed here in detail and the reader may refer to [LIB 80], [KUZ 90] and [POP 85]. Only the two main models will be considered here for simplicity.

The model by Dopazo and O'Brien (see [LIB 80, KUZ 90]) is based on an estimate of the shape of the PDF for the fluctuations in the variables vector c at two neighboring locations within the continuous medium (justifying its name "linear mean square estimation" (LMSE)). Using this, the term for micromixing can be written as:

$$+\sum_i \frac{\partial}{\partial \hat{Y}_i}\left(\rho_g(\hat{Y})\frac{C_{MM}}{\tau_t}(\tilde{Y}_i-\hat{Y}_i)\bar{\Phi}_g P_g(\hat{Y})\right)$$

τ_t is a characteristic time for the turbulence that creates mixing, thus here the turbulence in the gas phase, and C_D is a dimensionless constant to be determined through experimentation.

The second model was proposed earlier on heuristic bases in the field of chemical engineering by Curl (see [BOR 00] or [POP 85]): it represents mixing as an interaction between fluid particles, which coalesce before dispersing and provide the micromixing term in the form of an integral:

$$\frac{C_{curl}}{\tau_t}\bar{\Phi}_g\left[2\iint P_g(\hat{Y}')P_g(\hat{Y}'')\delta\left(\hat{Y}-\frac{\hat{Y}''+\hat{Y}'}{2}\right)d\hat{Y}''d\hat{Y}'-P_g(\hat{Y})\right]$$

However, drops are still surrounded by a non-turbulent gas layer in which the mass fraction gradients are controlled by molecular diffusion, and these gradients are the controlling factor (as mentioned in the beginning of section 16.3.1) for drops in the $\hat{Y}_V$ region where vaporization–condensation takes place, therefore close to $Y_{Vsat}(T_s)$. As a result, the classic turbulent models quoted above are altered. Here again, the assumption that local exchange occurs around the drops in a quasi-static regime shows that the profile around the drop is exponential, therefore $\partial Y_V / \partial n \propto 1/Y_V$. Durand *et al.* [DUR 99] then proposed a modified version of Curl's model using this property. The importance of this modification is still to be studied for other applications.

In a non-reactive single-phase fluid medium, time τ_t is nothing other than k/ε if the *k*-*ε* model is used to calculate turbulence characteristics, but if the Reynolds number is not large enough, this time may also be affected. As a consequence, the constants C_{MM}, or C_{curl}, are therefore a function of the Reynolds number tending toward a constant if the latter tends toward infinity. In the case of premixing turbulent flames, the time is also affected by other dimensionless parameters of the problem (see [BOR 00]). For the reactive two-phase media being considered here, the flamelets are mainly non-premixed flamelets, which should not significantly influence the micromixing time. On the contrary, the presence of drops in the medium imposes a different characteristic length scale, linked to the mean spacing between drops. The turbulence involved here is that of the gas phases, and, as seen in Chapter 6, the dissipation rate may take into account the characteristic length scale of the two-phase field if the turbulence is calculated by *k*-*ε* model. It can nonetheless be questioned whether or not the volume fraction of the liquid may influence the values of constants C_{MM}, or C_{curl}. This must also be studied more closely.

16.4.2. *Turbulent diffusion terms in PDF equations*

This term is also an important ingredient for modeling. It can be simply represented by a law containing a turbulent diffusion coefficient (which can be calculated with *k* and *ε*) by a simple analogy with the turbulent diffusion of a mean quantity. Therefore:

$$-\overline{\rho_g v'^{\alpha}_g \Phi_g \delta(\hat{Y} - c)} = \bar{\rho}_g \bar{\Phi}_g D_t \frac{\partial P_g(\hat{Y}; x, t)}{\partial x_\alpha}$$

Starting with the PDF equation, all mean values $\tilde{c}_i$ in the balance equation can be calculated, which clearly means that the diffusion fluxes of the mean values also follow a gradient law, with the same turbulent diffusion coefficient. Coefficient D_t defined here is thus the same as the one mentioned in Chapter 6 but is only relative to the gas phase, thus dependent on the kinetic energy of velocity fluctuations in this phase and of its scale.

Modeling of turbulent dispersion can be avoided if an interest is taken in the joint PDF for composition and velocities, as was done by Pope for reactive single-phase medium [POP 85].

16.5. Practical use of PDF equations

Even after modeling, use of the PDF equation [16.8] is still difficult to implement due to the fact that the PDF is a generalized function with a large number of dimensions: it can have sharp peaks, and even for steady regimes the number of dimensions it has is equal to the number of dimensions in the geometrical space plus the number of the random variables considered. As a result, the numerical processing of [16.8] involves a long calculation time and requires a lot of memory.

A simple way to by-pass this problem, when only one random variable is considered, consists of taking a "presumed PDF" which has a definite shape in the random variable space, a shape that depends on only three parameters. In these conditions, it is therefore only necessary to know the mean value and the variance of the variable considered, the PDF having been normalized, to calculate the shape, at any point in the medium. It has already been argued that an equation for the variance can also be found from the PDF equation by multiplying by $(\hat{Y}-\tilde{Y})^2$ and by integrating it over the $\hat{Y}$ space. This method is widely used for flames in gas media, see [BOR 00], for example. It was used for flames in a spray by Demoulin and Borghi [DEM 00]. The reader may also refer to [MUR 04].

The classic method for numerical calculation of the PDF equation consists of performing a Monte Carlo simulation, without dismissing the option of calculating very diverse shapes or of considering two or three random variables or more. The simulation is achieved by tracking a certain number of "stochastic particles" both in the physical space and in the random variable space. The tracking can be performed by meshing the physical space with a large number of small (yet finite) volumes and by recording the passage of various stochastic particles in these volumes (in which the PDF can be reconstituted by creating a histogram of all the particles that have gone through). More complete tracking of the particle trajectories can also be done in the physical space when turbulent dispersion is represented in a Lagrangian way. The first type of simulation was used by Durand *et al.* [DUR 99]. The second was developed and perfected by Pope and also uses velocity components among the random variables for single-phase turbulent flows. The reader may refer to [POP 85] and [MIN 01]. A simplified version of the Monte Carlo–Lagrangian simulation, in which the PDF for velocities is simply reconstructed from a turbulence model, can also be used beneficially as it was used for turbulent flames in a gas medium, allowing a complex representation of the kinetic chemical processes to be taken. An example is presented by Zamuner and Dupoirieux [ZAM 00], which studies the production of soot by a turbulent flame. This "Lagrangian" simulation by stochastic particles will be discussed further in Chapter 17.

Chapter 17

Temperature Fluctuations in Condensed Phases

17.1. Problems

In all the situations where a liquid phase is vaporized upon contact with a gas phase that does not only contain the vapor of the liquid, it was discussed in Chapter 5 that the mass flow rate vaporized per unit of interface area depends on the temperature of the surface and the mass (or volume) fraction of the vapors in the gas phase. For example, for the vaporization of droplets, it was seen in Chapter 5, section 5.6.2, that the mass flow rate vaporized per unit interface area is:

$$\rho_l v_{rl}^{\alpha} n_g^{\alpha} = \dot{m}_V = \frac{\rho_g D_V}{r_d} Ln\left(\frac{1-Y_{V\infty}}{1-Y_{Vsat}(T_s)}\right) \quad [17.1]$$

To use this result in the building of models for the mean medium, as studied in Chapter 5, the mass fraction of the vapor far from the interface was identified with the mean mass fraction in the gas and the temperature of the interface with the mean temperature of the liquid phase. For studying the probability density function (PDF) of the temperature fluctuations and the composition of the gas phase, the fluctuations of $Y_{V\infty}$ and of the variable Y_V were identified, without taking into account the eventual presence of an interface. The knowledge of the surface temperature T_s is another essential characteristic to be known before evaluating the mass flow rate vaporized with [17.1]: as the condensation–vaporization equilibrium is verified at the interface at any instant, the mass fraction of the vapor at the interface depends on it.

If it is assumed that the thermal conductivity of the liquid is large, the surface temperature is simply the uniform temperature of the droplet; this is a very realistic, if not perfect, assumption. Other approximations have been based on an approximate calculation of the internal temperature field of the droplet, but the possibility of movements of the fluid inside the droplet does complicate the calculation and makes the uniform internal temperature assumption more plausible (see [SIR 99]). In any case, even if the temperature is (practically) uniform inside the droplet, it depends on the considered droplet because the heat exchange with the gas medium is a function of the radius of the droplet, just like the mass flow rate vaporized, and consequently when the droplets have different radii their temperatures will evolve differently. In all rigorousness, it is therefore necessary to be able to know more than the mean temperature of the liquid phase to be able to estimate the surface temperature of each droplet. It is, therefore, very interesting to calculate the PDF of the temperature of the liquid or $P_l(\hat{T})$.

Formula [17.1], again identifying $Y_{V\infty}$ in Y_V (or rather supposing that the PDF's of the two variables are the same), gives the mass flow rate vaporized as a function of the two variables that can fluctuate, Y_V and $T_s = T_l$. Other non-fluctuating variables can be taken into account, like the radius of the droplets, the relative velocity of the flow (the possibility of varying radii of the droplets will be discussed in Chapter 18). This requires studying joint PDF of the fluctuations of these two variables. Here, the focus is on the surface temperature fluctuations, we will neglect in the following the correlations with the fluctuations of Y_V, and thus consider only the mean mass flow rate vaporized relative to the Y_V fluctuations, which is written as:

$$\dot{m}_V^{mg}(T_l) = \frac{\rho_g D_V}{3\bar{\Phi}_p} \bar{\sigma}_{lg}^s \int_{\hat{Y}} P_g(\hat{Y}) Ln\left(\frac{1-\hat{Y}}{1-Y_{Vsat}(T_l)}\right) d\hat{Y} \qquad [17.2]$$

An approximation of this formula, completely independent of the PDF of the gas phase studied in Chapter 16, would simply be due to the complete negligence of the fluctuations of this phase: $\dot{m}_V^{mg} = \frac{\rho_g D_V}{3\bar{\Phi}_p} \bar{\sigma}_{lg}^s Ln\left(\frac{1-\tilde{Y}_V}{1-Y_{Vsat}(T_l)}\right)$. Since the logarithmic form is not too "nonlinear", the use of this last formula is probably not too incorrect, especially in relation to the strongly nonlinear dependence of $Y_{Vsat}(T_l)$ and the use of $\tilde{T}_l$.

The correct way to find the PDF $P_l(\hat{T})$ will be discussed in this chapter.

The previous discussions can obviously be used again for liquid parcels of any shape, where the formula for the mass flow rate vaporized per unit interface area is different but depends on the same variables. This is also valid for the "liquid core" stretched and fluctuating at the exit of the liquid injector. However, in this case, the extra difficulty will be that the hypothesis stating that the temperature in the liquid is uniform will be a lot less likely, if not false.

Looking now at solid particles that can interact with the surrounding gas phase through chemical surface reactions, it is clear that it is important to know the temperature, and its fluctuations between particles, before evaluating the mean reaction rate. The developments of this chapter can be transposed onto these situations without much difficulty.

17.2. Instantaneous equation for the temperature of the liquid phase

First, the instantaneous balance equation is considered starting from:

$$\begin{aligned}&\frac{\partial}{\partial t}(\rho_i\Phi_i h_{ti})+\frac{\partial}{\partial x_\alpha}(\rho_i\Phi_i v_i^\alpha h_{ti})=\frac{\partial}{\partial x_\alpha}(\Phi_i(-j_i^{Q\alpha}+\tau_i^{\alpha\beta}v_i^\beta))+\Phi_i\frac{\partial}{\partial t}(p_i)\\&+\rho_i\Phi_i v_i^\alpha g^\alpha+\rho_{is}\dot{w}_i^{\mathrm{sech}}h_{ti}-\tau_i^{\alpha\beta}v_i^\alpha n_i^\beta\sigma_i^s-j_i^{Q\alpha}n_i^\alpha\sigma_i^s\end{aligned} \qquad [17.3]$$

where $\rho_{is}\dot{w}_i^{echs}=-\rho_{is}v_{ris}^{\alpha}n_i^\alpha\sigma_i^s=-\dot{m}_i\sigma_i^s$ and $-n_i^\alpha\sigma_i^s=\dfrac{\partial\Phi_i}{\partial x_\alpha}$.

The heat flux by conduction in the phase, $j_i^{Q\alpha}$, must be given by an adequate law of irreversible phenomenon for the phase that is considered, and this law calls upon temperature.

In the vast majority of situations where the temperature of the liquid phase plays a role, the kinetic energy of this liquid phase is perfectly negligible with respect to its internal energy. Moreover, the effect of gravity is only really important for meteorological applications.

The balance equation can thus be rewritten as:

$$\frac{\partial}{\partial t}(\rho_l\Phi_l h_l)+\frac{\partial}{\partial x_\alpha}(\rho_l\Phi_l v_l^\alpha h_l)=\frac{\partial}{\partial x_\alpha}(-\Phi_l j_l^{Q\alpha})+\Phi_l\frac{\partial p_l}{\partial t}+\rho_{ls}\dot{w}_l^{echs}h_l-j_l^{Q\alpha}n_l^\alpha\sigma_l^s \qquad [17.4]$$

The terms of power of the contact forces other than those due to pressure, i.e. viscosity for a liquid, have also been neglected.

By developing the first term of the equation:

$$h_l(\frac{\partial}{\partial t}(\rho_l\Phi_l)+\frac{\partial}{\partial x_\alpha}(\rho_l\Phi_l v_l^\alpha))+\rho_l\Phi_l\frac{\partial}{\partial t}(h_l)+\rho_l\Phi_l v_l^\alpha\frac{\partial}{\partial x_\alpha}(h_l)=$$

$$\frac{\partial}{\partial x_\alpha}(-\Phi_l j_l^{Q\alpha})+\Phi_l\frac{\partial p_l}{\partial t}+\rho_l\dot{w}_l^{echs}h_l-j_l^{Q\alpha}n_l^\alpha\sigma_l^s$$

However, the liquid mass gives:

$\frac{\partial}{\partial t}(\rho_l\Phi_l)+\frac{\partial}{\partial x_\alpha}(\rho_l\Phi_l v_l^\alpha)=\rho_l\dot{w}_l^{echs}$, so there remains:

$$\rho_l\Phi_l(\frac{\partial}{\partial t}(h_l)+v_l^\alpha\frac{\partial}{\partial x_\alpha}(h_l))=\frac{\partial}{\partial x_\alpha}(-\Phi_l j_l^{Q\alpha})+\Phi_l\frac{\partial p_l}{\partial t}-j_l^{Q\alpha}n_l^\alpha\sigma_l^s$$

If the pressure equilibrium between the gas and the liquid phases is assumed to take place at each instant, which would neglect the effects of capillarity, and the exchange of mass is assumed to be small (as in the case of vaporization–condensation), the pressure level in the liquid is that of the carrier gas phase, and in most cases its contribution to the enthalpy of the liquid can be neglected. It is, therefore, realistic to take the simple relationship for the thermal state equation: $h_l=C_lT_l+e_l^0$, where the calorific capacity and the formation energy are characteristic constants of the liquid considered. This is the same as neglecting the effects of the compression of the liquid with respect to the thermal effects, and it is therefore logical at the same time to neglect the temporal derivative of the pressure in the balance equation. An equation in terms of the liquid temperature can then be deduced:

$$\rho_l\Phi_lC_l(\frac{\partial}{\partial t}(T_l)+v_l^\alpha\frac{\partial}{\partial x_\alpha}(T_l))=\frac{\partial}{\partial x_\alpha}(-\Phi_l j_l^{Q\alpha})-j_l^{Q\alpha}n_l^\alpha\sigma_l^s \qquad [17.5]$$

If the temperature in a droplet is assumed to be uniform, the flux term in the liquid phase (the first term on the right side) is due to the exchange of heat between the droplets during collisions, coalescences and re-dispersions of these droplets. The final term on the right-hand side is the exchange of heat between the droplets and the surrounding gas phase.

Equation [17.5] is also valid even if the liquid phase is not simply in the form of droplets and is in no way dependent on the assumption that the temperature is uniform in the liquid. In the general case of liquid parcels of various sizes and shapes, even including the presence of very long tongues, the physical significance of the internal heat flux of the liquid phase is not limited to collisions between droplets, and takes into account conduction in the different regions of the liquid. Its local expression can be given simply by Fourier's law: $j_i^{Q\alpha} = -\lambda \dfrac{\partial T_l}{\partial x_\alpha}$, in every case, and if the droplets have a uniform temperature, it is only during contact between droplets, where the temperature gradient is very large, that the heat transfer takes place.

Moreover, the equation is also valid for solid phases, more or less dispersed as particles or parcels of various sizes and shapes. Obviously, in this case, the modeling of heat transfers in the solid or during contact between particles can also be done using Fourier's law for conduction, but radiation between these particles must be taken into account through more complex formulas.

If it is assumed, idealistically, that collisions between droplets are rare enough to be negligible (and the effect of gravity is also negligible), it can be deduced that along the trajectory of a droplet or a liquid parcel, the (homogeneous) temperature of the droplet only varies due to the heat flux exchanged with the gas. In general, the initial temperature of droplets is quite weak; the droplets are heated by the surrounding gas, and their temperature increases, but it cannot overtake the saturation temperature corresponding to the vapor content of their surrounding medium and in any case boiling temperature (for a medium made entirely of vapor from the droplet). In these conditions, the temperature of all the droplets is in between the injection temperature and boiling temperature. However, it must be taken into account that close to boiling temperature, small temperature differences result in large differences in partial pressure of the vapor in the gas phase and therefore of the vaporized mass flow rate.

17.3. Equation for the PDF of the temperature of the liquid

In the same way as in Chapter 16, the Lundgren method will be used. It is based on "fine-grained PDF", the probability that, at the point x and at the instant t, in *one single realization*, the temperature of the condensed phase $T_l(x,t)$ is equal to the random variable $\hat{T}$, at $d\hat{T}$ nearby. This probability is simply the Dirac distribution $\delta(\hat{T} - T_l(x,t))$. The probability density over all realizations is, by definition, the statistical average of the "fine-grained PDF", and the conditioned average in the liquid, which we are interested in, is such that:

$$\bar{\Phi}_l P_l(\hat{T}; x,t) = \overline{\Phi_l \delta(\hat{T} - T_l(x,t))}$$

It is then possible to find an equation for the PDF from balance equations for $T_l(x,t)$, in the form of [17.5]:

$\frac{\partial}{\partial t}(\delta(\hat{T}-T_l(x,t)))=\frac{\partial}{\partial \hat{T}}(\delta(\hat{T}-T_l)).(-\frac{\partial T_l}{\partial t})$, and the same for the partial derivative as a function of any spatial coordinate:

$$\frac{\partial}{\partial x_\alpha}(\delta(\hat{T}-T_l(x,t)))=\frac{\partial}{\partial \hat{T}}(\delta(\hat{T}-T_l)).(-\frac{\partial T_l}{\partial x_\alpha})$$

Thus, from the balance equation [17.5], it can be written:

$$\Phi_l\frac{\partial}{\partial t}(\delta(\hat{T}-T_l(x,t)))+\Phi_l v_l^\alpha\frac{\partial}{\partial x_\alpha}(\delta(\hat{T}-T_l(x,t)))=$$

$$-\frac{\partial}{\partial \hat{T}}(\delta(\hat{T}-T_l))(\frac{1}{\rho_l C_l}\frac{\partial}{\partial x_\alpha}(\Phi_l j_l^{Q\alpha})+\frac{1}{\rho_l C_l}j_l^{Q\alpha}n_l^\alpha\sigma_l^s)$$

After multiplication by the density and the use of the continuity equation for the liquid phase, the equation becomes:

$$\frac{\partial}{\partial t}(\rho_l\Phi_l\delta(\hat{T}-T_l))+\frac{\partial}{\partial x_\alpha}(\rho_l\Phi_l v_l^\alpha\delta(\hat{T}-T_l))-(\rho_l v_{rl}^\alpha n_l^\alpha\sigma_l^s)\delta(\hat{T}-T_l)=$$

$$-\frac{\partial}{\partial \hat{T}}(\delta(\hat{T}-T_l)(\frac{\partial}{\partial x_\alpha}(\Phi_l j_l^{Q\alpha}/C_l)-j_l^{Q\alpha}n_l^\alpha\sigma_l^s/C_l))$$

The average of the expression must now be taken. The density of the liquid is assumed to be constant here. If, moreover, the velocity fluctuation is again defined as the difference from its mean value, the following is obtained by taking the average of the equation above:

$$\frac{\partial}{\partial t}\left(\rho_l\bar{\Phi}_l P_l(\hat{T})\right)+\frac{\partial}{\partial x_\alpha}(\rho_l\bar{\Phi}_l P_l(\hat{T})\tilde{v}_l^\alpha)=\frac{\partial}{\partial x_\alpha}(-\overline{\rho_l\Phi_l v_l'^\alpha\,\delta(\hat{T}-T_l)})+\overline{\rho_l v_{rl}^\alpha n_l^\alpha\sigma_l^s\delta(\hat{T}-T_l)}$$

$$-\frac{\partial}{\partial \hat{T}}(\overline{\frac{\partial}{\partial x_\alpha}(\Phi_l j_l^{Q\alpha}/C_l)\delta(\hat{T}-T_l)}+\overline{(j_l^{Q\alpha}/C_l)n_l^\alpha\sigma_l^s\delta(\hat{T}-T_l)})\qquad[17.6]$$

Like in Chapter 16, the second last term can further be transformed, using Fourier's law:

$$\frac{\partial}{\partial \hat{T}}(\overline{\frac{\partial}{\partial x_\alpha}(\Phi_l j_l^{Q\alpha} / C_l \delta(\hat{T}-T_l))} = \overline{\frac{\partial}{\partial x_\alpha}(\Phi_l \frac{\lambda_l}{C_l}\frac{\partial T_l}{\partial x_\alpha})\frac{\partial}{\partial \hat{T}}(\delta(\hat{T}-T_l))}$$

$$= \overline{\frac{\partial}{\partial x_\alpha}(\frac{\lambda_l}{C_l}\Phi_l \frac{\partial T_l}{\partial x_\alpha}\frac{\partial \delta}{\partial \hat{T}})} - \overline{\frac{\lambda_l}{C_l}\Phi_l \frac{\partial T_l}{\partial x_\alpha}\frac{\partial}{\partial x_\alpha}(\frac{\partial \delta}{\partial \hat{T}})} =$$

$$\frac{\partial}{\partial x_\alpha}(-\overline{\frac{\lambda_l}{C_l}\Phi_l \frac{\partial \delta}{\partial x_\alpha}}) + \frac{\partial}{\partial \hat{T}}(\overline{\frac{\lambda_l}{C_l}\Phi_l \frac{\partial T_l}{\partial x_\alpha}\frac{\partial T_l}{\partial x_\alpha}\frac{\partial \delta}{\partial \hat{T}}}) =$$

$$\frac{\partial}{\partial x_\alpha}(-\overline{(\lambda_l / C_l)}\frac{\partial(P_l(\hat{T})\bar{\Phi}_l)}{\partial x_\alpha} + \overline{\frac{\lambda_l}{C_l} n_l^\alpha \sigma_l^s \delta})) + \frac{\partial^2}{\partial \hat{T}\partial \hat{T}}(\overline{\frac{\lambda_l}{C_l}\frac{\partial T_l}{\partial x_\alpha}\frac{\partial T_l}{\partial x_\alpha}\Phi_g \delta})$$

At the end of the calculation the fluctuations of the coefficient (λ_l / C_l), or rather their correlation with those of $\frac{\partial}{\partial x_\alpha}(\Phi_l \delta(\hat{T}-T_l))$, were neglected. Finally:

$$\frac{\partial}{\partial t}\left(\rho_l \bar{\Phi}_l P_l(\hat{T})\right) + \frac{\partial}{\partial x_\alpha}(\rho_l \bar{\Phi}_l P_l(\hat{T})\tilde{v}_l^\alpha) =$$

$$+\frac{\partial}{\partial x_\alpha}((\overline{(\lambda_l / C_l)}\bar{\Phi}_l \frac{\partial P_l}{\partial x_\alpha} - \overline{\rho_l \Phi_l v_l'^\alpha \delta(\hat{T}-T_l)} + \overline{(\lambda_l / C_l)}(P_l \frac{\partial \bar{\Phi}_l}{\partial x_\alpha} + \overline{n_l^\alpha \sigma_l^s \delta(\hat{T}-T_l)}))$$

$$-\frac{\partial^2}{\partial \hat{T}^2}(\overline{\frac{\lambda_l}{C_l}\Phi_l \frac{\partial T_l}{\partial x_\alpha}\frac{\partial T_l}{\partial x_\alpha}\delta(\hat{T}-T_l)}) - \frac{\partial}{\partial \hat{T}}((\overline{j_l^{Q\alpha} / C_l)n_l^\alpha \sigma_l^s \delta(\hat{T}-T_l)}) + \overline{\rho_l v_{rl}^\alpha n_l^\alpha \sigma_l^s \delta(\hat{T}-T_l)} \quad [17.7]$$

The physical interpretation of the terms is similar to that of Chapter 16. First, the equation concerns $\bar{\Phi}_l P_l(\hat{T})$ and not $P_l(\hat{T})$, and by this fact contains a term of exchange of the total mass, the last term on the right side. This ensures the conservation of the probability density. The first term on the right-hand side is a term of dispersion, with one part due to turbulence and another due to the presence of interfaces. The next term is a term of dissipation of the temperature fluctuations caused by conduction, analogous to "micro-mixing", and the second to last term on the right-hand side is a term due to the exchange of heat with the gas phase, which therefore provides the temperature fluctuation in the liquid. Apart from that which concerns the term of molecular diffusion, all these terms are not in a closed form and must be approximated so as to allow the integration of the equation.

17.4. Closure of the equation of the temperature PDF

Discussions regarding the terms of turbulent dispersion and dissipation of the fluctuations, that of "micro-mixing of the temperature", are identical in all points to those concerning these same terms for the case of the PDF of the fluctuations in composition of the gas phase, Chapter 16. This will not be discussed again, but it should be noted that Curl's "coalescence-redispersion" model has been suggested in conjunction with the scenario of droplets that come into collision, coalesce and therefore mix their constituents, and then re-disperse. Here, it is exactly the same thing, except looking at mixing internal energy and not the masses of constituents.

Nevertheless, it is important to ask, for the case of dispersed droplets, what the characteristic time of the fluctuations in velocity (of the droplets) that produce the micro-mixing must be. If the physical phenomenon is mixing between droplets that coalesce, it is reasonable to represent it as the ratio between a mean distance between droplets and an agitation velocity of the droplets. In the case of a dispersed medium, this is exactly what would provide the equation of the dissipation rate (see Chapter 6, section 6.2.3).

For the three other terms, it is significant to consider the situation where the liquid phase is principally dispersed in the form of droplets and where these droplets are small enough so that it is reasonable to assume that they have a uniform temperature. In this situation, there is no difference between the PDF of the temperature of the liquid and the PDF of the temperature of the liquid at the surface of this liquid.

First, this implies that $\overline{n_l^\alpha \sigma_l^s \delta(\hat{T}-T_l)} = \overline{n_l^\alpha \sigma_l^s} P_{ls}(\hat{T}) = \overline{n_l^\alpha \sigma_l^s} P_l(\hat{T}) = -\dfrac{\partial \overline{\Phi}_l}{\partial x_\alpha} P_l$; thus, the term of the dispersion due to interfaces is cancelled, quite simply.

Next, $+\overline{\rho_l v_{rl}^\alpha n_l^\alpha \sigma_l^s \delta(\hat{T}-T_l)}$ must be studied. Here again, the vaporized mass flow rate is modeled per unit interface area looking at the situation of an isolated droplet, as indicated in Chapter 5. This results in $\rho_l v_{rl}^\alpha n_l^\alpha = -m_V$, and its simplification as $\rho_l v_{rl}^\alpha n_l^\alpha = -m_V^{mg}(T_l)$ was mentioned at the beginning of this chapter. This leads simply to:

$$+\overline{\rho_l v_{rl}^\alpha n_l^\alpha \sigma_l^s \delta(\hat{T}-T_l)} = -m_V^{mg}(\hat{T}) \overline{\sigma}_l^s P_l(\hat{T}) \quad [17.8]$$

Next, the term of mean exchange of heat must be closed. It can be recalled that the term of exchange of heat with the gas phase can be modeled, with the help of study of the exchange around a droplet (Chapter 5, section 5.6.2), as a local function of the temperature of the liquid (at the surface):

$$j_l^{Q\alpha} n_l^\alpha = \dot{m}_V \left(\frac{C_g (T_{g\infty} - T_s)}{Y_{Vsat}(T_s) - Y_{V\infty}} (1 - Y_{Vsat}(T_s)) - L_{Vs} \right) \text{ with } \dot{m}_V = \frac{\rho_g D_V}{r_d} Ln\left(\frac{1 - Y_{V\infty}}{1 - Y_{Vsat}(T_s)} \right).$$

Still in accordance with the discussions at the beginning of this chapter, it can be formally written:

$$j_l^{Q\alpha} n_l^{\alpha} = \frac{\rho_g D_V}{3\bar{\Phi}_p} \bar{\sigma}_{lg}^s Ln\left(\frac{1-Y_V}{1-Y_{Vsat}(T_s)}\right)\left(\frac{C_g(T_g - T_s)}{Y_{Vsat}(T_s)-Y_V}(1-Y_{Vsat}(T_s)) - L_{Vs}\right) = q_l(T_s, T_g, Y_V, ..)$$

where other (assumed) non-fluctuating variables can also figure.

If the droplets are assumed to be uniform in temperature, $T_s = T_l$, it is consistent to assume that there is a decorrelation between the liquid temperature fluctuations and the fluctuations of the gas phase, so it is defined:

$$q_l^{mg}(T_l) = \frac{\rho_g D_V}{3\bar{\Phi}_p} \bar{\sigma}_{lg}^s \int_{\hat{Y}} P_g(\hat{Y}) Ln\left(\frac{1-Y_V}{1-Y_{Vsat}(T_l)}\right)\left(\frac{C_g(T_g(h_g, Y_V) - T_l)}{Y_{Vsat}(T_l)-Y_V}(1-Y_{Vsat}(T_l)) - L_{Vs}\right) d\hat{Y}$$

It can therefore be written:

$$\overline{(j_l^{Q\alpha} / C_l) n_l^{\alpha} \sigma_l^s \delta(\hat{T} - T_l)} = q_l^{mg}(\hat{T}) \bar{\sigma}_l^s P_l(\hat{T}) / C_l \qquad [17.9]$$

Formula [17.9] constitutes the closure hypothesis for the term of heat exchange with the gas phase. If it is impossible to assume the liquid parcels are homogenous in temperature, the problem becomes a lot more complicated. This requires a distinction between the PDFs of the temperature at the surface and in the liquid.

Chapter 18

Study of the PDF for Velocity Fluctuations and Sizes of Parcels

Lundgren's method, described in Chapter 16, can also be used to find the equation of the velocity probability density function (PDF) for any phase, no matter whether it is dispersed or not. In these conditions, the use of an equation for the kinetic energy of the turbulence's velocity fluctuations in this phase, or for the complete Reynolds tensor, is not necessary as the PDF can be used to find them. It will be seen that the method can be adapted and has already been used for single-phase flows, although a problem then arises relative to the connection of velocities in the other phase because the contact force depends on it.

In addition, it was shown in Chapter 4, section 4.5, that it was possible to obtain more information on the dispersed phase from the equations for "numbers per unit volume" of several classes of parcels of different volumes. Until now, the exchange terms between phases were included by a mean interface area per unit volume, but it has been shown that these terms are not simply expressed as proportional to this mean interface area. If more is known on the size distribution of parcels of each dispersed phase, then these exchanges between phases, for the case here, can be assessed more accurately. In principle, using these numerical density equations for classes enables the joint distribution of particle sizes and velocities to be found with, of course, a defined class resolution.

This chapter will discuss these two possible improvements. A link must then be made with the Euler–Lagrange approaches, which are widely used for low-density dispersed media and enable the detailed study of all these aspects. Finally, it will be shown how the stochastic simulation of the Lagrangian equations is very similar to PDF calculation.

18.1. Phase velocity PDF equation

As an example, a general liquid phase will be considered here, not necessarily able to disperse or vaporize, and the approach used can clearly be transposed for a gas phase or even a dispersed solid phase.

The basic equations are the instant mass and momentum equations for the phase, rewritten here by neglecting the effects linked to surface tension and gravity. For clarity, it is assumed that the liquid has a constant density:

$\frac{\partial}{\partial t}(\rho_l \Phi_l)+\frac{\partial}{\partial x_\alpha}(\rho_l \Phi_l v_l^\alpha)=\rho_l \dot{w}_l^{ec\,hs}$, with a vaporized mass flow rate $\rho_l \dot{w}_l^{ec\,hs}=$ $-\rho_l v_{rl}^\alpha n_l^\alpha \sigma_l^s$, and $\frac{\partial}{\partial t}(\rho_l \Phi_l v_l^\beta)+\frac{\partial}{\partial x_\alpha}(\rho_l \Phi_l v_l^\alpha v_l^\beta)=\frac{\partial(\Phi_l \sigma_l^{\alpha\beta})}{\partial x_\alpha}+\rho_l \dot{w}_l^{echs} v_l^\beta+\sigma_l^{\alpha\beta} n_l^\alpha \sigma_l^s$

The last term is the contact force with the gas phase.

The fact that the density of the liquid is assumed constant avoids the need to couple these equations with the energy equation. The momentum balance equation can then be rewritten as:

$$\rho_l \Phi_l \frac{\partial}{\partial t}(v_l^\beta)+\rho_l \Phi_l v_l^\alpha \frac{\partial}{\partial x_\alpha}(v_l^\beta)=\frac{\partial}{\partial x_\alpha}(\Phi_l \sigma_l^{\alpha\beta})+\sigma_l^{\alpha\beta} n_l^\alpha \sigma_l^s \qquad [18.1]$$

The "fine-grained PDF" will now be considered in the three-dimensional space created by the random variables corresponding to the three velocity components, and of which the conditioned average in the liquid phase provides the PDF of interest here:

$$\bar{\Phi}_l P_l(\hat{v};x,t)=\overline{\prod_{i=1,2,3} \Phi_l \delta(\hat{v}_i - v_l^i(x,t))}=\overline{\Phi_l \delta(\hat{v} - v_l)}\,.$$

In the same way as before, $\frac{\partial}{\partial t}(\delta(\hat{v}-v_l(x,t)))=\sum_i \frac{\partial}{\partial \hat{v}_i}(\delta(\hat{v}-v_l)).(-\frac{\partial v_l^i}{\partial t})$, and $\frac{\partial}{\partial x_\alpha}(\delta(\hat{v}-v_l(x,t)))=\sum_i \frac{\partial}{\partial \hat{v}_i}(\delta(\hat{v}-v_l)).(-\frac{\partial v_l^i}{\partial x_\alpha})$.

Then, using balance equation [18.1], it is written that

$$\Phi_l \frac{\partial}{\partial t}(\delta(\hat{v}-v_l(x,t)))+\Phi_l v_l^\alpha \frac{\partial}{\partial x_\alpha}(\delta(\hat{v}-v_l(x,t)))=$$
$$-\sum_i \frac{\partial}{\partial \hat{v}_i}(\delta(\hat{v}-v_l))(\frac{1}{\rho_l}\frac{\partial}{\partial x_\alpha}(\Phi_l \sigma_l^{\alpha i})+\frac{1}{\rho_l}\sigma_l^{\alpha i} n_l^\alpha \sigma_l^s)$$

and also:

$$\frac{\partial}{\partial t}(\rho_l \Phi_l \delta(\hat{v}-v_l))+\frac{\partial}{\partial x_\alpha}(\rho_l \Phi_l v_l^\alpha \delta(\hat{v}-v_l))=$$
$$-\sum_i \frac{\partial}{\partial \hat{v}_i}(\delta(\hat{v}-v_l)(\frac{\partial}{\partial x_\alpha}(\Phi_l \sigma_l^{\alpha i})+\sigma_l^{\alpha i} n_l^\alpha \sigma_l^s))-(\rho_l v_{rl}^\alpha n_l^\alpha \sigma_l^s)\delta(\hat{v}-v_l)$$

The average of the equation must now be taken. As a result, the following is obtained:

$$\frac{\partial}{\partial t}\left(\rho_l \bar{\Phi}_l P_l(\hat{v})\right)+\frac{\partial}{\partial x_\alpha}(\rho_l \bar{\Phi}_l P_l(\hat{v}) v_l^\alpha)=-\overline{\rho_l v_{rl}^\alpha n_l^\alpha \sigma_l^s \delta(\hat{v}-v_l)}$$
$$-\sum_i \frac{\partial}{\partial \hat{v}_i}(\overline{\frac{\partial}{\partial x_\alpha}(\Phi_l \sigma_l^{\alpha i})\delta(\hat{v}-v_l)}+\overline{\sigma_l^{\alpha i} n_l^\alpha \sigma_l^s \delta(\hat{v}-v_l)}) \quad [18.2]$$

By splitting the Cauchy stress tensor in the same way as in Chapter 4, section 4.2.2, i.e. by making the mean liquid pressure on the interface appear via $\sigma_l^{\alpha\beta}=-\bar{p}_l^s \delta_{\alpha\beta}+\tau_l^{\times\alpha\beta}$, the following relationship is found:

$$\frac{\partial}{\partial t}\left(\rho_l \bar{\Phi}_l P_l(\hat{v})\right)+\frac{\partial}{\partial x_\alpha}(\rho_l \bar{\Phi}_l P_l(\hat{v}) v_l^\alpha)=-\overline{\rho_l v_{rl}^\alpha n_l^\alpha \sigma_l^s \delta(\hat{v}-v_l)}$$
$$-\sum_i \frac{\partial}{\partial \hat{v}_i}(\overline{\frac{\partial}{\partial x_\alpha}(\Phi_l \tau_l^{\times\alpha i})\delta(\hat{v}-v_l)}-\overline{\frac{\partial}{\partial x_i}(\Phi_l \bar{p}_l^s)\delta(\hat{v}-v_l)}+\bar{p}_l^s\overline{\frac{\partial \Phi_l}{\partial x_i}\delta(\hat{v}-v_l)}+\overline{\tau_l^{\times\alpha i} n_l^\alpha \sigma_l^s \delta(\hat{v}-v_l)})$$

or
$$\frac{\partial}{\partial t}\left(\rho_l \bar{\Phi}_l P_l(\hat{v})\right)+\frac{\partial}{\partial x_\alpha}(\rho_l \bar{\Phi}_l P_l(\hat{v}) v_l^\alpha)=-\overline{\rho_l v_{rl}^\alpha n_l^\alpha \sigma_l^s \delta(\hat{v}-v_l)}$$
$$-\sum_i \frac{\partial}{\partial \hat{v}_i}(\overline{\frac{\partial}{\partial x_\alpha}(\Phi_l \tau_l^{\times\alpha i})\delta(\hat{v}-v_l)}-\bar{\Phi}_l \frac{\partial \bar{p}_l^s}{\partial x_i} P_l(\hat{v})+\overline{\tau_l^{\times\alpha i} n_l^\alpha \sigma_l^s \delta(\hat{v}-v_l)}) \quad [18.3]$$

Tensor $\tau_l^{\times\alpha\beta}$ is not only viscous, but also includes fluctuations in the liquid's pressure, relative to the mean interface pressure. This tensor must be taken into account to make the "micro-mixing" term appear responsible for the viscous dissipation of velocity fluctuations in the liquid itself. The following relationship is reused:

$$-\sum_i \frac{\partial}{\partial \hat{v}_i}\left(\overline{\frac{\partial}{\partial x_\alpha}(\Phi_l \tau_l^{\times \alpha i})\delta(\hat{v}-v_l)}\right) =$$

$$+\sum_i \frac{\partial}{\partial \hat{v}_i}\left(\overline{\Phi_l \frac{\partial p'_l}{\partial x_i}}\delta(\hat{v}-v_l)\right) - \sum_i \frac{\partial}{\partial \hat{v}_i}\left(\overline{\frac{\partial}{\partial x_\alpha}(\Phi_l \tau_l^{\alpha i})\delta(\hat{v}-v_l)}\right),$$

and by bringing in the previous relationship [18.3], we obtain:

$$\frac{\partial}{\partial t}\left(\rho_l \bar{\Phi}_l P_l(\hat{v})\right) + \frac{\partial}{\partial x_\alpha}(\rho_l \bar{\Phi}_l P_l(\hat{v}) v_l^\alpha) = -\overline{\rho_l v_{rl}^\alpha n_l^\alpha \sigma_l^s \delta(\hat{v}-v_l)}$$

$$+\sum_i \frac{\partial}{\partial \hat{v}_i}\left(\bar{\Phi}_l \frac{\partial \bar{p}_l^s}{\partial x_i} P_l(\hat{v}) + \overline{\frac{\partial p'_l}{\partial x_i}\Phi_l \delta(\hat{v}-v_l)} - \overline{\tau_l^{\times \alpha i} n_l^\alpha \sigma_l^s \delta(\hat{v}-v_l)}\right)$$

$$-\sum_i \frac{\partial}{\partial \hat{v}_i}\left(\overline{\frac{\partial}{\partial x_\alpha}(\Phi_l \tau_l^{\alpha i})\delta(\hat{v}-v_l)}\right) \qquad [18.4]$$

The standard structure of a PDF equation can be recognized. The first term of the second member is due to the fact that the equation deals with $\bar{\Phi}_l P_l(\hat{v})$, the integral of which is not constant and equal to 1, the others are transports in the random variable space. The second term is the contribution of the mean pressure gradient which increases or decreases the velocity component according to its value in the direction of this component. The third term shows that the fluctuating pressure gradient modifies the shape of the PDF. The fourth term includes the effect of viscosity in the liquid phase, which decreases fluctuations around the mean value.

The last term can be split into two parts by bringing in Newtonian viscosity:
$$\tau_l^{\alpha\beta} = \mu_l\left(\frac{\partial v_l^\alpha}{\partial x_\beta} + \frac{\partial v_l^\beta}{\partial x_\alpha}\right) = 2\mu_l \frac{\partial v_l^\beta}{\partial x_\alpha} + \mu_l\left(\frac{\partial v_l^\alpha}{\partial x_\beta} - \frac{\partial v_l^\beta}{\partial x_\alpha}\right)$$

Effectively, $\sum_i \frac{\partial}{\partial \hat{v}_i}\left(\overline{\frac{\partial}{\partial x_\alpha}(\tau_l^{\alpha i}\Phi_l)\delta(\hat{v}-v_l)}\right) =$

$$\sum_i \frac{\partial}{\partial \hat{v}_i}\left(\overline{\frac{\partial}{\partial x_\alpha}\left(\mu_l \Phi_l \frac{\partial v_l^i}{\partial x_\alpha}\right)\delta(\hat{v}-v_l)}\right) + \sum_i \frac{\partial}{\partial \hat{v}_i}\left(\overline{\frac{\partial}{\partial x_\alpha}\left(\mu_l \Phi_l \frac{\partial v_l^\alpha}{\partial x_i}\right)\delta(\hat{v}-v_l)}\right)$$

On the one hand,

$$\sum_i \frac{\partial}{\partial \hat{v}_i}\overline{(\frac{\partial}{\partial x_\alpha}(\mu_l \Phi_l \frac{\partial v_l^i}{\partial x_\alpha})\delta} = \sum_i \frac{\partial}{\partial \hat{v}_i}\overline{(\frac{\partial}{\partial x_\alpha}(\mu_l \Phi_l \frac{\partial v_l^i}{\partial x_\alpha}\delta)} - \sum_i \frac{\partial}{\partial \hat{v}_i}\overline{\mu_l \Phi_l \frac{\partial v_l^i}{\partial x_\alpha}\frac{\partial \delta}{\partial x_\alpha}} =$$

$$-\frac{\partial}{\partial x_\alpha}\overline{(\mu_l \Phi_l \sum_i \frac{\partial \delta}{\partial \hat{v}_i}.\frac{\partial v_l^i}{\partial x_\alpha})} + \sum_i \frac{\partial}{\partial \hat{v}_i}\overline{(\mu_l \Phi_l \frac{\partial v_l^i}{\partial x_\alpha}\sum_j \frac{\partial \delta}{\partial \hat{v}_j}.\frac{\partial v_l^j}{\partial x_\alpha})} =$$

$$-\frac{\partial}{\partial x_\alpha}\overline{(\mu_l \Phi_l \frac{\partial \delta}{\partial x_\alpha})} + \sum_i \sum_j \frac{\partial^2}{\partial \hat{v}_i \partial \hat{v}_j}\overline{(\mu_l \Phi_l \frac{\partial v_l^i}{\partial x_\alpha}.\frac{\partial v_l^j}{\partial x_\alpha}\delta)}$$

and on the other, because of the liquid's (assumed) incompressibility:

$$\sum_i \frac{\partial}{\partial \hat{v}_i}\overline{(\frac{\partial}{\partial x_\alpha}(\mu_l \Phi_l \frac{\partial v_l^\alpha}{\partial x_i})\delta(\hat{v}-v_l))} = \sum_i \frac{\partial}{\partial \hat{v}_i}\overline{(\mu_l \frac{\partial v_l^\alpha}{\partial x_i}\frac{\partial \Phi_l}{\partial x_\alpha}\delta(\hat{v}-v_l))} + 0 =$$

$$-\sum_i \frac{\partial}{\partial \hat{v}_i}\overline{(\mu_l \frac{\partial v_l^\alpha}{\partial x_i} n_l^\alpha \sigma_l^s \delta(\hat{v}-v_l))}$$

In total, the following is obtained: $\sum_i \frac{\partial}{\partial \hat{v}_i}\overline{(\frac{\partial}{\partial x_\alpha}(\tau_l^{\alpha i}\Phi_l)\delta(\hat{v}-v_l))} =$

$$-\frac{\partial}{\partial x_\alpha}\overline{(\mu_l \Phi_l \frac{\partial \delta}{\partial x_\alpha})} + \sum_i \sum_j \frac{\partial^2}{\partial \hat{v}_i \partial \hat{v}_j}\overline{(\mu_l \Phi_l \frac{\partial v_l^i}{\partial x_\alpha}.\frac{\partial v_l^j}{\partial x_\alpha}\delta)} - \sum_i \frac{\partial}{\partial \hat{v}_i}\overline{(\mu_l \frac{\partial v_l^\alpha}{\partial x_i} n_l^\alpha \sigma_l^s \delta(\hat{v}-v_l))}$$

First, a part for molecular diffusion is identified in the PDF, as in previous chapters, and finally:

$$\sum_i \frac{\partial}{\partial \hat{v}_i}\overline{(\frac{\partial}{\partial x_\alpha}(\tau_l^{\alpha i}\Phi_l)\delta(\hat{v}-v_l))} = -\frac{\partial}{\partial x_\alpha}(\mu_l \bar{\Phi}_l \frac{\partial P_l}{\partial x_\alpha}) - \frac{\partial}{\partial x_\alpha}(\mu_l (P_l - P_{ls})\frac{\partial \bar{\Phi}_l}{\partial x_\alpha})$$

$$+\sum_i \sum_j \frac{\partial^2}{\partial \hat{v}_i \partial \hat{v}_j}\overline{(\mu_l \Phi_l \frac{\partial v_l^i}{\partial x_\alpha}.\frac{\partial v_l^j}{\partial x_\alpha}\delta)} - \sum_i \frac{\partial}{\partial \hat{v}_i}\overline{(\mu_l \frac{\partial v_l^\alpha}{\partial x_i} n_l^\alpha \sigma_l^s \delta(\hat{v}-v_l))}.$$

In addition to the molecular diffusion of the PDF in physical space, a standard-form term for "micro-mixing" is found which decreases the velocity, in addition to two terms linked to the presence of interfaces, one that corrects molecular diffusion and the other that corrects micro-mixing.

By summarizing the PDF equation, the following is obtained:

$$
\begin{aligned}
&\frac{\partial}{\partial t}\left(\rho_l \bar{\Phi}_l P_l(\hat{v})\right)+\frac{\partial}{\partial x_\alpha}(\rho_l \bar{\Phi}_l P_l(\hat{v}) v_l^\alpha)=\frac{\partial}{\partial x_\alpha}(\mu_l \Phi_l \frac{\partial P_l}{\partial x_\alpha})+\frac{\partial}{\partial x_\alpha}(\mu_l (P_l \frac{\partial \bar{\Phi}_l}{\partial x_\alpha}-\overline{n_l^\alpha \sigma_l^s \delta(\hat{v}-v_l)}))\\
&+\sum_i \frac{\partial}{\partial \hat{v}_i}(\bar{\Phi}_l \frac{\partial \bar{p}_l^s}{\partial x_i} P_l(\hat{v})+\overline{\frac{\partial p'_l}{\partial x_i}\Phi_l \delta(\hat{v}-v_l)}-\overline{\tau_l^{\times\alpha i} n_l^\alpha \sigma_l^s \delta(\hat{v}-v_l)})\\
&-\sum_i \sum_j \frac{\partial^2}{\partial \hat{v}_i \partial \hat{v}_j}(\mu_l \overline{\Phi_l \frac{\partial v_l^i}{\partial x_\alpha}.\frac{\partial v_l^j}{\partial x_\alpha}\delta(\hat{v}-v_l)})+\sum_i \frac{\partial}{\partial \hat{v}_i}(\mu_l \overline{\frac{\partial v_l^\alpha}{\partial x_i} n_l^\alpha \sigma_l^s \delta(\hat{v}-v_l)})\\
&-\overline{\rho_l v_{rl}^\alpha n_l^\alpha \sigma_l^s \delta(\hat{v}-v_l)}
\end{aligned}
\qquad [18.5]
$$

It is noticeable that this equation has an identical structure to that of the PDFs in previous chapters. Some terms are closed, and some terms still need to be modeled. Among the latter, some terms are similar to those that appear in single-phase problems: micro-mixing, which decreases velocity fluctuations, and the effect of pressure fluctuations which, in contrast, increase them. The effect of the mean pressure gradient plays the same role as an external forcefield that accelerates, or decelerates, all the velocities that are oriented in the same direction, and is a term that requires closure hypotheses once the mean liquid pressure gradient is known (sometimes equal to the mean pressure gradient in the gas phase). Had gravitational effects been kept, it would have "*in fine*" appeared as a term similar to the mean pressure gradient. Finally, some terms represent the influence of the other phase: the last term is due to mass exchange and it explains the variation of the liquid volume fraction through evaporation or condensation. There is a term linked to viscous friction with the gas phase and two additional terms due to viscous action at interfaces. The first of these corrects the term for diffusion in physical space and the second corrects the term for micro-mixing. The first term has already been discussed in previous chapters and is proportional to the difference between the PDF for velocities in the liquid and the same PDF at the surface of liquid parcels. In the context of velocities, the interfaces do not play a significant part if the influence of surface tension and mass exchange on the velocities is neglected: velocity fields are continuous as well as the tangential stress and the pressure. There is, therefore, a tendency to assume that the PDF for velocities in the liquid or at its surface must be identical. This should be studied in more depth by performing direct numerical simulations. The second: $-\sum_i \frac{\partial}{\partial \hat{v}_i}(\mu_l \overline{\frac{\partial v_l^\alpha}{\partial x_i} n_l^\alpha \sigma_l^s \delta(\hat{v}-v_l)})$ is an additional contribution to dissipation linked to viscosity at the interface. According to its form, it is the effect of the gradient of the velocity component normal to the interface, different from the friction at the interface. When the liquid medium is highly turbulent, this term is expected to be negligible in comparison to the dissipation term itself, similar to the molecular diffusion term in comparison to all others. They can only be significant in the case of a medium that is only slightly turbulent.

In the following section, the way in which all these terms can be modeled will be discussed.

18.2. Modeling the exchanges between phases and the internal interactions

18.2.1. *Terms of exchanges between phases*

The term for *exchange by contact force* is given by $\overline{-\tau_l^{\times\alpha i} n_l^{\alpha} \sigma_l^{s} \delta(\hat{v}-v_l)}$, other than the generalized Archimedes force, and a problem, analogous to the term for heat and mass exchange in previous chapters, arises when it is modeled: the involved stress tensor, as well as the interfacial area per unit volume, does not appear among the variables in the PDF. However, the approximate relationships found from local configurations, presented in Chapter 5, can be of help for modeling. It will be assumed, for now, that the contact force between phases is restricted to a friction force.

If the liquid is assumed to be in the form of drops, for example, and we adopt the Stokes equation, the force exerted on a particle is given by: $F_l^{\alpha} = 6\pi v_f r_p \rho_f (v_f^{\alpha} - v_p^{\alpha})$. In the previous equation, the gas velocity (noted "f") is assumed to be far from the drop, the drop's velocity (noted "p") is the velocity of drop's center which does not distort or spin. By assuming that all drops are identical (an assumption that will be re-examined later), the following is obtained, per unit area of the liquid:

$$F_{ls}^{\alpha} = \frac{F_l^{\alpha}}{4\pi r_p^2} = \frac{\overline{\sigma}_p^s}{2\overline{\Phi}_p} \rho_f v_f (v_f^{\alpha} - v_p^{\alpha}) \qquad [18.6a]$$

For liquid–gas interfaces with a low curvature, a definition for the tangential force per unit area was discussed in Chapter 5:

$$F_{sl}^{ft} = \frac{\mu_l \mu_g}{(\mu_l d + \mu_g d')(d+d')}(V_g - V_l)$$

and with identical "friction elements", in projection in the direction α, the force per unit area of interface is calculated as:

$$F_{ls}^{\alpha} = \frac{\mu_l \mu_g}{(\mu_l \overline{\Phi}_g + \mu_g \overline{\Phi}_l)} \overline{\sigma}_l^s (V_g^{\alpha} - V_l^{\alpha}) \qquad [18.6b]$$

The velocity v_p in [18.6a] is nothing other than what is called fluid velocity v_l in the previous equations and V_l in [18.6b] is also a liquid velocity at a given distance from the interface, for which the fluctuation PDF is expected to be similar to $P_l(\hat{v})$, studied here. There is some doubt as to whether v_f and V_g can be identified to v_g, or whether they only have the same PDF as v_g. Given that the dynamic viscosity in the gas is usually larger than that in the liquid (because of density), the instant velocity gradient in the gas at the interface is larger. In any case, the following can be written:

$$\overline{-\tau_l^{\alpha i} n_l^\alpha \sigma_l^s \delta(\hat{v}-v_l)} = -\overline{F_{ls|v}^i}(\hat{v})\bar{\sigma}_l^s P_l(\hat{v}), \qquad [18.7]$$

where $\overline{F_{ls|v}^i}$ is the mean value conditioned by $v_l = \hat{v}_l$, the force F_{ls}^i per unit area of the surface given above. It can also be defined that from [18.6a] for droplets:

$$\overline{F_{ls|v}^i} = \frac{1}{2}\rho_f V_f \frac{\bar{\sigma}_l^s}{\bar{\Phi}_l}(\tilde{v}_g^i - \hat{v}^i) + \frac{1}{2}\rho_f V_f \frac{\bar{\sigma}_l^s}{\bar{\Phi}_l}\overline{v''^i_{f|\hat{v}}} \qquad [18.8a]$$

and from [18.6b]:

$$\overline{F_{ls|v}^i} = \frac{\mu_l \mu_g}{\mu_l \bar{\Phi}_g + \mu_g \bar{\Phi}_l}\bar{\sigma}_l^s(\tilde{v}_g^i - \hat{v}^i) + \frac{\mu_l \mu_g}{\mu_l \bar{\Phi}_g + \mu_g \bar{\Phi}_l}\bar{\sigma}_l^s \overline{v''^i_{f|\hat{v}}} \qquad [18.8b]$$

Velocity $\overline{v''^i_{f|\hat{v}}}$ is the mean fluctuation of this “far-away” gas velocity v_f, relative to $\tilde{v}_g$, but is restricted by the fact that the liquid’s instant velocity has a value $\hat{v}$. The zero value for $\overline{v''^i_{f|\hat{v}}}$ is questioned for two reasons: first, the PDF for v_f is not necessarily the same as $P_g(\hat{v}_g)$, and, in particular, $\tilde{v}_f \neq \tilde{v}_g$ is the justification of the definition of the Simonin’s drift velocity (see Chapter 5); second, one expects that fluctuations of v_g and v_l still correlate. If equation [18.7] is used for the model, an approximation for $\overline{v''^i_{f|\hat{v}}}$ is, therefore, required and the first terms in [18.8a] and [18.8b] are closed and are no longer a problem.

The presence of this term in [18.7] is often linked to an increase in the velocity fluctuations of the liquid, as revealed in the study of the kinetic energy equation for phase fluctuations (Chapter 6). The recent article by Jones and colleagues looked into this phenomenon (see [JON 10]). They proposed to model the effect by a stochastic Wiener process, using a Lagrangian simulation for the PDF. This corresponds to making the assumption that the PDF equation is of a Fokker–Planck type, and this will be discussed in more detail in the next section. Using [18.18], this comes back to modeling the entire term as:

$$+\sum_i \frac{\partial}{\partial \hat{v}_i}(-\overline{\tau_l^{\alpha i} n_l^\alpha \sigma_l^s \delta(\hat{v}-v_l)}) = -\sum_i \frac{\partial}{\partial \hat{v}_i}(\frac{1}{2}\mu_f \frac{(\bar{\sigma}_l^s)^2}{\bar{\bar{\Phi}}_l}(\tilde{v}_g^i - \hat{v}^i)P_l(\hat{v})) + \frac{1}{2}\sum_i \frac{\partial}{\partial \hat{v}_i}\frac{\partial}{\partial \hat{v}_i}(C_{F0}\frac{\mu_f}{\bar{\bar{\Phi}}_l}\varepsilon_g^*(\bar{\sigma}_l^s)^2 P_l(\hat{v})) \quad [18.9]$$

The latter term is influenced by turbulence and corresponds to the effect of $\overline{v''^i_{f|\hat{v}}}$, through its dissipation rate ε_g^*. We can wonder whether the characteristic time of particles should also be involved. C_{F0} is a constant. In order to predict the effect of the fluid's turbulence on the dispersion of heavier particles, it is already known that the correlation time for the disturbing effect also depends on the time during which the drops remain in a "fluid eddy", inducing what is termed a "crossing trajectory" correction. It will be seen later that this type of modeling will be used also to represent the effect of pressure fluctuations in the gas phase itself.

One could also use the modeling results obtained by O. Simonin and his colleagues [SIM 00] concerning the drift velocity, to get another model for this term, maybe in a form that is different from [18.9], and compare these two approaches. If the effect of the mean pressure gradient, i.e. generalized Archimedes force, on the contact force needs to be taken into account, then a force per unit volume must be added, $F_{sP}^i = -\bar{\Phi}_l \frac{\partial \bar{p}_l^*}{\partial x_i}$, the classical local pressure hypotheses allows the mean gas pressure to appear, and there is no direct coupling with the liquid's velocity fluctuations. This leads simply to add:

$$-\sum_i \frac{\partial}{\partial \hat{v}_i}(\bar{\Phi}_l \frac{\partial \bar{p}_g^*}{\partial x_i} P_l(\hat{v}))$$

It is also possible to take into account the lift force linked to the curl of the fluid mean velocity.

The term for *mass exchange* is given by $-\overline{\rho_l v_{rl}^\alpha n_l^\alpha \sigma_l^s \delta(\hat{v}-v_l)}$, and it was shown in the previous chapter that a similar term is involved in the PDF for the liquid's temperature. A similar modeling method can be applied here.

In the previous chapter, the vaporized mass flow rate per unit surface area was modeled by the mass flow rate of drops, thus providing:

$\rho_l v_{rl}^{\alpha} n_g^{\alpha} = \dot{m}_V = \dfrac{\rho_g D_V}{r_d} Ln\left(\dfrac{1-Y_{V\infty}}{1-Y_{Vsat}(T_s)}\right)$; the formula was then simplified to $\dot{m}_V^{mg}(T_l) = \dfrac{\rho_g D_V}{3\bar{\Phi}_p} \bar{\sigma}_{lg}^s Ln\left(\dfrac{1-\tilde{Y}_V}{1-Y_{Vsat}(T_l)}\right)$. In these conditions, the mass exchange no longer depends on the velocity fluctuations in the liquid or the gas, and it is simple to calculate that:

$$-\overline{\rho_l v_{rl}^{\alpha} n_l^{\alpha} \sigma_l^s \delta(\hat{v}-v_l)} = \int_{\hat{T}} P_l(\hat{T}) \dot{m}_V^{mg}(\hat{T}) d\hat{T} . \bar{\sigma}_l^s P_l(\hat{v}) \qquad [18.10]$$

If one does not want to take into account the PDF for the liquid's temperature, the formula can be simplified to:

$$-\overline{\rho_l v_{rl}^{\alpha} n_l^{\alpha} \sigma_l^s \delta(\hat{v}-v_l)} = \dot{m}_V^{mg}(\tilde{T}_l) \bar{\sigma}_l^s P_l(\hat{v}) \qquad [18.11]$$

It is also easily feasible to take into account the effect of relative velocities between drops and the gas by correcting the initial formula with a factor that depends on the drop's Reynolds number, which involves mean relative velocity as well as the gas turbulence via its kinetic energy and length scale, as was discussed in Chapter 5, sections 5.3.1 and 5.3.3. This does not change anything else. However, the fluctuations in relative velocity can also be taken into account more exactly by simply using the method used above for the friction term. This type of modeling is also used in the work by Jones *et al.*, discussed above, and who again model an increase of fluctuations by means of a Wiener process. It remains to be seen whether this significantly changes the final results, relative to the consideration of the mean velocity and turbulence alone.

18.2.2. *Internal dissipation and production of fluctuations*

Terms $P_{PDF} = \sum_i \dfrac{\partial}{\partial \hat{v}_i} \overline{\dfrac{\partial p'_l}{\partial x_i} \Phi_l \delta(\hat{v}-v_l)}$ and $D_{PDF} = -\sum_i \sum_j \dfrac{\partial^2}{\partial \hat{v}_i \partial \hat{v}_j} \overline{(\mu_l \Phi_l \dfrac{\partial v_l^i}{\partial x_\alpha} . \dfrac{\partial v_l^j}{\partial x_\alpha} \delta(\hat{v}-v_l))}$ also appear in the PDF equation for the velocity of a single-phase turbulent medium (with $\Phi_l = 1$), and it is, therefore, possible to use standard models. The second term brings into play pseudo-dissipation $\rho_l \varepsilon_{ij} = \mu_l \dfrac{\partial v_l^i}{\partial x_\alpha} . \dfrac{\partial v_l^j}{\partial x_\alpha}$, classically assumed to be isotropic and independent of velocity fluctuations at large scale so that the following is obtained:

$$-\sum_i \sum_j \frac{\partial^2}{\partial \hat{v}_i \partial \hat{v}_j} \overline{(\mu_l \Phi_l \frac{\partial v_l^i}{\partial x_\alpha} \cdot \frac{\partial v_l^j}{\partial x_\alpha} \delta(\hat{v} - v_l))} = -\sum_i \frac{\partial}{\partial \hat{v}_i} \frac{\partial}{\partial \hat{v}_i} (\frac{1}{3} \bar{\Phi}_l \rho_l \bar{\varepsilon}_l^* P_l(\hat{v})) .$$

The first term is interpreted as the source of velocity fluctuations, which also leads to their isotropization. Standard models for both these two terms adopt a form which provides a "Fokker–Planck"-type PDF equation; thus:

$$P_{PDF} + D_{PDF} = -\sum_i \frac{\partial}{\partial \hat{v}_i} (D_i \bar{\Phi}_l P_l(\hat{v})) + \frac{1}{2} \sum_i \frac{\partial}{\partial \hat{v}_i} \frac{\partial}{\partial \hat{v}_i} (C_0 \bar{\Phi}_l \bar{\varepsilon}_l^* P_l(\hat{v})) \qquad [18.12]$$

where

$$D_i = -\rho_l \frac{\hat{v}_i - \tilde{v}_l^i}{\tau_L} + \rho_l G_{ij}^a (\hat{v}_i - \tilde{v}_l^i) \text{ and } \frac{1}{\tau_L} = (\frac{1}{2} + \frac{3}{4} C_0) \frac{\bar{\varepsilon}_l^*}{k_l} \qquad [18.13]$$

This modeling allows standard models for the Reynolds tensor to be recovered (in the case of single-phase turbulent flows), taking into account that tensor G_{ij}^a has several variants. The reader may refer to the more in-depth discussions by Minier and Pozorski [MIN 97] and Pope [POP 85], and particularly the review article by Minier and Peirano [MIN 01].

It will also be remarked that the characteristic time τ_L of velocity fluctuations in the liquid, introduced in [18.13], linked to the dissipation rate $\bar{\varepsilon}_l^*$, can take into account both the length scale of turbulent velocity fluctuations in the liquid and the length scale of liquid parcel separation, as discussed in Chapter 6.

Standard models for single-phase flows have always been relevant to highly turbulent flows. The terms linked to molecular viscosity have been always neglected. In other cases, the term for the PDF's molecular diffusion in physical space, already closed, could be calculated, but the additional terms found are linked to interfaces that would have to be modeled. The problem then needs to be studied from the start.

18.3. Practical calculation of PDF

Similar to the PDF given in Chapter 16 for composition, the practical calculation of the PDF for liquid velocities is not simple due to the number of variables in the equation.

For this reason, it has already been mentioned that Monte Carlo stochastic simulations are generally used, moreover with the tracking of stochastic particles in a Lagrangian context.

It can be simply said that this method aims to follow a relatively large number of "stochastic particles" that "simulate" parcels of the phase being considered. The Lagrangian-type equations that control the motion of these particles, within the physical space as well as within the random variables' space, are built so that all the particles have a PDF at a given point (i.e. in a small region neighboring this point) of the physical space, which approximates the solution to the equation modeled above.

For example, in the case considered, assuming that a particle is without contact with the gas flow, the equations that enable this simulation for each stochastic particle (k) are written as:

$$dx_\alpha^k = U_\alpha^k dt \qquad [18.14]$$

$$dU_\alpha^k = -\frac{1}{\rho^k}\frac{\partial \overline{p}_l^s}{\partial x_\alpha} dt + D_\alpha dt + \sqrt{C_0 \overline{\varepsilon}_l^*}\, dW_\alpha \qquad [18.15]$$

D_α is the term defined above in section 18.2 and it results from physical modeling. dW_α is the increment of a stochastic Wiener process, forming a series of random values $W_\alpha(t_i)$ with Gaussian PDFs such that the mean of dW_α is zero, the mean of its square is dt and the moments of higher orders are infinitely small in comparison to dt, in particular. More details on these stochastic simulations, as well as very interesting developments, are discussed in [MIN 01]. In addition, consideration is given to the modeling of friction with the carrier gas flow.

Minier and Peirano also discuss another model for the dispersion of a particle cloud in a carrier flow, studied by Reeks, and propose another model in which a joint PDF is considered for the velocity fluctuations of particles and the carrier gas phase, or more accurately of the velocity field "seen by the particles", which has been called v_f here.

18.4. The study of the sizes of the dispersed phase parcels

From Chapter 5 onward, it has been shown that the size of liquid parcels (or the radius of drops) has a noticeable influence on their vaporization. This is also the case for the friction between particles or drops and the continuous flow. We have also noticed that the exchange terms are directly connected to the instant interface area per unit volume, and this quantity has a clear physical meaning. However, the exchange terms are not simply linear relative to the interface area and yet involve the size of the parcel. The statistical mean is, therefore, not calculated without some difficulty. It has been demonstrated that if the size distribution in the flow is always in a form that can be defined as a function with two variables that are then tracked, $\overline{\sigma}_p^s$ and $\overline{\Phi}_p$, this can be taken into account and does not complicate the formulas much. However, vaporization phenomena, as well as possible fracture and coalescence of parcels, can considerably change the size distribution at any point in the flow: it may depend on other parameters, and it is clear that it would be beneficial to know more about this distribution. It is also apparent that the joint distribution of drop velocities and radii must be found, and not simply the two distributions independently, as it is realistic to consider the evolution of drop velocity to be linked to their size, the velocity being important both for their movement and the vaporization rate. The question is how we take these variations into account.

Consider the general case for a multiphase flow without assuming that the particles, drops or bubbles, are carried everywhere within the same fluid phase. For such a case, it has been seen that the local exchange formulas are always presented in the form of a mass flow rate, a force or a flux per unit of interface area. In most cases, the PDF for fluctuations in the interface area per unit volume is, therefore, the one that needs to be calculated, or, better still, the joint PDF for interface areas, velocities and temperature in the two touching phases, as all these variables are involved. Can such a PDF be obtained?

The direct balance equation for the interface area per unit volume was presented in Chapter 2, although only its incomplete form was given because it is not known how to take into account the phenomenon of interface annihilation. Nonetheless, it is possible to write a general joint PDF for velocity fluctuations in each phase and for the interface density between these phases, using seven random variables. If all the variables mentioned above are included, this would make a total of 10. This equation is not closed for any of the PDFs considered until now, and some terms will, therefore, need to be modeled, in particular the ones corresponding to fracture and coalescence phenomena and, to the best of our knowledge, this remains to be done. Some studies have already investigated how the distribution function of particle or drop sizes can be calculated, using fracture and agglomeration processes, and should provide a source of inspiration.

This approach uses the instant interface density and it is, therefore, logical to consider some "local configurations" which include curved portions of the interface instead of particles, drops or bubbles. At the level of the interface, these are effectively the normal temperature gradients on either side, and the mass fraction gradient of the vaporized liquid on the side of the gas, which play a direct role. The study of local configurations enables these gradients to be found using information on velocities, temperatures and compositions in the vicinity. Can our current knowledge of the local situations of particles, drops or bubbles be transformed to situations for curved portions of the interface?

On the other hand, considering the simpler case of a medium with dispersed phase parcels, it was seen in Chapter 4 that it was possible to define a "number per unit volume" of phase parcels of different volumes. This consists of considering a few more different classes in each dispersed physical phase. These new classes consist of parcels with a value for the volume lying between several well-defined limits (this can also be performed using the surface area of parcels instead of their volume). The mass balance equations for these parcels' classes involve the mean velocity of parcels in the class, all of which also follow a balance equation for momentum. It is, therefore, quite possible to calculate the particle size (volumes) and velocity distribution simultaneously, without the assumption that there is a particular correlation between these two quantities.

This provides the possibility of providing a more complete Eulerian description of the dispersed phase (or phases) of the medium in terms of sizes and velocities. In fact, this approach has been studied for quite a long time for spherical droplets or particles. It is often called "sectional approach" in the literature, and the reader should, for example, refer to [MAS 07]. The mass balance equation for different classes brings into play source or sink terms that will take into account the parcels' change of class, caused by their fracture or coalescence (for example, refer to [LAU 04]).

It is easily conceived that the number of classes, and therefore, the number of balance equations to be solved simultaneously, will become significantly large to obtain a sufficiently accurate distribution. Several techniques have been proposed to represent the distribution and lower this number; for example, the reader may refer to [FOX 08] or [KAH 11].

The terms for exchange between phases in these new Eulerian equations remain and are always in the form of an average: the velocity of particles in a given class always exhibits fluctuations, as is the case for its temperature, and mainly the temperature and composition of the adjacent gas medium. The influence of these fluctuations is still impossible to study, unless it is studied via PDFs that have previously been discussed.

For situations in which the dispersed phase is not very dense, and for which interactions, collisions, coalescences and even fracture of parcels can be neglected, the Eulerian–Lagrangian simulations mentioned in Chapter 1 of this book allow these media to be modeled and computed while taking into account most of these questions, with the possible exception of those concerning fluctuations in the temperature or composition of the gas phase. This approach is not described here, nor it is compared with the approach discussed throughout this book. However, it is desired by the authors to disclose some of the similarities between the two approaches, primarily to highlight their coherence, and to possibly induce cross-improvement.

18.5. Eulerian–Lagrangian simulation of dispersed media

The definition of parcels is only valid in all cases as long as the dispersed multiphase medium is diluted enough for *coalescences or fractures to be neglected.* In these conditions, it is possible to follow the evolution of the parcels along their Lagrangian path, and to take advantage of a decrease in the amount of calculation by representing a large number of particles or drops by a small number of representative samples for them. It is initially only in these conditions that a so-called Eulerian–Lagrangian model has been proposed, where the continuous fluid medium is still described by Eulerian equations and the particles, or drops, are tracked in a Lagrangian way. This approach is nowadays widely used, even when coalescence and fracture can sometimes occur, assuming that phenomena occur suddenly and sporadically.

First, it is interesting to see how the Langrangian balance equations can be found directly from Eulerian balance equations for the piecewise continuous medium considered in Chapter 2. It will then be seen how stochastic simulations achieved via this method can be compared to calculations for the velocity PDFs previously mentioned.

18.5.1. *Lagrangian equations of the parcels*

The balance equations were written in Chapter 2 in Eulerian form, for the balance of mass, energy and momentum, for each phase in a given fixed volume of any size, as a starting point. They were then used to find balance equations in the form of partial derivatives by generalizing classic writing to take into account discontinuities introduced by the presence of different phases. These balance equations can clearly also be written for any finite size volume that moves continuously and that is chosen in advance. The only change to be made will be in the exchange flux between the volume and the rest of the medium, involved in the surface integrals at the volume's boundaries. In these integrals, it will then be necessary to replace the local convection velocity (of the total mass present at this point) relative to the fixed reference frame by the convection velocity relative to the surface of the moving volume considered. The other effects of exchange, species diffusion, heat and momentum from contact forces remain the same as they are defined relative to the movement of the total mass and are, therefore, objective. The balance equations are thus obtained more simply for each parcel of a dispersed phase (drop or particle) by tracking its movement.

The mass balance for a parcel of phase i, say a liquid parcel, defined by its interface at the initial time is given by the following relationship for any instant:

$$\frac{d}{dt}(\iiint_{V_p} \rho_i dx) = \frac{dm_{pi}}{dt} = -\iint_{Sp} \rho_{is}(v_{is}^{\alpha} - v_s^{\alpha})n_i^{\alpha} d\sigma_p \qquad [18.16]$$

where m_{pi} is the mass of a parcel (p) of phase i, V_p is its volume and S_p is its surface. The right-hand side term is only the total mass flow rate gained (or lost) by the parcel through mass exchange with one or more phases that it touches at that time. Let it be reminded that phenomena of parcel coalescence and fracture are neglected here.

Likewise, the momentum balance is written as:

$$\frac{d}{dt}(m_{pi}V_{Gpi}^{\beta}) = m_{ip}g^{\beta} + \iint_{Sp}(\sigma_{is}^{\alpha\beta}n_i^{\alpha} - \rho_{is}(v_{is}^{\alpha} - v_s^{\alpha})n_i^{\alpha}v_{is}^{\beta})d\sigma_p \qquad [18.17]$$

where V_{Gpi} is the velocity of the parcel's center of gravity. The effects of gravity, contact forces along the entire surface, pressure, friction forces and mass exchange appear on the right-hand side of the equation. It is already apparent that if the flow field and the stresses around a spherical parcel are spherically symmetric, then the surface integral is zero overall. However, this symmetry clearly does not exist due to the movement of the particle relative to the fluid, even if the particle is itself spherical and also in case of a mean pressure gradient in the medium.

The total energy balance is given by:

$$\frac{d}{dt}(m_{pi}e_{tpi}) = \iint_{Sp}(-j_{is}^{Q\alpha}n_i^{\alpha} + \sigma_{is}^{\alpha\beta}n_i^{\alpha}v_{is}^{\beta} - \rho_{is}(v_{is}^{\alpha} - v_s^{\alpha})n_i^{\alpha}e_{tis})d\sigma_p + m_{pi}V_{Gpi}^{\beta}g^{\beta} \qquad [18.18]$$

The exchange of heat flux via conduction (or radiation), the exchange of mass, etc. can be identified here. The particle's total energy is usually mainly internal energy as kinetic energy is often negligible. Temperature can be calculated using this internal energy although it only has a clear meaning if the temperature is homogeneous in the particle. Otherwise, it is a "mean particle temperature" and the internal distribution needs to be known to calculate the surface temperature, which is of direct interest for heat transfers.

It is of importance to look at the consequences of the parcel's energy balance piece by piece. If friction is neglected, as well as the parcel's kinetic energy and the gravitational effects on the parcel, this energy balance is written as:

$$\frac{d}{dt}(m_p e_p) = \iint_{Sp}(-j_{ps}^{Q\alpha}n_p^{\alpha} - p_{ps}n_p^{\alpha}(v_{ps}^{\alpha} - v_s^{\alpha}) - \rho_{ps}(v_{ps}^{\alpha} - v_s^{\alpha})n_p^{\alpha}e_{ps} - p_{ps}v_s^{\alpha}n_p^{\alpha})d\sigma_p \qquad [18.19]$$

The integral of the second member is the overall energy flux received by the parcel from the surrounding medium. It is calculated here by using the liquid's characteristics at the parcel's surface, although it can also be calculated using the characteristics of the surrounding gas at the surface if the phenomenon of surface tension is neglected because the total local energy balance at the interface, seen in Chapter 2, implies that the two kinds of characteristic are directly related in this case.

Only part of the flux will actually be used to increase the parcel's temperature if there is an exchange of mass. The other part compensates the latent heat for the liquid to be vaporized. Indeed, the previous expression can be rewritten as:

$$\frac{d}{dt}(m_p e_p) = \varphi_p^Q + \frac{dm_p}{dt}\overline{h}_p^s - \iint_{Sp} p_{ps} v_s^\alpha n_p^\alpha d\sigma_p$$

where $\varphi_p^Q = \iint_{Sp} (-j_{ps}^{Q\alpha} n_p^\alpha) d\sigma_p$ and $\overline{h}_p^s$ is the parcel's mean enthalpy on the surface such that: $\iint_{Sp} (-p_{ps} n_p^\alpha (v_{ps}^\alpha - v_s^\alpha) - \rho_{ps}(v_{ps}^\alpha - v_s^\alpha) n_p^\alpha e_{ps}) d\sigma_p = \overline{h}_p^s \iint_{Sp} \rho_{ps}(v_{ps}^\alpha - v_s^\alpha) n_p^\alpha d\sigma_p$

If the parcel is spherical and has homogeneous pressure and density, then the last term is given by $-\frac{\overline{p}_p}{\rho_p}\frac{dm_p}{dt}$, and the following equation is obtained:

$$m_p \frac{d}{dt}(e_p) + h_p \frac{dm_p}{dt} = \varphi_p^Q + \frac{dm_p}{dt}(\overline{h}_p^s - h_p)$$

When the drop can be assumed to have homogeneous temperature, only $m_p \frac{d}{dt}(e_p) = \varphi_p^Q$ remains, which clearly shows that only flux by conduction in the liquid can increase the temperature.

Besides, by neglecting kinetic energy and friction effects (see Chapter 2, equation [2.12b]), the interface balance relation for energy can be written as follows:

$$\varphi_p^Q + \frac{dm_p}{dt}\overline{h}_p^s - \iint_{Sp} p_{ps} v_s^\alpha n_p^\alpha d\sigma_p = -\varphi_f^Q + \frac{dm_p}{dt}\overline{h}_f^s + \iint_{Sp} p_{fs} v_s^\alpha n_f^\alpha d\sigma_p$$

If the pressure at the parcel's surface in both phases (index p for the parcel and f for the carrier fluid, i.e. the gas) is the same, by neglecting the surface tension, and the temperature is also the same, then the following equation is obtained:

$\varphi_p^Q - \frac{dm_p}{dt}(\overline{h}_f^s - \overline{h}_p^s) = -\varphi_f^Q$, which can also be written as $\varphi_p^Q + (-\frac{dm_p}{dt})\overline{L}_{Vs} = -\varphi_f^Q$, where factor $\overline{L}_{Vs}$ is the mean, along the parcel's surface, of the latent heat of vaporization.

Similar to the equations in Chapter 2, it is not possible to use equations [18.16]–[18.18] or [18.19] in their current form. The state of the liquid inside the parcel and the gas outside needs to be known in order to express the surface integrals of the right-hand side terms. A Eulerian–Lagrangian method, therefore, requires these integrals to be modeled, and they represent exchanges with the surrounding medium. To achieve this modeling, it is useful to consider local exchange configurations, as was done in Chapter 5, although only the configurations with particles are of significance. The difficulty linked to calculating the average for numerous parcels, likely to pass at a given point in the medium, does not appear here because these equations are not statistical averages. The exchange terms can be formally written as: $F_{sp}^\beta, \varphi_p^Q, \dot{m}$ and are functions of certain variables for parcel's characteristics, r_p, V_{Gp}, T_p, as well as certain variables for the gas medium (located at a given distance from the parcel, or "experienced by the parcel"), $V_f^\beta, T_{f\infty}, Y_{V\infty}$. These variables need to be modeled according to what is known about the gas medium, therefore, most often according to this medium's mean characteristics.

18.5.2. *Stochastic simulations*

The Eulerian–Lagrangian simulation method consists of following the path and Lagrangian evolution of some "stochastic particles" that are considered to be representative of the whole population of the parcels of the dispersed phase. For each stochastic particle, it is simply a case of numerically integrating:

– the kinetic equation

$$\frac{dx_p^\beta}{dt} = V_{Gp}^\beta \qquad [18.20]$$

– balance equations [18.16]–[18.18], conveniently modeled.

First, it is clear that by identifying $V_f^\beta = \tilde{v}_f^\beta$, the particle's trajectories do not exhibit truly turbulent dispersion. Only the dispersion induced by the scatter of initial positions and velocity conditions will be observed. It is necessary to take into account the fluctuations of the fluid velocity "experienced by the particle" to reproduce a turbulent dispersion effect.

The gas phase in which particles move is always represented by Eulerian balance equations and only its mean characteristics are provided by the equations described in Chapter 4. If the actual number of parcels per unit volume is sufficiently low, then their presence can be assumed to have no effect on the movement of the fluid. However, this is scarcely true, and Chapter 4 discussed how the parcels contributed to balance equations for mass, mean momentum, mean energy, kinetic energy of velocity fluctuations, for the gas phase. The formulas are written as mean values on the interface between the fluid and parcels. In the case of the dispersed parcels being studied here, the equations can clearly be written as mean values for one parcel:

Formally:

$$\overline{F^{\beta}_{fs}} = -\overline{\tau^{\alpha\beta}_{p} n^{\alpha}_{p} \sigma^{s}_{p}} = -\iiint \left\langle F^{\beta}_{sp} \right\rangle^{f} f_p(r_p, V_{Gp}, T_L; x, t) dr_p dV_{Gp} dT_L \qquad [18.21]$$

$$\overline{q_{fs} \sigma^{s}_{p}} = -\overline{j^{Q\alpha}_{f} n^{\alpha}_{f} \sigma^{s}_{p}} = \iiint \left\langle \varphi^{Q}_{f} \right\rangle^{f} f_p(r_p, V_{Gp}, T_L; x, t) dr_p dV_{Gp} dT_L \qquad [18.22]$$

$$\overline{\rho_f \dot{w}^{\sec h}_{f}} = +\overline{\rho_p v^{\alpha}_{rp} n^{\alpha}_{p} \sigma^{s}_{p}}^{s} = \iiint 4\pi r_p^{2} \left\langle \dot{m} \right\rangle^{f} f_p(r_p, V_{Gp}, T_L; x, t) dr_p dV_{Gp} dT_L s \qquad [18.23]$$

where $f_p(r_p, V_{Gp}, T_L; x, t)$ is the probability of particles being present, with radius r_p, velocity V_{Gp} and temperature T_L at the point and time being considered, such that: $\iiint f_p dr_p dV_{Gp} dT_p = \overline{\Phi}_p$. The previous formulae do not take into account the effect of non-homogeneity of the medium, then of the particles volume fraction gradient, as discussed in Chapter 5.

Probability $f_p(r_p, V_{Gp}, T_L; x, t)$ can, in practice, be calculated by a Lagrangian simulation (and an ergodic hypothesis). During this simulation, the number of passing particles can be calculated, and their radius can be found, for any small given volume of space in each short given time interval. An estimate of the distribution is thus found, which will be especially accurate for a higher number of particles and smaller time intervals.

Notation $\left\langle \dot{m} \right\rangle^{f}$, and those similar to it, means that the exchange terms concerning the fluctuations of the surrounding fluid characteristics have to be averaged. Until now, an approximate method has been used for this, which consists of replacing the instantaneous variables by their average in the instantaneous relations, and so any effects of fluctuations are completely neglected. This is another problem that still remains to be studied.

Even if only the mean contact force between particles and the carrier fluid is enough for this carrier fluid motion, it has previously been said that fluctuations in the fluid velocity had to be taken into account in one way or another in order to calculate the turbulent dispersion of particles. The procedure that is classically used is the following: using known fluid turbulence, k_f, l_{tf}, and maybe all the components of the Reynolds tensor R_f, a random choice for V_f is introduced into the simulation which satisfies all these known properties. The choice is random and must be done periodically on all the particle trajectories, which depends on the characteristic Lagrangian turbulence time, effectively on $l_{tf}/(k_f)^{1/2}$. The problem is, in fact, a little more complex than this because the particle's inertia must be considered and is represented by the time l_{tf}/V_{Gp}, giving the so-called "crossing trajectory effect". The example provided by Berlemont *et al.* [BER 90] and Crowe [CRO 01] can be referred to for more detail.

Some models use a separate PDF model to calculate the carrier fluid turbulence which is mainly solved by fluid stochastic particle simulation. It is then a simple question of transposing the developments discussed in sections 18.1–18.3 of this chapter to calculate the PDF for fluid velocities, which is then used to represent the PDF for the velocity V_f "seen by the particles". For these models, the influence of dispersed parcels is often neglected if their numerical density is low. It would be possible to directly model the PDF for parcels velocity, proposed above, and this was the method used in the study by Jones and colleagues [JON 10]. However, standard stochastic simulations, which choose the carrier fluid velocity randomly, are able to represent the Lagrangian time correlation of fluid fluctuations more realistically, which is important for dispersion over a small time or at small distances from the point of origin.

As far as $T_{f\infty}, Y_{V\infty}$ are concerned, they can only be identified to the gas carrier phase's mean values $\tilde{T}_f, \tilde{Y}_V$ if nothing is known about their fluctuations. This has been the case for nearly all simulations up to now. However, a random choice can be implemented for these two variables using the PDF that could be determined from Chapter 16. Without calculating this PDF, it is also possible to retrieve the equations for variances $\overline{\rho_f \Phi_f Y_V'^2}$ and $\overline{\rho_f \Phi_f T_f'^2}$, which are easier to solve, from PDF equations. The problem concerning the calculation of the fluid averages, such as $\langle \dot{m} \rangle^f$, can thus be solved.

As has been shown, stochastic Eulerian–Lagrangian simulation is directly adapted to situations in which parcels are of different sizes. Stochastic particles representative of all sizes simply need to be tracked. When vaporization or condensation occurs, the assumption that droplets are spherical and have a known homogeneous density is enough to follow the variation of droplets' radii. The question related to how fracture or coalescence of drops can be taken into account discontinuously is now being studied. However, it requires quite a lot of calculations, because it is always better to use a lot of particles and many Eulerian fluid meshes. However, we see that it is not in an especially favorable position to deal with the general problems of the effects of fluctuations in the Eulerian fluid medium, as only the effect of velocity fluctuations is taken into account. In multiphase media where such effects are important, it would then be necessary to associate the modeling of the PDF for sizes and velocity fluctuations, maybe used in the Lagrangian framework, with the PDF of temperature, composition and velocity of the carrier phase.The approach described throughout this book shows all these possibilities, but their full implementation is a rather huge practical work. We have also shown how to build models with partial coupling with a balance equation of the interface area per unit volume, treated simply in an Eulerian way. They are able to give approximations that can be adapted to different applications, which will remain attractive for a while at least.

Bibliography

[ALB 03] ALBRECHT H.E., DAMASCHKE N., BORYS M., *et al.*, *Laser Doppler and Phase Doppler Measurement Techniques*, Springer-Verlag, Berlin-Heidelberg, 2003.

[AND 11] ANDREOTTI B., FORTERRE Y., POULIQUEN O., *Les milieux granulaires*, EDP Sciences/CNRS Editions, Paris, 2011.

[AUT 88] AUTON T.R., HUNT J.C.R., PRUD'HOMME M., "The force exerted on a body in inviscid unsteady non-uniform rotational flow", *Journal of Fluid Mechanics*, vol. 197, pp. 241–257, 1988.

[BAE 86] BAER M., NUNZIATO J., "A two-phase mixture theory for the deflagration-to-detonation transition (ddt) in reactive granular materials", *International Journal of Multiphase Flow*, vol. 12, no. 6, pp. 861–889, 1986.

[BAG 54] BAGNOLD R.A., "Experiments on a gravity-free dispersion of large solid spheres in a Newtonian fluid under shear", *Proceedings of the Royal Society of London Series A*, vol. 225, no. 1160, pp. 49–63, 1954.

[BAR 99] BARRAU E., RIVIÈRE N., POUPOT C., *et al.*, "Single and double optical probes in air-water two-phase flows", *International Journal of Multiphase Flow*, vol. 25, no. 2, pp. 229–256, 1999.

[BAS 06] BASU P., *Combustion and Gasification in Fluidized Beds*, CRC Press, Taylor & Francis, London, 2006.

[BEE 97] VAN BEECK J. RIETHMULLER M.L., "Rainbow interferometry with wire diffraction for simultaneous measurement of droplet temperature, size and velocity", *Particle and Particle Systems Characterization*, vol. 14, pp. 186–192, 1997

[BEH 07] BEHESHTI N., BURLUKA A.A., FAIRWEATHER M., "Assessment of $\Sigma-Y_{liq}$ model predictions for air-assisted atomisation", *Theoretical and Computational Fluid Dynamics*, vol. 21, pp. 381–397, 2007.

[BEI 07] BEISHUIZEN N.A., NAUD B., ROEKAERTS D., "Evaluation of a modified Reynolds stress model for turbulent dispersed two-phase flows including two-way coupling", *Flow, Turbulence and Combustion*, vol. 79, pp. 321–341, 2007.

[BER 90] BERLEMONT A., DESJONQUÈRES P., GOUESBET G., "Particle Lagrangian simulation in turbulent flows", *International Journal of Multiphase Flow*, vol. 16, no. 1, pp. 19–34, 1990.

[BLO 02] BLOKKEEL G., DEMOULIN F.-X., BORGHI R., "An Eulerian model to improve the primary break up modeling of atomizing jet", *Proc. Conf. ILASS-Europe*, Zaragoza, Spain, 2002.

[BOR 00] BORGHI R., CHAMPION M., *Modélisation et théorie des flammes*, Technip, Paris, 2000.

[BOR 07] BORGHI R., BONELLI S., "Towards a constitutive law for the unsteady contact stress in granular media", *Continuum Mechanics and Thermodynamics*, vol. 19, pp. 329–345, 2007.

[BOR 08a] BORGHI R., *Thermomécanique et modélisation par systèmes*, Cépadues, Toulouse, 2008.

[BOR 08b] BORGHI R., *Les milieux continus multiphysiques hors d'équilibre et leur modélisation*, Cépadues, Toulouse, 2008.

[BOU 12a] BOUTIER A. (ed.), *Vélocimétrie laser pour la mécanique des fluides*, Hermès-Lavoisier, Paris, 2012.

[BOU 12b] BOUTIER A. (ed.), *Métrologie laser pour la mécanique des fluides: Granulométrie et techniques spectroscopiques*, Hermès-Lavoisier, Paris, 2012.

[BRI 12] BRIARD P., SAENGKAEW S., MEUNIER-GUTTIN-CLUZEL S., *et al.*, "Caractérisation de la structure 3D d'un nuage de gouttes par Imagerie Interférométrique de Fourier (FII)", *13th Congress AFVL*, CORIA, Rouen, 2012.

[BUR 00] BURLUKA A.A., BORGHI R., "Stretch effect on the vaporization rate in Fendell's problem", *Combustion and Flame*, vol. 122, no. 3, pp. 227–232, 2000.

[CAM 98] CAMBOU B., *Behaviour of Granular Materials*, Springer-Verlag, Berlin-Heidelberg, 1998.

[CAM 01] CAMBOU B., JEAN M., *Micromécanique des matériaux granulaires*, Hermes, Paris, 2001.

[CAO 96] CAO J., AHMADI G., MASSOUDI M., "Gravity granular flows of slightly frictional particles down an inclined bumpy chute", *Journal of Fluid Mechanics*, vol. 316, pp. 197–221, 1996.

[CAR 91] CARTELLIER A., ACHARD J.-L., "Local phase detection probes in fluid/fluid two-phase flows", *Review of Scientific Instruments*, vol. 62, pp. 279–303, 1991.

[CAR 92] CAREY V.P., *Liquid Vapor Phase Change Phenomena*, Taylor & Francis, London, 1992.

[CAR 95] CARRICA P.M., LEONARDI S.A., CLAUSE A., "Experimental study of the two-phase flow dynamics in nucleate and film pool boiling", *International Journal of Multiphase Flow*, vol. 21, no. 3, pp. 405–418, 1995.

[CHA 05] CHAOUAT B., SCHIESTEL R., "A new partially integrated transport model for subgrid-scale stresses and dissipation rate for turbulent developing flows", *Physics of Fluids*, vol. 17, no. 6, pp. 065106/1–065106/19, 2005.

[CHA 07] CHAOUAT B., SCHIESTEL R., "From single-scale turbulence models to multiple-scale and subgrid-scale models by Fourier transform", *Theoretical and Computational Fluid Dynamics*, vol. 21, pp. 201–229, 2007.

[CHA 09a] CHAOUAT B., SCHIESTEL R., "Progress in subgrid-scale transport modelling for continuous hybrid non-zonal RANS/LES simulations", *International Journal of Heat and Fluid Flows*, vol. 30, no. 4, pp. 602–616, 2009.

[CHA 97] CHAOUKI J., LARACHI F., DUDUKOVIC M.P.P., *Non-Invasive Monitoring of Multiphase Flows*, Elsevier, Amsterdam, 1997.

[CHA 60] CHAPMAN S., COWLING T.G., *The Mathematical Theory of Non-Uniform Gases*, Cambridge University Press, Cambridge, 1960.

[CHA 09b] CHAREYRON D., Développement de méthodes instrumentales en vue de l'étude Lagrangienne de l'évaporation dans une turbulence homogène isotrope, PhD Thesis, Ecole Centrale de Lyon, 2009.

[CHA 02] CHASSAING P., ANTONIA R.A., ANSELMET F., *et al.*, *Variable Density Fluid Turbulence*, Kluwer Academic Publisher, Dordrecht, 2002.

[CHA 75] CHAVARIE C., GRACE J.R., "Performance analysis of a fluidized bed reactor. Part I: visible flow behavior", *Industrial and Engineering Chemistry Fundamentals*, vol. 14, no. 2, pp. 75–78, 1975.

[CHE 05] CHEN J.C., GRACE J.R., GOLRIZ M.R., "Heat transfer in fluidized beds: design methods", *Powder Technology*, vol. 150, no. 2, pp. 123–132, 2005.

[COL 11] COLOMBET D. *et al.*, "Experimental study of mass transfer in a dense bubble swarm", *Chemical Engineering Science*, vol. 66, no. 14, pp. 3432–3440, 2011.

[COQ 02] COQUEL F., GALLOUËT T., HÉRARD J., *et al.*, "Closure laws for a two-fluid two-pressure model", *Comptes Rendus de l'Académie des Sciences, Série I*, vol. 334, pp. 927–932, 2002.

[CRO 01] CROWE C.T. (ed.), *Multiphase Flows Handbook*, Taylor & Francis, London, 2001.

[DAR 77] DARTON R.C., LA NAUZE R.D., DAVIDSON J.F., *et al.*, "Bubble growth due to coalescence in fluidized beds", *Transactions of the Institution of Chemical Engineers*, vol. 55, pp. 274–280, 1977.

[DE 09] DE LUCA M., VALLET A., BORGHI R., "Pesticide atomization modeling for hollow-cone nozzle", *Atomization and Sprays*, vol. 19, no. 8, pp. 741–753, 2009.

[DEL 81] DELHAYE J.-M., GIOT M., REITHMULLER M.L., *Thermohydraulics of Two-Phase Systems for Industrial Design and Nuclear Engineering*, Hemisphere, New York, 1981.

[DEM 00] DEMOULIN F.-X., BORGHI R., "Assumed PDF modeling of turbulent spray combustion", *Combustion Science and Technology*, vol. 158, pp. 249–271, 2000.

[DEM 07] DEMOULIN F.X. *et al.*, "A new model for turbulent flows with large density fluctuations: application to liquid atomization", *Atomization and Sprays*, vol. 17, no. 4, pp. 315–345, 2007.

[DIH 91] DIHR V.K., "Nucleate and transition boiling heat transfer under pool and external flow conditions", *International Journal of Heat and Fluid Flow*, vol. 12, no. 4, pp. 290–314, 1991.

[DIH 98] DIHR V.K., "Boiling heat transfer", *Annual Review of Fluid Mechanics*, vol. 30, pp. 365–401, 1998.

[DRE 83] DREW D.A., "Mathematical modeling of two-phase flow", *Annual Review of Fluid Mechanics*, vol. 15, pp. 261–291, 1983

[DUR 99] DURAND P.P., GOROKHOVSKI M., BORGHI R., "An application of the probability density function model to diesel engine combustion", *Combustion Science and Technology*, vol. 144, pp. 47–78, 1999.

[ELG 94] ELGOBASHI S.E., "On predicting particle-laden turbulent flows", *Applied Scientific Research*, vol. 52, no. 4, pp. 309–329, 1994.

[EME 96] EMERIAULT F., CAMBOU B., "Micromechanical modelling of anisotropic non-linear elasticity of granular medium", *International Journal of Solids and Structures*, vol. 33, no. 18, pp. 2591–2607, 1996.

[ENW 96] ENWALD H., PEIRANO E., ALMSTEDT A.E., "Eulerian two-phase flow theory applied to fluidization", *International Journal of Multiphase Flow*, vol. 22, pp. 21–66, 1996.

[ERG 52] ERGUN S., "Fluid flow through packed columns", *Chemical Engineering Progress*, vol. 48, pp. 89–94, 1952.

[ERO 91] EROGLU H., CHIGIER N.A., FARAGO Z., "Coaxial atomizer liquid intact lengths", *Physics of Fluids A*, vol. 3, no. 2, pp. 303–308, 1991.

[FAE 90] FAETH G.M., "Structure and atomization properties of dense turbulent sprays", *23rd International Symposium on Combustion*, Pittsburgh, pp. 1345–1352, 1990.

[FAL 09] FALCONE G., HEWITT G.F., ALIMONTI C., *Multiphase Flow Metering: Principles and Applications*, Elsevier, Amsterdam, 2009.

[FOX 08] FOX R.O., LAURENT F., MASSOT M., "Numerical simulation of spray coalescence in an Eulerian framework: direct quadrature method of moments and multi-fluid method", *Journal of Computational Physics*, vol. 227, no. 6, pp. 3058–3088, 2008.

[FOX 12] FOX R.O., "Large-eddy-simulation tools for multiphase flows", *Annual Review of Fluid Mechanics*, vol. 44, pp. 47–76, 2012.

[FRE 02] FRENETTE R., ZIMMERMAN T., EYHERAMENDY D., "Unified modeling of fluid or granular flows on dam-break case", *Journal of Hydraulic Engineering*, vol. 128, no. 3, pp. 299–305, 2002.

[FRO 66] FROST V.A., "Mathematical model of turbulent combustion", *Transactions of the Third All-Union Conference on Combustion Theory*, Izd-vo AN SSSR, Moscow, pp. 121–125, 1966.

[GAT 83] GATIGNOL R., "The Faxén formulae for a rigid particle in an unsteady non-uniform Stokes flow", *Journal de Mécanique Théorique et Appliquée*, vol. 1, no. 2, pp. 143–160, 1983.

[GAV 02] GAVRILIUK S., SAUREL R., "Mathematical and numerical modeling of two-phase compressible flows with micro-inertia", *Journal of Computational Physics*, vol. 175, no. 1, pp. 326–360, 2002.

[GID 94] GIDASPOW D., *Multiphase Flow and Fluidization*, Academic Press, London/New York, 1994.

[GOD 86] GODDARD J.D., "Dissipative materials as constitutive models for granular media", *Acta Mechanica*, vol. 63, pp. 3–13, 1986.

[GOR 08] GOROKHOVSKI M., HERRMANN M., "Modeling primary atomization", *Annual Review of Fluid Mechanics*, vol. 40, pp. 343–366, 2008.

[GRA 00] GRACE J.R., "Reflections on turbulent fluidization and dense suspension upflow", *Powder Technology*, vol. 113, no. 3, pp. 242–248, 2000.

[GRE 93] GREENBERG J.B., SILVERMAN I., TAMBOUR Y., "On the origins of spray sectional conservation equations", *Combustion and Flame*, vol. 93, no. 1–2, pp. 90–96, 1993.

[GUA 12] GUAZELLI E., MORRIS J.F., *A Physical Introduction to Suspension Dynamics*, Cambridge University Press, Cambridge, 2012.

[GUY 97] GUYON E., HULIN J.-P.P., *Granites et Fumées, Un peu d'ordre dans le mélange*, O. Jacob Editions, Paris, 1997.

[HAM 10] HAMAD F.A., HE S., "Evaluation of hot-film, dual optical and Pitot tube probes for liquid-liquid two-phase flow measurements", *Flow Measurement and Instrumentation*, vol. 21, pp. 302–311, 2010.

[HIR 54] HIRSCHFELDER J.O., CURTISS C.F., *et al.*, *Molecular Theory of Gases and Liquids*, John Wiley & Sons, Hoboken, 1954.

[HOG 97] HOGSETT S., ISHII M., "Local two-phase flow measurements using sensor techniques", *Nuclear Engineering and Design*, vol. 175, no. 1–2, pp. 15–24, 1997.

[HON 04] HONG M., CARTELLIER A., HOPFINGER E.J., "Characterization of phase detection optical probes for the measurement of the dispersed phase parameters in sprays", *International Journal of Multiphase Flow*, vol. 30, no. 6, pp. 615–648, 2004.

[HRE 97] HRENYA C.M., SINCLAIR J.L., "Effects of particle-phase turbulence in gas-solid flows", *American Institute of Chemical Engineers Journal*, vol. 4, pp. 853–869, 1997.

[HUN 02] HUNT M.L., ZENIT R., CAMPBELL C.S., *et al.*, "Revisiting the 1954 suspension experiments of R.A. Bagnold", *Journal of Fluid Mechanics*, vol. 452, pp. 1–24, 2002.

[HUT 94] HUTTER K., RAJAGOPAL K.R., "On flows of granular materials", *Continuum Mechanics and Thermodynamics*, vol. 6, pp. 81–139, 1994.

[ISH 75] ISHII M., *Thermo-Fluid Dynamics Theory of Two-Phase Flow*, DER-EDF, Eyrolles, Paris, 1975.

[ISH 90] ISHII M, "Two-fluid model for two-phase flow", *Multiphase Science and Technology*, vol. 5, pp. 1–63, 1990.

[JAC 00] JACKSON R., *The Dynamics of Fluidized Particles*, Cambridge University Press, Cambridge, 2000.

[JAC 97] JACKSON R., "Locally averaged equations of motion for a mixture of identical spherical particles and a Newtonian fluid", *Chemical Engineering Science*, vol. 52, no. 15, pp. 2457–2469, 1997.

[JAF 04] JAFARI R., SOTUDEH-GHAREBAGH R., MOSTOUFI N., "Performance of the wide-ranging models for fluidized bed reactors", *Advanced Powder Technology*, vol. 15, no. 5, pp. 533–548, 2004.

[JAK 08] JAKOBSEN H.A., *Chemical Reactor Modelling*, Springer-Verlag, Berlin-Heidelberg, 2008.

[JON 10] JONES W.P.P., LETTIERI C., MARQUIS A.J., *et al.*, "Large-eddy simulation of the two-phase flow in an experimental swirl-stabilized burner", *Proceedings of the 8th International Symposium on Engineering Turbulence Modelling and Measurements-ETMM8*, Marseille, vol. 2, pp. 498–503, 2010.

[JOP 06] JOP P., FORTERRE Y., POULIQUEN O., "A constitutive law for dense granular flows", *Nature*, vol. 441, pp. 727–730, 2006.

[JOU 01] JOU D., CASAS-VAZQUEZ J., LEBON G., *Extended Irreversible Thermodynamics*, 3rd ed., Springer-Verlag, Berlin-Heidelberg, 2001.

[KAD 08] KADEM N., TCHIFTCHIBACHIAN A., BORGHI R., "Experimental and numerical modelling study for irrigation gun water jet", *Atomization and Sprays*, vol. 18, no. 4, pp. 315–341, 2008.

[KAH 11] KAH D. *et al.*, "Eulerian polydisperse spray modelling and validation for I.C.Engines", *ILASS Europe 2011*, Estoril, Portugal, 2011.

[KAT 86] KATAOKA I., "Local instant formulation of two-phase flow", *International Journal of Multiphase Flow*, vol. 12, no. 5, pp. 745–758, 1986.

[KRU 03] KRUYT N.P.P., "Contact forces in anisotropic frictional granular materials", *International Journal of Solids and Structures*, vol. 40, no. 13–14, pp. 3537–3556, 2003.

[KUM 10] KUMARA W.A.S., HALVORSEN B.M., MELAAEN M.C., "Single-beam gamma densitometry measurements of oil-water flow in horizontal and slightly inclined pipes", *International Journal of Multiphase Flow*, vol. 36, no. 6, pp. 467–480, 2010.

[KUT 50] KUTATELADZE S.S., "Hydrodynamic model of heat transfer crisis in free-convection boiling", *Journal of Technical Physics*, vol. 20, no. 11, pp. 1389–1392, 1950.

[KUZ 90] KUZNETSOV V.R., SABEL'NIKOV V.A., *Turbulence and Combustion*, Hemisphere Publishing Corporation, New-York, 1990.

[LAB 12] LABERGUE A., DELCONTE A., LEMOINE F., "Fluorescence induite par laser à trois couleurs appliquée à la mesure de température de gouttes dans les sprays", *13th Congress AFVL*, CORIA, Rouen, 2012.

[LAG 88] LAGUÉRIE C., "Techniques de mise en contact entre phases solides et gazeuses", *Techniques de l'Ingénieur*, vol. A 5850, pp. 1–9, 1988.

[LAM 12] LAMADIE F., BRUEL L., HIMBERT M., "Holographie numérique de particules en ligne appliquée à la mesure de gouttes liquides millimétriques", *13th Congress AFVL*, CORIA, Rouen, 2012.

[LAN 91] LANCE M., BATAILLE J., "Turbulence in the liquid phase of a uniform bubbly air-water flow", *Journal of Fluid Mechanics*, vol. 222, pp. 95–118, 1991.

[LAN 87] LANDAU L.D., LIFSHITZ E.M., *Fluid Mechanics*, MIR/Butterworth-Heinemann, Amsterdam/London, 1987.

[LAP 95] LAPORTA A., BERTOGLIO J.P.P., "A model for inhomogeneous turbulence based on two-point correlations", in BENZI R. (ed.), *Advances in Turbulence V*, Kluwer Academic Publisher, Dordrecht, pp. 286–297, 1995.

[LAS 00] LASHERAS J.C., HOPFINGER E.J., "Liquid jet instability and atomization in a coaxial gas stream", *Annual Review of Fluid Mechanics*, vol. 32, pp. 275–308, 2000.

[LAU 04] LAURENT F., MASSOT M., VILLEDIEU P.P., "Eulerian multi-fluid modeling for the numerical simulation of coalescence in polydisperse dense liquid sprays", *Journal of Computational Physics*, vol. 194, no. 2, pp. 505–543, 2004.

[LAV 95] LAVIÉVILLE J., DEUTSCH E., SIMONIN O., "Large eddy simulation of interactions between colliding particles and a homogeneous isotropic turbulence field", *Proceedings of 6th International Symposium on Gas-Particle Flows ASME FED*, vol. 228, pp. 347–357, 1995.

[LEB 09] LEBAS R. *et al.*, "Numerical simulation of primary break-up and atomization: DNS and modelling study", *International Journal of Multiphase Flow*, vol. 35, no. 3, pp. 247–260, 2009.

[LEC 03] LE CORRE J.-M., HERVIEU E., ISHII M., *et al.*, "Benchmarking and improvements of measurement techniques for local-time-averaged two-phase flow parameters", *Experiments in Fluids*, vol. 35, no. 5, pp. 448–458, 2003.

[LEG 98a] LEGENDRE D., BORÉE J., MAGNAUDET J., "Thermal and dynamic evolution of a spherical bubble moving steadily in a superheated or subcooled liquid", *Physics of Fluids*, vol. 10, no. 6, pp. 1256–1272, 1998.

[LEG 98b] LEGENDRE D., MAGNAUDET J., "The lift force on a sperical bubble in a viscous linear shear flow", *Journal of Fluid Mechanics*, vol. 368, pp. 81–126, 1998.

[LEM 05] LEMAITRE P.P. *et al.*, "Global rainbow refractometry for droplets temperature measurement", *Conférence NURETH-11*, Avignon, 2005.

[LHU 09] LHUILLIER D., "Migration of rigid particles in non-Brownian viscous suspensions", *Physics of Fluids*, vol. 21, no. 2, pp. 023302/1–023302/6, 2009.

[LIB 80] LIBBY P.P.A., WILLIAMS F.A., *Turbulent Reacting Flows*, Springer-Verlag, Berlin-Heidelberg, 1980.

[LUN 84] LUN C.K.K., SAVAGE S.B., JEFFREY D.J., *et al.*, "Kinetic theories for granular flow: inelastic particles in Couette flow and slightly inelastic particles in a general flowfield", *Journal of Fluid Mechanics*, vol. 140, pp. 223–256, 1984.

[MAG 00] MAGNAUDET J., EAMES I., "The motion of high-Reynolds number bubbles in inhomogeneous flows", *Annual Review of Fluid Mechanics*, vol. 32, pp. 659–708, 2000.

[MAN 09] MANERA A. *et al.*, "Comparison between wire-mesh sensors and conductive needle-probes for measurements of two-phase flow parameters", *Nuclear Engineering and Design*, vol. 239, no. 9, pp. 1718–1724, 2009.

[MAR 67] MARLE C., "Ecoulements monophasiques en milieux poreux", *Revue de l'Institut Français du Pétrole*, vol. XXII, no. 10, pp. 1471–1509, 1967.

[MAR 04] MARMOTTANT P.P., VILLERMAUX E., "On spray formation, *Journal of Fluid Mechanics*, vol. 498, pp. 73–111, 2004.

[MAS 02] MASHAYEK F., TAULBEE D.B., "Turbulent gas-solid flows, part I: direct simulations and Reynolds stress closures", *Numerical Heat Transfer, Part B,* vol. 41, no. 1, pp. 1–29, 2002.

[MAS 07] MASSOT M., "Eulerian multi-fluid models for polydisperse evaporating sprays", *Multiphase Reacting Flows: Modeling and Simulations*, CISM Course and Lectures Series, vol. 492, Marchisio D.L., Fox R.O. (eds), Springer-Verlag, Berlin-Heidelberg, 2007.

[MAT 04] MATAS J.P., MORRIS J.F., GUAZELLI E., "Lateral forces on a sphere", *Oil and Gas Science and Technology*, vol. 59, no. 1, pp. 59–70, 2004.

[MEI 96] MEI R., "Velocity fidelity of flow tracer particles", *Experiments in Fluids*, vol. 22, no. 1, pp. 1–13, 1996.

[MÉN 07] MÉNARD T., TANGUY S., BERLEMONT A., "Coupling level set/VOF/ghost fluid methods: validation and application to 3D simulation of the primary break-up of a liquid jet", *International Journal of Multiphase Flow*, vol. 33, no. 5, pp. 510–524, 2007.

[MER 05] MERLE A., LEGENDRE D., MAGNAUDET J., "Forces on a high-Reynolds-number spherical bubble in a turbulent flow", *Journal of Fluid Mechanics*, vol. 532, pp. 53–62, 2005.

[MEU 12] MEUNIER-GUTTIN-CLUZEL S., SAENGKAEW S., GARO A., *et al.*, "Sur la mesure de la taille de nano et microparticules par Imagerie Interférométrique de Fourier", *13th Congress AFVL*, CORIA, Rouen, 2012.

[MID 04] MiDi GDR, "On dense granular flows", *European Physical Journal E*, vol. 14, pp. 341–365, 2004.

[MIN 01] MINIER J.P., PEIRANO E., "The PDF approach to turbulent polydispersed two-phase flows", *Physics Reports*, vol. 352, no. 1–3, pp. 1–214, 2001.

[MON 09] MONTOUT M., Contribution au développement d'une approche prédictive locale de la crise d'ébullition, PhD Thesis, INP Toulouse, 2009

[MOR 07] MOREL C., "On the surface equations in two-phase flows and reacting single-phase flows", *International Journal of Multiphase Flow*, vol. 33, no. 10, pp. 1045–1073, 2007.

[MUR 04] MURA A., DEMOULIN F.X., BORGHI R., "Turbulent reaction rate in two-phase flow flames", *Proc. ECCOMAS 2004*, Jyväskylä, Finland, 2004

[MUS 96] MUSTE M., PATEL V.C., PARTHASARATHY R.N., "Discriminator laser Doppler velocimetry for measurement of liquid and particle velocities in sediment-laden flows", *Experiments in Fluids*, vol. 22, no. 1, pp. 45–55, 1996.

[NIG 91] NIGMATULIN R.I., *Dynamics of Multiphase Media*, vol. 1–2, Hemisphere, New York, 1991.

[NOR 87] NOREM H., IRGENS F., SCHIELDROP B.A., "A continuum model for calculating snow avalanches", in SALM B., GUBLER H. (eds), *Avalanche Formation, Movement and Effects*, IAHS Press, Wallingford, UK, pp. 363–379, 1987.

[NOZ 87] NOZIÈRES P., "A local coupling between sedimentation and convection: application to the Beenakker-Mazur effect", *Physica A*, vol. 147, no. 12, pp. 219–237, 1987.

[ODA 64] ODAR F., HAMILTON W.S., "Forces on a sphere accelerating in a viscous fluid", *Journal of Fluid Mechanics*, vol. 18, pp. 302–314, 1964.

[ODD 04] ODDIE G., PEARSON J.R.A., "Flow rate measurement in two-phase flows", *Annual Review of Fluid Mechanics*, vol. 36, pp. 149–172, 2004.

[OES 06] OESTERLÉ B., *Ecoulements multiphases*, Hermès Lavoisier, Paris, 2006.

[ONO 11] ONOFRI F. *et al.*, "Near-critical-angle scattering for the characterization of clouds of bubbles: particular effects", *Applied Optics*, vol. 50, no. 30, pp. 5759–5769, 2011.

[ONO 12] ONOFRI F., RADEV S., BARBOSA S., SENTIS M., "Modèle d'optique-physique de la diffusion au voisinage de l'angle critique de bulles elliptiques", *13th Congress AFVL*, CORIA, Rouen, 2012.

[PEI 02] PEIRANO E. *et al.*, "Numerical simulation of the fluid dynamics of a freely bubbling fluidized bed: influence of the air supply system", *Powder Technology*, vol. 122, no. 1, pp. 69–82, 2002.

[PEI 98] PEIRANO E., LECKNER B., "Fundamentals of gas-solid flows applied to circulating fluidized bed combustion", *Progress in Energy and Combustion Science*, vol. 24, no. 4, pp. 259–296, 1998.

[PEZ 11] PEZO M., STEVANOVIC V., "Numerical prediction of critical heat flux in pool boiling with the two-fluid model", *International Journal of Heat Mass Transfer*, vol. 54, no. 15–16, pp. 3296–3303, 2011.

[PLE 54] PLESSET M.S., ZWICK S.A., "The growth of vapor bubbles in superheated liquids", *Journal of Applied Physics*, vol. 25, no. 4, pp. 493–500, 1954.

[POP 85] POPE S., "PDF methods for turbulent reacting flows", *Progress Energy and Combustion Science*, vol. 19, pp. 119–192, 1985.

[PU 05] PU S.L. *et al.*, "Particle field characterization by digital in-line holography: 3D location and sizing", *Experiments in Fluids*, vol. 39, no. 1, pp. 1–9, 2005.

[RAS 09] RASTELLO M., MARIÉ J.-L., GROSJEAN N., *et al.*, "Drag and lift forces on interface-contaminated bubbles spinning in a rotating flow", *Journal of Fluid Mechanics*, vol. 624, pp. 159–178, 2009.

[REI 82] REITZ R.D., BRACCO F.V., "Mechanism of atomization of liquid jets", *Physics of Fluids*, vol. 25, no. 10, pp. 1730–1740, 1982.

[REN 05a] RENSEN J., LUTHER S., LOHSE D., "The effects of bubbles on developed turbulence", *Journal of Fluid Mechanics*, vol. 538, pp. 153–187, 2005.

[REN 05b] RENSEN J., LUTHER S., VRIES DE J., *et al.*, "Hot-film anemometry in bubbly flow I: bubble-probe interaction", *International Journal of Multiphase Flow*, vol. 31, no. 3, pp. 285–301, 2005.

[REV 05] RÉVEILLON J., VERVISCH L., "Analysis of weakly turbulent dilute-spray flames and spray combustion regimes", *Journal of Fluid Mechanics*, vol. 537, pp. 317–347, 2005.

[SAE 09] SAENGKAEW S., GODARD G., BLAISOT J.B., *et al.*, "Experimental analysis of global rainbow technique", *Experiments in Fluids*, vol. 47, no. 4–5, pp. 839–848, 2009.

[SAE 12] SAENGKAEW S., MEUNIER-GUTTIN-CLUZEL S., BRUNEL M., *et al.*, "Mesure du taux d'évaporation de gouttes par imagerie en défaut de mise au point", *13th Congress AFVL*, CORIA, Rouen, 2012.

[SAG 06] SAGAUT P.P., *Large Eddy Simulations for Incompressible Flows*, Springer-Verlag, Berlin-Heidelberg, 2006.

[SAV 98] SAVAGE S.B., "Analyses of slow high-concentration flows of granular materials", *Journal of Fluid Mechanics*, vol. 377, pp. 1–26, 1998.

[SCA 99] SCARDOVELLI R., ZALESKI S., "Direct numerical simulation of free-surface and interfacial flow", *Annual Review of Fluid Mechanics*, vol. 31, pp. 567–603, 1999.

[SHE 05] SHEN X., SAITO Y., MISHIMA K., *et al.*, "Methodological improvement of an intrusive four-sensor probe for the multi-dimensional two-phase flow measurement", *International Journal of Multiphase Flow*, vol. 31, no. 5, pp. 593–617, 2005.

[SHE 06] SHEW W.L., PONCET S., PINTON J.-F., "Force measurements on rising bubbles", *Journal of Fluid Mechanics*, vol. 569, pp. 51–60, 2006.

[SIM 00] SIMONIN O., "Statistical and continuum modeling of turbulent reactive particulate flow", Part I and Part II, *VKI Lecture Series*, Von Karmann Institute, Rhode-St-Genèse, Belgium, 2000.

[SIM 07] SIMONNET M., GENTRIC C., OLMOS E., *et al.*, "Experimental investigation of the drag coefficient in a swarm of bubbles", *Chemical Engineering Science*, vol. 62, no. 3, pp. 858–866, 2007.

[SIM 12] SIMONINI A., VETRANO M.R., STEELANT J., *et al.*, "Application de la PLIF à deux couleurs à un jet en état de surchauffe", *13th Congress AFVL*, CORIA, Rouen, 2012.

[SIM 96] SIMONIN O., "Continuum modelling of dispersed turbulent two-phase flows", *VKI Lecture Series*, vol. 2, Von Karmann Institute, Rhode-St-Genèse, Belgium, 1996.

[SIR 05] SIRIGNANO W.A., "Volume averaging for the analysis of turbulent spray flows", *International Journal of Multiphase Flow*, vol. 31, no. 6, pp. 675–705, 2005.

[SIR 92] SIRIGNANO W.A., "Fluid dynamics of sprays", Freeman scholar lecture, *ASME Winter Annual Meeting*, 1992 [published in *Journal of Fluids Engineering*, vol. 115, no. 3, pp. 345–378, 1993].

[SIR 99] SIRIGNANO W.A., *Fluid Dynamics and Transport of Droplets and Sprays*, Cambridge University Press, Cambridge, 1999.

[SOO 89] SOO S.L., *Particulates and Continuum-Multiphase Fluid Dynamics: Multiphase Fluid Dynamics*, CRC Press/Taylor & Francis, London, 1989.

[STÉ 12] STÉVENIN C., Étude de l'atomisation d'un jet d'eau haute vitesse: Application à l'irrigation par aspersion et à la pulvérisation, PhD Thesis, Ecole Centrale Marseille, 2012.

[SYA 93] SYAMLAL M., ROGERS W., O'BRIEN T.J., Mfix documentation theory guide, Tech. Note, U.S. Department of Energy, Office of Fossil Energy, 1993.

[TCH 47] TCHEN C.M., Mean value and correlation problems connected with the motion of small particles suspended in a turbulent fluid, PhD Thesis, University of Delft, 1947.

[THE 02] THEOFANOUS T.G., TU J.P., DINH A.T., *et al.*, "The boiling crisis phenomenon. Part I: nucleation and nucleate boiling heat transfer", *Experimental Thermal and Fluid Science*, vol. 26, no. 6–7, pp. 775–792, 2002.

[TRO 11] TROPEA C., "Optical particle characterization in flows", *Annual Review of Fluid Mechanics*, vol. 43, pp. 399–426, 2011.

[TRY 11] TRYGGVASON G., SCARDOVELLI R., ZALESKI S., *Direct Numerical Simulations of Gas–Liquid Multiphase Flows*, Cambridge University Press, Cambridge, 2011.

[VAL 01] VALLET A., BURLUKA A.A., BORGHI R., "Development of a Eulerian model for the "atomization" of a liquid jet", *Atomization and Sprays*, vol. 11, no. 6, pp. 619–642, 2001.

[VIL 07] VILLERMAUX E., "Fragmentation", *Annual Review of Fluid Mechanics*, vol. 39, pp. 419–446, 2007.

[VIS 12] VISENTINI R., Etude expérimentale des transferts thermiques en ébullition transitoire, PhD Thesis, INP Toulouse, 2012.

[WAC 01] VAN WACHEM B.G.M. *et al.*, "Comparative analysis of CFD models of dense gas-solid systems", *American Institute of Chemical Engineers Journal*, vol. 47, pp. 1035–1051, 2001.

[WAN 11] WANG J. 007A*et al.*, "Numerical study of global rainbow technique", *Experiments in Fluids*, vol. 51, no. 1, pp. 149–159, 2011.

[WEN 66] WEN C.Y., YU Y.H., "Mechanics of fluidization", *Chemical Engineering Progress Symposium Series*, vol. 62, pp. 100–111, 1966.

[WES 13] WESTERWEEL J., ELSINGA G.E., ADRIAN R.J., "Particle image velocimetry for complex and turbulent flows", *Annual Review of Fluid Mechanics*, vol. 45, pp. 409–436, 2013.

[WIL 58] WILLIAMS F.A., "Spray combustion and atomization", *Physics of Fluids*, vol. 1, no. 6, pp. 541–545, 1958.

[YAT 83] YATES J.G., *Fundamentals of Fluidized Bed Chemical Processes*, Butterworth-Heinemann, Amsterdam/London, 1983.

[ZAM 00] ZAMUNER B., DUPOIRIEUX F., "Numerical simulation of soot formation in a turbulent flame with a Monte-Carlo PDF approach and detailed chemistry", *Combustion Science and Technology*, vol. 158, pp. 407–438, 2000

[ZHA 94] ZHANG D.Z., PROSPERETTI A., "Averaged equations for inviscid disperse two-phase flow", *Journal of Fluid Mechanics*, vol. 267, pp. 185–219, 1994.

[ZUB 58] ZUBER N., "On the stability of boiling heat transfer", *ASME Journal of Heat Transfer*, vol. 80, no. 2, pp. 711–720, 1958.

Index

G, H, I

J, K, L

M, N, O

P, Q

R, S

T, V, W

Printed and bound by CPI Group (UK) Ltd, Croydon, CR0 4YY

13/09/2026

14961078-0006